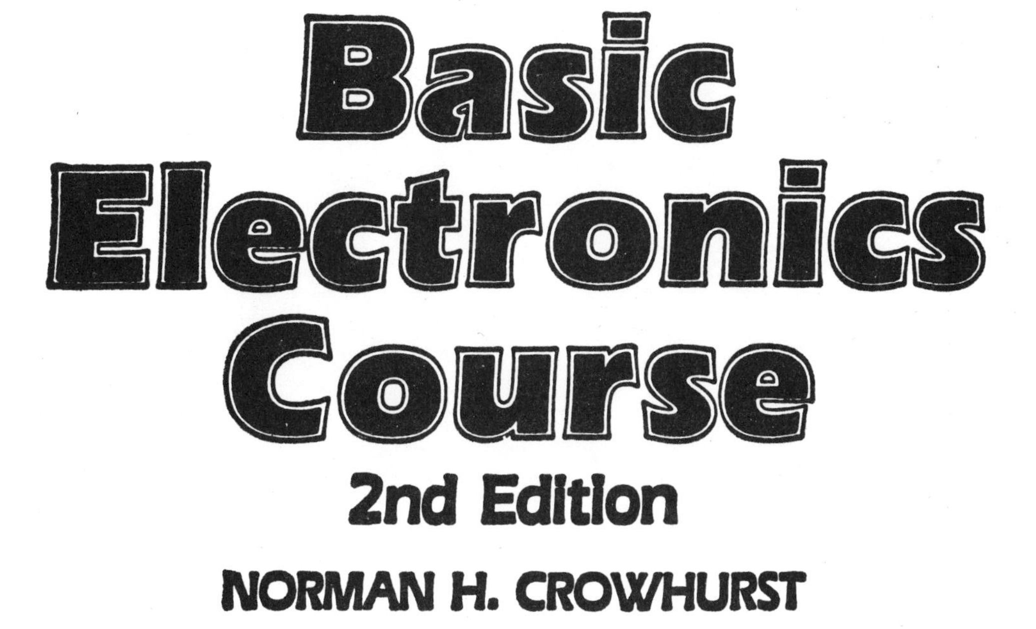

2nd Edition

NORMAN H. CROWHURST

TAB Books
Division of McGraw-Hill, Inc.
New York San Francisco Washington, D.C. Auckland Bogotá
Caracas Lisbon London Madrid Mexico City Milan
Montreal New Delhi San Juan Singapore
Sydney Tokyo Toronto

pbk 151617 QPFQPF 0210
hc 3456789 QPFQPF 0210

Library of Congress Cataloging-in-Publication Data

Crowhurst, Norman H.
Basic electronics course / by Norman H. Crowhurst.—2nd ed.
p. cm.
Includes index.
ISBN 0-8306-0913-X ISBN 0-8306-0413-8 (pbk.)
1. Electronics. I. Title.
TK7816.C69 1987 87-19411
621.381—dc19 CIP

OTHER BOOKS IN THE TAB HOBBY ELECTRONICS SERIES

This series of five newly updated and revised books provides an excellent blend of theory, skills, and projects that lead the novice gently into the exciting arena of electronics. The Series is also an extremely useful reference set for libraries and intermediate and advanced electronics hobbyists, as well as an ideal text.

The first book in the set, *Basic Electronics Course—2nd Edition*, consists of straight theory and lays a firm foundation upon which the Series then builds. Practical skills are developed in the second and third books, *How to Read Electronic Circuit Diagrams—2nd Edition* and *How to Test Almost Anything Electronic—2nd Edition*, which cover the two most important and fundamental skills necessary for successful electronics experimentation. The final two volumes in the Series, *44 Power Supplies for Your Electronic Projects* and *Beyond the Transistor: 133 Electronics Projects*, present useful hands-on projects that range from simple half-wave rectifiers to sophisticated semiconductor devices utilizing ICs.

How to Read Electronic Circuit Diagrams—*2nd Edition*

ROBERT M. BROWN, PAUL LAWRENCE, and JAMES A. WHITSON

Here is the ideal introduction for every hobbyist, student, or experimenter who wants to learn how to read schematic diagrams of electronic circuits. The book begins with a look at some common electronic components; then some simple electronic circuits and more complicated solid-state devices are covered.

How to Test Almost Everything Electronic—*2nd Edition*

JACK DARR and DELTON T. HORN

This book describes electronic tests and measurements—how to make them with all kinds of test equipment, and how to interpret the results. New sections cover logic probes and analyzers, using a frequency counter and capacitance meter, signal tracing a digital circuit, identifying unknown ICs, digital signal shaping, loading and power supply problems in digital circuits, and monitoring brief digital signals.

44 Power Supplies for Your Electronic Projects

ROBERT J. TRAISTER and JONATHAN L. MAYO

Electronics and computer hobbyists will not find a more practical book than this one. A quick, short, and thorough review of basic electronics is provided along with indispensable advice on laboratory techniques and how to locate and store components. The projects begin with simple circuits and progress to more complicated designs that include ICs and discrete components.

Beyond the Transistor: 133 Electronics Projects

RUFUS P. TURNER and BRINTON L. RUTHERFORD

Powerful integrated circuits—both digital and analog—are now readily available to electronic hobbyists and experimenters. And many of these ICs are used in the exciting projects described in this new book: a dual LED flasher, an audible continuity checker, a proximity detector, a siren, a pendulum clock, a metronome, and a music box, just to name a few. For the novice, there is information on soldering, wiring, breadboarding, and troubleshooting, as well as where to find electronics parts.

Contents

Preface viii

1 What Is Electronics? 1

2 Properties of Resistance 9

Ohm's Law • Other Resistance Values • Conductor Resistance • Series and Parallel Resistors • Effect of Tolerances • Temperature Coefficients

3 Kirchhoff's Laws 30

The Bridge Circuit • Thevenin's Theorem • Low-Resistance Measurement • Instrument Multipliers • Potentiometer Measurements • Voltage and Current Inputs and Outputs

4 Power Calculations 53

Power Dissipation • Heating Effects • Negative Temperature Coefficient Materials • Measuring Resistance Change • Effect of Pulsing on Dissipation

5 From DC to AC Circuits 73

Sinusoidal Waveform • AC Measurements • The Origin of RMS Readings • Rectified Average Readings • Effect of Other Waveforms on Readings • Synthesizing Waveforms • Frequency Ranges

6 Analog and Digital Bases of Operation 98

Sound • Video • Distinction between Digital and Analog • Conversion Between Digital and Analog • Understanding the System

7 Magnetism 115

Some Basic Quantities • Permanent Magnets • Demagnetization • Early Permanent Magnet Concepts and Calculations • Optimizing a Magnet for its Application • Soft Magnetic Materials • Magnetic Calculations

8 **Electromagnetism** 142

Spatial Relationships • Force Due to a Current in Magnetic Field • Moving a Conductor in a Magnetic Field • Electromotive Force Generated • Mechanical Force Generated • Energy Transfer and Balance • Transducers • Reversibility • Efficiency • Inductance • Leakage • Electromagnetic Radiation

9 **Reactance** 166

Inductance • Capacitance • Reactance • Time Constants • Inductor and Capacitor Types • Temperature Coefficients • Skin Effect

10 **Reactance In Circuits** 185

Power Flow in Circuits with Reactance • Impedance • J as an Operator • Equivalent Series and Parallel Values • Resonance • Bridge Circuits • Frequency Independent Bridges • Effects of Spurious Elements • Complicated Nature of Inductors • Q Values • Frequency-Dependent Bridges • Reactance Filters • Some Practical Features

11 **Diodes and Simple Semiconductors** 210

Basic Characteristics • Logarithmic Property • Zener Voltage • Instrument Rectifiers • Calibration • Low-Voltage Scales • Input Resistance • Half-Wave Rectifier • Multimeters • Temperature Coefficients • Ratings

12 **Compound Semiconductors** 233

The Current-Amplifying Transistor • Grounded (Common) Emitter • Grounded (Common) Base • Grounded (Common) Collector • Variety of Current-Amplifying Transistors • Transistor Data

13 **Linear Amplification** 249

Achieving Linearity • Input: Current, Voltage, or Combination • AC Collector Loading • Bias • Removing the AC Feedback • A Voltage-Gain Stage • Split Load Stages • Darlington Connection • Op Amps

14 **Solid-State Switching Functions** 272

How Switching Types Differ • High-Frequency Cutoff • Switching Characteristics • Test Circuits • Practical Circuits • Duty Cycle Problems • SCR Circuits • Two-State Logic Systems

15 **Thermionic Tubes** 296

Indirectly Heated Types • Limits of Operation • From Triode to Pentode • Parameters • Dissipation Problems • Biasing • Gas-Filled Tubes • Beam-Type Tubes • Accelerator Devices

16 **Other Devices** 322

Photosensitive Devices • Piezoelectric Devices • Hall-Effect Devices • Newer Developments in Solid-State

17 **Instrumentation** 334

Functions of Negative Feedback • Digital Instrumentation • Frequency Measurement • Other Measurements

18 **System Design** 346

Conditions of Stability • Open and Closed Systems • Design Approaches • System Building • What If? • Interfacing

19 **Successive "Generations"** 366

Electronics in Music • Video Games • Sensory Projection • Successive Generations • Speed

20 **Hybrid Systems** 379

In Nature • In Machines • Time or Frequency Based • Efficiency Aspects

21 **On Keeping Up with Change** 395

Touch Dialing • Rules for Keeping Up

Glossary 404

Answers to Examination Questions 416

Index 424

Preface

The second edition of this book has been a long time coming. It was first published in 1972. A lot has happened since then in electronics. A few years back, the publishers asked me to do an update; to make this book "in step" with today's electronics. That request made me feel as if I must have known Noah!

In 1972 ICs had just made their first, limited appearance. They were still expensive, though those who understood knew the price would come down. Modern semiconductors had not been around long, although radio reception started with the crystal and catwhisker, the primitive forerunner of today's diode. For a long while, virtually everything was done with "tubes."

Aware of how much had changed, I didn't feel equal to the task of such a revision, being no longer active in the electronic technology forefront. For the longest time I tried to find a coauthor to work with me on it. The publisher kept nudging me. "Have you found anyone yet?" Such "leads" as I had kept fizzling out. Nobody seemed anxious to do it, I didn't know why.

Eventually I realized that I must do it myself, that I must tackle what looked like a mammoth "catch-up" job. Perhaps I could get work in the field again . . . But nobody wanted me, I was too "out of date." So I'd go back to school. I enrolled in a class for fifth term students, sure that everything would be "over my head," but by struggling I'd get enough to serve my purpose.

My instructor knew me, from the old days—he could have been one of my students. He is a good instructor. But his students are something else! He kept telling them they were doing fine—compared to others, that is. But today's students would never rate with those of my day.

So this proved to be the best way I could have chosen to get myself updated so I could revise this book. As well as insight into "state of the art," I also got insight into the state of education in the art! I was not nearly so far behind as I imagined. Maybe Noah wouldn't have been, either! The basics haven't changed.

I had kept telling myself that, but I hadn't really believed that was enough. I needed to know more about where basics have "got to" since I left the scene.

Well, basics haven't gone far—maybe even backwards! Microminiaturization just makes it look impressive: so much can happen in such a small space. What is actually possible hasn't changed that much. If in the old days, you could have rented the Empire State Building to house what you can now put in a pocket computer—which nobody could, of course—you could have done it all then.

Microminiaturization has, more than anything else, made possible things we knew were possible but didn't have the resources to do. I must say I wish I was born later: you can have so much fun with electronics these days! But then, looking at the younger students, I wonder: where would I have got the ground work that served me so well that when I came back I found I was not really out of date at all?

About all I lacked was knowledge of the part numbers for the latest ICs and other hardware items. They compress so much into a single chip, they do a lot more than anything we had then, but really very little that's actually new.

Before concluding this preface, I must say what this leads up to. Modern electronics seems to be intelligent—well almost! It is not; that's an illusion. It will just faithfully do (unless it breaks down) exactly what you or someone else programs it to do: no more, no less. Well, maybe less—if it breaks down in the process.

Years ago, science fiction writers projected a time when computers would become intelligent enough to run the world, at which time they would "turn on us." Now some think that day is practically here. That scenario needs changing a little. Computers keep changing, as new designs come out. Who knows how to fix them? Who fixes them? You don't, you just buy a later model!

So who designs these new models? Virtually nobody: that's the scary part. You track down a guy credited with having designed a particular chip. Actually he didn't, he just did part of it. Nobody knows who did the rest. He doesn't remember anything about the part he designed—he's done a dozen things since then—he only understood about half of anything he did. Do you see where this kind of thing leads?

We need people who understand what's happening. That is what this book aims to prepare you for. Else we're going to land up back with the horse and buggy, the candle, the "little house" at the end of the garden, and whatever else we associate with centuries gone by, simply because nobody knows how to do anything and computers don't really have the intelligence we credit them with.

So, dear reader, approach this book with that attitude in mind. Have fun, but don't forget, God made you and I with brains, which nobody has been able to pass on to computers, although they may sometimes give that impression!

1
What Is Electronics?

Historically, the study of electronics is always preceded by the studies of electricity and magnetism. In the earliest times, various experiments revealed phenomena that are well-known today, but the facts uncovered in these experiments seemed only partially related at the time, so that decades—even centuries—elapsed before the relationships between them were established as they are today.

There seems to be some evidence that electro-chemical devices were used by the ancient Greeks, centuries before Christ, presumably to produce electric currents. Whether they used it for some form of electric light or what has not been established, but the remains of primitive electric batteries have been found.

Many such things, apparently known to the ancients, have been rediscovered much more recently. The fact that an electric current flowing in a wire could be detected by means of a magnetic compass needle (Fig. 1-1) was demonstrated by Oersted (Copenhagen, 1819).

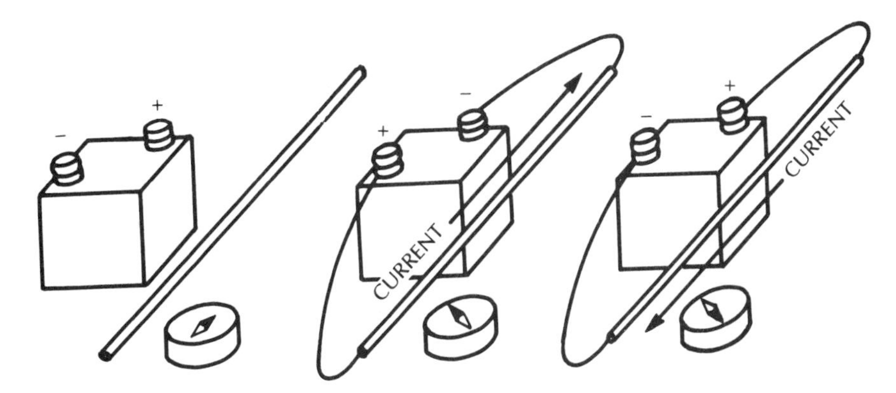

Fig. 1-1. An electric current produces a magnetic field.

When no current flows in the wire over the compass needle, the needle points to magnetic North. When current flows *upward* (middle) the needle points to the left. When current flows *downward* the needle points to the right. From this discovery came the earliest meters, called galvanometers, which detected what was then called galvanic current. Of course, the galvanometer was the forerunner of modern electrical instruments.

In 1831, Faraday discovered that moving a magnet in proximity to a wire or coil could induce a current flow of the same nature as that produced with an electro-chemical cell (Fig. 1-2). When the magnet is stationary, relative to the wire or coil, no current is induced (left). But when the magnet is moved within the coil, the current thereby induced causes the magnetic compass in the other coil to point in a direction opposite to the movement of the magnet. Faraday also continued to experiment with electrical charges and induction, a subject which was the objective of many experiments that led to the design of "influence" machines for half a century or so before his time.

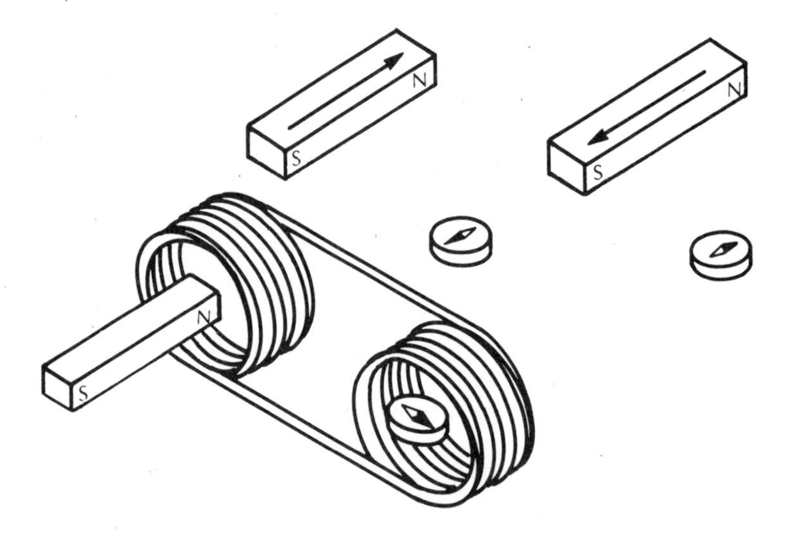

Fig. 1-2. Moving a magnet relative to an electrical circuit generates a current or electromotive force.

In addition to these relatively recent experiments with what was termed *current electricity* and *static electricity*, the presence of magnetism had been known for centuries. Mariners have used magnetic compasses for navigation since the dawn of history.

Now that the electron theory has linked these various phenomena together, their inter-relationship seems obvious, whether or not we know or understand the theory involved. But in those early days, each phenomenon was experimented with and explored as a separate entity. Although most experimenters undoubtedly suspected that an inter-relationship existed, its precise identification eluded them.

Two problems were responsible for hindering the earlier discovery of the inter-relationship: (a) the curious three-right-angle juxtaposition of the interaction; and (b) the fact that the velocity of light is involved in conversion from one form of inter-related reaction to another.

Thus, the reason early experimenters had difficulty relating static electricity to current electricity was because the charges encountered in static electricity run into thousands, even millions of volts (although there were no such things as "volt meters" in those days to measure it) accompanied by charges that represent quite small currents, while current electricity concerns a steady flow of appreciable magnitude with only a few volts of potential.

The concept of voltage, or potential, as an electrical *pressure* and current as an electrical *flow* (using an analogy based on the physical behavior of liquids), was a big step forward. This led to the enunciation and proof of what became known as *Ohm's law*, but still a precise definition of static electricity evaded physical scientists. The great difference in relative values made it difficult to prove that the momentary currents which seemed to change electrical charges at the enormous voltages that accompanied them were, in fact, of the same form as the steady currents flowing at much lower voltages.

To further frustrate efforts, it was noticed that voltages in the thousands and millions (and there was no way to measure them in those days, except by the "inches of spark" they would generate) also produced strong electric fields, yielding attraction and repulsion, breaking down or ionizing gases (such as air) in the form of a discharge; the latter phenomenon is impossible with only a few volts such as are generated by an electrochemical cell. However, experiments with an induction coil began to show these inter-relationships. It was learned that if a current flowing through a relatively few turns wound on a magnetic core (Fig. 1-3) was interrupted, the magnetization of the core changed very rapidly. This changing magnetization also was used to provide the interruption by opening (vibrating) the contacts (as shown).

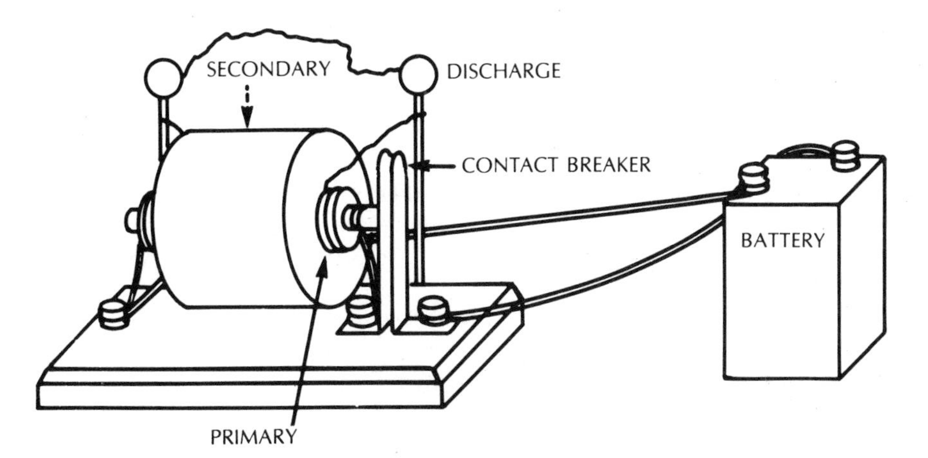

Fig. 1-3. This induction coil setup associated current with static electricity.

The experiment was based on a theory developed from the discoveries: that an electric current can produce magnetism (the invention of the electromagnet) and that a moving or changing magnetic field generates an electromotive force or voltage. Utilizing these facts, experimenters soon discovered that by putting another winding on the same magnetic core, with many times the number of turns in the first winding (using much finer wire to get more turns into the space), a much higher voltage could be induced. By the process of electromagnetic induction, working both ways in this case, the larger number of turns generated a voltage correspondingly greater than the interrupted voltage supplied to the lower number of turns. Experiments then showed that this higher voltage exhibited properties similar to those already discovered with static electricity: it could produce a long spark through air, dependent on the voltage generated by the coil, of course.

Such a historic look, in bare outline, at the way these bits of theory became associated and the uses that were made of them as those associations were proved, helps us appreciate what has repeatedly happened in the development of electronics—what is still happening, in fact. For example, in those early days, the electrostatic generator, which is a machine that uses electrostatic generation (better called electric than electrostatic, because if it was truly "static" it couldn't happen), was merely a novelty intended to produce some rather spectacular demonstrations (simulated lightning, etc.). The machines that first achieved practical wide use were of the electromagnetic induction type, in the form of motors and dynamos (generators) dc and ac. These were the practical machines of the day, because they could produce power, do work, and so forth.

The principle of the "influence" machine was to apply the fact that like charges repel each other and unlike charges attract by moving the charge carriers physically (and thus putting work into the machine) to increase the magnitude of the charges in relation to each other. This principle was first shown in quite simple experiments, many of which had been performed for centuries.

If a glass rod is rubbed with dry fur, it becomes charged and will attract uncharged objects light enough to exhibit the attraction by movement, such as pieces of paper, pithballs (a favorite item in early experiments) etc. But if, or when, such charged particles touch the rod, they are then repelled from it. Figure 1-4 illustrates the swinging pithball experiment. A charged glass rod is brought toward the uncharged pithball (suspended by a piece of silk thread; nowadays nylon would serve). The pithball develops an induced charge, whereby the side nearer to the charged rod is attracted more strongly than the opposite side is repelled, so the overall effect is attraction. But when the two touch (center), the charge on the rod is shared, so they now have like charges and the pithball is repelled (right).

A more elaborate experiment involves a device known as an electrophorus (Fig. 1-5). The base is a dry insulating material (dielectric in electrostatic parlance) which is charged by rubbing. On it is placed a conducting (e.g., brass copper, etc.) disc, which can be moved by means of an insulating handle (Fig. 1-5B). At this stage in the experiment, the charge on the base induces an opposite charge of the face of the disc nearest the base and a like charge on the upper face of the disc. The next step is to discharge (e.g., by touching) the upper face of the disc and to place a small chargeable object (e.g., a pithball) on it (Fig. 1-5C). At this stage in the experiment, although the base is still charged, the upper surface of the disc is neutral. Finally, when the disc is lifted by its insulating handle (Fig. 1-5D), the charge on its underside spreads over its entire surface, including the pithball in contact with it, and causes repulsion, making the pithball jump off the disc.

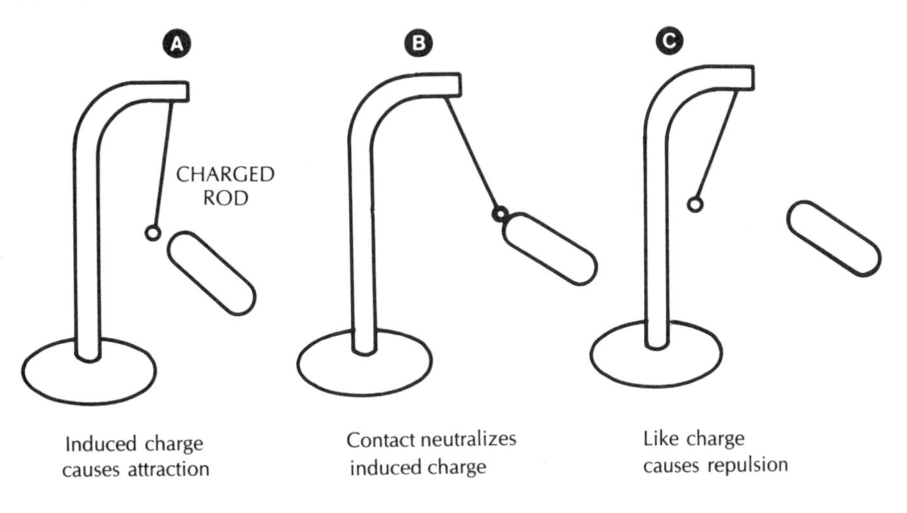

Fig. 1-4. The swinging pithball experiment shows the nature of static electricity.

These experiments explain the phenomena that unlike charges attract and like charges repel and that conduction equalizes charges. Today, the Van de Graaff machine, a variety of influence generator perfected as a result of new synthetic materials made available by modern technology, is widely used to generate the extremely high voltages necessary for the accelerator tubes associated with experiments in atomic physics and with processes that use such accelerators. The machine that was long only a novelty now has practical uses.

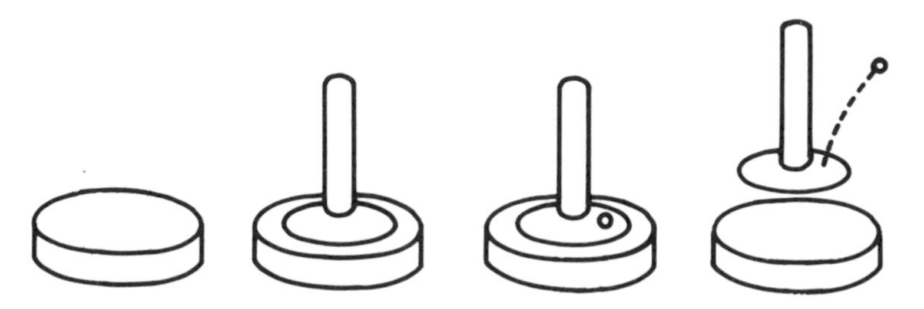

Fig. 1-5. Another demonstration of static electricity: the electrophorus.

So far, we have explored a variety of phenomena. Their practical use developed, first, into various uses of electricity for lighting, heat and power. In the early 20th century, this was the emphasis taught to young engineers. Meanwhile, young experimenters like Marconi were working with a related novelty, called *radio* or *wireless telegraphy*, that utilized electromagnetic radiation for transmission over distances. Today, radio transmission is a central fact of life. Associated with its development were a number of devices needed to make it work, first for transmission of any intelligence at all, then for transmission over successively greater distances. Thus were born the coherer (now extinct, except as an historic novelty), the crystal detector or diode, and finally a wide range of electronic tubes, not to mention the transistor.

This order of development was somewhat an accident of the way men made discoveries and theorized about what they found. At that stage, the crystal and catwhisker detector was just a useful (but largely unexplained) phenomenon. Because, in the form in which it was used, it was a somewhat "hit and miss" device; it was regarded as primitive, so the practical-minded industrial designers moved into the use of thermionic devices.

But some of the men who follow research—knowledge for knowledge's sake—theorized and developed, then verified the existence of a different category of device or material called semiconductors. Conductors had been known for centuries as materials that freely conduct electricity. Insulators, also known as dielectrics (although the properties thus named are not the same), had also been known. However, semiconductors proved to be a whole new range of materials, when constructed in appropriate devices, that can conduct one way and insulate the other.

So, at this point in time, the part of the field for which this book is aimed is known as "electronics." At one time, a course in "electricity and magnetism" would have been specified as a prerequisite. But the world is changing. A few decades ago, the basic course dealt with the generation and basic ways of using electricity: from static electricity, magnetism and current electricity, through electromagnetics and other basic application modes into various uses for electricity or the bases for them.

Today, most of those electromechanical applications are grouped under the heading of electrical engineering. Devices using any of the more sophisticated components that use electrons in some way other than merely conducting them along wires to various electromechanical or heating appliances come under the heading of electronic engineering.

It has not always been this way. When the author started, members of the profession were just beginning to allow that radio engineering was a distinct profession in itself, rather than a particularly sophisticated offshoot or application, still with some very "experimental" connotations to it, within electrical engineering. At that time, the people who dabbled in what is today known as electronics were out in a field of their own, as the radio experimenters of a generation earlier had been. The electrical engineers, content to continue with nice familiar things like electric motors, generators, heaters, etc., felt that the radio and electronics people were playing with toys, using a very imprecise technology. They forgot that the early electrical experimenters had no voltmeters and ammeters as we know them.

At the time, that viewpoint was true. Nobody was more conscious of the problem this imprecision brought than the people who worked with it. But the leading workers in the field labored to change all that, so that, today, electronic measurements are far more precise than electrical engineering measurements ever were. For example, electrical engineers, using electric clocks and devices derived from them, could measure time to a fraction of a second, about one fiftieth—an accuracy previously unheard of. But now electronic devices can measure time intervals to the order of one thousand-millionth of a second, which makes a fiftieth of a second seem like an eternity.

When electronics first started, as an offshoot of radio, which was an offshoot of electricity, it used vacuum tubes and other components in much the same way that radio did, in those days. Things have changed, several times over, since then. First, transistors replaced vacuum tubes with considerable space saving, and much greater versatility as to what could be done.

But it did not end there. Computers and other equipment, which today would be called "processors" (although today's equivalent are "microprocessors") began to introduce new concepts. Electronics could be used to automatically do all kinds of things that people had to think how to do. Computers and processors could be programmed to do those things automatically.

Before electronics had made such advances, men had designed mechanical machines to do all sorts of things. But electronics enabled a great many more things to be done. Anything that could be controlled humanly could now be automatically controlled by electronics.

But it didn't end there. That was possible, using a lot of the little transistors and other components, that man had to design in the first place. The next step was putting hundreds of such components onto a "chip" in a preprogrammed arrangement, that would enable that chip to do more than formerly a whole roomful of equipment could do.

In fact the real "working parts" of electronics have become so small that the connections needed to do all the wonderful things these "chips" can do are vastly bigger than the whole assembly of its working parts. Which means that the more functions that can be put on a single chip, the smaller the overall piece of equipment can be made.

And it doesn't end there. As its human designers put more into such individual parts, they began to think in terms of what assemblages of such parts could do, instead of how all the myriad individual electronic elements inside each part did its job. There were many accompanying changes.

For instance, in the old days, if an individual tube, and later an individual transistor, went bad, then a serviceman would change it. But when hundreds of such extremely miniature components were compressed into a chip that can sit on the tip of your little finger, nobody is going to change one of those components when it goes bad, even if he could see it, with the aid of a microscope.

Cost was another change that affected this. Individual transistors came down in price. At first they cost hundreds of dollars each. They came down in price, as they were made by the millions, to around a dollar each. But now, many of these microchips, each containing hundreds of transistors and other parts, cost no more, or even less, than a single transistor cost a few years ago.

And perhaps the biggest change has been that now electronic equipment can be made self-programming. In effect, it can design itself, and tell itself how to do its job.

Will the time come, when human beings will become unnecessary, and electronic machines will "do it all?" For now you want to come abreast of the science as quickly as possible, so you can "catch up" with what's going on. There is something else you need, that was not formerly seen as well as it is today: you need to be able to "keep up," as well as catching up.

This means you need to develop a way of thinking about what electronics does in more basic terms than has been customary recently, a way that the earlier edition of this book pioneered. At the rate at which electronics is advancing, where the latest device becomes "obsolete" often in less than two years, it is surprising that a book would not become obsolete in much less than ten years.

Yet the previous edition of this book was being used for that long before the publishers decided it needed updating. We mention this, because part of the purpose of this book is to convey that capability to you.

2

Properties of Resistance

One thing common to all electrical and electronic circuits is resistance, and the component in a circuit most obviously possessing resistance is a *resistor*. The term, *resistance,* originally was derived from the notion that such a component "resists" current flow. In modern circuits, resistors are usually found in small (although sometimes they are larger) cylindrical packages, with color-coded markings to indicate the resistance value in *ohms*. These markings use a basic 3-digit color code, sometimes with a fourth color to indicate tolerance.

Seldom does a resistor have the actual value indicated by the color code in fact, the value can vary as much as 20 percent, if there is no tolerance marking. If there is a tolerance marking, the deviation may vary from 5 to 10 percent. Making a resistor with the exact value designated by the code would be very costly. An exact value is not critical to the proper operation of most circuits. So long as the value is fairly close to its stated value, the circuit will work.

The 3-digit code uses selections from ten colors in a sequence, with a different meaning for each digit of the code. The first two color bands represent the first two digits of the number that states the component's resistance in ohms. The third digit of the code tells how many zeroes follow the first two digits to make up the number that states the value.

The numerals associated with the colors in the code are as follows:

BLACK: 0 BROWN: 1 RED: 2 ORANGE: 3 YELLOW: 4

GREEN: 5 BLUE: 6 PURPLE: 7 GRAY: 8 WHITE: 9

Applying the rules just given, a resistor with color bands, in order, BLUE, GRAY and YELLOW, will have a value of 680,000 ohms. The BLUE identifies 6, the GRAY indicates 8, and the YELLOW tells that four zeroes follow those numerals.

The coded digits are marked on resistors in a variety of ways. Some of the earlier resistors were painted all over with the first digit color, the second digit was marked by a spot on one end, and the third digit appeared as a spot on the side (Fig. 2-1A). More modern resistors use color bands in sequence along the body of the resistor, which usually has a neutral color, such as tan, off-white, or gray (a different shade of gray from the one used in the code). In this case, the three digits are placed in order, starting from the end where the ring or band is nearest to the end of the resistor (Fig. 2-1B).

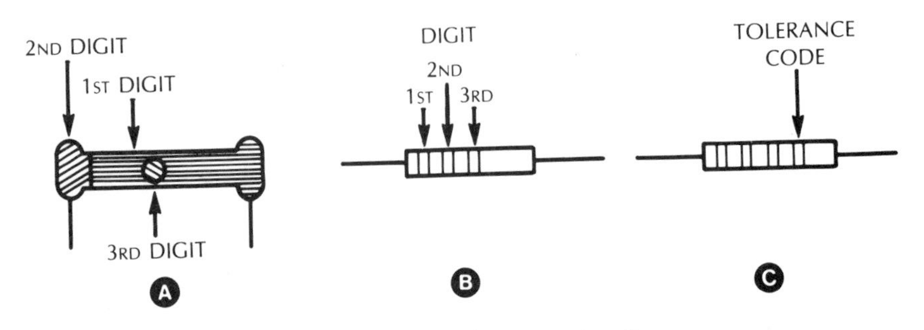

Fig. 2-1. Typical resistor construction.

Thus, we can identify resistors from 10 ohms (BROWN BLACK BLACK) to 99,000,000,000 ohms (WHITE WHITE WHITE) with the color code. Since the latter is 99,000 megohms (a megohm is a million ohms), it is far larger than is ever used as a resistor value. At the lower end, however, resistors are needed below 10 ohms. One way to mark lower values would be to use the black code for the first digit. Thus, BLACK BROWN BLACK would represent 1 ohm. But this procedure would limit the selection of values to exact integral values: 1, 2, 3, 4, 5, 6, 7, 8, 9 ohms, with no in-between values. So it is not used. With the method used, the first two colors still indicate two digits and GOLD is the third code element, which means the second digit is a decimal fraction. Thus, a resistor coded YELLOW PURPLE GOLD has a value of 4.7 ohms. This gold band should not be confused with its use to represent 5 percent tolerance, which we shall discuss presently.

Resistors can be constructed in many ways. A resistance composition may be sprayed on a glass or other insulating rod, with bands, open-ended cylinders or other contacting devices to make connections to the resistance material at the ends. Or they may be wound of fine wire, which also has resistance, especially if it is made of one of the alloys developed for that purpose.

This introduces the fact that every conductor of electricity possesses resistance, however small. The smaller the resistance, the better a conductor of electricity it is. Copper, silver, gold and aluminum are good electrical conductors, with very low resistance. They are not used to make resistors.

Resistors also are made of alloys including metals like nickel, chromium and other metals, sometimes even iron, although iron, along with some others, is used for making special kinds of resistors. Resistors are also made of carbon and compositions made up with carbon as one element. Graphite, the form of carbon used in so-called "lead" pencils, was much used in earlier days for making resistors.

Components made deliberately to be resistors are usually manufactured by a process that yields values within the general range wanted. The production run is then sorted by measurement into a series of bins, each resistor being put into the group to which its value is closest, after which they are coded with the value found by measurement (Fig. 2-2).

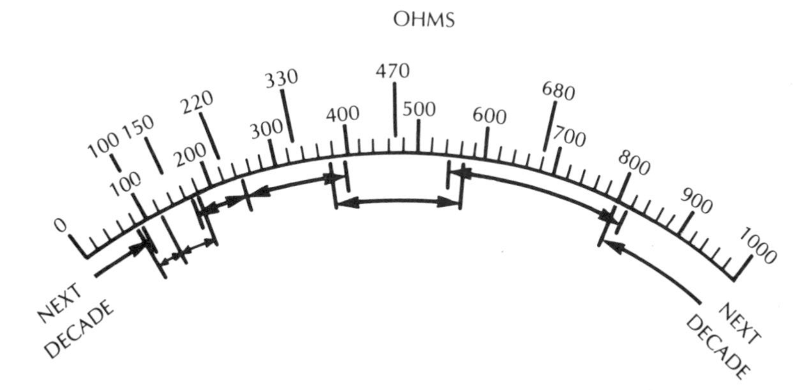

Fig. 2-2. Resistors over a decade range of values divide in 20-percent tolerance values that were at one time standard.

This method leads to the tolerance system. Sorting resistors into values costs more, the more closely they are sorted, because that involves a bigger number of bins and more critical measurement. The coursest sorting, of which a decade is indicated in Fig. 2-2, allows a 20-percent tolerance on values. This means that a resistor coded as having a value of 100 ohms may actually measure between 80 and 120 ohms. The next value in the 20-percent scale is 150 ohms, and one thus coded could have a resistance between 120 and 180 ohms, because 20 percent of 150 is 30, then 150 − 30 equals 120 and 150 + 30 equals 180.

The next value is nominally coded at 220 ohms. 20 percent of 220 is 44 ohms, so its actual value could vary between 220 − 44 or 176 ohms and 220 + 44 or 264 ohms. The following 20-percent code value is 330 ohms, of which 20 percent is 66 ohms, yielding tolerance limits for this coding of 330 − 66 or 264 ohms and 330 + 66 or 396 ohms.

Going up the scale of preferred values, as this succession is called, the next one is rated at 470 ohms, of which 20 percent is 94 ohms, yielding limits of 470 − 94 or 376 ohms and 470 + 94 or 564 ohms. One more value completes the decade scale of 20-percent resistor values. This one is rated at 680 ohms, of which 20 percent is 136 ohms, making limits of 680 − 136 or 544 ohms and 680 + 136 or 816 ohms.

The next value starts the next decade at 1000 ohms. Like the 100-ohm rating, which can vary between 80 and 120 ohms, a 1000-ohm resistor will range between 800 and 1200 ohms.

Notice that the six values just discussed cover a decade of resistance values with 20 percent of each other, so that every resistance value in that decade (in this case from 100 to 1000 ohms) will fall into at least one of the value slots. Between 100 and 150, as well as between 220 and 330, there is no overlap between adjacent tolerance ranges—a value must fall in one or other. Below 120, it goes in the 100 bin. Above 120, it goes in the 150 bin. Between all the other adjacent ratings, there is a little overlap. Between 176 and 180 ohms, a resistor could be coded either 150 or 220. Between 376 and 396 ohms, a resistor could be coded either 330 or 470. Between 544 and 564 ohms, a resistor could be coded either 470 or 680. And between 800 and 816 ohms, a resistor could be coded either 680 or 1000.

Nowadays, 20-percent tolerance resistors are not widely used, because production techniques have made it possible to produce resistors within 10 percent tolerances at no higher cost than those sorted within 20 percent tolerances. To fill a 10-percent range of preferred values, intermediate values of 12, 18, 27, 39, 56, 82 percent are inserted between the sequence of 10, 15, 22, 33, 47, 68 already established for the 20 percent range.

Table 2-1 lists the preferred values for the 20-, 10- and 5-percent ranges, with the tolerances on each. In the 5-percent range, there are two median or nominal values that are slightly different from the coding number because three digits are required for the number (before the number of zeroes are added). The actual middle value, on which the tolerances are based, is shown in parenthesis.

Between adjacent values in the limit or tolerance column, symbols are added to indicate whether the tolerance ranges exactly adjoin, overlap or leave a gap. An x indicates that adjacent tolerances exactly coincide: the upper tolerance of one is the same as the lower tolerance of the next. A − sign indicates there is a slight gap between the lowest value of the higher coded value and the highest value of the lower one. A + sign indicates that there is a slight overlap.

Resistors coded within a 20-percent tolerance have no mark to indicate the tolerance, only the three value code bands. Resistors coded with a 10-percent tolerance have a SILVER tolerance band, in addition to the three value coding bands (Fig. 2-1C). Resistors coded within a 5-percent tolerance have a GOLD tolerance band. Do not confuse gold as a 5-percent tolerance band when it is used to signify a low resistance (between 1 and 10 ohms). A resistor marked BROWN GREEN GOLD has a resistance of 1.5 ohms, tolerance unstated and thus presumed to be 20 percent. A color between the green and gold changes it. If that color is BROWN, the resistor would be 150 ohms, 5 percent.

A few more comments will make clear how the color code identifies possible variations in resistance value. A 33,000-ohm resistor is color coded ORANGE ORANGE ORANGE—three orange bands. Without a fourth band, its tolerance would be 20 percent, indicating its value could be anywhere between 26,400 and 39,600 ohms. With a SILVER fourth band, the tolerance is 10 percent and the

TWENTY PERCENT	LIMITS	TEN PERCENT	LIMITS	FIVE PERCENT	LIMITS	TWENTY PERCENT	LIMITS	TEN PERCENT	LIMITS	FIVE PERCENT	LIMITS
1000	800 1200	1000	900 1100 +	1000	950 1050 +	3300	2640 3960	3300	2970 3630 +	3300	3135 3465 +
				1100	1045 1155 +					3600	3420 3780 +
	×	1200	1080 1320 –	1200	1140 1260 -		+	3900	3510 4290 +	3900	3705 4095 +
				1300 (1340)	1275 1409 -					4300	4085 4515 +
1500	1200 1800	1500	1350 1650 +	1500	1425 1575 +	4700	3760 5640	4700	4230 5170 +	4700	4465 4935 +
				1600 (1640)	1558 1722 +					5100	4845 5355 +
	+	1800	1620 1980 ×	1800	1710 1890 -		+	5600	5040 6160 +	5600	5320 5880 -
				2000	1900 2100 +					6200	5890 6510 +
2200	1760 2640	2200	1980 2420	2200	2090 2310 +	6800	5440 8160	6800	6120 7480 +	6800	6460 7140 +
		–		2400	2280 2520 -					7500	7125 7875 +
	×	2700	2430 2970 ×	2700	2565 2835 -		+	8200	7380 9020 +	8200	7790 8610 -
				3000	2850 3150 +					9100	8645 9555 +

Table 2-1. Standard Preferred Resistor Values for One Decade.

value could be between 29,700 and 36,300 ohms. With a GOLD fourth band, the tolerance is 5 percent and the value could be between 31,350 and 34,650 ohms.

Nothing about the 20-percent tolerance indicates that its value could not be between 26,400 and 39,600 ohms, or even between 31,350 and 34,650 ohms; it is just less likely to be. Since each group is made up by sorting an even number of resistors (of great quantity) uniformly into the range slots that encompass the precise value for the tolerance of sorting, a large number of resistors of a given coding are likely to be distributed uniformly throughout the tolerance range. Thus, in a large batch of 20-percent tolerance resistors, half of them are likely to be within the 10-percent tolerance range, although they are coded 20 percent, and one fourth of them will probably be within the 5-percent tolerance. Obviously, if you need 10- or 5-percent tolerance, it would be less expensive to buy them to that tolerance in the first place, rather than having to buy twice as many, or four times as many, of the 20-percent variety, then test every one of them to find the proportion you can use.

A 33,000-ohm resistor could be coded at any of the standard tolerances. However, a 39,000-ohm resistor, coded orange white orange, would never be rated at a 20-percent tolerance, because it is not a 20-percent value but one of the 10-percent range. With a SILVER fourth band, its value could be between 35,100 and 42,900 ohms. With a GOLD fourth band, it could be between 37,050 and 40,950 ohms. Similarly, a 36,000-ohm resistor, coded ORANGE BLUE ORANGE, will always be a 5-percent value, bearing a GOLD fourth band, with a value between 34,200 and 37,800 ohms.

OHM'S LAW

So much for being able to recognize the ohmic value of a resistor from its coding. The next thing to know is how much a resistor affects voltages and currents in a circuit. The basic law that determines this, in every circumstance (although it does get modified from the simple relationship we discuss at first), is called "Ohm's law," named after the man who first discovered the relationship.

Ohm's law can be stated in a number of ways. Possibly the simplest to understand is to say that resistance controls the current flowing, in proportion to the voltage applied across the resistor (resistance). The basic relationship is between units of current, voltage and resistance: 1 amp, 1 volt and 1 ohm, respectively.

- **If a voltage of 1 volt is applied across a resistance of 1 ohm, a current of 1 amp will flow.** That is the starting point for calculations.
- Raising the resistance value cuts down the current flow for the same applied voltage.
- Raising the voltage applied increases the current in the same proportion, if the resistance remains unchanged.

- Raising the current flowing through a resistance causes the voltage appearing across it to rise proportionately.

Reverse statements to each of the above can easily be made. If voltage, E, is given in volts, current I in amps and resistance R in ohms, then Ohm's law can be stated in the following three ways:

$$I = \frac{E}{R} \quad (1)$$

$$E = IR \quad (2)$$

$$R = \frac{E}{I} \quad (3)$$

Given any two quantities relative to a particular circuit, the third can always be calculated by the appropriate formula.

Example 1

A resistance of 10,000 ohms (often written 10 K, with K standing for "thousands") has 55 volts across it. Find the current flowing:

$$I = \frac{E}{R} = \frac{55}{10{,}000} = 0.0055 \text{ amps}$$

or 5.5 milliamps. Milliamps is the more convenient unit here. A milliamp is one thousandth of an ampere, since an ampere consists of 1000 milliamps. To convert a figure in amps to a figure in milliamps, multiply by 1000. Thus,

0.0055 amps equals 0.0055×1000 or 5.5 milliamps.

Here is a useful secondary reference or starting point for Ohm's law calculations: If the voltage is in volts, resistance in kilohms (thousands of ohms), then the current will be given directly in milliamps:

$$I = \frac{E}{R} = \frac{55}{10} = 5.5 \text{ milliamps}$$

Example 2

A circuit controls the current through a resistor of 47 K so that it is 1.2 milliamps. Find the voltage across it. Here we can use the kilohm-milliamp reference direct:

$$E = IR = 1.2 \times 47 = 56.4 \text{ volts}$$

In practice, allowance should be made for the tolerance on the resistor.

Example 3

An unknown resistance value has a measured voltage of 45 volts across it and is passing a current of 3.2 milliamps. Find its resistance value. Again, we can use the kilohm-milliamp reference:

$$R = \frac{E}{I} = \frac{45}{3.2} = \begin{array}{c}\text{14.06 kilohms, or}\\ \text{14,060 ohms}\end{array}$$

(Divided out to the nearest 4th figure, the 6).

Had this been a coded resistor of either 20- or 10-percent tolerance, it should have been coded 15 K (BROWN GREEN ORANGE) with the appropriate tolerance band (SILVER if 10 percent). Alternatively, it could have been a 13 K, 5 percent, coded BROWN ORANGE ORANGE GOLD. However, the wording of the question suggests that its color code was missing for some reason or other, or that it was some kind of resistance with no indication of value.

OTHER RESISTANCE VALUES

In addition to the much-used preferred-value resistance ranges in Table 2-1, some other values are used for special purposes. Also, sometimes we need to know the resistance value of a component whose main purpose is not to serve as a resistor.

Close-tolerance resistors may be rated at values held to closer limits than 5 percent. Also, sometimes 5-percent values are encountered at other than the preferred resistances because the choice available does not suit the need as well as some in-between value. For example, an ideal value may be 9500 ohms, with a variation between 9000 and 10,000 ohms being acceptable. A value of 9500 ohms, plus or minus 5 percent, would serve and would be coded WHITE GREEN RED GOLD.

Also, coils or inductors intended to provide magnetic fields (as in magnetic relays, for example), are designed to work with predetermined voltages or currents, which requires that they be wound of a suitable wire gauge and number of turns to serve the purpose. (We discuss this more in Chapter 9.)

Other components, such as transistors, diodes, tubes, etc., are not primarily intended to function as resistance at all, but they are involved in relationships between voltage and current that may be expressed as a resistance for the particular combination used at a particular time. In other words, there is an effective resistance at that particular time or measurement condition.

When we calculate such resistances later in the book, be sure to remember that giving something a resistance value for a purpose like this does not necessarily mean it behaves like a resistance, unless it happens to obey Ohm's law. For example, a diode operating as a zener (which is discussed fully in Chapter 11) passes virtually no current until a certain voltage is reached, after which the voltage ceases to rise, while current rises rapidly with no accompanying rise in voltage (Fig. 2-3). If the zener voltage is, say, 10 volts, and other components in the circuit limit the current to, say, 5 milliamps, the effective resistance (for that condition only) could be stated as 2,000 ohms.

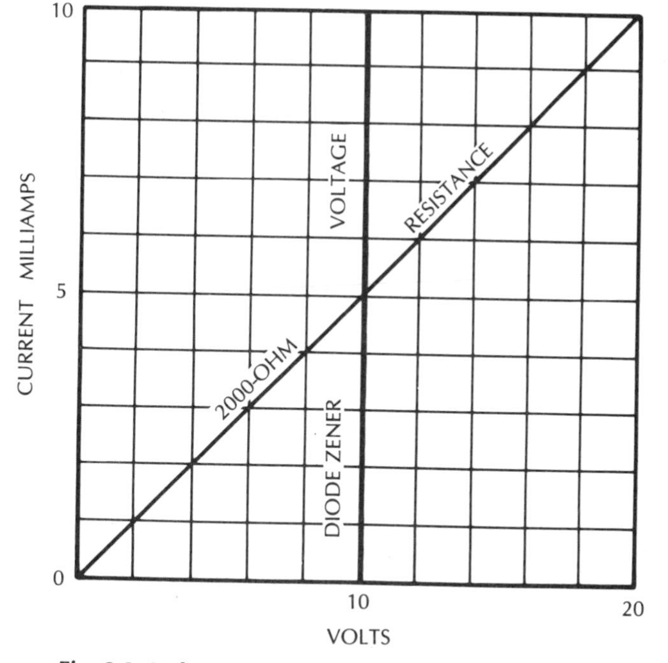

Fig. 2-3. Voltage-current relations for a zener diode.

But Ohm's law says that a resistance of 2000 ohms would pass 5 milliamps at 10 volts (as in this case) and also 10 milliamps at 20 volts, or 2 milliamps at 4 volts, for example. In this case, Ohm's law does not hold because the zener diode will pass no current at 4 volts, and no amount of current flowing would make it drop 20 volts before the diode was destroyed. The voltage is fixed at close to 10 volts, while current can vary from zero up to the maximum the device will safely handle. Saying it has an *effective* resistance value of 2000 ohms merely applies to that strict condition of 10 volts at 5 milliamps—nothing else.

A resistor may be needed at a value of 8000 ohms within a 5-percent tolerance. This means it can vary 400 ohms above or below the 8000-ohm rating, or from 7600 to 8400 ohms. Such a resistor would be so marked (GRAY BLACK RED GOLD).

A relay coil may take 1 volt-amp (i.e., 1 watt) to operate the relay on a 24-volt circuit. This means the resistance of the coil must be designed to draw one twenty-fourth amp at 24 volts. Dividing 24 by one twenty-fourth according to Ohm's law formula (3), gives a resistance of 576 ohms. This value could be put in code (as 570, which is GREEN PURPLE BROWN, or 580, which is GREEN GRAY BROWN), but is usually printed or otherwise marked on the coil in numerals. Alternatively, it may be marked with the operating voltage or current, instead of a resistance value. This is discussed more in Chapter 9.

CONDUCTOR RESISTANCE

In electronic circuits, we form the habit of thinking that resistances (or resistors) are things so marked, and that wires of various forms that connect them, whether wrapped in insulation or etched as strips of copper on circuit boards, are "conductors." In thinking of conductors we tend to forget that they, too, possess resistance. This is because such a resistance is usually very small compared to the values marked on components designed as resistors, and thus conductor resistance can usually be ignored since it will not affect circuit operation.

But for some purposes, conductor resistance can become important. Every substance that conducts electricity has a *specific resistance or resistivity*. This means that a piece of it having a given shape, usually a centimeter cube, will measure a specific resistance value when connections are made to opposite faces of the cube (Fig. 2-4). Thus, a heavy strip of this conductor, 1 centimeter square in cross-section by 10 centimeters long, will have 10 times the specific resistance of the substance of which it is made. If the strip is 1 centimeter wide by 1 millimeter thick, its resistance will increase by 10 again. If the strip is cut to 1 millimeter wide as well as thick, so its cross-section is 1 millimeter square, instead of the original 1 centimeter square, the resistance is multiplied by 10 again, or it is 100 times the piece 1 centimeter square in cross-section.

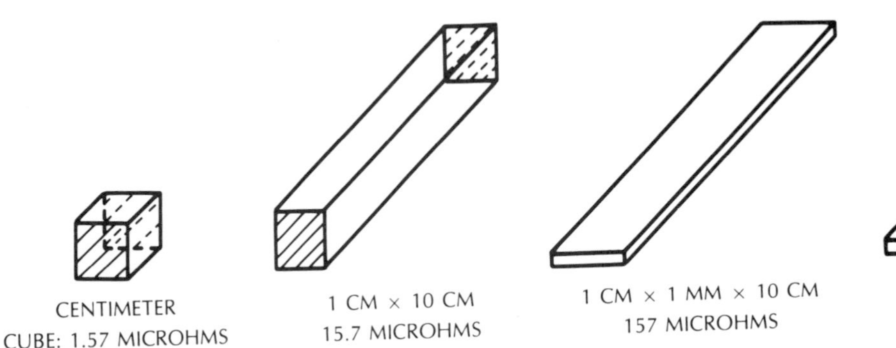

Fig. 2-4. Basic resistivity is measured across a centimeter cube.

Copper, the conductor metal used on most etched-circuit boards, has a specific resistance of 1.57 microhms (millionths of an ohm) for a centimeter cube. It may be plated on the board as thin as about 50 microns (a micron is one one-thousandth of a millimeter, or one ten-thousandth of a centimeter; thus, a micron is about one twenty-five thousandth of an inch, so 50 microns is about 0.002 inch) in strips as narrow as about 1 millimeter (0.04 inches wide). Thus, a 1-centimeter length of such a strip of etched circuit would measure 200 × 10 or 2000 times 1.57 microhms, or 0.00314 ohms. A 100-centimeter length of the same strip would measure 0.314 ohm, or nearly one third of an ohm.

Annealed silver has a slightly lower resistance than annealed copper, but not sufficiently lower to make the change worth the increase in cost for most purposes. Annealed aluminum has a resistivity of 2.89 microhms, as compared with copper's 1.57. Against this, copper has a specific gravity of 8.88 (times the weight of the same volume of water) while aluminum has a specific gravity of only 2.68.

Assume that a given length of conductor is needed for some purpose. Changing from copper to aluminum, to retain the same resistance value, the increase in resistivity would require an increase in cross-section (one dimension only, such as width or thickness, not diameter) of 2.89 divided by 1.57, which is 1.84 times. However, because aluminum is so much lighter, changing from one to the other, without changing measurements, would reduce weight by a factor of 8.88 divided by 2.68, or 3.32, while raising resistance by 1.84 times. So increasing the cross-section by 1.84 to keep resistance the same will result in a weight reduction factor of 3.32 divided by 1.84, which is 1.8. Figure 2-5 shows some actual figures to illustrate this.

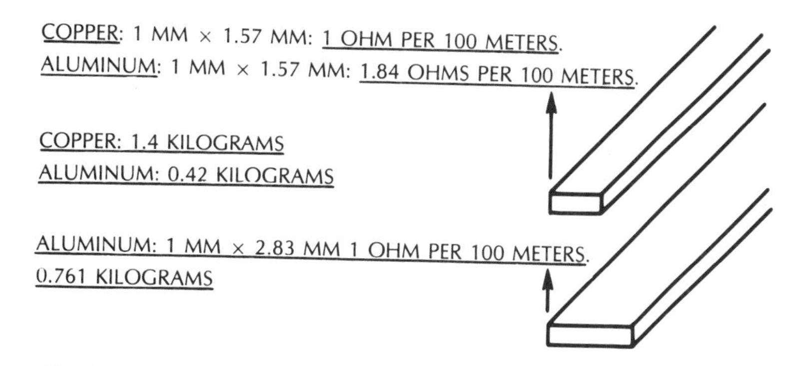

Fig. 2-5. A comparison between copper and aluminum as conductors.

So changing a conductor from copper to aluminum and increasing the size to keep the same resistance value results in a weight reduction of approximately one half (a division by 1.8). This is why overhead power lines use aluminum conductors. There may be applications in electronics where keeping weight down is more important than keeping size down.

However, another factor in the choice is the ease of working. Copper can be silver or tin-plated, and is easily soldered for making connections. Aluminum requires either a special solder process or crimping (which is also used for copper connections sometimes) and is more subject to corrosion in some circumstances than copper. We discuss this matter of conductor dimensions more in the next chapter, connected with dissipation and current-carrying capacity.

SERIES AND PARALLEL RESISTORS

Increasing the cross-section of a conductor reduces its resistance because the increase provides additional or parallel current paths. On the other hand, a conductor's resistance goes up with each added portion of length. The added portions each pass a specified current and drop a share of voltage, and these voltage drops add for successive pieces of length (Fig. 2-6).

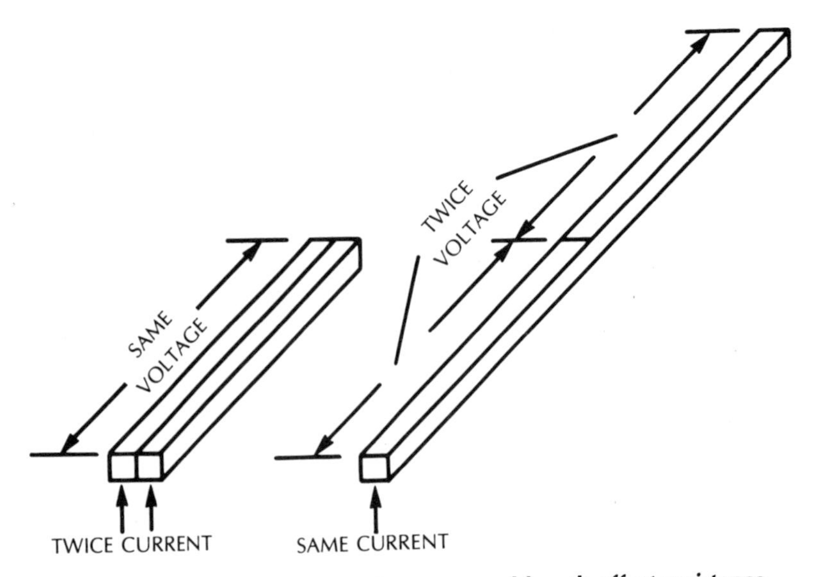

Fig. 2-6. Another way of seeing how area and length affect resistance.

The same is true if individual resistors are connected in series or in parallel. With two resistors in series, the same current must flow through both of them, simply because it has no other place to go. Therefore, the voltage drop across each will be proportional to its respective resistance and the voltage drops will add. The combined resistance of two resistors in series is the sum of their individual values. Connecting a 10-ohm resistor in series with a 100-ohm resistor produces a combined resistance of 110 ohms (Fig. 2-7).

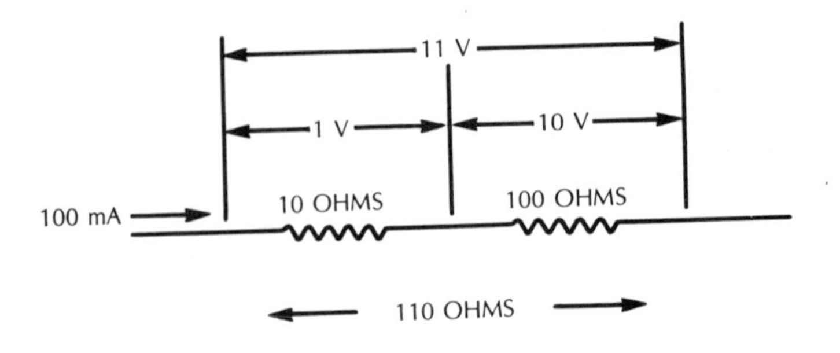

Fig. 2-7. Resistances in series.

When two or more resistors are connected in parallel, they have the same terminal voltage, and the total current is the result of adding together the current through each one separately. Thus, if 1 volt is applied across a 10-ohm and a 100-ohm resistor connected in parallel, the 10-ohm resistor will carry 100 milliamps and the 100-ohm resistor will carry 10 milliamps, for a total current of 100 + 10 or 110 milliamps. A resistance that will carry 110 milliamps at 1 volt represents 1 divided by 0.11, or 9.1 ohms. (Fig. 2-8).

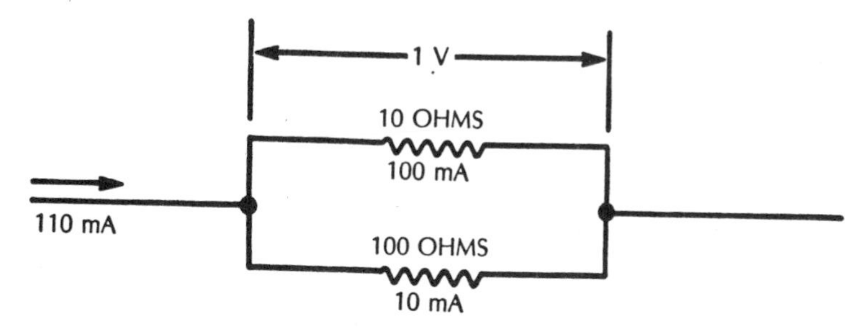

Fig. 2-8. Resistances in parallel.

Various formulas can be used to determine the value of each resistor in a parallel combination. The most direct and easiest to understand, because it shows how the result is derived practically, is the use of reciprocals.

$$\frac{1}{R_S} = \frac{1}{R_1} + \frac{1}{R_2} + \frac{1}{R_3} + \frac{1}{R_4} + \cdots \quad (4)$$

This form is a little unhandy to use, because each value must be converted to its reciprocal before adding, and then the final result is again converted back to its reciprocal. For two resistors in parallel, which is a calculation more frequently encountered, a simpler formula is:

$$R_S = \frac{R_1 \times R_2}{R_1 + R_2} \quad (5)$$

Since a more complicated combination (using more than two in parallel) can be treated successively, using only two values at a time, this formula is frequently used for more complicated arrangements. As a check, let's work in the following example both ways.

Example 4

Find the combined resistance of 1000 ohms, 2200 ohms, 3300 ohms and 4700 ohms in parallel. First, using the reciprocal method:

The reciprocal of 1000 is 0.001
The reciprocal of 2200 is 0.000455
The reciprocal of 3300 is 0.000303
The reciprocal of 4700 is 0.000213
Adding the reciprocals: 0.001971
The reciprocal of 0.001971 is 507 ohms.

Now, take the values two at a time. First combine 1000 and 2200 by formula (5):

$$\frac{1000 \times 2200}{1000 + 2200} = \frac{2200000}{3200} = 688$$

Combine 688 with 3300 by formula (5) again:

$$\frac{688 \times 3300}{688 + 3300} = \frac{2270000}{3988} = 570$$

Combine 570 with 4700 by formula (5) once again:

$$\frac{570 \times 4700}{570 + 4700} = \frac{2678000}{5270} = 507 \text{ ohms, as before}$$

The results agree.

EFFECT OF TOLERANCES

The above calculations assume that each resistance value is known accurately. More frequently, what we know is the *rated* values of the resistors, which are subject to tolerances, as discussed earlier. Take the example of the series connection in Fig. 2-7: 100 ohms in series with 10 ohms. The nominal total is 110 ohms. Now suppose the 100-ohm resistor has a tolerance of 10 percent: it could vary from 90 to 110 ohms, thus somewhat swamping the effect of the smaller value, whatever its tolerance.

In fact, one reason for using such a small resistor in series with a much larger one might be to correct for a deviation (within tolerance) in the larger value. If the actual value of the nominal 100-ohm resistor is 90 ohms, a 10-ohm resistor in series will correct it. And if the 10-ohm resistor has a 10-percent tolerance, this will correct the overall result to within 1 percent, as shown in Fig. 2-9.

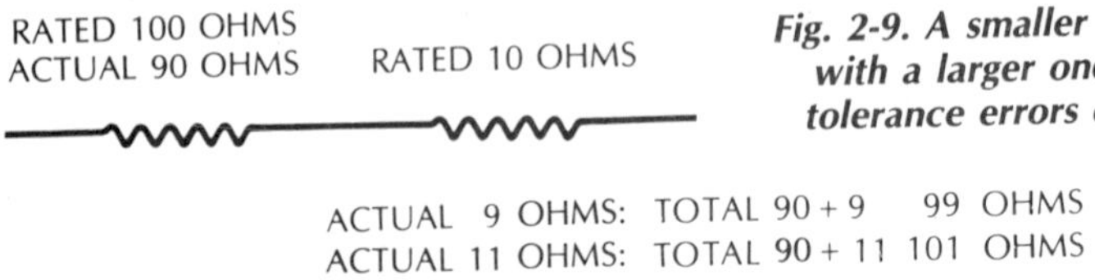

RATED 100 OHMS
ACTUAL 90 OHMS
RATED 10 OHMS

ACTUAL 9 OHMS: TOTAL 90 + 9 99 OHMS
ACTUAL 11 OHMS: TOTAL 90 + 11 101 OHMS

Fig. 2-9. A smaller resistance in series with a larger one can correct for tolerance errors of the larger one.

Now take a similar example with the parallel connection (Fig. 2-8): 1000 ohms in parallel with 100 ohms. Nominally, this figures to 91 ohms, as calculated earlier. Again, if the 100-ohm resistor has 10 percent tolerance, it can vary from 90 to 110 ohms, which swamps the effect of the parallel 1000-ohm resistor (Fig. 2-10). Suppose the nominal 100-ohm resistor has an actual value of 110 ohms.

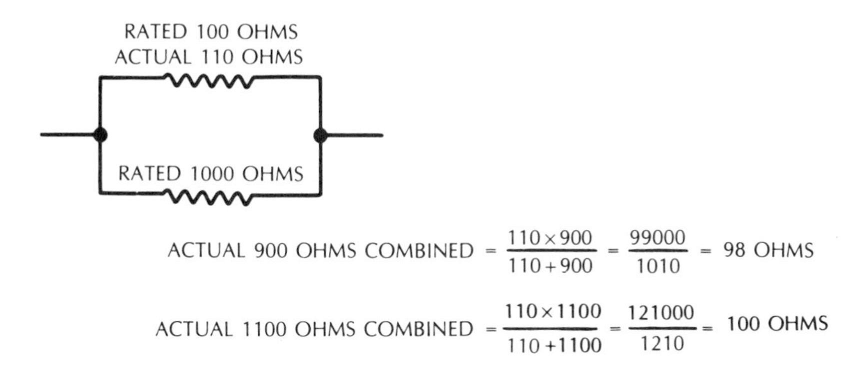

Fig. 2-10. A larger resistance in parallel can correct for an error in the opposite direction from the series addition.

Connecting 1000 ohms in parallel will produce a resulting resistance of:

$$\frac{110 \times 1000}{110 + 1000} = \frac{110000}{1110} = 99 \text{ ohms}$$

which is within 1 percent of the desired result.

Now suppose the nominal 1000-ohm resistor varies over a 10 percent tolerance range, from 900 to 1100 ohms; the possible results are:

$$\frac{110 \times 900}{110 + 900} = \frac{99000}{1010} = 98 \text{ ohms}$$

and

$$\frac{110 \times 1100}{110 + 1100} = \frac{121000}{1210} = 100 \text{ ohms}$$

Thus, a 10-percent variation in the larger of the parallel resistors produces about a 1-percent variation of the overall resistance. Of course, this is because its value is 10 times as high.

TEMPERATURE COEFFICIENTS

The resistance of a conducting material is affected in various ways by its temperature. The resistance value of most pure metals is almost perfectly proportional to its absolute temperature (measured in degrees Centigrade above −273 degrees C and called "degrees Kelvin "). Thus, at normal temperatures, for each degree Centigrade change in temperature the resistance varies a little less than 0.4 percent (Fig. 2-11).

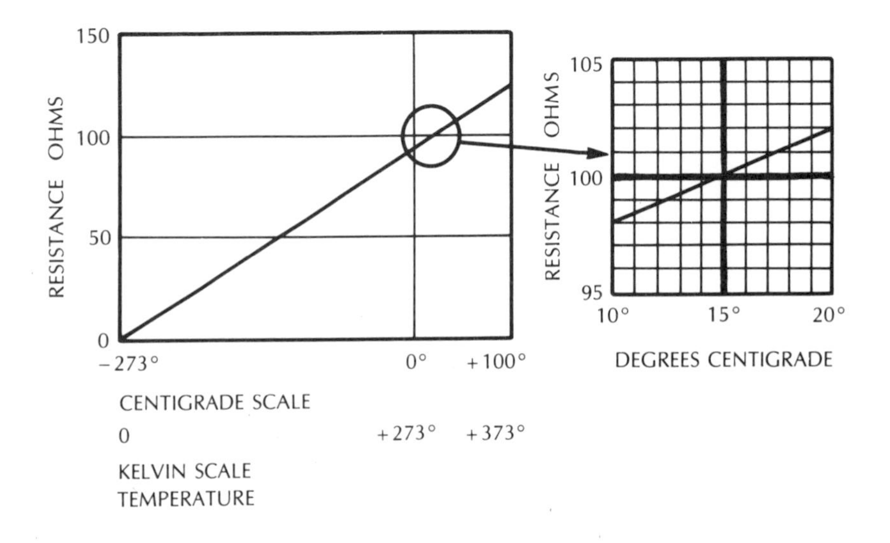

Fig. 2-11. The normal kind of temperature coefficient for resistors made of conventional conducting material (not composition material).

At temperatures approaching absolute zero, a different behavior shows up: a conductor suddenly becomes a super-conductor, rather than gradually diminishing in resistance all the way. At 4 or 5 degrees above absolute zero, where according to the rule given in the previous paragraph the resistance should be a little more than one one-hundredth of the normal-temperature resistance, it suddenly conducts about a million times as well—all resistance virtually disappears (Fig. 2-12). This phenomenon is the basis of the relatively new science of cryogenics, in which conductors are operated at temperatures close to absolute zero.

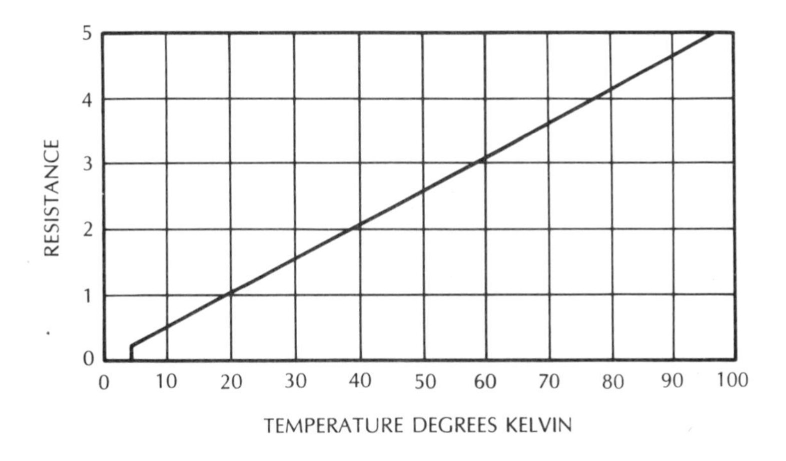

Fig. 2-12. Close to absolute zero, a superconductive region exists.

In more common use than cryogenic effects are other irregularities that occur in more normal temperature ranges. Many alloys have temperature coefficients that depart widely from that of pure metals over a specific temperature range. A typical temperature-resistance chart is shown in Fig. 2-13. Notice that over the wide range, the resistance change follows the general proportional relationship for all conductors, but that over a short portion, the temperature coefficient is close to zero. Resistance hardly changes at all over this range. This characteristic is useful for certain applications.

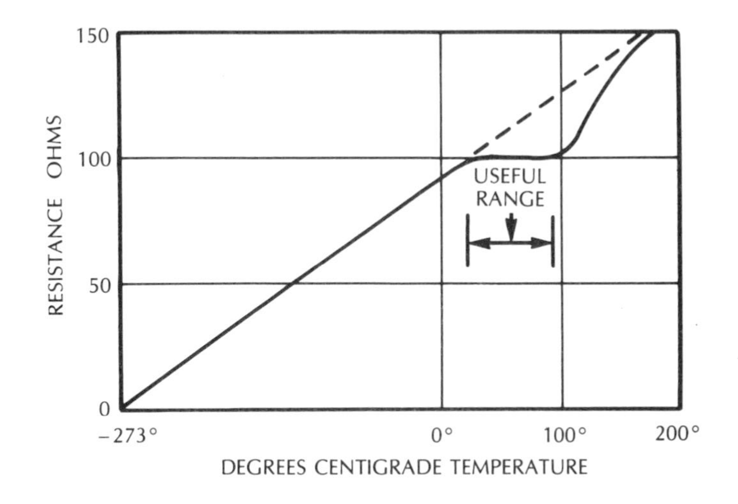

Fig. 2-13. Special alloys can almost eliminate resistance temperature coefficient over a limited range.

Iron has another useful characteristic. Under ordinary temperatures, its coefficient is about 0.0005 (instead of the more normal 0.004, approximately). Then, at 700 or 800 degrees, the temperature coefficient suddenly rises (Fig. 2-14), so that by appropriate design a component using this kind of "resistance" can hold current practically constant as the voltage across it is changed (Fig. 2-15).

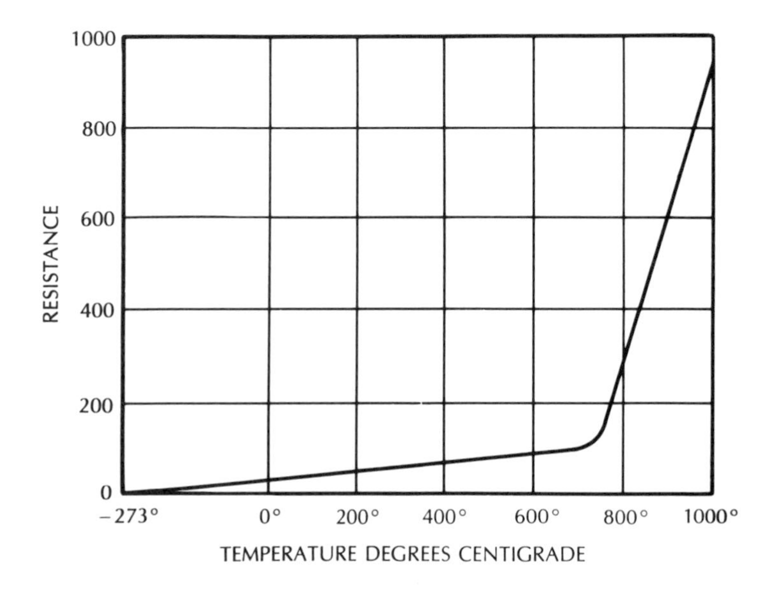

Fig. 2-14. Temperature characteristic typical of iron and some of its alloys.

To achieve this effect, the special resistance must be enclosed in vacuum or inert gas, which serves the dual purpose of controlling the heat dissipation and thus the current value at which this control is effected. The vacuum or gas protects the material from oxidation, which would quickly disintegrate the iron when raised to that temperature in a normal atmosphere.

Carbon, as well as most of the compositions used for resistors (other than wirewound types), has a negative temperature coefficient—the resistance drops as temperature rises. Also, carbon is not a very good conductor of heat, not as good as the metallic electrical conductors. For this reason, earlier resistors made from a solid stick of carbon, such as graphite, made poor resistors when it came to dissipating heat. The inner part of the resistor would naturally reach a higher temperature than its surface, which was cooled by the surrounding air. This meant that the inner part possessed a lower resistivity than the surface, causing the current to concentrate nearer the middle, thus increasing the dissipation there.

Because of this effect, resistors made of carbon had an unstable point in their operation, beyond which the central part rapidly heated to a sort of explosion point. This was the main reason for adopting designs in which the resistance composition is sprayed on a surface, forcing the current to flow only at the surface.

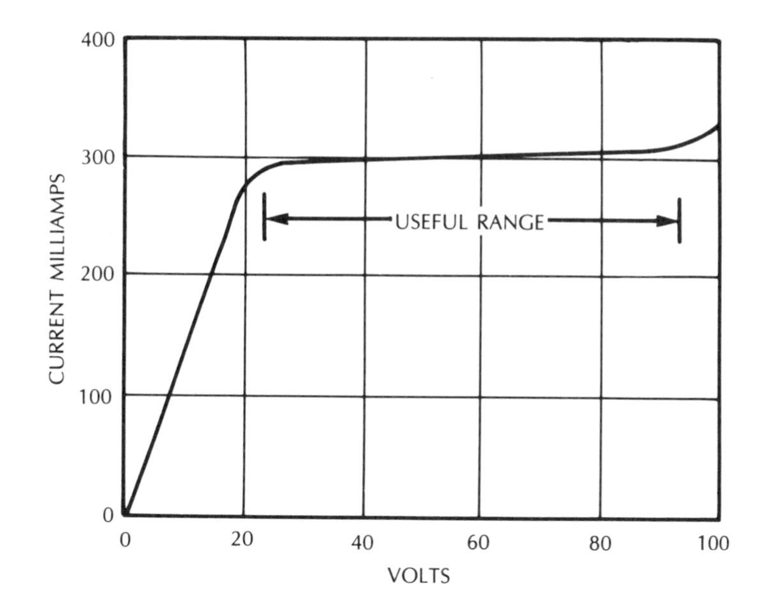

Fig. 2-15. In a controlled environment (vacuum or inert gas) a resistance made of iron alloy wire will maintain almost constant current over a range of voltage values.

Resistors constructed in this way will dissipate far more power than the solid type for the same size or, conversely, can be made much smaller for the same rating. (Dissipation in various types of components is discussed more fully in Chapter 3.)

Thus far, this chapter has followed the original edition of this book. While applications for resistance calculations have extended, as we shall find later in the book, uses for actual resistors are little different. The main difference is that there are far fewer of them—many are built right into the integrated circuits we will find later in the book. Where actual discrete resistors are required in circuitry, everything that has been said so far still applies.

The main difference is that resistors of a given wattage rating, which we don't get to until Chapter 4, are much smaller, through improved design that enables heat to be dissipated more readily, and that production of resistors to any degree of precision has been made easier, so that use of the preferred value sets discussed in this chapter, which was virtually a "must" for many years, is no longer necessary.

But preferred values are still available and in use. While they play a much reduced role, simply because discrete resistors (by which is meant ones that are wired into the circuit as separate entities from the integrated circuits) are used in fewer places. However, the principles learned in this chapter are still basic to what you will learn later.

Examination Questions

1. Resistors are needed to build an attenuator. They should have exact values, within 5 percent of the design value. The values below are required. State the color codes for each resistor.

29	4300
51	2200
110	1000
350	350
410	100
480	10

2. A revision in the attenuator design requires different resistor values, tabulated as below. State the new color codes.

300	2.0
170	4.0
76	8.1
33	24
21	85
17	890

3. The following resistors were found in a box marked "special values" in a laboratory. Identify the value and tolerance of each.

BLUE GREEN YELLOW GOLD	WHITE GREEN BLUE GOLD
RED PURPLE GREEN SILVER	GRAY RED GOLD SILVER
ORANGE BLUE RED GOLD	BLUE GRAY GOLD
YELLOW PURPLE GOLD	YELLOW PURPLE YELLOW
GRAY YELLOW BLACK GOLD	BROWN BLACK BLUE GOLD

4. The following also are found in the same box as those in Question 3.

PURPLE RED GREEN GOLD	GRAY BLUE BLACK GOLD
BLUE BROWN YELLOW GOLD	GREEN BLUE GOLD SILVER
GREEN YELLOW ORANGE GOLD	ORANGE WHITE GOLD SILVER
YELLOW ORANGE RED GOLD	ORANGE ORANGE GOLD
ORANGE RED BROWN GOLD	BROWN BLACK GOLD

5. List the value variations allowed by the tolerance in Question 1.
6. List the value variations allowed by the tolerance in Question 2.
7. List the tolerances represented by the resistor codes in Question 3.
8. List the tolerance represented by the resistor codes in Question 4.
9. A resistance of 1.5 megohms (a megohm is a million ohms) has 45 volts across it. Find the current flowing in microamps (millionths of an amp).

10. A resistance of 62 K is connected across 125 volts. What is the current flow in the resistor?
11. The current flowing in the resistor in Question 10 also passes through a resistor of 510 ohms. What is the voltage across this resistor?
12. This same current is passed through a 1 K resistor. What is the voltage across it?
13. Calculate the resistance of copper wire, measuring 0.1 millimeter each way, 15 meters long, wound into a coil.
14. Calculate the resistance of a loudspeaker voice coil, using rectangular copper wire 0.1 millimeter by 0.25 millimeter by 2.5 meters long.
15. What is the resistance of the coil in Question 13 if wound with aluminum wire?
16. What is the resistance of the voice coil in Question 14 if wound with aluminum wire?
17. Two resistors are connected in series, rated at 1.5 K, 10 percent and 2.2 K, 20 percent. Find the limits of the resistance variations in the combination.
18. The tolerance of the resistors of the previous question are reversed. Find the new limits in the combination.
19. Two resistors connected in parallel have ratings of 3.9 K, 10 percent and 22 K, 20 percent. Find the limits of the resistance variations in the combination.
20. How much closer (give the new limits) will the combined resistance in Question 19 be controlled if the 22 K resistor has a tolerance of 5 percent?
21. Find the overall resistance of the following resistors connected in parallel: 15 K, 27 K, 39 K, 56 K.
22. Find the overall resistance of resistors rated at 510, 560, 620 and 750 connected in parallel.
23. Assuming all the values in Question 21 have 10-percent tolerances, what are the variation limits for the combined resistance?
24. Assuming all the values in Question 22 have 5-percent tolerances, what are the limits? What are they if the 560-ohm resistor has a 10-percent tolerance?

3
Kirchhoff's Laws

The discussion of resistors in series and parallel in the previous chapter paves the way for a consideration of the laws of branching circuits and loops (also called meshes), known as Kirchhoff's laws. Since they are based on the simple facts stated in Ohm's law, they are really derived from it. In a simple circuit, both Kirchhoff's laws seem obvious enough not to need clarification. But in more complicated circuits, they provide a basis which is not immediately obvious for solving problems in those circuits.

Take the simple parallel connection (Fig. 3-1). Current reaches the parallel point and divides (at another point it may come together again): the current reaching that branching point is equal to the total current leaving it (or vice versa at the other end).

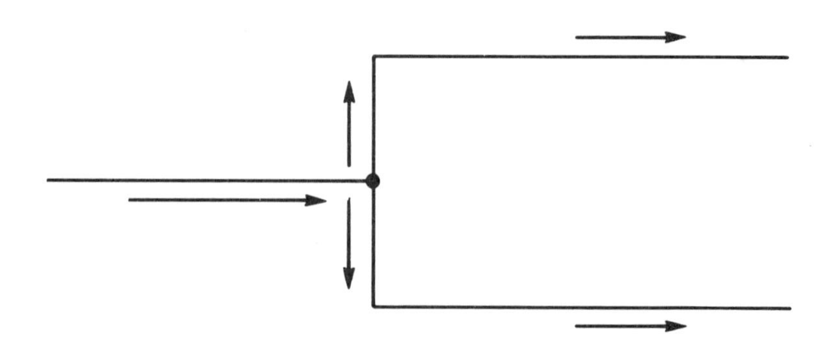

Fig. 3-1. In a branching circuit, the current divides (as shown here) or combines.

In slightly more complicated circuits, more than one current may enter the branching point and more than one current may leave the same point (Fig. 3-2). From the above statements it is obvious that the total current entering that point is equal to the total current leaving it.

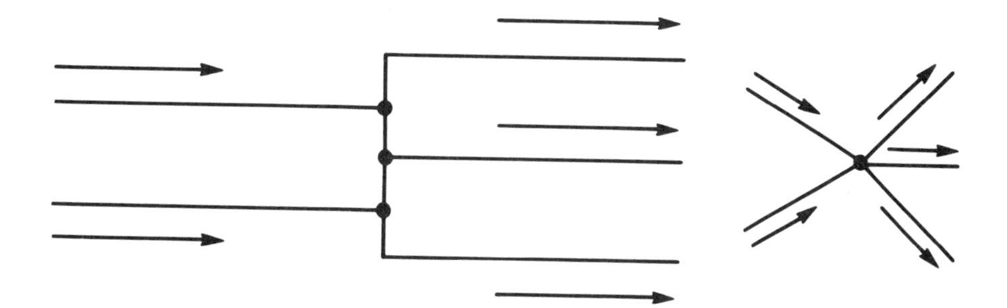

Fig. 3-2. Two ways of illustrating a more complicated branching circuit. The diagram on the right illustrates the Kirchhoff notion of a "point."

Kirchhoff, who had a mathematical way of thinking about things, said that the *algebraic* sum of the currents entering or leaving a point is always zero.This means that if we consider currents as entering the point, then those leaving it are negative, while if we consider current as leaving the point, those entering it are negative. It is merely a more sophisticated way of saying what we said in the previous paragraph: when the positive numbers total the same as the negative numbers, the sum of all of them is zero.

Kirchhoff's second law relates to the voltages, potentials or electromotive force around a loop or circuit. Voltage, potential and electromotive force are names for essentially the same kind of quantity or thing. Roughly speaking, electromotive force describes its means of generation, for example, by a battery, generator or coil, while potential describes its distribution and use. In any case, this quantity is called voltage. But the words often are used almost interchangeably.

If a battery produces 3 volts (Fig. 3-3) and is connected to two resistors in series, the circuit or loop consists of three components—the battery and two resistors.

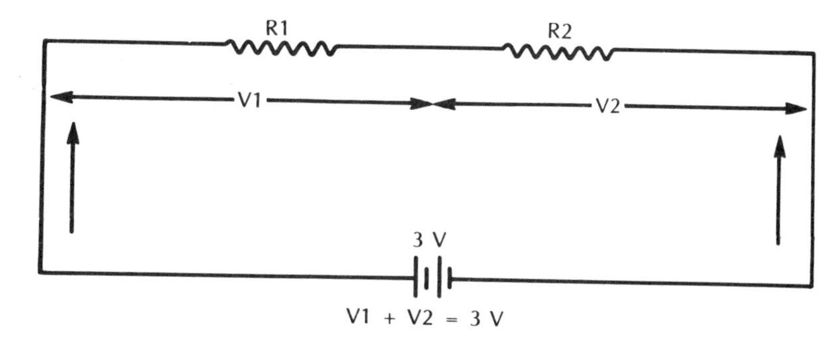

Fig. 3-3. Schematic illustrating a simple application of Kirchhoff's second law.

The voltage produced by the battery is equal to the sum of the voltage drops across the resistors. That is merely repeating the facts about series resistors from the previous chapter.

If more components or elements are included in the circuit or loop, the same holds true. And if you consider the voltage changes, as you follow the circuit through the battery, you come back to the same starting point, or potential, so the sum of the voltages around the complete circuit, back to the starting point is again zero (like the currents at a point). As the current goes through the battery, it recovers the voltage it lost going through the resistors.

Kirchhoff's laws may be stated briefly:

1. **The algebraic sum of the currents meeting at a point is always zero.**
2. **The algebraic sum of the voltages, taken in order round a loop or circuit, back to the starting point, is always zero.**

We have considered them as obvious in dc circuits. They may be less so, when we consider ac circuits, but the same laws still apply. However, in more complicated dc circuits we need a little more understanding, too, so the rest of the chapter is devoted to that.

Each of these laws is simple when you consider circuits that involve only one or other of them: a number of currents converging on and leaving a point, or a loop with voltages acting around it. It is where you have both acting at once that problems in understanding are apt to strike.

THE BRIDGE CIRCUIT

Consider the simple Wheatstone Bridge circuit (Fig. 3-4). The major use of this circuit is to find the value of an unknown component by adjusting the values of the components in the arms of the bridge so that no current flows in the middle arm (R5). This means that the voltage between the ends of this arm is zero; in other words, for the bridge to balance (zero current through R5),the same voltage must exist at both ends.

From this, we know that the voltage drops across R1-R2 and R3-R4 are the same. This represents the null or balanced condition. Since the voltage across R1 and R3 divides in proportion to their resistance values, because the same current flows through both (which must be so when none flows either way through R5), and since the voltage across R2 and R4 also divides in proportion to their resistance values, we have a simple statement in proportion,

$$\frac{R1}{R3} = \frac{R2}{R4} \quad \text{or} \quad \frac{R1}{R2} = \frac{R3}{R4} \qquad (6)$$

From this, if we know any three values, we can calculate the fourth:

$$R1 = \frac{R2R3}{R4}\,;\; R2 = \frac{R1R4}{R3}\,;\; R3 = \frac{R1R4}{R2}\,;\; R4 = \frac{R2R3}{R1} \qquad (7)$$

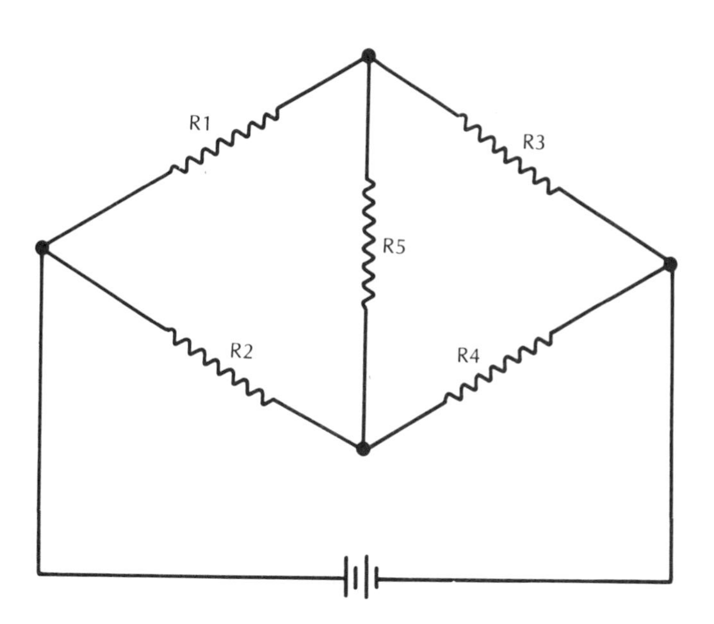

Fig. 3-4. Simple Wheatstone Bridge circuit.

Important Note: For these formulas to apply, the bridge must be adjusted to a null condition, so that no current flows through R5, which also means that R5 has no voltage across it or that the voltage at junctions R1-R3 and R2-R4 is the same.

This is only one form of bridge circuit. Other devices of a similar nature are used, both for dc and ac work, and we shall study them later. But the ratio derived in equation (6) and the dependent solutions (7) relate to a special condition of that circuit—balance or null. Sometimes, for various reasons, either in this or other bridge circuits, we are also concerned with what happens when null or balance does not exist in the bridge: then, current flows in R5 and there is a voltage across it.

Solving the condition, then, provides an exercise in the use of both Kirchhoff's laws. We will show how this is done, since it helps to understand circuit theory, but we do not suggest that you try to use the equations we develop as a means of solving the condition, because there are easier ways, as a rule.

Take the circuit in Fig. 3-4, in a condition of unbalance. Call the battery voltage V_b and the voltages across each of the resistors and the currents in them by V and I, using the same numerical subscript as the resistor with which the voltage or current is associated. Now we can write an entire sequence of facts. First, using Kirchhoff's first law:

At the left side junction: $I_b = I_1 + I_2$ (8a)

At the top junction: $I_1 = I_3 + I_5$ (8b)

Note: in that equation, we have assumed that the current in R1 divided between R3 and R5. If, in fact, the current in R5 flows the opposite way, which it may do, the value calculated for I_5 under this assumption would be negative. The important thing, if we are to arrive at a correct solution, is to make the same assumption for all equations.

At the bottom junction, following the same consistent assumption,

$$I_2 + I_5 = I_4 \tag{8c}$$

and at the right side junction: $$I_3 + I_4 = I_b \tag{8d}$$

In addition to these current-division facts, we can determine some voltage facts, according to Kirchhoff's second law. Over the top:

$$V_b = V_1 + V_3 \tag{9a}$$

Notice that, except in balance, I_1 and I_3 are not the same, as in a simple series circuit. Under the bottom of the bridge:

$$V_b = V_2 + V_4 \tag{9b}$$

Finally, taking the left side triangle, and remember our assumption about current flowing through R5,

$$V_2 = V_1 + V_5 \tag{9c}$$

and in the right side triangle:

$$V_3 = V_5 + V_4 \tag{9d}$$

That is a total of eight equations, all of which must be true at once (as well as the Ohm's law equation for each resistor), to determine what is actually happening in that circuit. The equations can be rearranged or recombined to get different equations, but it is still rather a lot of variables to handle for a rather small circuit. If you have a computer handy capable of this kind of solution, and if you know how to program it to solve such a problem, it will tell you the answers.

Mathematicians like that kind of "set," because it looks complicated or impressive, and because (they will tell you) it "generalizes," which is their way of saying that those equations will solve any problem of that form, which can include quite a variety. Although I, too, like mathematics, I also like things I can understand better than those that give me difficulty and, frankly, solving a bunch of equations with that many unknowns is a bit much for me to comprehend at one go. If I have my book handy, I can use the method of determinants, which goes much faster by computer, but that does not help me visualize what is happening in the circuit.

Alternatively, I can use another method that also goes much faster by computer, which is programmed to finish the job and present the answers. This method involves making a first guess at, say, I_5, finding how much this is wrong, then working to reduce the error until the correct answer emerges. For example,

example, I assume a current in R5 and, based on that assumed current, calculate everything else, only to find that I get a discrepancy at the end. So I adjust the assumed current in R5 for another try and see whether or how much that reduces the error, and so on, until I arrive at a correct answer.

When I went to school, that kind of working was called "cut and try," and teachers frowned on it. But now computers can do the work so much faster, it's a different story. There are problems for which this is the easiest method now. Still another method, and one that best helps to visualize what happens in this situation, is called:

THEVENIN'S THEOREM

The current in any impedance, Z_L (for which you can read resistance, R_L, with equal validity), connected to two terminals of a network consisting of any number of impedances and generators (or voltage sources) is the same as though Z_L were connected to a simple generator, whose output voltage is the open-circuit voltage at the terminals in question, and whose impedance is the impedance of the network looking back from the terminals, with all generators replaced by impedances equal to the internal impedances of these generators.

If you read that twice and still do not know what it means, welcome to the club. You may come back to it some day, and find it much more enlightening. However, it is much easier to understand than its language implies. And the best way to understand it is to see how it is used.

In a case like the bridge in Fig. 3-4, there is no problem if you leave out resistor R5 (Fig. 3-5).

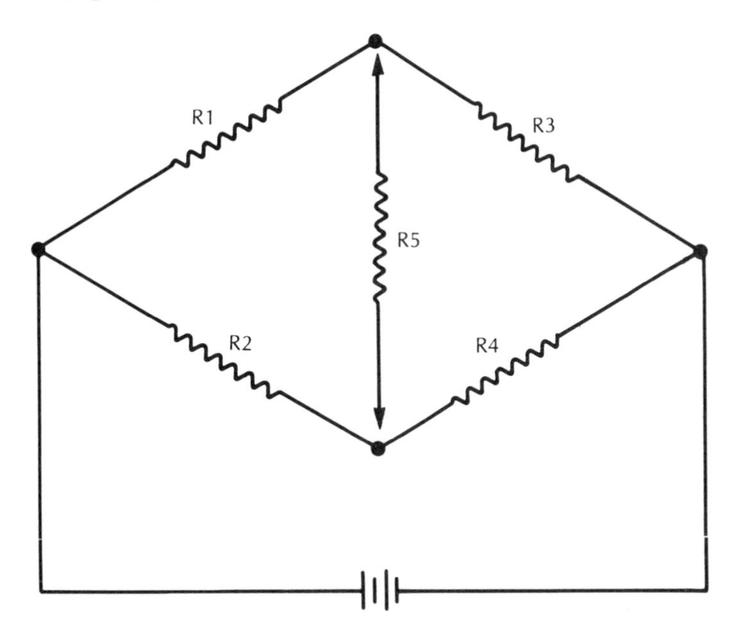

Fig. 3-5. Imagining the middle resistor, R5, to be disconnected at first provides an easier approach to solving an unbalanced bridge problem.

Now, you can calculate the voltages at the junctions of R1 and R3 and of R2 and R4 fairly easily. Taking the total of R1 and R3, you calculate the current through them and the voltage across each. With the total of R2 and R4, you calculate the current through them and the voltage across each. Let's put in some values to try it. Assume that R1 is 6 ohms, R2 is 10 ohms, R3 is 12 ohms, R4 is 10 ohms and R5 is 11 ohms.

R1 and R3 add up to 18 ohms, which, across 3 volts, will pass one-sixth (or 0.167) amp. One-sixth amp in 6 ohms produces a drop of 1 volt and in 12 ohms produces a drop of 2 volts, which adds up to 3, the battery voltage. R2 and R4 add up to 20 ohms, which, across 3 volts, pass 0.15 amp (or 150 milliamps, if you prefer). 0.15 amp in 10 ohms (each of them) produces a drop 1.5 volts each side.

Now, we are ready to apply Thevenin. What the Theorem says, in effect, is that each of these points, in addition to a voltage, has something we can view as an "internal resistance." In this kind of circuit, the value of the internal resistance is equivalent to the two resistances that produce each voltage, considered as in parallel. Thus, the internal resistance (Fig. 3-6) at the junction of R1 and R3 is equal to those two values in parallel, which calculates to 6 × 12 divided by 6 + 12 is 4 ohms. Likewise, the internal resistance at the junction of R2 and R4 is equal to those two resistors in parallel, which calculates to 10 × 10 divided by (10 + 10) or 5 ohms.

Now, the voltage difference between the points to which R5 will be connected is 0.5 volt, which is easily deduced either from the difference between V1 and V2 or between V3 and V4. And the total effective resistance that this 0.5 volt drives current through (internal and external, from the Thevenin viewpoint) is 4 + 11 + 5 or 20 ohms. The first and last are internal, the middle one being R5. This will allow a current of 0.5 divided by 0.02 or 25 milliamps.

From this we can calculate all the values in the circuit as follows. Connecting R5 will increase the current in R1 and reduce it in R3. To find out how much, we calculate the voltage first. The change in voltage at this junction will be the same as if current (25 milliamps) were drawn through 4 ohms. This yields a voltage change of 4 × 0.025 or 0.1 volt, so the voltage across R1 rises to 1.1 volt and the voltage across R3 drops to 1.9 volt. The current in R1 is 1.1 divided by 6 equals 0.1833 amp or 183.3 milliamps. And in R3 it is 1.9 divided by 12 equals 0.1583 amp or 158.3 milliamps. The difference is 183.3 − 158.3 or 25 milliamps, which confirms our calculation by Kirchhoff's first law.

The voltage across R5 is 0.025 × 11 or 0.275 volt, making the voltage at the junction of R2 and R4 equal to 1.1 + 0.275 or 1.375 volt, which should also be the voltage across R2. Looking at R2 and R4, the internal resistance is 5 ohms and the current 25 milliamps, which means the voltage shifts 5 × 0.025 or 0.125 volt from the 1.5 it started with before R5 was connected. 1.5 − 0.125 equals 1.375, which confirms the result according to Kirchhoff's second law. In effect, Thevenin serves as a direct tool to solve equations in many unknowns by a step-at-a-time procedure. At the same time, it helps us visualize what happens, in this case by considering the effect of making that last "connection": putting R5 across the bridge.

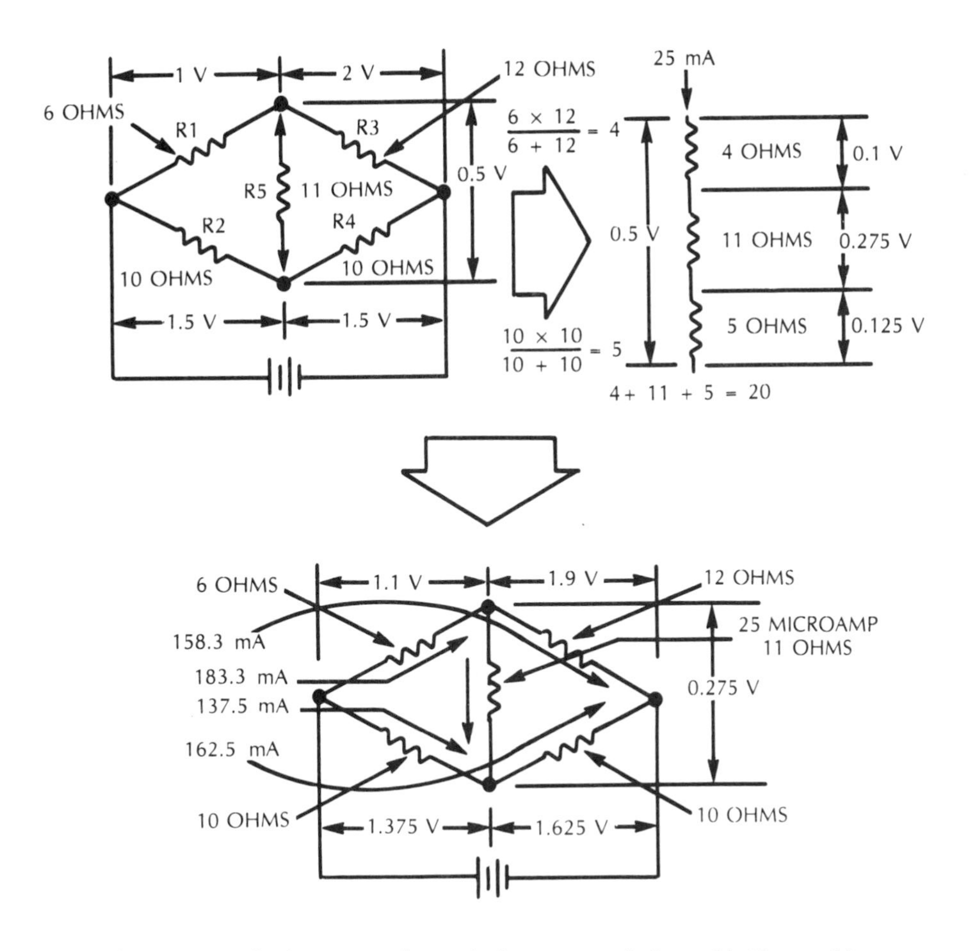

Fig. 3-6. Developing a complete solution to an unbalanced bridge problem.

Bridge circuits are found in many electrical and electronic applications based on variations of the simple Wheatstone variety. In most instances, the bridge is used to find a null or balance. But some form of instrument, such as a meter, must be connected between the null points to determine when a null has been obtained.

Figure 3-7 shows one form of bridge. It consists of precision resistors, arranged in decade units, so a reference value can be set precisely to balance, using (in this case) up to four significant figures, marked as thousands, hundreds, tens and units.

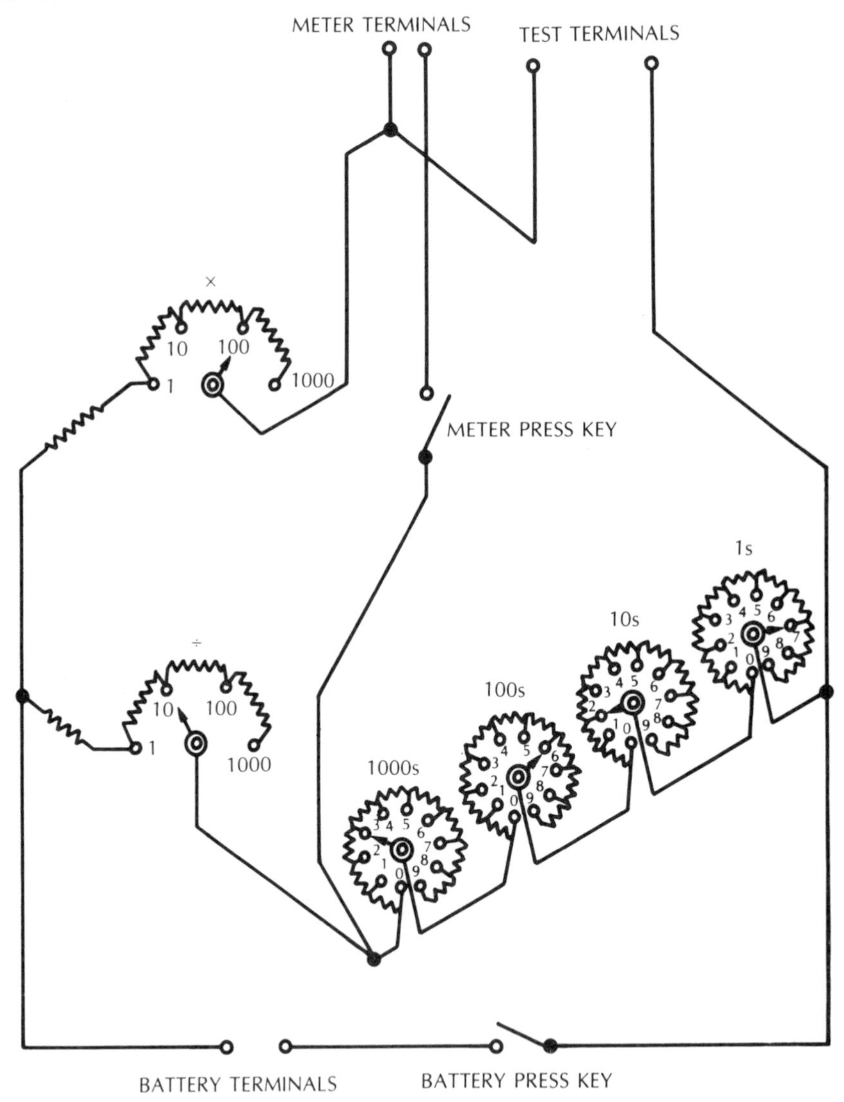

Fig. 3-7. Schematic layout of a typical dial-type Wheatstone Bridge. If the positions of the dials shown represent balance, the resistor connected to the test terminals will be 36.27K.

If the ratio arms of the bridge use identical resistors (both either 1s, 10s, 100s or 1000s), then balance will occur when the resistance connected to the test terminals is matched by the calibrated arm reading in thousands, hundreds, tens and units, directly.

With the appropriate ratio arm, it is possible to multiply or divide the range of resistance being measured by 10, 100 or 1000. Therefore, the bridge can measure, to a 4-figure accuracy, resistors from 1.000 ohm (the zeroes indicate the precision) up to 9,999,000 ohms (which is pretty close to 10 megohms). It will also give a close indication, based on a progressively reduced possible accuracy, down to 0.001 ohm. For the higher values, larger battery voltages are necessary to obtain accurate readings because the currents in the bridge are very small if a low voltage is used. For lower value measurements, use low battery voltages to avoid currents that might overheat the resistors.

Bridges of this type are usually provided with press keys to take readings. This is to conserve battery current and to facilitate observing any deflection of the meter that occurs. Always, and especially if you are measuring the resistance of a coil or anything else that possesses inductance, press the battery key first, allowing time for the current to become steady in any inductive element, before pressing the meter key. Then release the meter key before you release the battery key.

This is important, especially when a sensitive meter is used. The bridge will show balance only for resistance values after currents have reached a steady value, which will take a moment or two, due to inductive kicks, depending on the amount of inductance involved, of course. Breaking the battery circuit will produce a second inductive kick, which is why the meter circuit should be broken first (before the inductive kick occurs), although a pause before releasing the battery key is not necessary.

The best kind of meter is one with a center zero, so it can deflect either way. The meter press key helps you to see the exact null, by finding the setting of the bridge where no movement of the meter needle occurs when the battery key is held down and the meter key is repeatedly pressed and released.

LOW-RESISTANCE MEASUREMENTS

Measuring low-resistance values requires a somewhat different technique. If a standard bridge is used to measure a very low resistance, it may well be that the contact resistance or the connecting wire resistance, where the resistance being measured is connected to the bridge, will invalidate the reading, since it may be comparable to the resistance to be measured or even exceed it. For example, if you wanted to measure the resistance of a centimeter cube of copper, mentioned in the previous chapter, how would you go about it? Did you wonder that when you read about it? Any wires you used to connect the piece to be measured to the bridge would have many times the resistance of the block of copper, if that is how you tried to do it.

The best method is to use separate terminals to introduce the current to the resistor (or in this case, conductor, although it is resistance you want to measure) and to measure the voltage developed by its resistance.

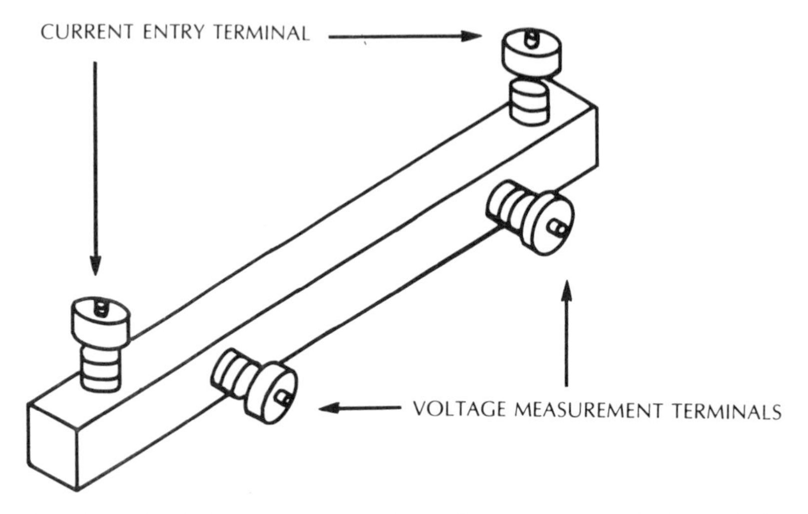

Fig. 3-8. A method of connecting to low-value resistors or test conductors for measurement purposes.

Basically, two pairs of terminals are used, one pair to lead the current in and out and another pair to read the voltage dropped in the resistance (Fig. 3-8).

INSTRUMENT MULTIPLIERS

An instrument multiplier will convert a sensitive meter to read high current values. But first we need to explain how instrument multipliers work. Voltmeters of the conventional moving-coil variety can be made to read higher voltages by adding resistors in series with the instrument, which is simple to do (Fig. 3-9).

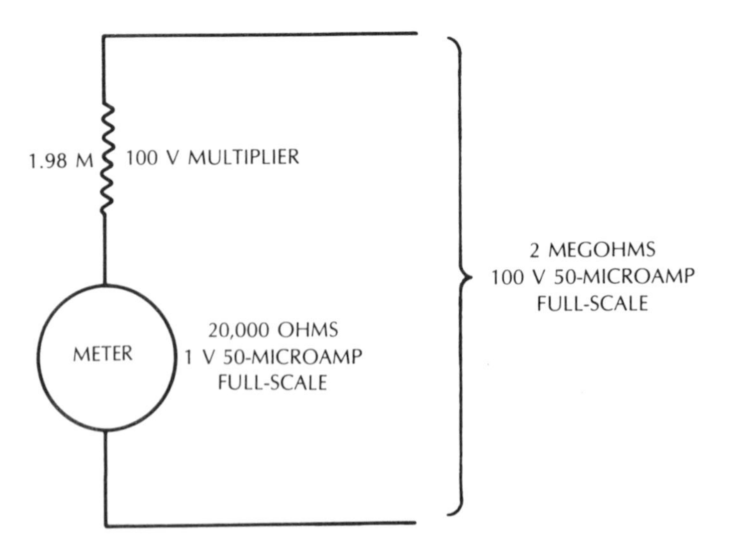

Fig. 3-9. How voltmeter ranges were extended, before the advent of digital instruments.

If a voltmeter, as a basic movement, reads 1 volt full scale and has an internal resistance of 20,000 ohms, it is said to have a sensitivity of 20,000 ohms per volt. It takes one twenty-thousandth amp, or 50 microamps of current for full-scale reading. To make the instrument read 100 volts full scale, the total resistance of the meter should be built up to 100 × 20,000 ohms, or 2 megohms, requiring a series resistor of 2 − 0.02 or 1.98 megohms. The meter takes the same current, 50 microamps in this case, to produce a full-scale reading, whatever the voltage of the circuit associated with the source of that current. Therefore, a half-scale reading would indicate 50 volts, instead of 0.5 volt, because half the current, or 25 microamps, flows through the meter, and so on.

Such a meter could not be used for reading heavy currents, because it requires only 1 volt for a full-scale reading. At 10 amps, say, this would mean the measurement would absorb 10 watts (volts × amps, see the next chapter) from the circuit being measured, which could not usually be allowed. Better use a different instrument coil (which would be supplied as a whole different movement with, perhaps, the same internal assembly, except for the coil) with a resistance of, say, 100 ohms.

As a meter with a coil of 20,000 ohms takes 1 volt, 50 microamps for a full-scale reading, one of 100 ohms, designed for the same assembly, would probably produce a full-scale reading with about 70 millivolts, 0.7 milliamps (700 microamps), in place of 1 volt, 50 microamps. Both of these combinations multiply to about 50 microvoltamps.

Now, suppose you want this meter to read 10 amps full scale. Since the meter will still take 70 millivolts to make it read full scale, the resistance, to pass 10 amps at that voltage needs to be 0.07 divided by 10 equals 0.007 ohm. With a resistance of 0.007 ohm connected in parallel with 100 ohms, the current should divide correctly. At full scale, this would be 10 amps through the 0.007-ohm resistor and 0.7 milliamps through the 100-ohm instrument movement.

But this is not as simple as it sounds. It can be difficult to achieve precisely because connections can easily have more than a 0.007-ohm resistance. To overcome this problem, use more than 0.007 ohm for the shunt (as the resistor that bypasses a current meter, to increase its range of reading, is called) and then connect the meter across precisely 0.007 ohms of that larger resistance, as a sort of voltage measurement (Fig. 3-8 again). Of course, the shunt is made of special, low-temperature coefficient resistance material, rather than copper bar, but the method is the same.

If the two pieces, the 0.007-ohm shunt, built like an ordinary resistor, and the instrument movement of 100 ohms, were connected in parallel directly, contact resistance would need to be kept very much smaller than 0.007 ohms. A contact resistance of 0.0007 ohms could cause a 10-percent reading error, and it could be much worse than that! But by doing it this way, the connections that can make the reading incorrect are those between the shunt and the meter, whose resistance is 100 ohms. The connections to the external circuit may increase the drop across the external meter terminals, but they will not change the shunt value as the meter sees it. So the resistance of the voltage connection to the 0.007-ohm shunt must be small, only in comparison with 100 ohms, which is much easier to achieve.

Using this method, it is simple to find the resistivity of a copper bar of 1 centimeter square cross-section: Current is passed through a length of it and the voltage measured along a central portion (theoretically 1 centimeter in length). However, a more accurate result would be attained by measuring along 10 centimeters and dividing the result by 10 (where the current is uniformly distributed throughout the conductor, which it may not be where it enters and leaves by some type of terminal). Measuring such a small voltage is a little difficult to do with a conventional meter, so this may be achieved by means of some kind of amplifier voltmeter. Such sensitive instruments are developed in Chapter 17.

POTENTIOMETER MEASUREMENTS

A potentiometer is a kind of bridge used a little differently. Perhaps the only use it really has in common with a bridge is null detection. Any kind of bridge is used for measuring resistance or conductance, or the associated ac counterparts (which are studied beginning with Chapter 4)—impedance or admittance. A potentiometer is used to measure voltages by employing a similar null technique.

Look at a bridge from a little different viewpoint. When the bridge arm is situated across null points, the voltage at both ends of the arm is the same and no current flows into or out of the circuits to which the arm is connected. If you measure the voltage at a junction point with an indicator voltmeter (Fig. 3-10), the voltmeter takes current to give a reading.

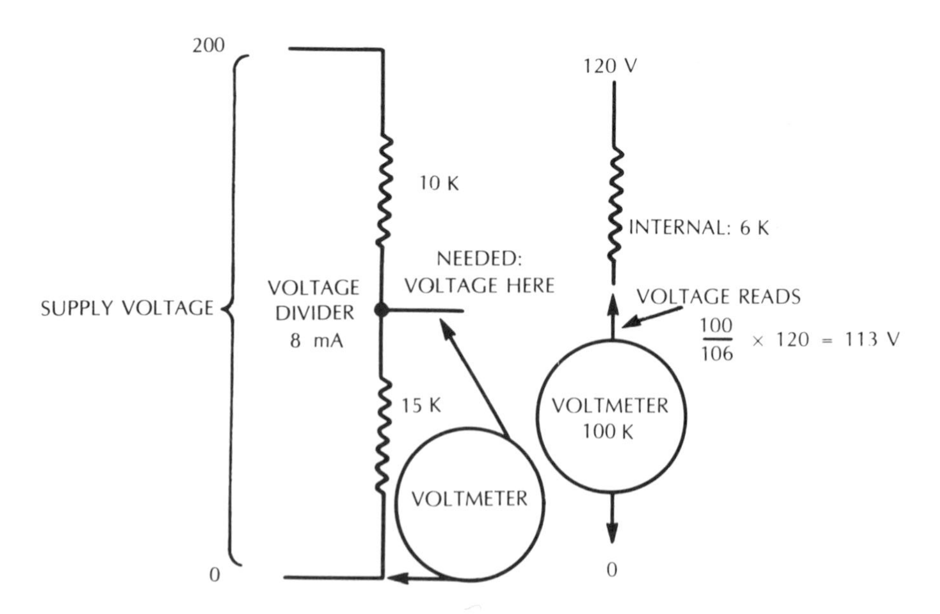

Fig. 3-10. Using Thevenin's Theorem to calculate the effect of connecting a voltmeter to a voltage divider.

Inaccuracy of Voltmeter Measurements

Maybe the current drawn by the meter is quite small, not enough to noticeably affect the voltage it is measuring. But the best way to see the effect of the current drawn by the meter is to look at some examples. Consider two resistors across a 200-volt supply, forming a voltage divider. The values of the resistors are 10,000 ohms and 15,000 ohms (with 10-percent tolerances). The current flowing in the resistors, in the absence of any load connection at the junction point, is 200 divided by (10 + 15) or 8 milliamps. With a load connected, the current in each resistor will change, as will the voltage at the junction. What you want to measure is the voltage at the junction.

If the voltmeter has a sensitivity of 20,000 ohms per volt, and uses a scale that reads 0-200, its resistance will be 200 × 20,000 ohms which is 4 megohms. Assuming the voltage is 60 percent of 200 (across the 15 K resistor) or 120 volts, then 4 megohms will take 120 divided by 4 or 30 microamps, which is quite a small fraction of 8 milliamps, less than half of 1 percent. So the meter loading would not affect the reading accuracy by as much as the probable internal inaccuracy in the meter itself.

But if the meter had a sensitivity of 500 ohms per volt, or if the resistors in the voltage divider where the voltage is to be measured were much higher, a different situation would prevail. At 500 ohms per volt, a meter to read 0-200 will have a resistance of 200 × 500 or 100,000 ohms. According to Thevenin, the resistance at the junction is 10 K in parallel with 15 K, or 10 × 15 divided by (10 + 15) or 6 K.

So the actual effective resistance across which the working voltage (that is present when the meter is not connected) is measured at 106 K, rather than only the meter's 100 K. Therefore, the current delivered to the meter coil will be 6 percent lower than it would be if the voltage remained stable when the meter was connected. Not only is the current lower. The actual voltage at the meter terminals drops this much, too. The drop is in the 6 K, as it delivers a reading to the 100 K that you know about.

This change in the actual voltage you read may be difficult to comprehend, because you tend to believe that an instrument is telling you the truth unless you know it to be "off." This is not a matter of meter error. It measures the voltage actually there when you connect the meter. The trouble is, connecting the meter changes the voltage, and you cannot measure it without the meter!

You can verify part of what is happening by using two meters, preferably with different sensitivities, both of which are in themselves accurate (or reasonably so). With the high sensitivity meter (20,000 ohms per volt) measure the voltage (Fig. 3-11). It is, say, 115 volts. (The reason for it not being the theoretical 120 volts could be due to resistor tolerances, or a load you have not measured, other than the meter.)

Now, without disconnecting that meter, also connect the low sensitivity meter. It probably will read 108.5 volts and at the same time the high sensitivity meter drops to a reading of 108.5 volts. Of course, there may be a slight difference between the readings, because one or both meters probably are not exactly accurate. We are assuming they are both equally accurate, or equally in error, as indicators of the voltage actually at their terminals. Finally, if you remove the more sensitive meter, the reading on the other may rise from 108.5 to 109 volts. You see it, you believe!

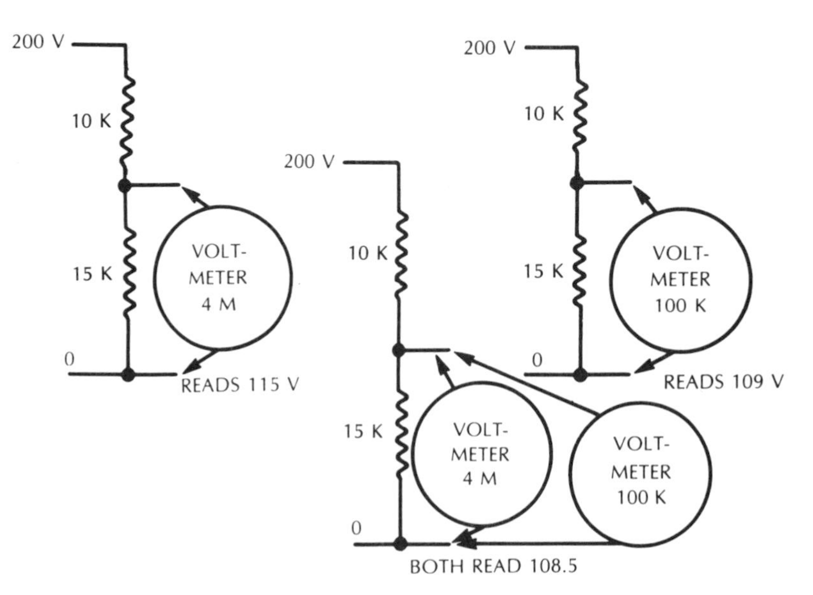

Fig. 3-11. Showing the different loading effect of different meters connected across a voltage-divider circuit.

But now, suppose that the resistors are for a much lower current circuit, where values of 1 megohm and 1.5 megohm replace those of 10 K and 15 K (Fig. 3-12).

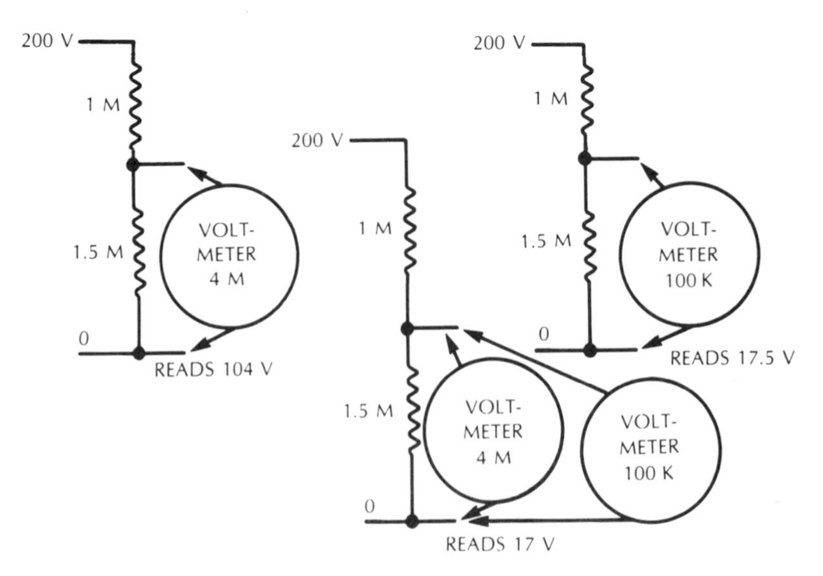

Fig. 3-12. Effect of the same two voltmeters on a voltage-divider circuit that uses much higher resistance values.

Now, the source resistance at the junction is 600 K instead of 6 K. The meter with 4-megohm resistance will use 4.6 megohms to measure the open-circuit voltage, but will read as if the resistance is that in the meter—4 megohms. So it will read about 13 percent low. But the meter with 100 K resistance will measure the open-circuit voltage with 600 K + 100 K or 700 K, reading as if the resistance is only 100 K, so it will indicate only one-seventh of the voltage that was there before it was connected—which is about 86 percent low!

Assuming the true voltage, with no meter connected, is 120 volts, the 4-megohm instrument will indicate 104 volts, both of them together will indicate 17 volts, and the 100 K instrument alone about 17.5 volts. If you change the range on this instrument, because 17.5 volts is so low on the 200-volt scale, the reading will go even lower. For example, if you switch to a 20-volt scale, which will have a resistance of 10 K, the reading will drop to about 2 volts. We have made the point somewhat graphically: voltage cannot be measured with complete accuracy by an instrument that takes current from the circuit to make the measurement.

A potentiometer avoids disturbing the circuit, because at null the measurement takes no current from the circuit being measured. The meter or indicator is used to determine that no current is taken, and thus that the circuit is, in fact, undisturbed. In the bridge circuit, two legs function as a voltage dividing device, while the other two yield an identical division. In a potentiometer, two legs again produce a controlled voltage within the potentiometer. The other two are in the circuit providing the voltage to be measured, or maybe there is only one leg, producing a single actual voltage to be measured (Fig. 3-13).

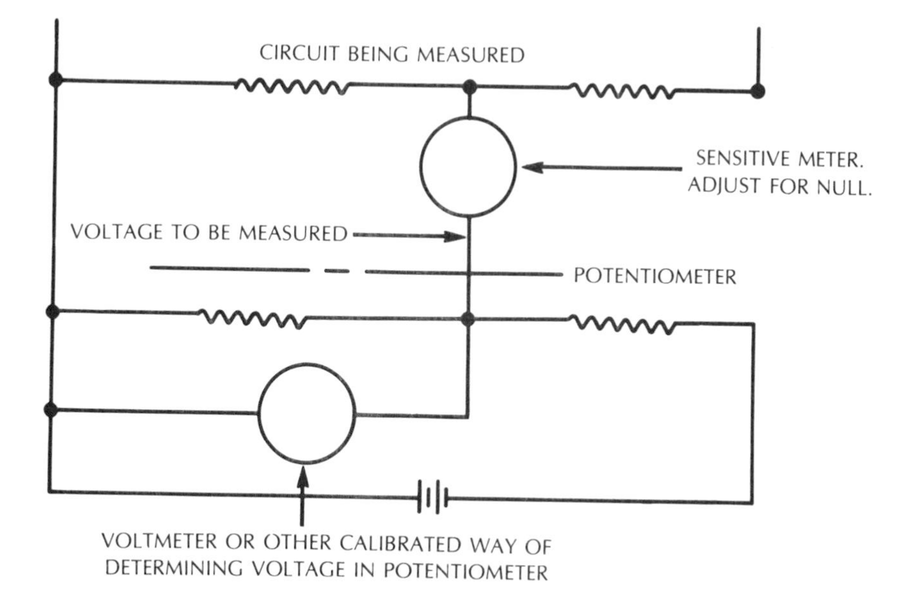

Fig. 3-13. Use of the basic potentiometer circuit, showing its similarity to a bridge, using a null of current flow to avoid disturbing the circuit being measured.

Let us use the potentiometer method to measure the voltage at the junction of the 1-meg and 1.5-meg resistors across the 200-volt supply. A crude but effective way would be to put much lower, heavier, current-carrying resistors across the same supply (Fig. 3-14). Say the resistors in the lash-up potentiometer circuit are 1 K and 1.5 K, with a 500-ohm potentiometer resistor in the middle. The current in this voltage divider is 200 volts across a total of 3 K, yielding 100 milliamps. Neither meter will load this voltage divider appreciably, and the 20,000 ohm per volt job will certainly give an adequate reading.

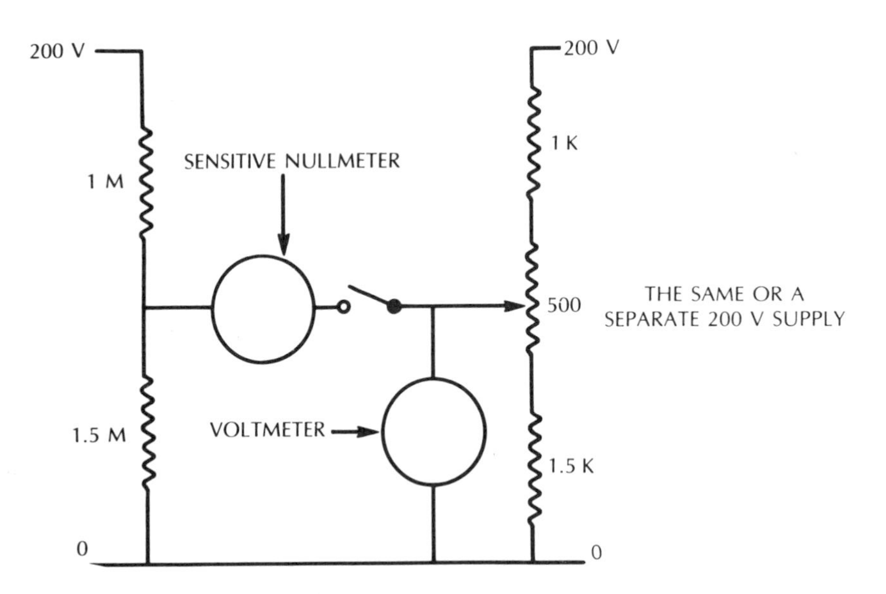

Fig. 3-14. A potentiometer used in a crude rig to measure the true voltage from the divider in Fig. 3-12.

If the 200-volt supply cannot deliver the extra 100 milliamps in addition to whatever else it supplies, or if the 100 milliamps would load it down (you can check this by putting a voltmeter on the supply as you connect the 3 K resistor chain) then a separate supply could be used for the potentiometer. Now, we connect a sensitive meter (possibly the 20,000 ohm per volt instrument, on a low voltage range) between the slider of the 500-ohm potentiometer and the voltage to be checked, and adjust the slider until the needle of the instrument does not move when the connection is made or broken.

Now, the voltage at the slider is the same as that at the junction of the two resistors (1 meg and 1.5 meg) and no current is flowing either way. Measure the voltage at the slider, where 100 milliamps is flowing through the pot, and you know that the voltage at the junction of the two high-value resistors is the same.

Alternatively, instrument potentiometers are made with precision parts (which was why we referred to the arrangement in Fig. 3-14 as crude but effective; it does not use precision or calibrated parts, except for the meter itself). The potentiometer is then calibrated so the setting of the slider, along with the resistors selected in series with it (all precision parts), tell what the voltage being measured is, without directly measuring it with a meter. Other types of instrumentation are developed later in this book, based on this principle.

VOLTAGE AND CURRENT INPUTS AND OUTPUTS

Something else, based on Kirchhoff's laws, that we will need later in this course, is an understanding of these basic forms of interconnection. Whenever two devices or circuits are connected together, and a signal whether dc or ac (which we will come to in Chapter 5) flows from one to the other, the one from which it flows is the source, and the one into which it flows is the output, or load.

For the purpose of illustrating this in a form that is greatly used in many different contexts, refer to Fig. 3-15. Here, in the center, is a four terminal network.

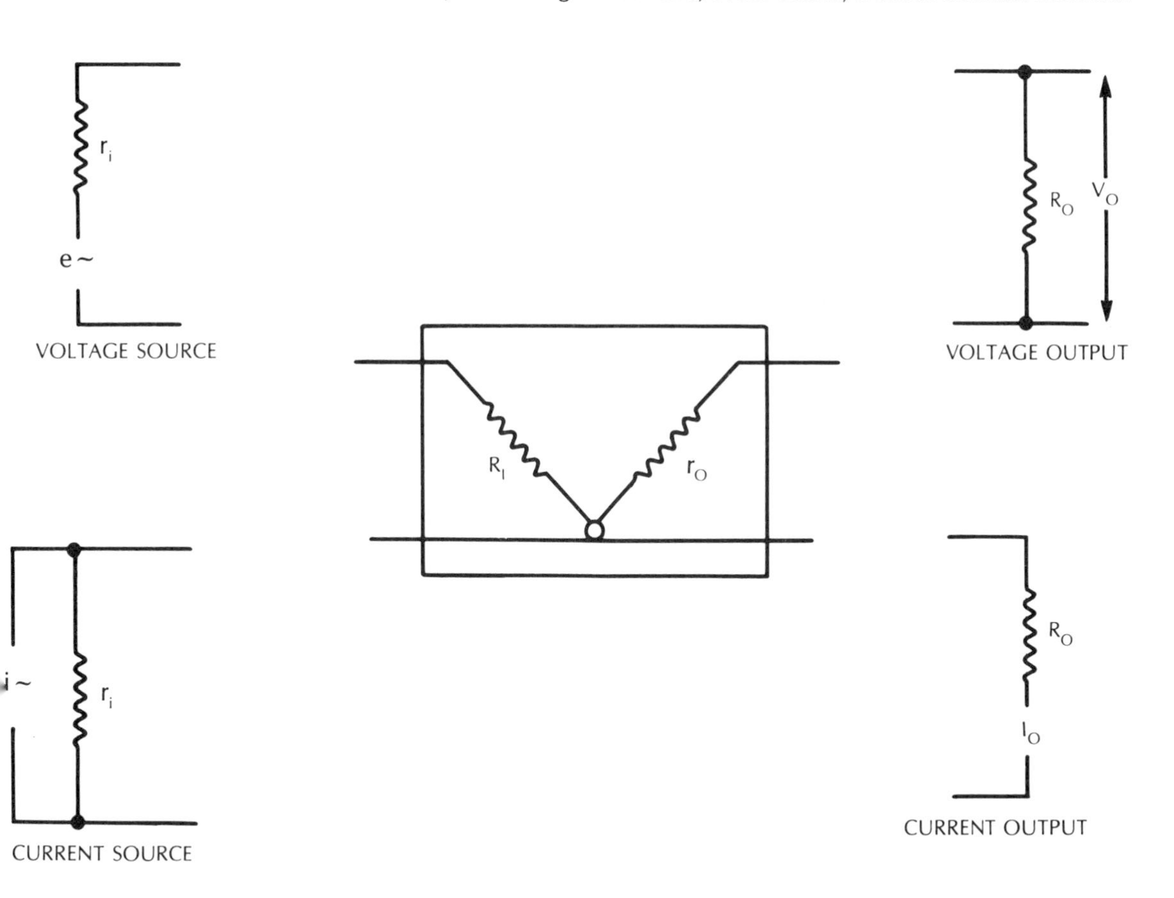

Fig. 3-15. How to apply Kirchhoff's Laws to interconnection problems.

Two of those terminals are common, so it is sometimes called a three-terminal network. The point is, whatever the circuit in the "box" is designed to do, the accuracy with which it does it can be affected by the connections at the input and output.

We have shown a small circle to indicate some function the box is designed to perform, which we shall get into later in the book. That could be simple attenuation, which makes the output some definite fraction of the input, or frequency discrimination, which lets some frequencies through and stops others, or something to do with digital signals, which are pulsed in form. Suppose, since we are far enough to understand this, that its function is attenuation.

Thevenin's theorem says that whatever resistances, etc. are in the network, these can be reduced to a simple equivalent. Later on, we will see that this may include reactance elements as well as resistance, but for the moment, we consider only resistance. So the resistance "looking in" we call R_i.

What this says is that when it is connected to any source, the terminals of that source behave exactly as if a resistance of that value were connected to them. Similarly, at the output send, "looking back," the source presented by this network behaves exactly like a resistance we have indicated as r_o. Now, the circuits to which this may be connected are of 3 basic kinds: voltage, current, or matched. Let us pursue each in turn. However, for the network in the middle to function as it was designed to do, it should operate on the kind of connections for which it was designed.

First take the voltage source input. If the voltage at the source is measured with a high resistance voltmeter of appropriate kind, it will measure e~ —whatever that is, because r_i will produce negligible change in it. However, when the network is connected, this voltage will be divided between r_i and R_i so that the voltage at the input terminals changes from e~ to $e\sim \left[\frac{R_i}{(R_i + r_i)}\right]$.

If it is designed as a voltage circuit throughout, then e~ going into the network is supposed to produce, let's say k • e~ out, where k is some definite fraction. But now you connected it to the output circuit, that has a resistance (or as we shall say later, an impedance) of R_o. So instead of getting k • e~ out, that's assuming we had e~ going in, we shall have $k \bullet e\sim \left[\frac{R_o}{(R_o + r_o)}\right]$.

So there are two sources of error there. Let's see what they mean. At the input, if r_i is so small, compared to R_i that the expression in brackets is essentially the same as R_i/R_i, nothing is changed by r_i. But if r_i becomes large, that expression may attenuate the signal more than the network that is supposed to do the attenuating.

At the output, for the expression in the brackets to be close to 1 requires R_o to be very large compared to r_o. Otherwise the same thing happens here. Thus we can see that the conditions for "voltage transfer" as it is called, to be accurate, are that r_i be close to zero, and R_o be close to infinity, always compared to the values inside the network.

Now take the current cases. Here the input is i ~. If r_i is infinitely large (like being zero, this is a theoretical ideal) then there would be infinite voltage accompanying i ~ before anything is connected, just as there would be zero current before anything was connected to the voltage source. But infinite voltage is a little more difficult to conceive of, than zero current! What makes it conceivable, is that everything is relative. In a current source, r_i must be very large, compared to R_i. Now the actual current input, instead of being i ~, will be $i\sim\left[\frac{r_i}{(r_i + R_i)}\right]$.

Similarly, at the output, assuming that with i ~ in, the network is supposed to deliver k • i ~ out, the actual output will be $k \bullet i\sim\left[\frac{r_o}{(r_o + R_o)}\right]$. So we have the opposite requirement for current connections that we had for voltage connections. The source resistance must be infinitely (or very) large compared to the network resistance, and the output load resistance must be as close to zero as possible.

Now, practical networks may be designed for either voltage in, voltage out, current in, current out, or combinations, such as voltage in, current out, or current in voltage out. With voltage in voltage out, the transfer function is a ratio. That is "k" as we have referred to it, is just a fraction, or a numerical quantity. But with voltage in, current out, a certain voltage in causes a corresponding current out, both of which are the properties of a conductance, so it is called a *transconductance*. Similarly, where a current in causes a corresponding voltage out, it would be called a *transresistance*.

Now in many electronics circuits, neither voltage nor current is used as the exclusive reference. These are the matched circuits, which result from making $r_i = R_i$ and $r_o = R_o$. If these equalities are departed from, the transfer will change. When this basis for design is used, the inputs are not e ~ and i ~ respectively (according to what type of circuitry is used in their design), but half of those values. These will be the quantities measured, by testing the sources with matched loads (not necessarily the network, but something of equal resistance value) connected. The same is true at the output.

Thus, while in practice, changing from matched to voltage in and out would result in 4 times as much transfer, doubling it at each end, that is not the basis for design, so you never need to take that into account.

If you think about this last section, you will realize that all of these cases are somewhat special applications of Kirchhoff's laws. We have introduced them before you have learned about ac, because they are easier to think about when you don't have to think, "Is this ac or dc, or are they mixed?" The same rules apply to both, but not always in the same way, because ac and dc don't always "see" the same circuits, as we shall see later on.

Examination Questions

1. In the circuit in Fig. 3-16, calculate the voltages and currents associated with each resistor (by number) and verify Kirchhoff's first law in your result.

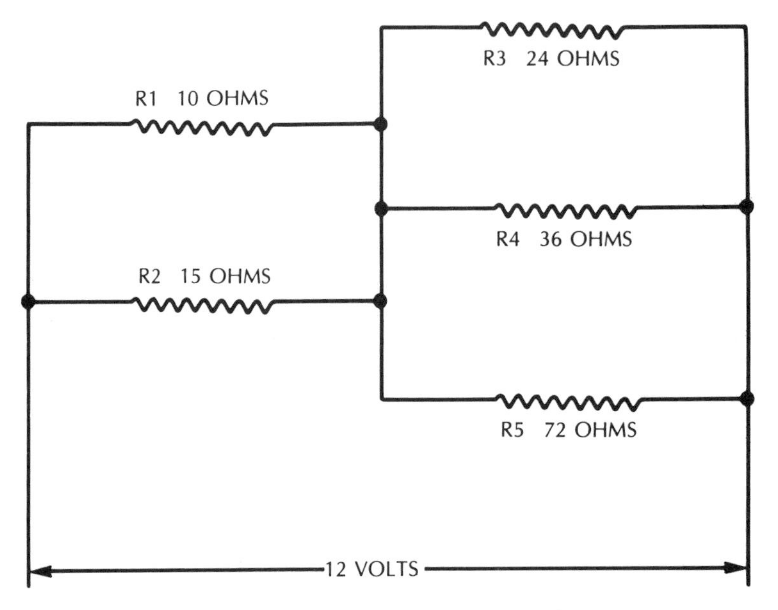

Fig. 3-16. Circuit for Examination Question 1.

2. Do the same thing for the circuit in Fig. 3-17.

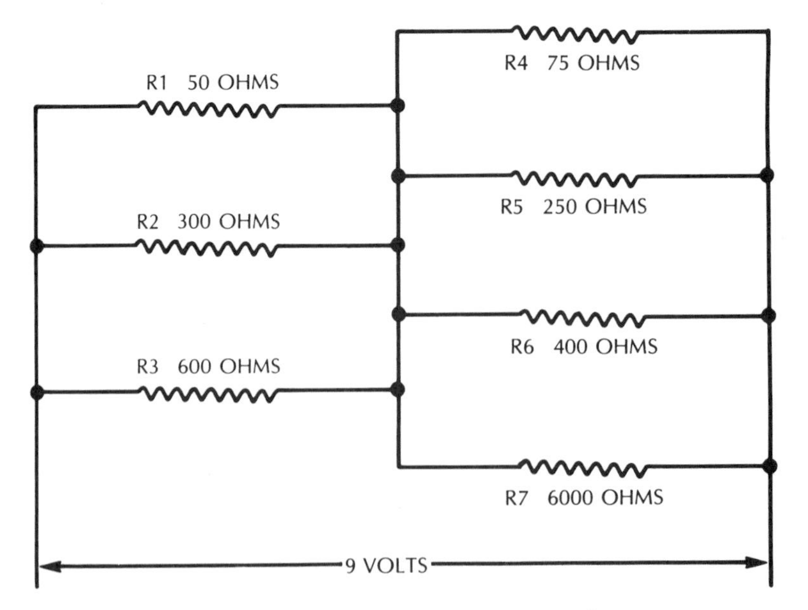

Fig. 3-17. Circuit for Examination Question 2.

3. When a certain 3-volt battery is connected to a load of 1.5 ohms, its voltage drops from 3 volts, which it measures without the load connected, to 2.5 volts when the load is connected. How would you explain this voltage drop?
4. Based on your conclusion in Question 3, what would you expect to be the terminal voltage of the battery when a load of 1.2 ohms is connected?
5. A load of 1.5 ohms in Question 3 is made up of resistors that measure 0.82 and 0.68 ohms, respectively. Find the current flowing and the voltage across each resistor. Verify Kirchhoff's second law in this result.
6. How much current would the battery deliver for the last three questions if it were short-circuited through an ammeter with very small volt drop?
7. In the Wheatstone Bridge in Fig. 3-4, R1, R2 and R4 are 4 K, 9 K and 6.75 K, respectively, as measured. R5 is a sensitive instrument that indicates when no current flows either way. Find the value of R3.
8. In Question 7, R4 is a variable resistor. It is found that making its value 6.9 K instead of 6.75 K results in a just discernible deflection of the meter in one direction. What value would you expect to produce to achieve just discernible deflection in the other direction?
9. Listed below are three columns. The first is a voltage. The second and third are resistance values which are connected in series across the voltage in the first column. Find the voltage appearing across the second resistor named in each case and the internal resistance at the junction.

6 volts	25 ohms	50 ohms
12 volts	40 ohms	50 ohms
15 volts	100 ohms	25 ohms
27 volts	120 ohms	150 ohms
45 volts	220 ohms	330 ohms

10. In Question 8, a value of 6.7 K makes the reading just discernible in the opposite direction. What adjustment would you make to your calculated value for R3 in Question 7?
11. In the circuit in Fig. 3-5, the supply is 15 volts and the following resistor values are used: R1 is 40 ohms, R2 is 90 ohms, R3 is 60 ohms, R4 is 60 ohms and R5 is 140 ohms. Find the voltage and currents in the circuit before and after R5 is connected. Check the application of Kirchhoff's first law at the top and bottom points of the bridge, and of the second as it relates to the left and right triangles.
12. In the arrangement described in Question 11, R2 and R4 are both exchanged for resistors of 80 ohms. Find the new set of answers applicable to these values.
13. An instrument of the type represented in Fig. 3-9 requires 50 millivolts at 1 milliamp to produce a full-scale reading. What is the resistance of its coil? What series resistance will be needed to enable it to read 100 volts full scale? What shunt resistance is needed for 1 amp full scale?

14. If an instrument with a movement that gives a full-scale reading at 1 volt at 50 microamps is used to make voltage and current scales that read 100 volts and 1 amp full scale, what series and shunt resistors are needed? Does not the fact that this shunt has a higher value than the one in answer to Question 13 make it a more practical solution? If not, why not?
15. A voltage divider across a 100-volt supply uses 27 K and 39 K resistors. A carefully conducted potentiometer measurement (using the method in Fig. 3-14) determines that the voltage at the junction (measured across the 39 K resistor) is precisely 59 volts. Using a meter on its 0-100 volt range, the voltage reads 58 volts, and the potentiometer checks that this is, in fact, the voltage with the meter connected. Estimate the sensitivity of the meter, in ohms per volt.
16. Two meters, each with a 100-volt scale, are used to measure a voltage at an unknown internal resistance. One of the meters is known to have a sensitivity of 20,000 ohms per volt, the sensitivity of the other is unknown. One meter reads 73 volts (the 20,000 ohms per volt one), the unknown one reads 71 volts. When the 20,000 ohms per volt meter is removed from connection, the other one registers a rise to 75 volts. What is the internal resistance of the circuit being measured?
17. Two resistors with coded values of 22 K and 33 K are connected across a 100-volt supply to derive a voltage (across the 33 K resistor) of 60 volts. With the divider connected the voltage at the tap is 69 volts. What value resistor should be connected across the 33 K resistor to bring the voltage closer to 60 volts?
18. In Question 16, removing the other meter causes the reading on the one with 20,000 ohms per volt sensitivity to rise from 73 to 79.35 volts. What is the sensitivity of the other voltmeter?
19. A simple bridge circuit of the type shown in Fig. 3-4 is to be used to check resistor tolerances. The resistor under test is put in the position of R3. Each of the other resistors, R1, R2 and R4, have values of precisely 100 ohms. The supply is 10 volts, and the meter is a center-zero instrument with full-scale readings of 0.5 milliamp either way and a coil resistance of 75 ohms. What resistance should be connected in series with the movement, across the position occupied by R5 (the coil + series resistance equals R5), to get an approximate full-scale reading each way for a 20-percent tolerance?
20. What resistance should be used in the question above to realize a tolerance of 10 percent read full-scale?

4
Power Calculations

Since electricity is a form of energy, it can be converted into other forms of energy. However, during any conversion process, all changes and transfers must conform to the well-established principle of the conservation of energy. Whatever form energy takes, it must come from or be derived from somewhere, and it must "go somewhere," too.

When electricity drives a motor or an electromechanical transducer of some kind (which we consider in Chapter 8), some of the electrical energy supplied to the motor or transducer is converted into mechanical energy. Any of the electrical energy that is not converted into mechanical energy must appear in some other form, notably heat. When electricity is used to generate light, some of the energy appears as light, but (except in the more exotic light sources, such as lasers) most of it is converted into heat. In an electrical resistor, all of the electrical energy dissipated appears as heat, which has to be conducted away at a rate sufficient to prevent the temperature from rising to a danger point or high enough to change the value of the resistor so that the circuit no longer operates correctly.

POWER DISSIPATION

Electrical energy is measured in watts, a quantity determined by multiplying voltage in volts by current in amperes. Electrical energy is converted to heat wherever there is a product of voltage and current, wherever current is flowing with an accompanying voltage difference or drop (Fig. 4-1). In dc circuits, the statement made in the previous paragraph is simple to apply. In ac circuits (which we discuss more fully beginning in the next chapter), voltages and currents both can be present in a way that may not be directly related to energy dissipation in the form of heat.

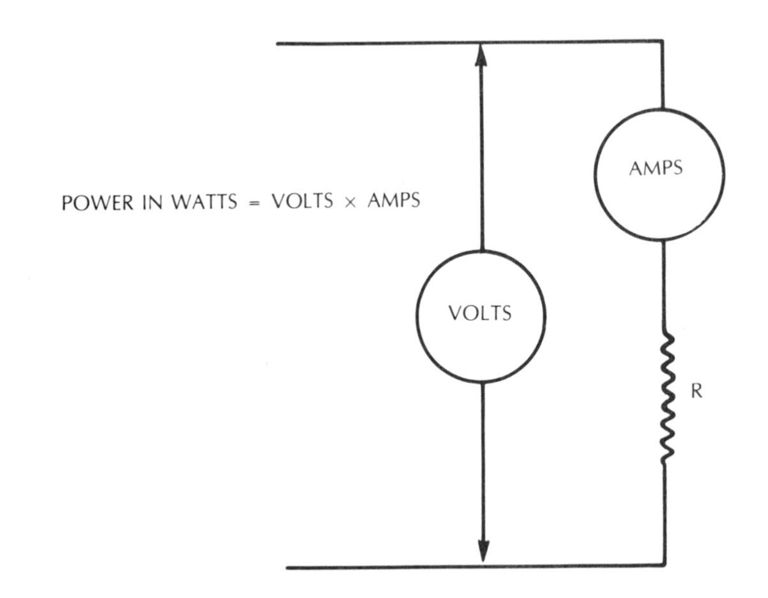

Fig. 4-1. Power in watts is the product of voltage across and current (amps) flowing in a resistance.

To better understand how all this works, we first consider dc circuits, where a steady state of voltage and current exists for a long enough period to provide a measurable indication on a voltmeter and an ammeter. Calculating the heat generated, which must be dissipated by conduction, convection or whatever means is available, is simple: multiply volts by amps and the answer is in watts.

$$W = EI \qquad (10)$$

To perform the calculation, you must know both volts and amps or voltage and current in whatever units are appropriate.

Volts and amps are not the only units for calculating power. As well as millivolts, microvolts and kilovolts, milliamps and microamps, power can be measured in kilowatts, watts, milliwatts, microwatts, or even smaller units like millimicrowatts and micromicrowatts. Table 4-1 gives some combinations of units that can be used.

If only voltage or current and resistance are known, the third quantity, current or voltage, can be calculated, and from the voltage and current thus obtained, formula (10) gives the wattage dissipated. But this calculation can be shortened by combining the formulas. If voltage and resistance are known, formula (1) gives the current in terms of voltage and resistance. Combining (1) and (10) gives:

$$W = E \times I = \frac{E}{R} \times E = \frac{E^2}{R} \qquad (11)$$

Or if current is the quantity known in addition to resistance, the voltage is given by formula (2) and combining (2) and (10) gives:

$$W = I \times E = I \times IR = I^2R \qquad (12)$$

In using formulas (11) and (12), a variety of units can also apply, some of which also are listed in Table 4-1.

VOLTAGE	CURRENT	POWER
VOLTS	AMPS	WATTS
KILOVOLTS	MILLIAMPS	WATTS
KILOVOLTS	MICROAMPS	MILLIWATTS
VOLTS	MICROAMPS	MICROWATTS
MILLIVOLTS	MILLIAMPS	MICROWATTS
MICROVOLTS	AMPS	MICROWATTS
VOLTAGE	**RESISTANCE**	**POWER**
VOLTS	OHMS	WATTS
KILOVOLTS	MEGOHMS	WATTS
VOLTS	MEGOHMS	MICROWATTS
CURRENT	**RESISTANCE**	**POWER**
AMPS	OHMS	WATTS
MILLIAMPS	MEGOHMS	WATTS
MILLIAMPS	OHMS	MICROWATTS

Table 4-1. Units for Calculating Powers.

These formulas make it evident that changing the voltage or the current associated with a specific resistance changes the dissipation, and thus the heat generated, in proportion to the square of the voltage or current change. Doubling voltage or current quadruples the power if resistance is unchanged, thus heat generation also quadruples.

HEATING EFFECTS

From a practical viewpoint, we should look at what can happen to heat. If the resistance consists of a conducting material in which the current distributes itself uniformly, the heat will be generated uniformly throughout the body of the conductor. However, the surface of the conductor, which has contact with the air or is in thermal contact with a heat conductor, is cooled because the heat is carried away, while the heat from the center part has to travel to the surface of the conductor before it can leave (Fig. 4-2).

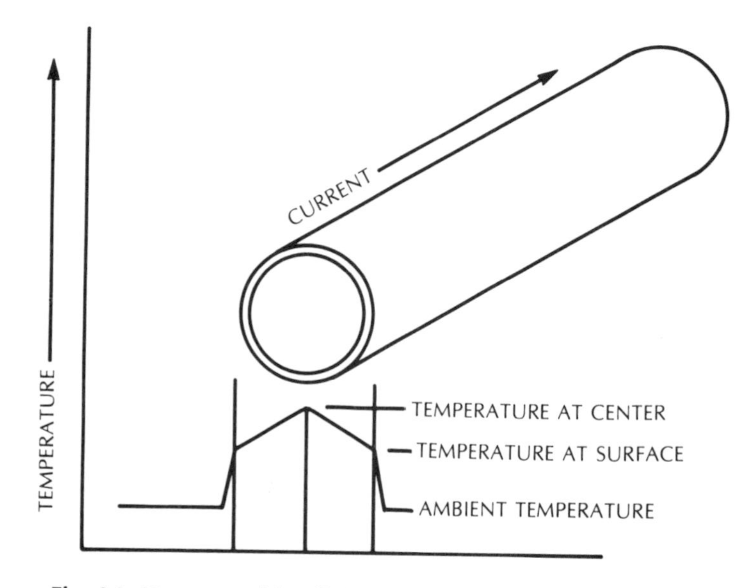

Fig. 4-2. Heat caused by dissipation results in temperature rise.

Heat, like electricity, flows from points of higher temperature (which is heat potential, and thus like voltage in electrical terms) to points of lower potential. Thus, before heat can move away from the center of a conductor toward its surface, the internal temperature must be higher than the surface temperature.

To understand how heat moves, we need to know that different substances have different heat conductivities, just as they possess specific electrical conductivities. In general, you will find that substances, like copper, silver, etc., which are good electrical conductors, are also good heat conductors. The poor electrical conductors, such as ceramics, are generally poor heat conductors. However, there are some notable exceptions, and these prove very useful in electronics. Particularly useful are certain materials, mostly synthetic, that conduct heat much better than electricity. This enables them to be used to conduct heat away, while providing electrical insulation at the same time (Fig. 4-3).

Another factor in getting rid of heat is the use of a large area with a short conducting path. Thus, two conducting surfaces (in both the electrical and heat sense) may be separated by a very thin but adequate layer of electrical insulation, such as mica or anodized and filled aluminum (which, in fact, would be the surface of one of the conductors). The heat conducting path may be only a few thousandths of an inch in "length" (more commonly regarded as thickness, under the circumstances, but it is the length the heat has to travel), with a considerable "cross-section" (the area over which the heat may escape). However, what counts is the total area of contact between the two conductors with the insulation between (Fig. 4-4). Then there are other facilitators, such as silicones or other compounds, which provide good heat conduction and contact, while maintaining sufficient electrical insulation.

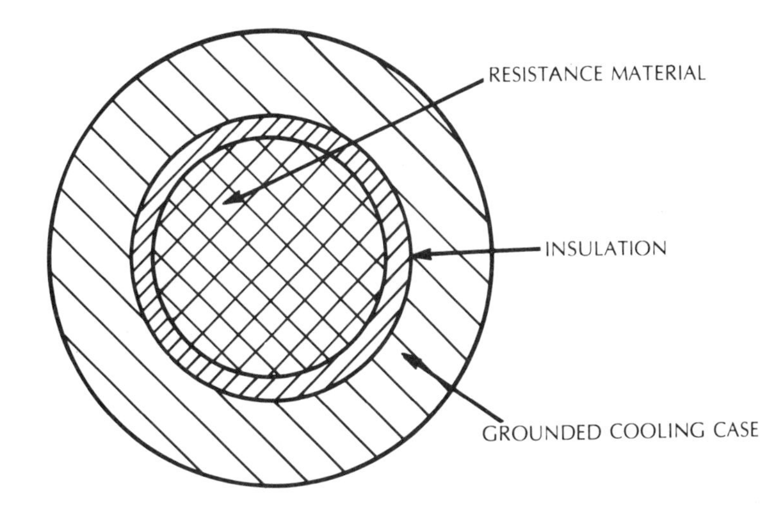

Fig. 4-3. Heat flow through thin insulation to a good heat conductor can help cooling.

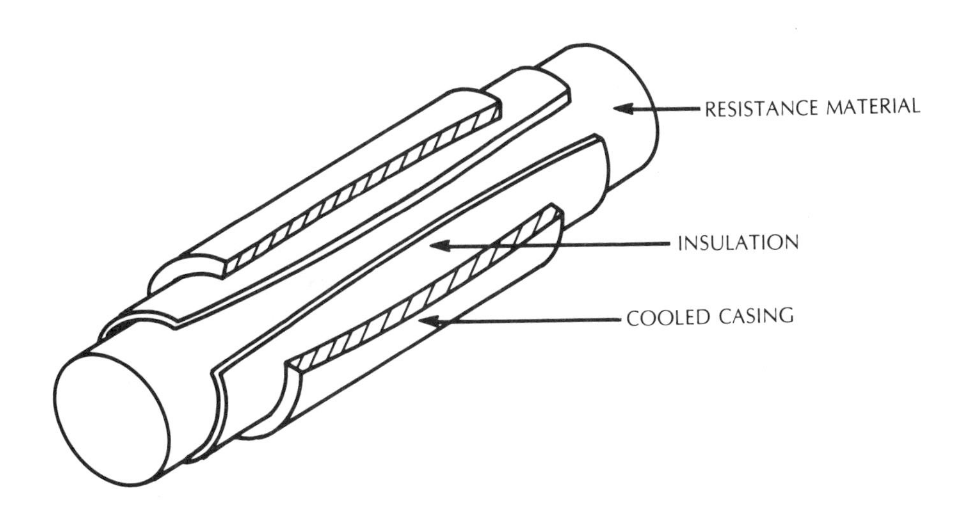

Fig. 4-4. Increasing the area where heat can flow improves cooling further.

Earlier textbooks list such substances, as are known at the time. But the state of the art is advancing so rapidly today that the best substance for the purpose now may be obsolete tomorrow. So unless you are in the business of making components that use these substances or are involved in applying the same substances on the assembly line, the name of the latest product will not help you much. Their existence is mentioned here to illustrate the problem that component designers face and the approach to solution. The better the ways found to carry heat away, with a lower temperature drop for the amount of heat conveyed, the more liberally can components using the improved methods of heat removal and the associated materials be rated (Fig. 4-5).

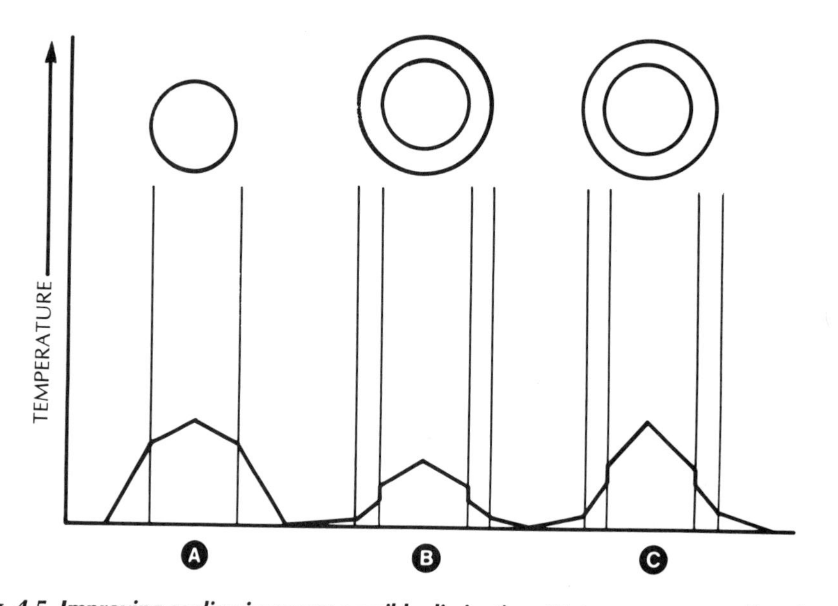

Fig. 4-5. Improving cooling increases possible dissipation: (A) temperature gradient from a resistor in air; (B) reduction in temperatures when a cooling casing is used; (C) increasing dissipation brings temperature back to the same limits as (A).

For most readers and students, a resistance or other component is accepted as a ready-made product. The component manufacturer attaches a dissipation specification, and it is this specification that we need to understand. As in electricity, where current flow is proportional to the voltage difference driving it, heat flows in proportion to the temperature difference moving the heat. So doubling the power dissipation doubles the temperature rise caused between the two points that heat must flow between.

Relation of Temperature Rise to Voltage and-or Current

Assuming that a change in temperature does not change the resistance (which it almost invariably does, but we assume this as a starting point for dealing with the problem), doubling the voltage or current will quadruple the temperature rise. Suppose the ambient (room or adjoining equipment) temperature is 70 degrees F. In this environment, let's say that a resistor or other component reaches a surface temperature 15 degrees higher than this, or 85 degrees F, and a maximum internal temperature 15 degrees higher than that, or 100 degrees F, when 10 volts is applied across it. Then, doubling the voltage will result in four times the heat rise in both instances: the surface temperature will reach 130 degrees (70 + 60) and the internal temperature will reach 190 degrees (130 + 60).

The smaller temperature rise that accompanies the lower dissipation may not significantly alter the value of the resistor, but the higher one almost definitely will do so unless the resistor is made of a special material that holds its resistance constant over such a wide temperature range.

Change in Current Distribution Due to Heating

In most metallic conductors the resistance rises with temperature, so the internal current paths, where the temperature rise is higher, will attain a higher specific resistance than those nearer the surface (Fig. 4-6). This means a smaller current will flow in the center than at the surface. Consequently, that current will no longer distribute uniformly throughout the material.

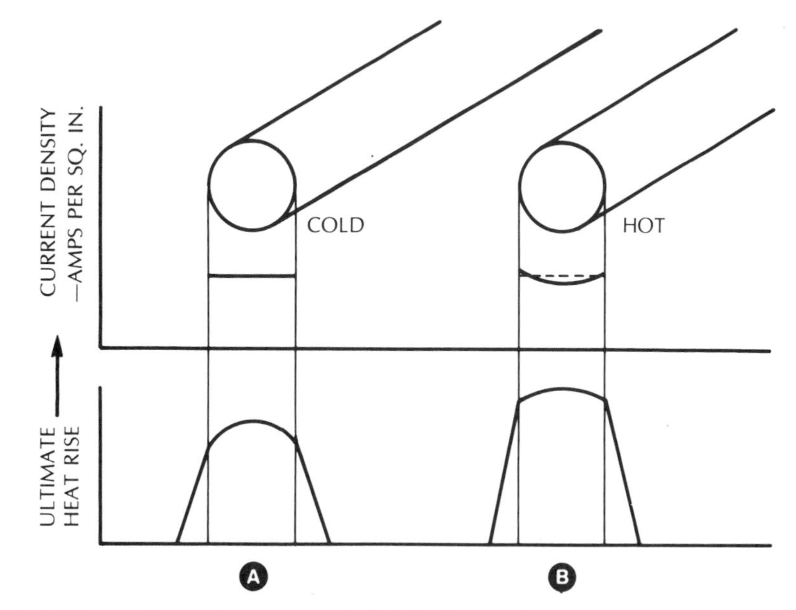

Fig. 4-6. Effect of heat rise on current distribution and temperature gradient: (A) cold; (B) after steady temperature is reached.

Therefore, the temperature rise from the center to the surface may be less than four times the original 15 degrees (say 55 degrees) when dissipation quadruples in the entire conductor. This reduction of current at the center means that more current must flow at the surface; therefore, the heat rise is likely to be increased by more than the 5 degrees "saved," making, say, a 75-degree surface rise in place of the 60-degree figure, calculated on the assumption that the resistance will remain constant. So the surface temperature may rise from 85 degrees to 145 degrees instead of 130 degrees, while the center temperature will rise from 100 degrees to 200 degrees, instead of 190 degrees.

Notice that, although there is only a 5-degree differential between the inside and outside, as compared with the 60 degrees based on uniform current distribution, the higher surface rise causes the actual internal temperature to be higher than the 190 degrees it would be if the current were uniformly distributed.

If the voltage across a resistor, rather than the current through it, is the quantity the circuit is intended to hold constant, a resistance with a positive temperature coefficient increases in value when the temperature rises. As a result, doubling the voltage will no longer quadruple the dissipation, because the resistance value rises as dissipation goes up. Thus, current does not double when voltage does if the heat has risen to correspond with the increase in voltage.

Suppose the increase in temperature causes the resistance to rise 20 percent. Doubling voltage will increase current to twice at first, before the resistance rises due to increased heat. As the temperature rises, the current will fall to five-sixths (100 to 120 percent) of this double value, or 1.67 times its original value before the voltage is doubled. Thus, doubling the voltage will increase the ultimate dissipation by 2×1.67 or 3.33 times, instead of the four times that happens initially, before the resistance has had time to change due to the cumulative heat rise.

On the other hand, if the circuit is such that the current is held constant and the voltage adjusts according to current and resistance, it is a different story. Increasing the resistance by 20 percent means that voltage, instead of doubling when the current doubles, will increase by 2×1.2 or 2.4 times, making the dissipation increase by a factor of 2.4×2 or 4.8, instead of four times.

Further, if increasing the dissipation by 3.33 times raises the resistance by 20 percent, increasing it by 4.8 times will raise the resistance by 20 percent $\times$ 4.8 divided by 3.33 or 28.8 percent. Voltage then rises by 2×1.288 or 2.576 times, and dissipation by twice that, or 5.15 times. The more the resistance rises, the more the dissipation rises, causing a further resistance rise.

This calculation can be carried out a step at a time like this in an actual circuit although the actual change will be gradual until a steady state is reached. If you calculate in this way, making corrections a step at a time, you will be able to tell if the circuit will be stable or not by noticing whether the steps diminish or increase. In the example just quoted, the first step increased dissipation by 4.8 times, instead of the expected four times; the next step increased it to 5.15, instead of 4.8 times. To pursue this one more step, the resistance at 5.15 times dissipation will go up in value by 31 percent, resulting in a dissipation of 4×1.31 or 5.24 times. Each increment is smaller: $4.8 - 4$ equals 0.8; $5.15 - 4.8$

is 0.35; 5.24 − 5.15 is 0.09. This indicates that a stable condition is being approached.

But now suppose that the increase in dissipation causes the resistance to go up by 100 percent at the temperature which the resistor reaches as a result of that dissipation. This will double the dissipation, which will again increase the resistance by another 100 percent, resulting in three times the initial dissipation, and so on. This rate of growth represents a high temperature rise—an unstable condition that will quickly burn up the resistor if supply power is not removed from the circuit.

A more direct formula can be derived to give the steady temperature where a component will stabilize, according to the rate of temperature rise with dissipation. But there are usually extraneous factors, such as the fact that when voltage rises, there comes a point where current ceases to stay constant and begins to fall off, or that the temperature coefficient ceases to be linear as the physical state changes due to heat. So the more practical procedure is to test the circuit and allow some latitude below the unstable point to be sure that a little extra supply voltage will not cause the component to blow.

This same effect explains why a fuse blows quickly, once its temperature starts to rise. A fuse is in series with the circuit it protects. When the circuit current reaches a critical point for the gauge and method of mounting the fuse wire (usually determined by the cartridge in which it is mounted), its resistance starts to rise, causing the voltage across the fuse to rise, thus accelerating the resistance rise, until it melts and disrupts the circuit.

Slow-blow fuses merely have a greater thermal capacity (more ability to absorb heat before the temperature rises). Thus, they may be capable of carrying two or three times the blow current for a short (but appreciable) time without blowing. Fast-blow fuses have a much smaller thermal capacity, so their temperature rises very rapidly when the critical point is exceeded by however little (Fig. 4-7).

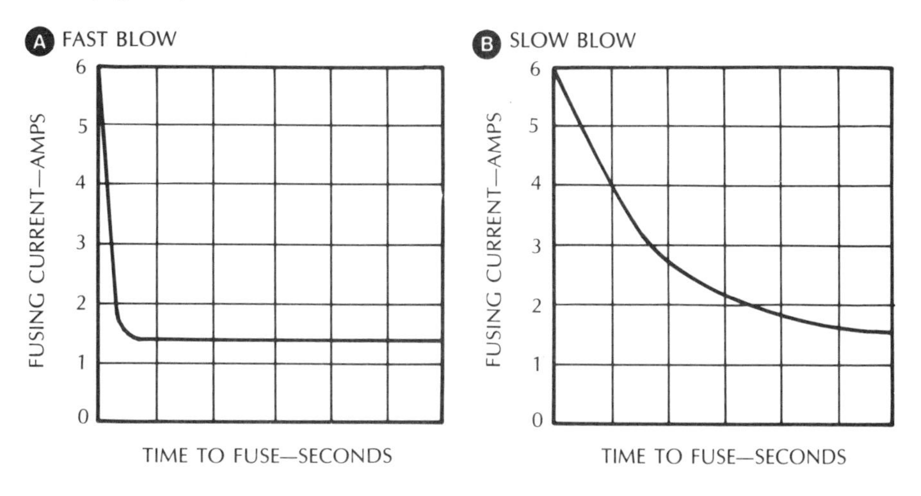

Fig. 4-7. Comparison between fusing current and time: (A) fast blow fuse; (B) slow blow fuse.

Effect of Cooling Arrangement

A factor sometimes overlooked in modern electronic circuits is that etched circuits can have properties very similar to fuses. Normally, the conductor stays cool enough and has a low enough resistance not to affect circuit operation. But if excessive current flows through part of a circuit board, due to some abnormal condition, that portion of the etched circuit may behave exactly like a fuse (except that it may not be so easy to replace!).

In considering the capability of any given size conductor to carry a specified current, an important factor is the method by which heat escapes from the conductor. For example, an etched circuit board mounted vertically (Fig. 4-8), with air space to allow cooling currents to circulate past the board, will probably emit heat in proportion to the area exposed, or close to that.

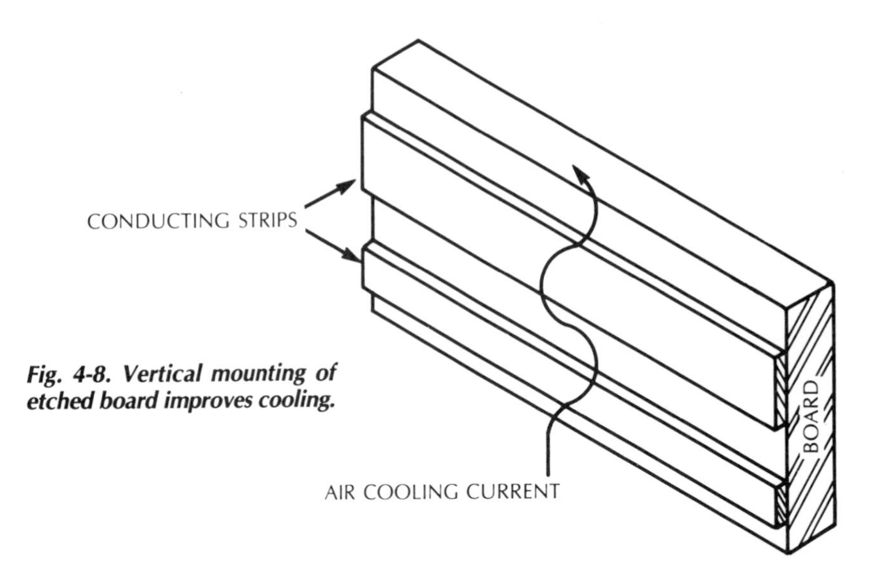

Fig. 4-8. Vertical mounting of etched board improves cooling.

Thus, doubling the width of etched strip will almost double its current-carrying capacity. But if the board is mounted horizontally, air currents will remove little heat. About the only way heat can escape is to pass through the insulation of the board (Fig. 4-9). In this case the relationship of conductor width to current-carrying capacity depends on the relative thickness of the board and its ability to distribute the heat by conduction, in a relatively poor heat conductor (which most circuit board materials are).

If the board is thin, heat will not disperse very readily in a lateral direction through the board, since doubling the width of the conductor will double the amount of heat to be dispersed from that immediate area if the current is also doubled. So the current-carrying capacity may not be increased proportionately by doubling the width of conductor, but it will be increased.

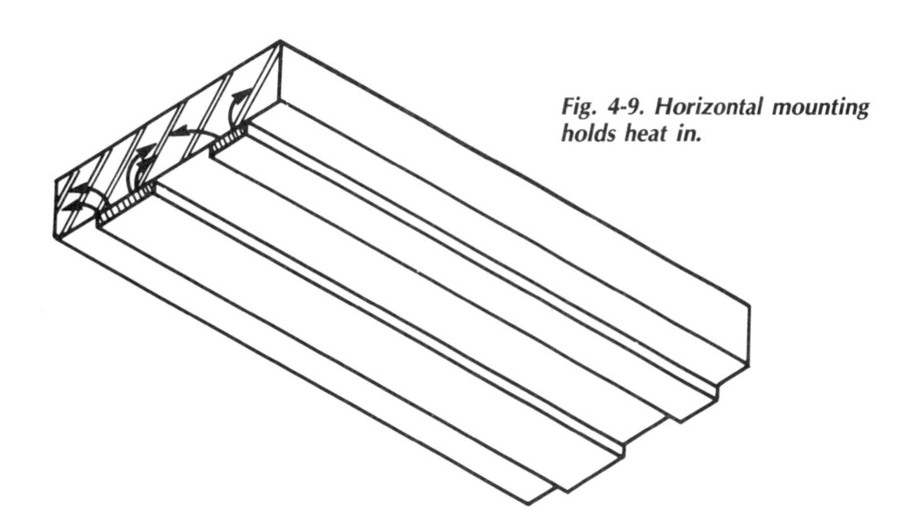

Fig. 4-9. Horizontal mounting holds heat in.

If the width of the strip is doubled, its resistance is halved. Thus, at the same current, dissipation is halved. If current is increased by a factor of root 2 (1.414) the dissipation will be the same. But the area of conductor in contact with the board is doubled, so a little better heat escape is provided, if it is not doubled. The current capacity should be increased by a factor of at least 1.5 by doubling its width.

Change in Voltage Distribution Due to Heat

In coil and transformer windings, whether resistance wire is used or not, current also generates heat (Fig. 4-10). A suitable current density must be figured, based on the size of the winding and the means by which heat can escape from it, and the winding must be operated within that rating to avoid overheating.

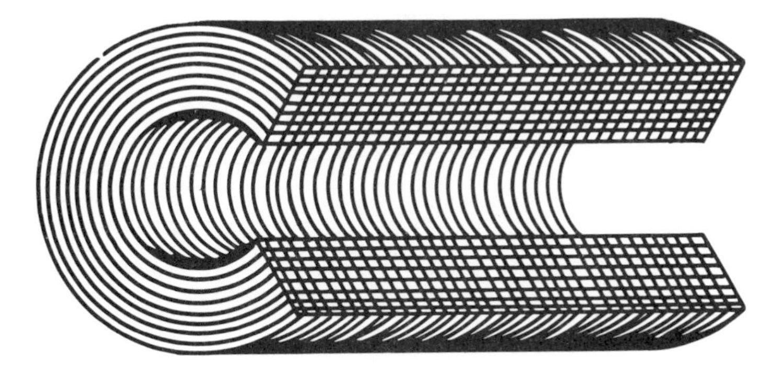

Fig. 4-10. Section through a cylindrical coil.

In this instance, the current has to flow through the entire winding, so the current in the turns of the winding near its center of cross-section will be the same as that near the outside or surface of the coil. But because the heat from the center has to escape through the turns at the surface, the central part of the winding will get hotter than the outside. As a result the turns at the center develop a higher resistance per unit length of wire. This means that more heat will be generated for a given bulk of wire in the central body than will occur near the surface, producing an accentuated rise in temperature on the inside of the coil. This effect can be cumulative, too, and constitutes a danger when the critical rating of the coil is approached.

The critical rating also depends to some extent on the outside or ambient temperature, because this is the point from which the internal temperatures start to rise. The higher the ambient temperature, the higher the internal temperature, so that the critical point where the effect becomes dangerously cumulative is reached with a smaller rise from ambient.

NEGATIVE TEMPERATURE COEFFICIENT MATERIALS

This discussion has related to the effect of temperature rise in changing the current or voltage distribution (and thus the relative heating) of conductors and resistors that have a positive temperature coefficient. More important, for some purposes, is the situation that prevails when the resistance material has a negative temperature coefficient, such as carbon, or the compositions usually used for resistors.

A solid resistor of this type will distribute current uniformly only when there is zero dissipation, or when the current is so small that the power dissipated is so little as to generate a negligible amount of heat. As soon as enough power is dissipated to generate appreciable heat, such a resistor starts heading for an unstable condition. Making the current paths non-uniform always increases total heating. In this case, regardless of the voltage-current relationship across the entire resistor, as determined by the external circuit, the resistor voltage is uniform across all the paths, inside and outside (Fig. 4-11).

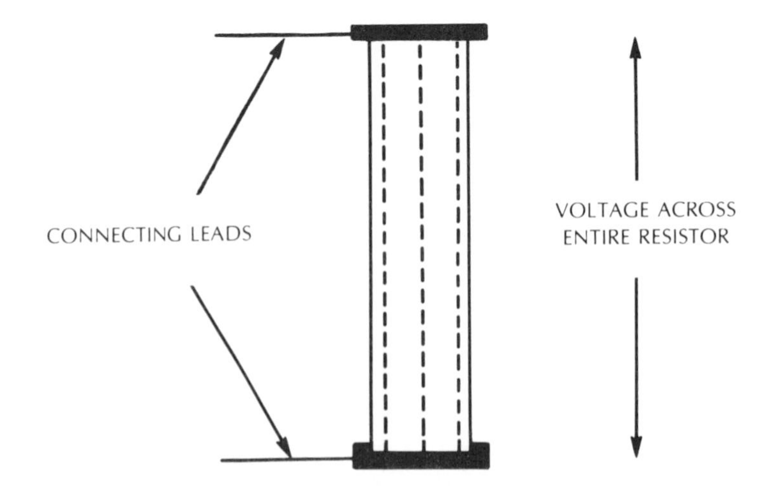

Fig. 4-11. Structure of a solid composition resistor (little used today), showing possible current paths.

So when the resistor starts to heat, its interior develops a lower resistance than its exterior, because the inside gets hotter and the temperature coefficient is negative.

If the central resistivity drops 20 percent, the current density there will be 20 percent higher than at the surface, which will increase the relative heat generated at the center by some 40 percent (it varies as current squared) as compared with the exterior. But this extra heat has to escape through the exterior, thus the temperature of the entire resistor rises, but the interior rises faster than the exterior, for two reasons:

1. Because a greater concentration of current passes along inner paths, in direct proportion to the temperature rise;
2. Because the heat thus generated still has to reach the surface to escape, and, as the inner part starts to heat more rapidly, the rise in surface temperature does not keep pace with the inner temperature.

Thus, an ever increasing proportion of the current concentrates toward the middle, the inside literally burns up, usually causing the resistor to explode.

This was the problem with resistors made from solid carbon or composition rods. A resistor constructed this way, rated at 1 watt dissipation for normal ambient temperatures, was ¼ inch in diameter by two inches long. Even then, without adequate cooling so the ambient couldn't rise because of poor ventilation, it could blow up when called upon to handle its rated 1-watt dissipation. That 1-watt figure was based on a location where the resistor was well ventilated on all sides.

A modern 1-watt resistor, in which the composition material consists of a thin layer deposited on a ceramic or glass cylinder, is a fraction of the size and is far more stable. It still has the negative temperature coefficient, which may affect its operation in circuits where voltage or current changes differ in accordance with resistance value changes, but it will not explode because of this unstable effect.

Warm-Up Changes in Equipment

However, a change in resistance can alter voltage or current distribution as the components warm up, and this is important in much electronic equipment. Suppose resistors are connected in series across a voltage supply (Fig. 4-12) to provide a voltage division. If these resistors have the same wattage rating but different resistance values,the one with the larger value will generate more heat, and thus, as the resistors warm up, the one having the larger value will drop in value more than the one of smaller value. Two resistors connected in this way tend to cause the voltage to drift toward center as they warm up.

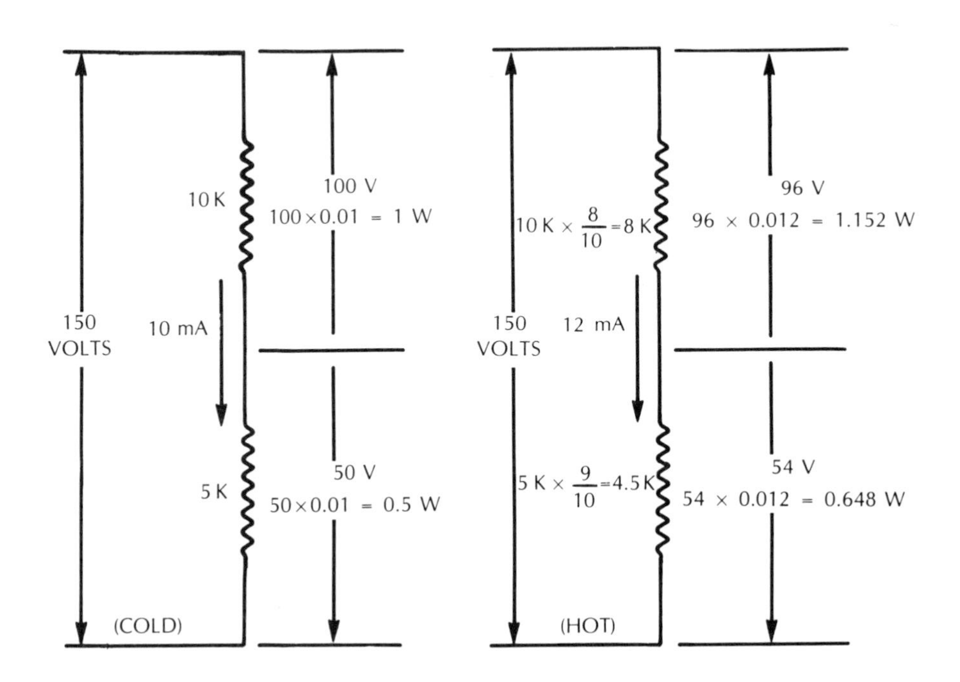

Fig. 4-12. Schematics showing how temperature change results in change of voltage division in a voltage divider.

For example, if one resistor is 10 K and the other 5 K across 150 volts, and both are rated at 1 watt the 10 K resistor may drop in value by 20 percent and the 5 K resistor by 10 percent, changing their values to 8 K and 4.5 K, respectively. Cold, the current through 15 K (total) across 150 volts is 10 milliamps, which produces 100 volts across the 10 K resistor and 50 volts across the 5 K resistor. The 10 K resistor dissipates 100 × 0.01 or 1 watt and the 5 K resistor 50 × 0.01 or 0.5 watt.

Hot, the resistors change to values of 8 K and 4.5 K, a total of 12.5 K. Across 150 volts, the current is 12 milliamps instead of 10. The voltage across the 8 K resistor is 96 volts and 54 volts across the 4.5 K. Now, the 10 K 1-watt resistor has become an 8 K resistor, dissipating 96 × 0.012 or 1.152 watts, and the nominal 5 K resistor, now actually 4.5 K, dissipates 54 × 0.012 or 0.648 watts.

With the old solid type, this would have been enough to make the nominal 1-watt resistor explode, without question. With modern resistors, it may be OK, although it is over the rating. But the value would be safe at the working temperature, for certain.

If resistors are connected in parallel to achieve a certain value, the one with the lower resistance will dissipate more. Suppose a 1-watt resistor has a value of 1.5 K, and the circuit needs a resistor of value 1.2 K (Fig. 4-13). Our formula for parallel resistors determines that the parallel value needs to be 1.2 K × 1.5 K divided by (1.5 K − 1.2 K) or 1.2 K × 1.5 K divided by 0.3 K is 6 K.

If a 6 K and a 1.5 K resistor are connected in parallel, they will have the same terminal voltage and the 6 K will pass ¼th the current passed by the 1.5 K resistor. So if the 1.5 K resistor dissipates 1 watt, the 6 K resistor will dissipate ¼ watt. If resistors of these respective ratings are used together, each will probably get equally hot and the values will stay in the same proportion, although heating may cause the combined value to drop somewhat. But if two resistors of the same rating (say 1 watt) are used, the 1.5 K resistor will get about four times as hot as the 6 K resistor. As a result, the 1.5 K resistor will drop in value more than the 6 K resistor will. Perhaps the 1.5 K resistor will drop by 20 percent to 1.2 K, so that its dissipation will now be five times that of the 6 K resistor.

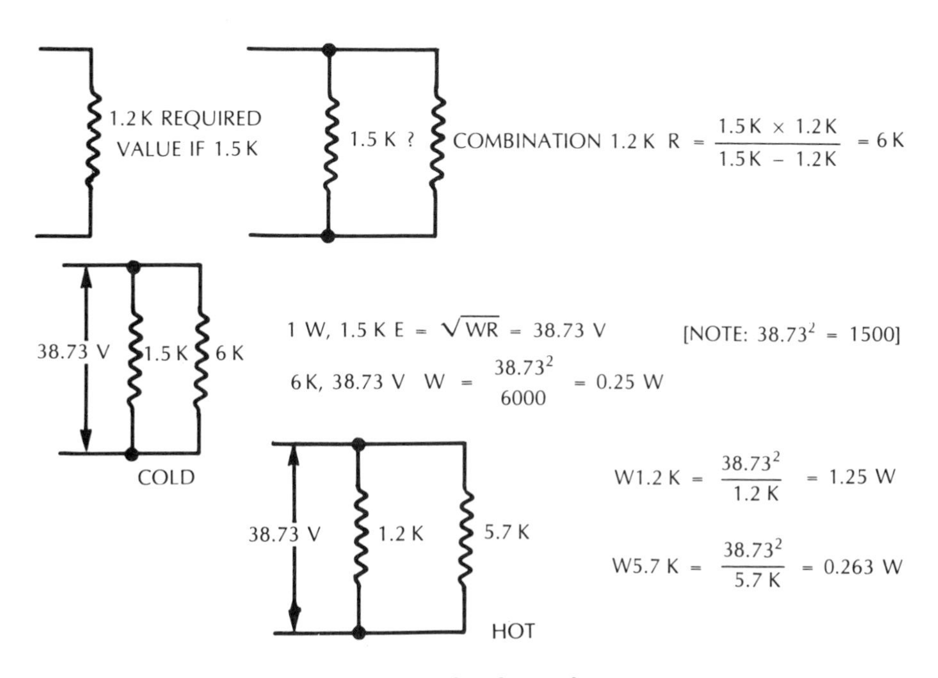

Fig. 4-13. Calculations showing current distribution between two resistors in parallel.

Unless your job is to design resistors, all this is somewhat a matter of guesswork although the principles stated above apply. If you need to know exactly what happens, you can measure the changes in resistance. Resistors that have a known positive temperature coefficient provide a means of reading the average change in temperature by measuring their resistance before and after warm up. For resistors with negative temperature coefficients, there is no reliable formula. The important thing, though, is that the resistance value does not change too much (whatever that means). This is subject to interpretation, according to the circuit; in any event, it should never change beyond its rated percentage of tolerance.

MEASURING RESISTANCE CHANGE

Measuring a resistor's value cold is no problem. But if any alternative circuit exists for the resistance-measuring current from the bridge (such as other resistors in parallel, or effectively so), which can affect the accuracy of the reading, you will need to make a correction or allowance, or else disconnect one end of the resistor you are measuring. But making the disconnection with a solder gun will heat the resistor through its connecting lead. You must then wait for it to cool before measuring it.

Measuring the value after warm up, which means after all voltages in the equipment have reached a steady value, takes a little more care. The measurement must be made quickly after the power is disconnected. If the resistor did not have to be disconnected on one end to measure it cold, this is merely a matter of making the measurement with utmost speed before the resistor has time to cool.

If you have to disconnect one end of the resistor to measure it, the result will be unreliable, unless you can avoid using a solder gun, because two things happen: the resistor will cool during the time you take to disconnect it; it will also heat, due to the effect of the solder gun. You will have no way of knowing which effect is the greater, and thus which way your resultant error in measurement is.

However, if you can persuade the resistor to make a good connection temporarily without soldering while the equipment warms up, and then quickly disconnect it for measurement again, you may get reliable figures. You can then make a secure soldered joint after you have made the measurement.

EFFECT OF PULSING ON DISSIPATION

In the next chapter, we shall apply much of the material from the first four chapters to circuits that use alternating currents (ac). However, one thing can be treated here that will partly pave the way for that. It is not strictly an ac consideration, but rather more related to pulse techniques, which are not quite the same as conventional ac.

Suppose a dc supply, which could be either rectified ac (which we discuss later) or a battery, is used intermittently to charge a capacitor through a resistor. The load which the capacitor supplies for a while after the dc supply is removed takes a fairly steady current (Fig. 4-14). But every time the supply is connected, the capacitor charges quickly.

For example, suppose the capacitor load is equivalent to a resistor of 10 times the value of the resistor through which the capacitor charges when the supply is connected; let's make these values 100 and 1000 ohms respectively. Assume that the voltage across the load and the capacitor is 100 when the supply is disconnected, and 90 just before it is connected again. The supply needs to be connected for one-tenth of the time to maintain this condition. (This kind of relationship is discussed more fully in the study of time constants.) Now, the average voltage across the 1000-ohm resistor is 95 volts, because it fluctuates, fairly linearly, between 90 and 100 volts. So the average current flowing in the

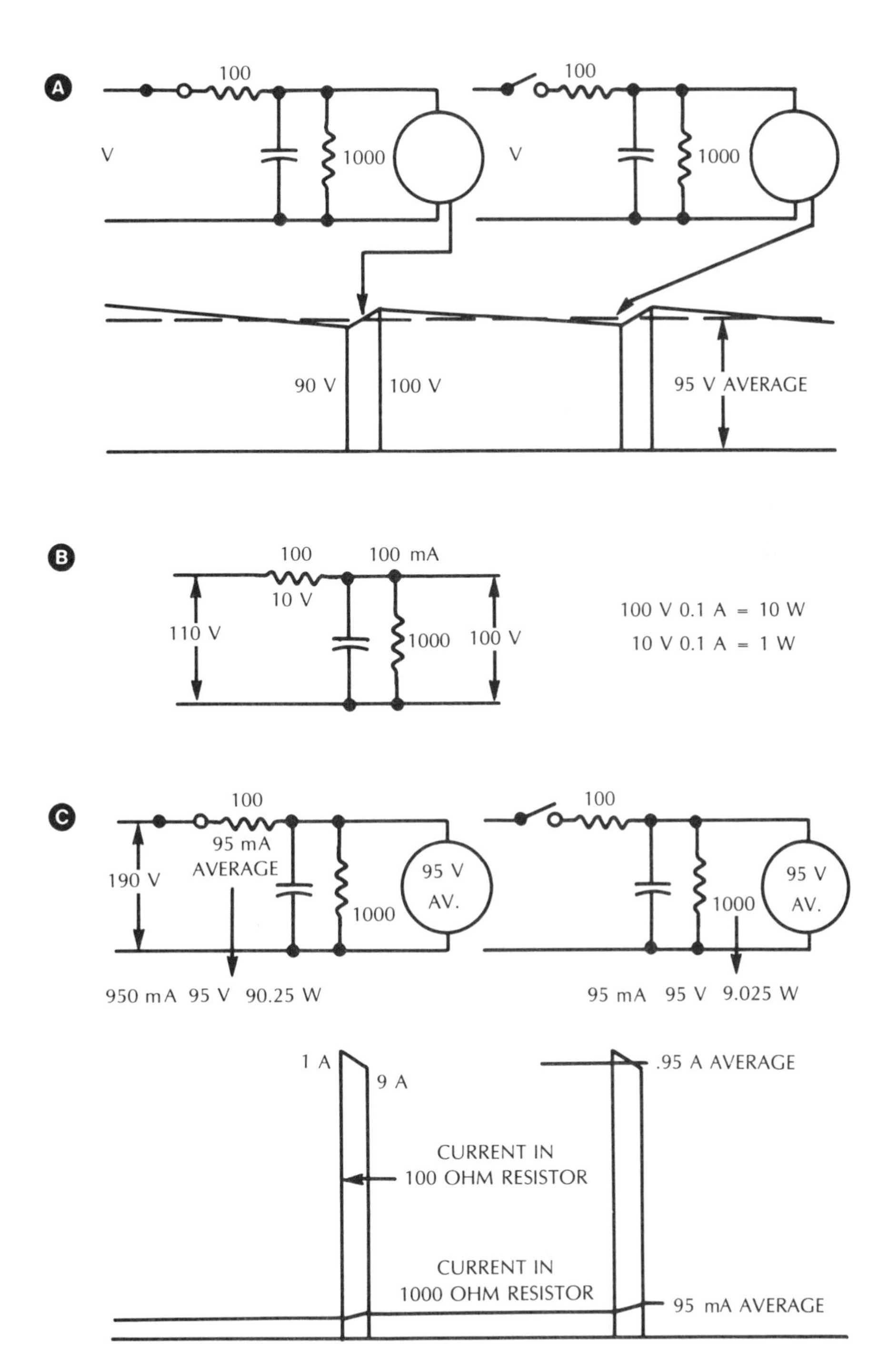

Fig. 4-14. Step-by-step examination of a simple pulse circuit: (A) sequence of charge and discharge; (B) the same circuit as a steady-state connection; (C) typical current conditions in the pulse arrangement.

1000-ohm resistor will be 95 milliamps, resulting in an average dissipation of 95 × .095 or 9.025 watts.

If the supply (of a different voltage, as we shall see in a moment) remained connected all the time, and the voltage across the 1000-ohm resistor is required to be 100, the 1000-ohm resistor would dissipate 100 × 0.1 or 10 watts. The drop in the 100-ohm resistor would be 10 volts, meaning the total supply needs to be 110 volts. And the 100-ohm resistor would dissipate 10 × 0.1 or 1 watt. Remember this. But when the 100-ohm resistor is connected to the supply for only one-tenth of the total time, its average current must be 10 times as high as the average discharge current to maintain that discharge current for the rest of the time.

This is a first step toward understanding Kirchhoff's laws as they apply to other than strict dc circuit conditions. In this case, the quantity that serves as a reference is not simply current, but the quantity of electricity flowing over time. The traditional electrical unit for this, corresponding with the volt, amp and ohm system, is the coulomb. A coulomb is the amount of electricity conveyed by 1 amp flowing for 1 second. So if 100 milliamps flow for 1 second, the amount of electricity flowing is 0.1 × 1 or 0.1 coulomb. A current of 1 amp flowing for 0.1 second also conveys 0.1 coulomb—the same amount in electrical charge.

If the average current flowing through the 1000-ohm resistor is 95 milliamps, then the average current that must flow through the 100-ohm resistor for one-tenth of the total time to maintain a repetitive condition, must be 10 times 95 milliamps, or 0.95 amp. Now the dissipation in the 100-ohm resistor during that one-tenth of the time, will be I^2R, or $0.95^2 \times 100$ or 90.25 watts. But as this dissipation occurs only for one-tenth of the time (for the other nine-tenths that no current is flowing in the 100-ohm resistor), the average dissipation is 90.25 divided by 10, or 9.025 watts. Because the effective current (average value) is, in fact, concentrated into one-tenth of the time, the dissipation it causes is 10 times as high (it is 9.025 watts instead of 10, because the voltage output is 95 instead of 100).

To complete the calculations, since the average current is 0.95 amp, the average volt drop across the 100-ohm resistor must be 95 volts, not the 10 volts of the steady-state condition, so the supply now needs to be about 190 volts to maintain such a pulsing condition. During the one-tenth charge time, there will be 100 volts across the 100-ohm resistor and 90 volts across the capacitor and 1000-ohm resistor at the beginning of charge, changing to 90 volts across the 100-ohm resistor and 100 volts at the output at the end of charge.

This step-by-step consideration is shown in Fig. 4-14. There are other kinds of pulse circuits which we consider later. But this is a relatively simple one to follow, so it forms a useful basis for the notions we shall explore in more depth later. The actual times of the charge and discharge depend on the value of the capacitor, and the time constants thus resulting. For now, it is enough to see that the relative times depend on the current flow, just so that all the current flowing into the capacitor also flows out—no more and no less, when averaged over time.

Examination Questions

1. The following voltages and currents are values measured with certain resistances. For each pair of values, find the resistance and the wattage represented by the voltage and current combination: 20 volts, 50 microamps; 25 volts, 200 microamps; 100 volts, 30 milliamps; 23 millivolts, 17 microamps; 490 millivolts, 37 milliamps.
2. Make the same calculations as in Question 1 for the following values: 22 millivolts, 500 milliamps; 1000 volts, 220 milliamps; 500 volts, 10 microamps; 290 microvolts, 43 milliamps; 66 volts, 22 milliamps.
3. State the formulas that give (a) voltage and (b) current when the resistance value and required wattage are known.
4. From the formulas given in answer to Question 3, calculate the voltages and current equivalents to the following resistance-wattage combinations: 1000 ohms, 10 watts; 270 ohms, 300 milliwatts; 1800 ohms, 2 watts; 750 ohms, 3 milliwatts; 3600 ohms, 10 milliwatts.
5. An electronic component is rated to dissipate a maximum of 10 watts when mounted with two flat, dry surfaces (each of which is insulated), in contact to conduct the heat away. At this rating, the internal temperature reaches 100 degrees C, the maximum safe temperature. The two surfaces reach temperatures of 80 degrees C and 45 degrees C when the ambient temperature is 15 degrees C, the ambient for which the rating applies. If a silicone lubricant, applied to carefully fill the dry space between the surfaces, increases the heat conduction between them seven times, find the maximum temperature reached within the component with the same 10-watt dissipation.
6. In the component mentioned by Question 5, how much could the rating be increased when the silicone is used, without causing the internal temperature to exceed the safe limit?
7. In the component described by Question 5, a different mount, which has heat-radiating fins attached results in a temperature reduction from 45 degrees C to 25 degrees C. How much does this reduce the maximum internal temperature at 10 watts dissipation, and what dissipation is likely to be possible without exceeding the safe limit?
8. What dissipation is likely to be possible if the different mounting in Question 7 is used in addition to the silicone suggested by Question 6?
9. A voltage-divider chain consisting of two fixed resistors, values 180 K 2 watts, and 22 K ½ watt, with a 50 K potentiometer between them, is used to divide a voltage supply of 840 volts so that a point of exactly 168 volts from one side can be obtained. Assuming all values are precisely what they are rated, at how many ohms from the end of the 50 K potentiometer nearest to the 22 K resistor will the 168-volt be obtained, assuming no current is drawn at this voltage?

10. Assuming the 50 K potentiometer is low in value by its maximum tolerance of 10 percent, and that the two fixed resistors each have a 10-percent tolerance, find the range of voltages that the potentiometer will provide when (a) both fixed resistors have values 10 percent high; (b) when the 180 K is 10 percent high and the 22 K is 10 percent low; (c) when the 180 K is 10 percent low and the 22 K is 10 percent high.
11. If, after warming up due to the current flowing, the 180 K resistor drops in value by 20 percent, and the drop in resistance is assumed to be approximately proportional to the ratio of the actual to the rated dissipation of the resistor, by how much will the 22 K resistor drop in value (percent)?
12. The 50 K potentiometer in the previous questions is wirewound, so it increases in value by 10 percent when the entire circuit has warmed up. Assuming values were "on" when all cold (as in Question 9), at how many ohms from the same end will the 168-volt tap be obtained after the circuit has warmed up?
13. A coil has a measured resistance of 500 ohms, cold (15 degrees C). After being connected to its supply for more than an hour, it measures 600 ohms, carefully checked immediately after it was disconnected. Using the rule that resistance rises 0.4 percent for every degree C, what is the average temperature of the coil after it has warmed up?
14. To measure temperature rise, a contact thermometer is strapped to the outside of the coil in Question 13, which registers 45 degrees C at the end of the test. From this, what would you predict to be the lowest possible maximum temperature reached inside the coil?
15. In the arrangement in Fig. 4-14, the supply is connected for only one twentieth of the time, and the voltage across the 1000-ohm resistor varies between 90 and 110. Find the current and dissipation in the 1000-ohm resistor, the 100-ohm resistor, and the voltage of the supply connected for one-twentieth of the time.
16. In the same arrangement, the voltage used is 150, and a voltmeter across the 1000 ohms gives a steady reading of 100 volts. For what proportion of the time is the capacitor being charged through the 100-ohm resistor?

5
From DC to AC Circuits

The simplest way to think of an alternating current is to visualize a current that is switched alternately in opposite directions. However, this simple concept does not represent the way the majority of ac circuits actually behave. The reason is that no natural movement of an alternating nature can be regarded as a 2-position or 2-state condition. A swinging pendulum has two basic "positions," left and right from center. But the pendulum does not change its position suddenly from extreme left to extreme right (Fig. 5-1). The same is true for a vibrating string, tuning fork, or any other device to which a natural frequency of oscillation or alternation may be assigned.

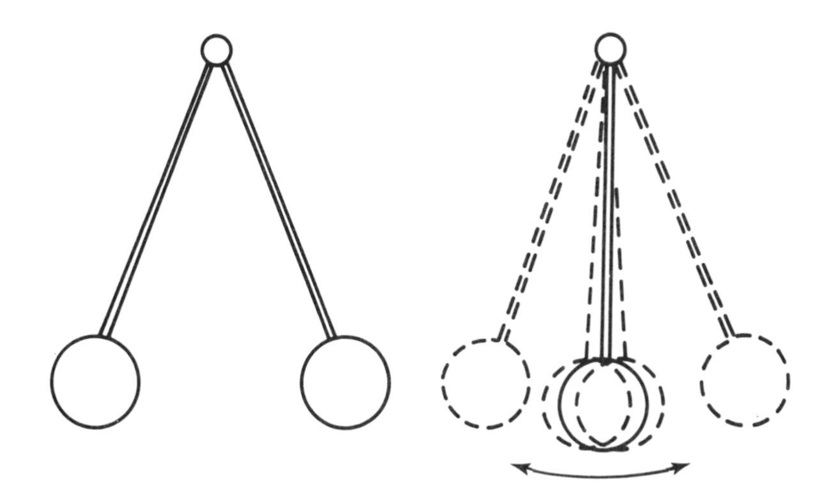

Fig. 5-1. A pendulum illustrates the distinction between 2-state and smooth fluctuation. At left, a pendulum can be considered in only two positions, but this is unrealistic, because it has to move back and forth (right).

Later in this book, we revert to a consideration of 2- and 3-position or -state devices, but the more common devices deal with a continuous change that grows in one direction (say, positive) to a maximum, then dies to zero, passes through zero into the negative region, then grows to a maximum in the negative direction, finally falling to zero from that direction.

The time taken for such an alternation to go through one complete cycle of change, from zero, through maximum in one direction, back through zero to maximum in the other direction and finally back to zero again, ready to start repeating that cycle, is called a *period* of alternation. It is also called a *cycle*. The rate at which such alternations, periods or cycles occur within a specified time period is called the *frequency*. The commonly used unit of frequency measures the number of cycles or periods occurring every second, and the figure obtained is given the name *Hertz* (abbreviated Hz).

SINUSOIDAL WAVEFORM

The smooth, sinusoidal waveform, just described in general terms, but in which the precise shape known as a sine wave is the important characteristic of the wave, is a "pure" frequency, or waveform. Such a wave represents pure simple harmonic motion (Fig. 5-2).

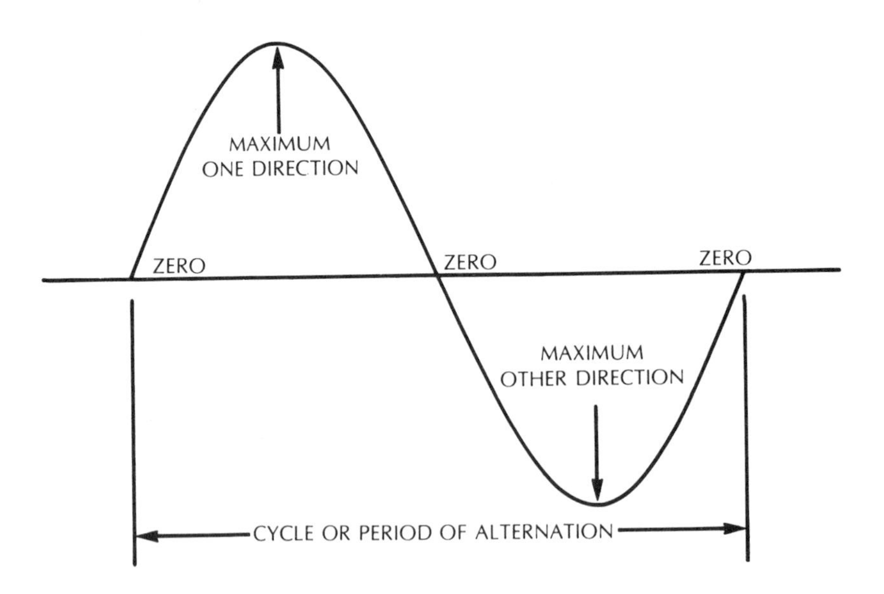

Fig. 5-2. A plot of position against time for any simple harmonic alternating movement, which could be the pendulum, or else voltage or current.

One way to see this in physical terms is to consider a vibrating string, which has the capability of vibrating in more than one mode at a time. In each mode, the string vibrates with this characteristic motion. In each, the motion occurs at a maximum rate as it crosses the center position (where it comes to rest when not vibrating) and, moving more and more slowly, it is momentarily stationary at maximum deflection from the position of rest, before its starts to move back again.

The characteristics that control string movement in such a natural vibration are the tension of the string, which always tends to bring it back to its position of rest, and the mass (weight per unit length) of the string, which tends to keep it moving until the tension retards it and pulls it back again.

These two qualities about a stretched string determine the rate, and thus the frequency, at which it vibrates. At the fundamental vibration rate, the entire string moves and the maximum movement is in the middle of the string's length (Fig. 5-3). But a string can vibrate in a number of modes, called harmonics of the fundamental, where each has its own symmetry of movement.

Fig. 5-3. A string vibrating in its fundamental mode.

In each mode of vibration, the string moves as if its length had been divided by an integral number, with each part vibrating independently at the same frequency; however, the movement of all segments is related, being controlled by the same constants. The string is free to move, under the control of the restrictions mentioned above, throughout its length (except at the ends, where it is fixed). But at the nodes (points that do not move in a particular mode of vibration) the movement of the string on either side is such that all the string does at the node is pivot, without moving (Fig. 5-4). When one segment is deflected up, the next to it is deflected down.

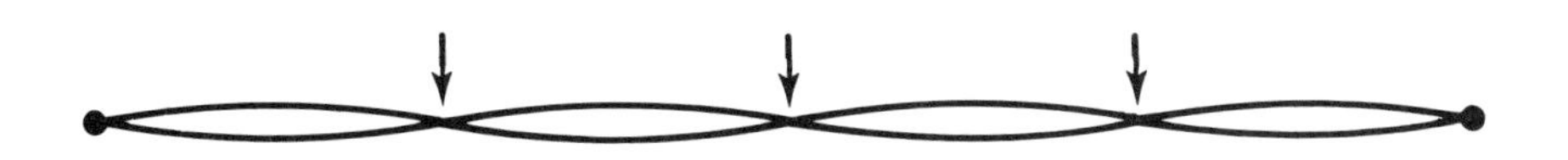

Fig. 5-4. A string vibrating in a harmonic mode: the fourth is shown here.

When a string vibrates in more than one mode at the same time, which is quite possible and happens more often than not, certain points along its length possess movements made up of a composite of the individual wave movements (Fig. 5-5).

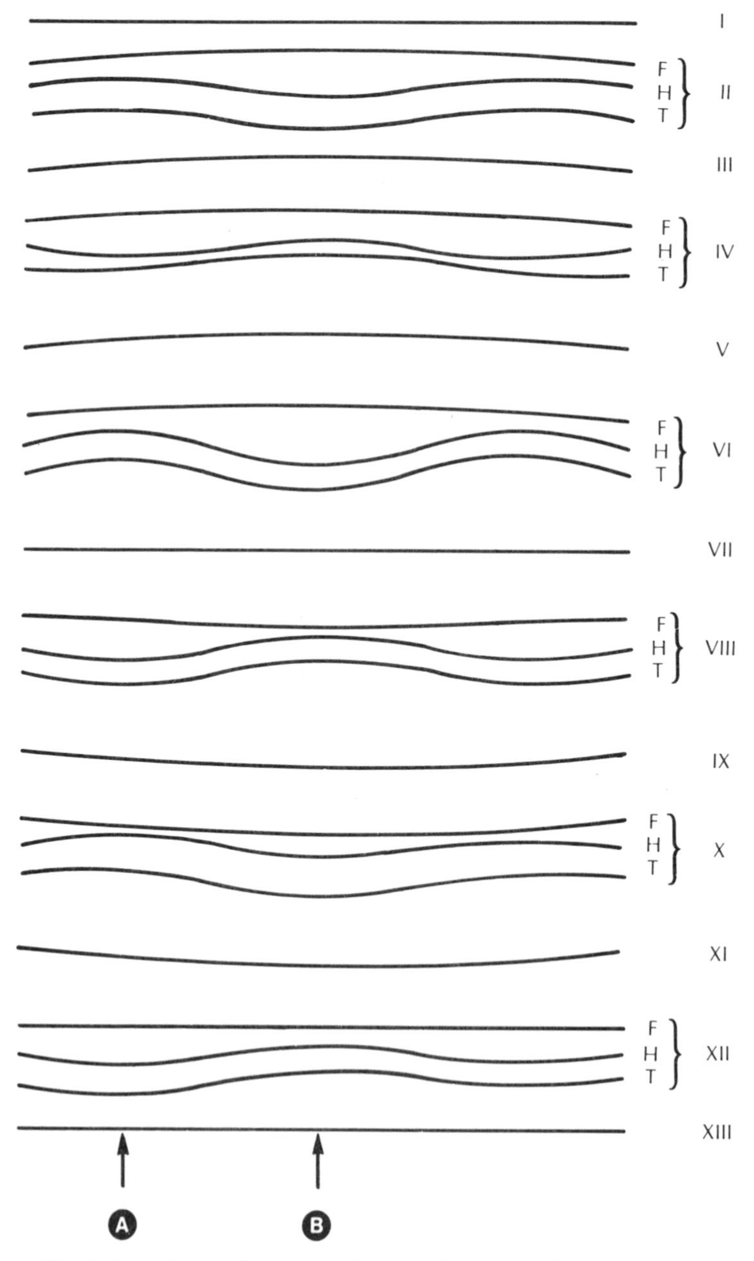

Fig. 5-5. Analysis of a string vibrating in two modes at once.

At a particular point, the total movement is the sum of the movements in the individual modes composing that vibration. And this may be different at various points along the string (Fig. 5-6).

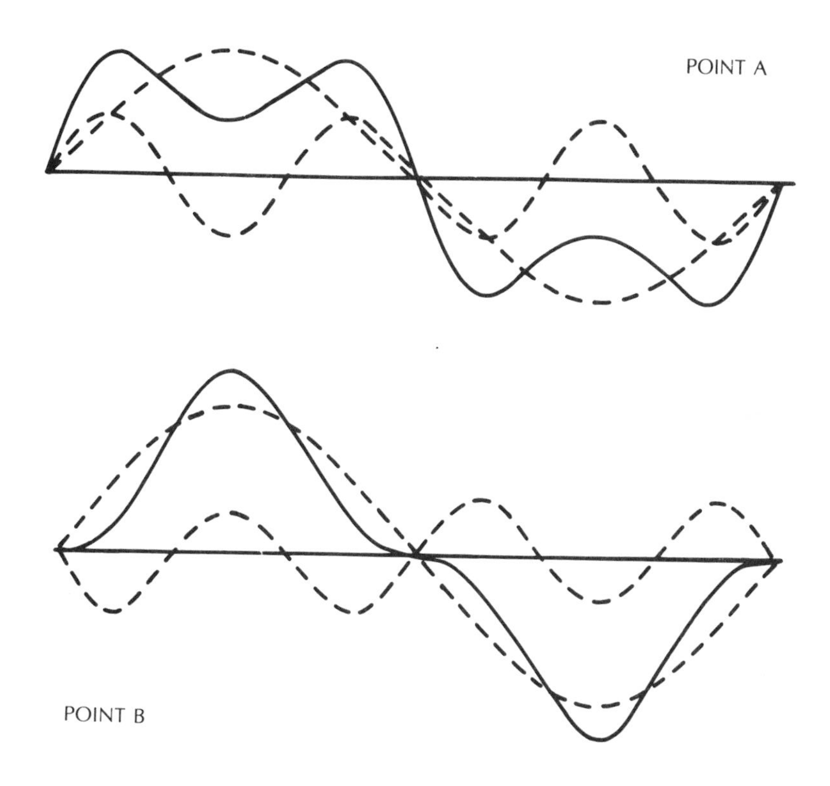

Fig. 5-6. Representation of string movement at two different positions along its length, indicated by A and B on Fig. 5-5.

Current and voltage in electrical circuits or electronic circuits can be compared to the movement of a vibrating string. It may consist of a single frequency or of a combination of frequencies. Thus, both voltage and current may possess waveforms made up of more than one sine-wave frequency.

The best way to observe an electrical waveform, and one used very commonly in modern electronics shops, is with an oscilloscope. It has a special type of tube (which is discussed later) that produces an electron beam focused to a point on a screen. By using voltages or currents to manipulate the spot on the screen, the traveling spot can be made to draw a graphical representation of any electronic quantity, voltage or current.

To observe a waveform that occurs through time (the vibrations in the alternation take time to occur) a special form of wave, called a "time base" is used. Applied to the electrodes that move the spot horizontally, this wave makes the spot traverse the screen at a steady rate from left to right, and then return quickly to the left again, ready to start another "trace." (Fig. 5-7).

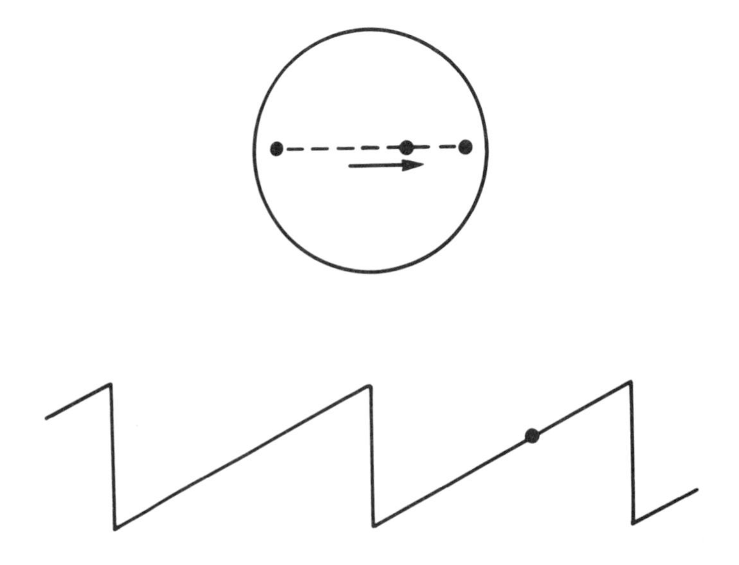

Fig. 5-7. A sawtooth wave, called a time-base, is used with an oscilloscope to make the spot travel from left to right at a uniform rate, returning it quickly to start another trace.

If, at the same time, a voltage waveform is applied to the electrodes that move the spot vertically, the spot will move up and down as it goes from left to right, thus tracing the waveform on the screen. The tube screen produces a glowing image of the waveform for a long enough period to allow it to be seen as a plot of the voltage against time (Fig. 5-8).

A true sinusoidal waveform illustrates the principle that displacement and rate of movement have the same form, merely being displaced in time. When displacement (distance from a position of rest) is at a maximum in one direction or the other, the rate of movement is momentarily zero. When displacement is zero, the rate of movement is maximum in one direction or the other. The inter-relationship between displacement and rate of movement is only perfectly true for a sinusoidal waveform. For the sine waveform, this relationship extends over every part of the waveform. Particularly useful reference points to show this are the one-twelfth intervals along the cycle. By regarding the cycle as a circle "opened up," which contains 360 degrees, these one-twelfth intervals correspond to 30 degrees each (360 divided by 12).

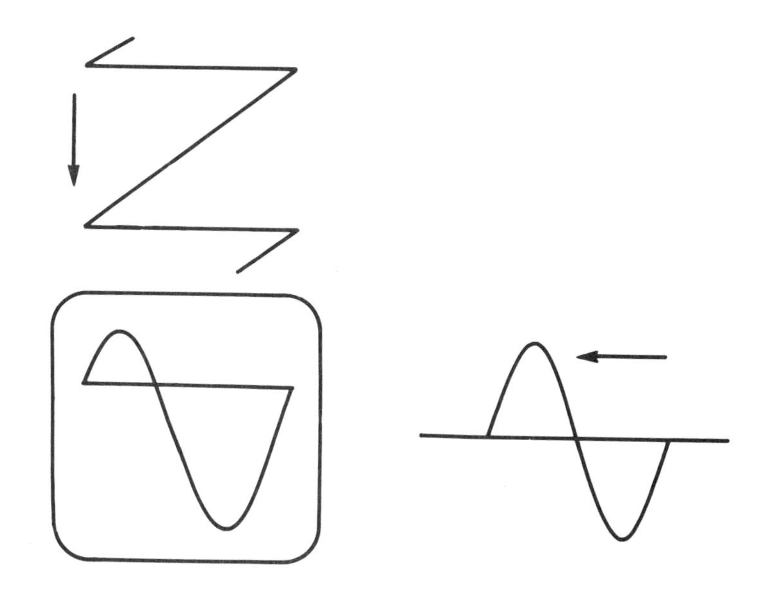

Fig. 5-8. A time base horizontal input, with a waveform vertical input will cause the oscilloscope to display the waveform graphically.

In fact, as we shall find when we study the relationships between waveforms further, relating time intervals within a waveform period or cycle to angles has an almost real significance when associated with the behavior of electronic circuits. And when an understanding of the behavior of such circuits is aided by the use of vectors, the meaning of angles helps, too. Consider that the instantaneous value of a sinusoidally varying voltage (a true sine-wave ac waveform) is represented by projecting the end of a vector along the axis of the waveform (Fig. 5-9).

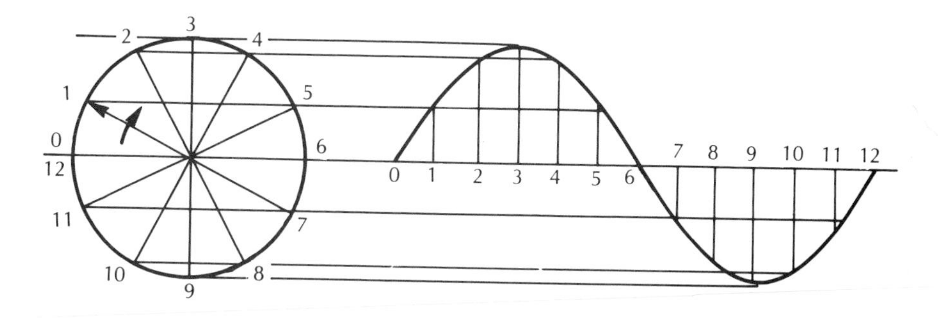

Fig. 5-9. A rotating vector used for conceptual generation of a sine wave.

As a vector rotates at a uniform speed, the instantaneous values plot a sine wave. If, by drawing tangents to the curve at individual points, the slope of the curve is determined at each point and calculated as a fraction of the maximum slope that occurs when the curve crosses the zero line, the values of slope produce another sine wave, displaced by a quarter wave, or 90 degrees (Fig. 5-10).

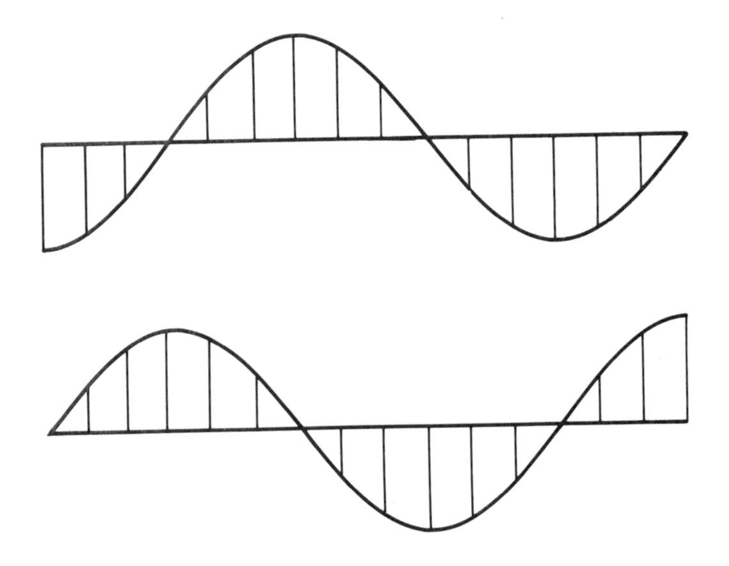

Fig. 5-10. If the slopes of a sine wave are used to plot another curve (the lower curve plots the slope of the upper curve) the result is another sine wave, displaced a quarter of a period from the original.

AC MEASUREMENTS

In an ac circuit, the instantaneous values of voltage and current follow both of Kirchhoff's laws perfectly. Ohm's law must be modified a little, because the instantaneous relationship between voltage and current is affected by components in the circuit other than resistance. But for the resistances in the circuit, the Ohm's law relationship holds as perfectly on ac as on dc.

But voltages and currents in ac circuits are not measured as instantaneous values. It is quite impractical to state or read what voltage and current are present at every instant, even throughout a single period of waveform, without attempting to do this on a continuous basis through all periods. For this reason, alternating waveforms are measured and specified in terms of certain constants.

A constant we have already referred to is frequency, which specifies how often the cycle is repeated per second (or other unit of time). Voltage and current each could be specified by the peak value in each direction, from which the instantaneous values throughout the rest of each period could be calculated (if needed for any reason) by a knowledge of the properties of the sine wave, if the cycle in question has a pure sine waveform. However, until relatively recently, with the advent of the oscilloscope and electronic peak-reading meters, it was not very easy to measure the peak value of a waveform. For this reason other reference values came into use, and they still have much validity and usefulness.

The earlier, more primitive meters were divided into two main types: polarized meters, such as the moving-coil type, in which the needle deflects in the *direction* of the voltage or current it is used to measure; and nonpolarized types, such as the moving-iron voltmeter or ammeter and most electrostatic voltmeters, which give a reading in the same direction regardless of the polarity of dc that is applied to them.

Basically, the nonpolarized meter is capable of reading ac just as well as dc, both voltages and currents. Accuracy may vary, particularly as the frequency of the ac rises. But the polarized meter tries to read in alternate directions very rapidly (as the half periods come by). So that type of instrument does not have time to register the voltage or current in either direction, and the needle usually just vibrates near the zero mark, from which there is no good way to judge the voltage or current that makes it vibrate.

Nonpolarized meters usually work that way because the voltage or current being measured provides both parts of the moving force that causes deflection of the instrument needle. With a polarized meter, one of these elements is fixed. Take the moving coil instrument as an example: A permanent magnet provides a field that is fixed, against which the current in the coil reacts to produce a reading. With a nonpolarized meter, the current generates the magnetic field and reacts with it.

Because of this distinction, in a perfect polarized meter of the simplest form, half of the voltage or current required for full-scale reading produces a deflection exactly half of full-scale. And in a perfect (which is purely a theoretical thing), nonpolarized meter of the simplest form, half of the voltage or current required for a full-scale reading will produce a deflection exactly one fourth of full-scale (Fig. 5-11). This distinction between linear and square-law will readily tell you when a meter is definitely of one type or the other.

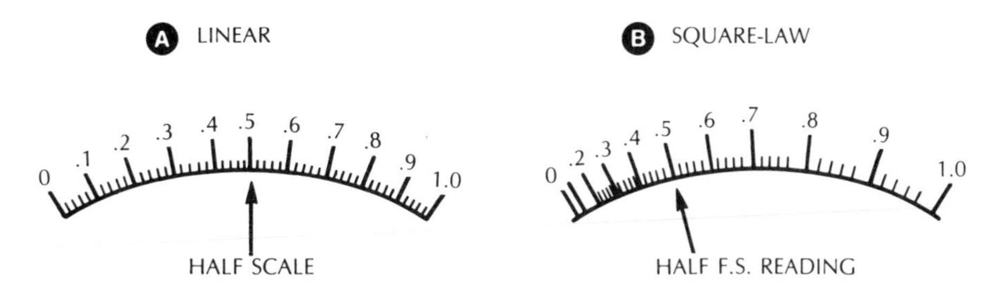

Fig. 5-11. Comparison of linear (left) and square-law (right) scales.

Just one caution about that: Some nonpolarized meters are specially designed so that the central portion of the scale more closely resembles a linear type (that of a polarized meter), which makes it a little more difficult to tell (Fig. 5-12).

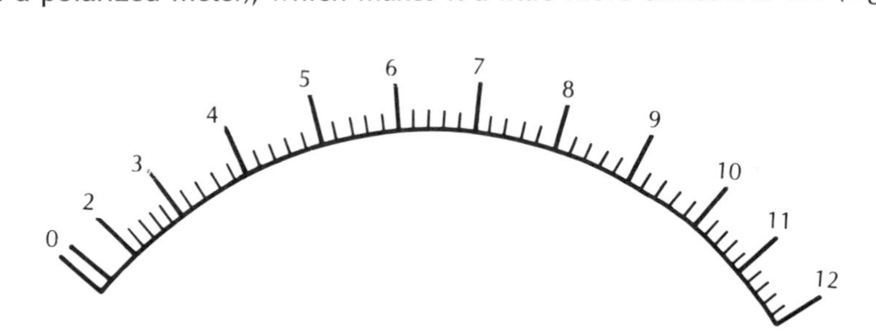

Fig. 5-12. A modified scale, derived by using a device that would be square-law, but has its shape changed by other considerations.

The foregoing relates to earlier electrical meter types. Most modern meters used for electronic work, when used for reading ac, employ a polarized meter with some kind of rectifier that converts all the current to be measured so it flows the same way through the meter itself. This results in yet another kind of reading on the same set of waveforms.

THE ORIGIN OF RMS READINGS

In a nonpolarized meter, the drive on the instrument needle follows the instantaneous force developed by each point on the waveform, current reacting with itself, or voltage with voltage. Thus, the instantaneous drive applied to the needle throughout the waveform is proportional to the squares of the instantaneous values of the current or voltage being measured.

When the meter is calibrated with a dc voltage or current, the scale is marked with the actual voltage or current producing that deflection. This deflection is produced by the steady dc voltage or current reacting with itself in the movement. In a perfect, simple instrument, such a scale would have the square-law appearance shown in Fig. 5-11B. But when the same meter is used to read ac, the steady position taken by the needle will average the square-law force provided by the instantaneous values throughout the cycle. Since the instrument is calibrated with simple readings, not the forces that produce them, the reading obtained on ac is the square root of the average of the instantaneous squares, called the *root-mean-square* or rms of the ac wave.

This means that the meter squares every instantaneous value and then averages the squares. But because the reading indicated on the scale is not the square of the simple dc applied but the simple dc that produced that square-law reading, that is where the needle points. For example, if the meter read 10 volts full scale on dc, and an ac waveform of 10 volts peak (in each direction) is applied, the needle will get a pull at each peak that would make it read 10 volts, if that peak were maintained, instead of coming and going. At the point where voltage crosses over from one direction to the other, the force is momentarily zero. The average force on the needle, taken over the whole sine wave, is half the maximum.

This is the same as the force produced by a steady dc voltage, of either polarity, whose square is half the square of 10, the voltage for a full-scale reading. The square of 10 is 100; half 100 is 50, and the square root of 50 is 7.07. This is the dc voltage that would produce the same reading as a sinusoidal ac voltage with a peak value of 10 volts (each way). So a sine wave has a root-mean-square value that is its peak value divided by root 2, which is 1.414. This is an important factor to remember.

Waves other than sinusoidal (a perfect sine wave) have a different relationship between the root-mean-square and peak values. This can be seen by considering two specific shapes: a square wave, in which both values, peak and rms, are the same, and a triangular wave.

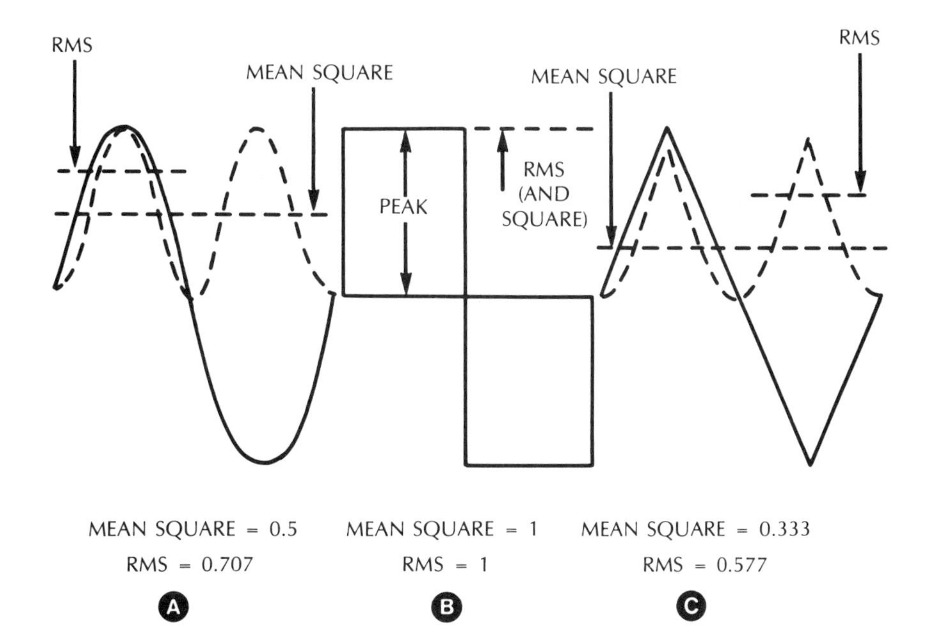

Fig. 5-13. Comparison of waveforms to show relationship between waveform and mean square, and rms values: A, sine wave; B, square wave; C, triangular wave.

Deriving the square-wave relationship is obvious, because that wave only has one value: peak. Average and rms are the same as peak. To derive the triangular relationship squaring each half wave's instantaneous values produces a spiky, parabolic waveform, the average value of which is one-third the peak. The square root of one third is 0.577, the rms value as a fraction of peak.

RECTIFIED AVERAGE READINGS

The other way of measuring alternating waveforms becomes a little more complicated. It uses the average, or mean value, after each portion of the wave—the positive-going and the negative-going—has been changed to go the same way. With this form of measurement, obviously a square wave still has the same value for mean and peak value, since it is at all times at one peak or the other. A triangular waveform has an average reading of one-half the peak value, which is less than the reading of 0.577 given by a root-mean-square meter. On a sine wave, the half wave is averaged (Fig. 5-14) and has a value of 2 over pi or 0.637 times the peak value, which is less than the rms 0.707.

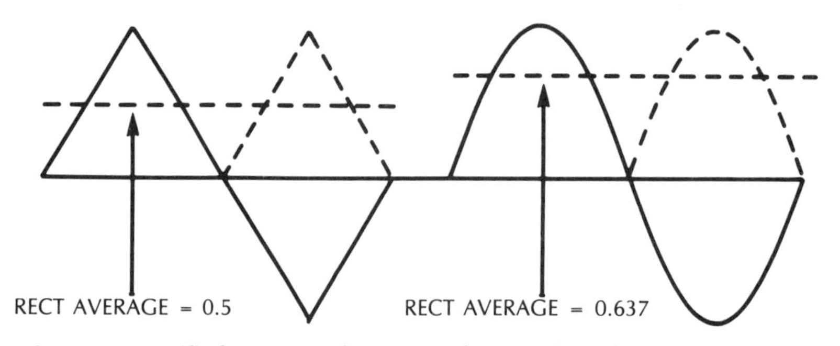

Fig. 5-14. Rectified average of two waveforms: triangular and sine wave.

In transferring from the older rms meters to the modern rectified average, alternating waveforms, which were predominantly sinusoidal, continued to be specified by the rms value. The newer instruments were calibrated to indicate the rms value of a sine wave. To see what this means, let's use an example to see what the calibration really means.

A sine wave with a true value of 10 volts rms will have a peak value of 14.14 volts. The current flowing through a rectifier meter will be the same as it would be from a dc voltage of 0.637 times the ac peak of 14.14: 0.637 × 14.14 is 9.007 (Fig. 5-15).

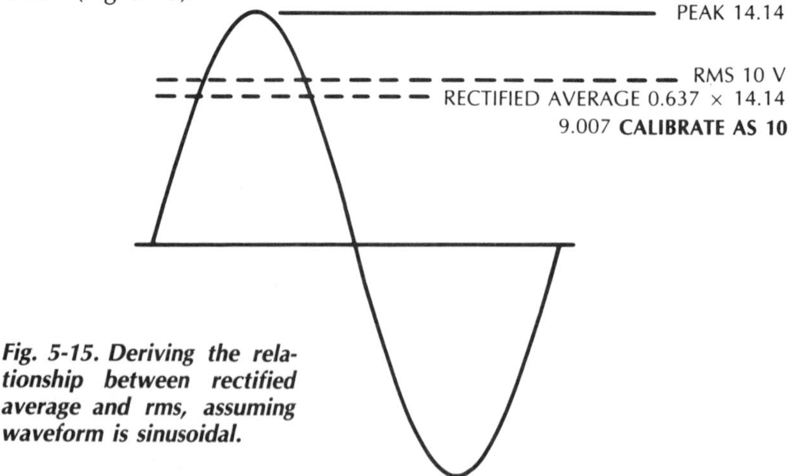

Fig. 5-15. Deriving the relationship between rectified average and rms, assuming waveform is sinusoidal.

So a sine wave with an rms value of 10 volts will read almost exactly 9 volts on such a meter calibrated on dc. The usual practice is to provide the instrument with a circuit change that will cause it to read 10 volts when used to measure ac, so it agrees with the established rms reading.

Form Factor

The ratio between the rms and the rectified average value of a waveform is known as the *form factor*. Dividing 9, the value of the rectified average for a wave with a peak 14.14, into 10, its rms value, yields 1.11, which is the accepted form factor for a sine wave. As we shall see presently, the form factor does not identify the shape of a wave. Quite different waves may possess the same form factor. The term is not used much in electronics, but finds considerable use in power engineering.

EFFECT OF OTHER WAVEFORMS ON READINGS

Using the relationship where an rms reading is equal to the rectified average multiplied by 1.11 is neat, so long as the waveform is sinusoidal; that is, the ac voltage or current being measured is a true sine wave. If it is not, the reading has a dubious value. Take some of the examples we have already used.

Suppose a square wave has an actual peak of 10 volts. Its true rectified average value will also be 10 volts. But a rectified average reading meter, calibrated to read rms on a sine wave, will read 10 multiplied by 1.11, or 11.1 volts. A similar thing happens to a triangular waveform, for a slightly different reason.

If the triangular waveform has a peak of 10 volts, its rectified average *value* will be 5 volts. As shown earlier, its rms value will be 5.77 volts, from which the form factor calculates to 5.77 divided by 5 or 1.155. So the meter calibrated to read rms on a sine wave will read 5.55 volts, instead of the true 5.77 volts (Fig. 5-16).

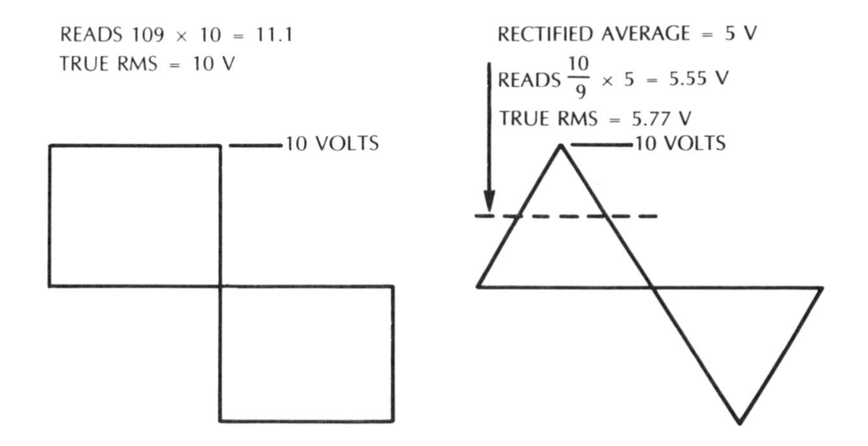

Fig. 5-16. Rectified average readings on square or triangular waveforms.

SYNTHESIZING WAVEFORMS

To arrive at more variants, as well as to cover possible variations in waveform, we recall the demonstration used at the beginning of this chapter to show that waveforms can be made up of more than one sine-wave frequency. The simplest waveforms are made by combining the sine waves where the extra frequencies are exact multiples of a first or fundamental frequency. In these forms, the waveform exactly duplicates itself every cycle of the fundamental frequency.

First, suppose that a frequency exactly twice that of the fundamental, but of smaller amplitude, is added to a sine wave (Fig. 5-17). The exact effect on the resulting waveform depends on the relationship or phase between the waves that are added together. If alternate tops of the second harmonic (as the higher frequency is called) coincide with the tops of the fundamental, the other tops of the second harmonic will coincide with the bottoms of the fundamental. The bottoms of the second harmonic will coincide with the points where the fundamental crosses the zero line. In this relationship, the tops of the resultant wave are sharpened and the bottoms blunted. If the alternate tops coincide with the points where the fundamental crosses the zero line, while the bottoms correspond with top and bottom of the fundamental, an exactly reverse modification of the wave occurs; the tops are blunted and the bottoms sharpened.

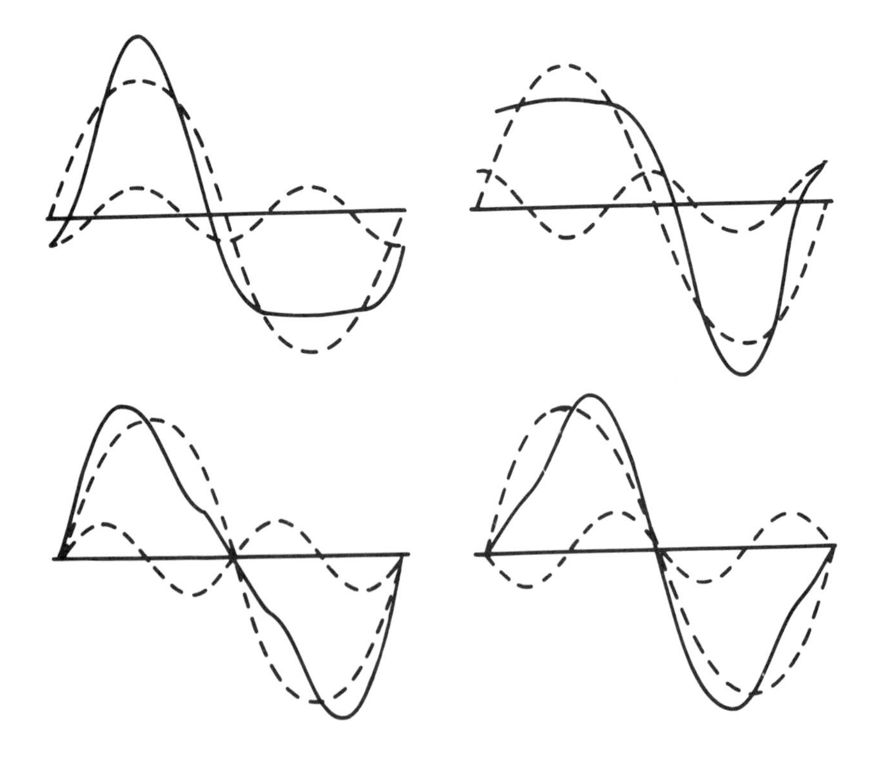

Fig. 5-17. Effects of adding second harmonic to fundamental when the phase relationship of the two waves is changed.

There are two more major possibilities (as well as an infinite number of in-betweens) where the zero points of the second harmonic coincide with the crossovers and tops and bottoms of the fundamental. This results in a change in the up and down slopes of the resultant wave, rather than an alteration in the tops and bottoms. One way the ups are steeper than the downs, the other way it is vice versa.

Now suppose that the wave added to the fundamental is three times the fundamental frequency, known as the third harmonic. This, too, can produce a variety of effects, according to the phase relationships. If the tops of the third harmonic wave coincide with tops of the fundamental and the bottoms with the bottoms, the tops and bottoms of the resultant wave are sharpened. At the same time, the points where the wave crosses the zero line find the fundamental and harmonic crossing in opposite directions. This characterizes the sharpening as being spiky, because the crossover points are less steep than the fundamental by itself. (Fig. 5-18).

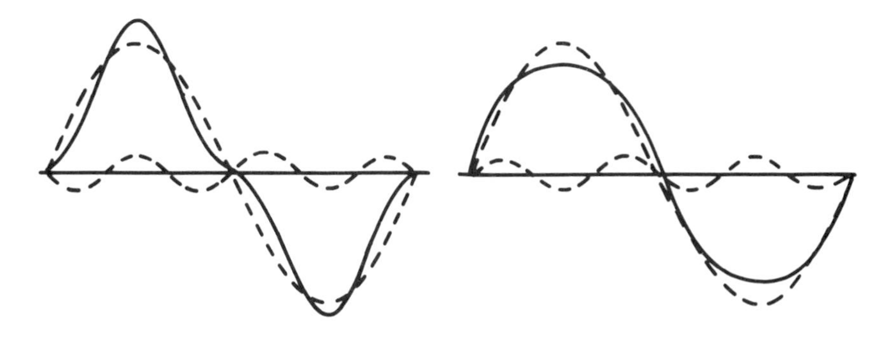

Fig. 5-18. A third harmonic can modify its fundamental in two basic ways.

An opposite effect is produced if the third harmonic bottoms coincide with the tops of fundamental and vice versa. Now, both waves cross the zero line at the same time and in the same direction. So at this point on the waveform, their effects are additive. But at the top of the fundamental the harmonic opposes it, so this produces a wave in which the tops and bottoms are flattened, while the zero-line crossovers are steepened. Notice the similarity between the derivations in Figs. 5-18 and 5-6, where the harmonic is larger in comparison with the fundamental.

Other phase relationships produce a more complicated wave. All of these forms, as well as many more that include more than two waveforms, are encountered in electronic circuits due to various causes, both in the generation and in the handling of various signal waveforms.

To pursue the effect of changing the amplitude of the harmonic in relation to the fundamental, notice what happens in the case of the second harmonic, where the tops or bottoms coincide with the tops and bottoms of the fundamental. There comes a point where the blunting or flattening effect reaches a maximum, beyond which a further increase in the harmonic amplitude causes that end of the wave to break into a double peak (Fig. 5-19). This occurs when the amplitude of the second harmonic exceeds one fourth the amplitude of the fundamental.

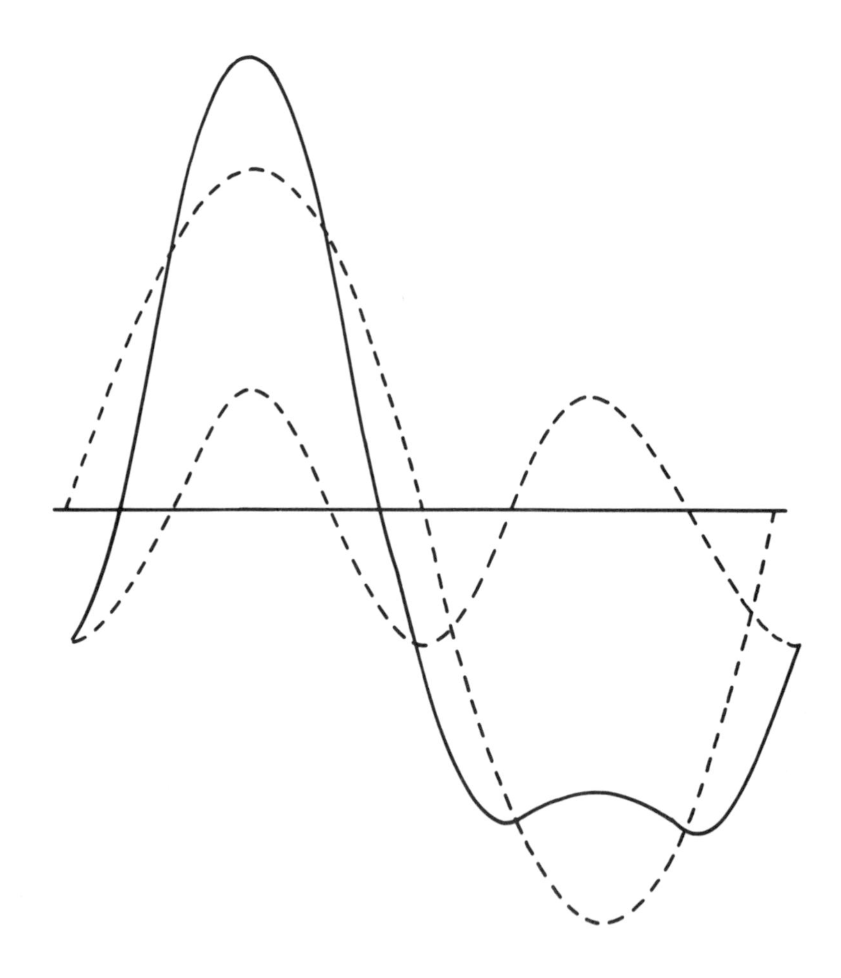

Fig. 5-19. When second harmonic exceeds one fourth of fundamental, the flattening of one peak goes into a double-peaked effect.

If the second harmonic is added so that the zero crossing points coincide, the steepness of the resultant crossing is always accentuated when they both cross the same way. But at the other crossing, where they cross in opposite directions, as soon as the steepness of the second harmonic at that point exceeds the steepness of the fundamental, there will be a triple crossing in this vicinity (Fig. 5-20). This occurs as soon as the second harmonic exceeds one half the amplitude of the fundamental.

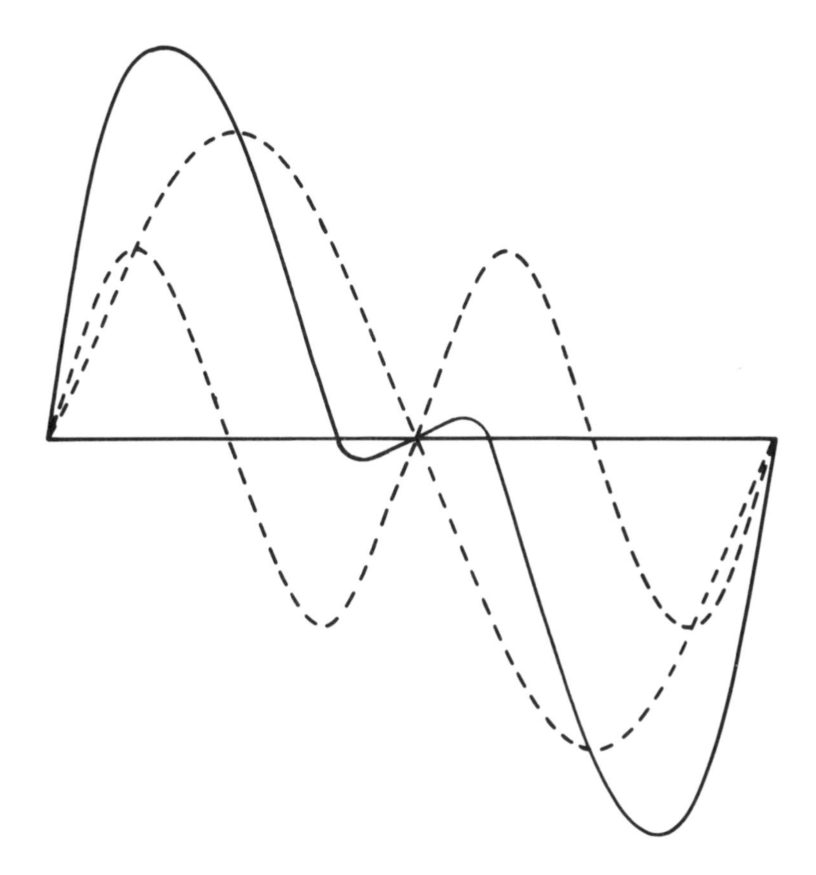

Fig. 5-20. In another relationship, when second harmonic exceeds half the fundamental, double crossing of the zero line occurs in one direction.

In the case of the third harmonic, the limit to the flattening effect is reached when the third harmonic is one-ninth the amplitude of the fundamental, after which the tops and bottoms develop double peaks (Fig. 5-21).

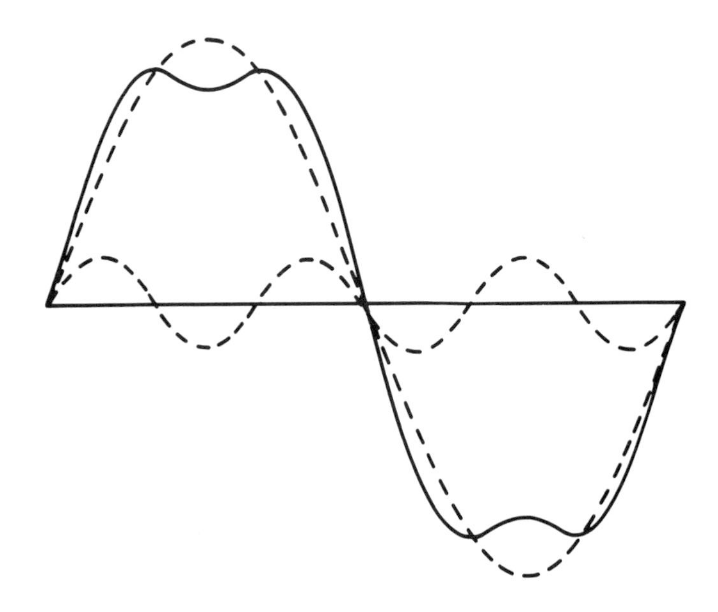

Fig. 5-21. Third harmonic exceeding one ninth of fundamental goes from flattening to double peaking of both half waves.

If the phase of the harmonic is inverted to sharpen the resultant wave, each zero crossing will become a triple crossing when the third harmonic exceeds one-third of the fundamental amplitude with that particular phase relationship (Fig. 5-22).

This is a good place to take another look at form factor. The average value of a wave can be found by taking the average over each half of the larger (fundamental) wave, noting phase effects. When the harmonic reverses from the fundamental, it subtracts instead of adding. The rms is always the root mean square of the sums of the squares of the instantaneous values of both waves.

Thus, adding the second harmonic to the fundamental will never change the average rectified value: it does not affect the peak, and it adds the same as it subtracts to each half wave of the fundamental, because each will contain exactly a whole wave of the second harmonic. If the second harmonic is strong enough to double-peak one side of the waveform, it does change the peak amplitude of the resultant, but by using the dip as the peak-to-peak reference, the same is true. On the other hand, if the second harmonic causes triple crossings at one zero crossover, the average is affected, and what is said here does not apply.

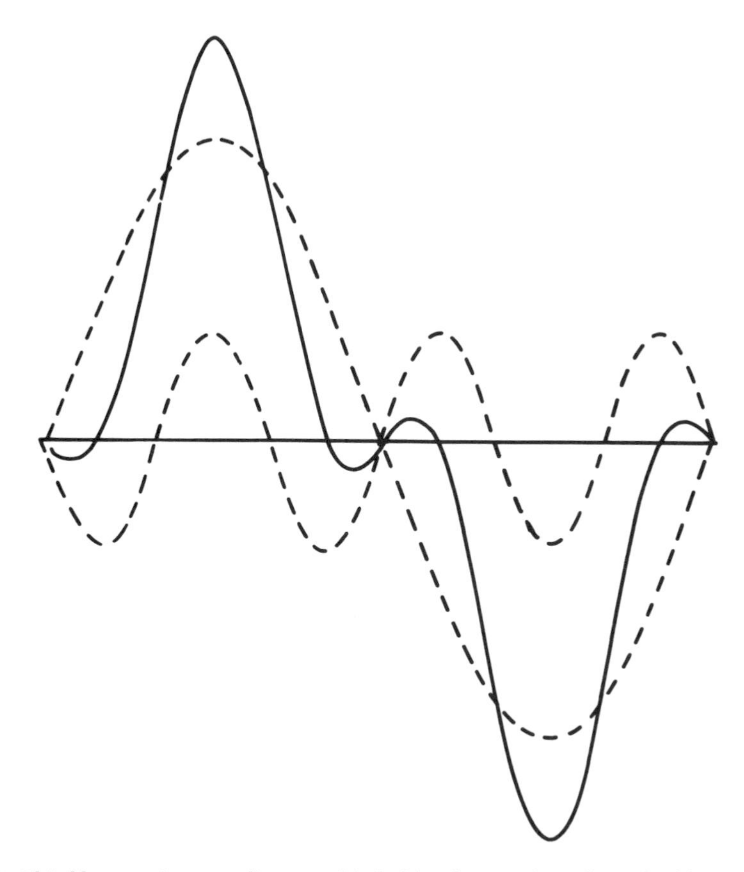

Fig. 5-22. Third harmonic exceeding one-third of fundamental results in double-crossing of the zero line both ways.

Suppose a second harmonic of one-fourth the fundamental amplitude is added. The average value is unchanged, but the rms is increased by the square root of 1 + one-sixteenth. So the form factor is increased very slightly.

Now, take a look at the third harmonic. Assume the added harmonic is one-third the fundamental amplitude. The rms value is increased, using the peak amplitude of the original fundamental rather than that of the resultant wave as a reference, by the square root of 1 + one-ninth which changes it from 0.707 to 0.745.

Now for the average value. With the flattened wave, each half of resultant wave contains two half-waves of the third harmonic in phase and one in opposition. The area of each half wave is one-ninth that of the fundamental: one third the amplitude, one third the duration. So the net result is an increase in the average of one ninth, from 0.637 to 0.708. Dividing 0.745 by 0.708 yields a form factor of 1.05 for this wave, which shows it is approaching a square.

With the sharpened wave, each half of the resultant wave contains one half-wave of the third harmonic in phase and two in opposition, resulting in a net decrease of one-ninth, from 0.637 to 0.566. Dividing 0.745 by 0.566 yields a form factor of 1.316.

One more thing, before we leave this matter of waveforms. Take the case where the second harmonic steepens one side of the resultant wave, up to the point where putting a "kink"—a triple crossing—limits the slope the other way. Now, add a third harmonic of one-third the original fundamental amplitude, so that it steepens the already steep side and offsets the leveling at crossover. This restores the slope at this point to its original value, because all three component sine waves have the same slope at their zero point (Fig. 5-23).

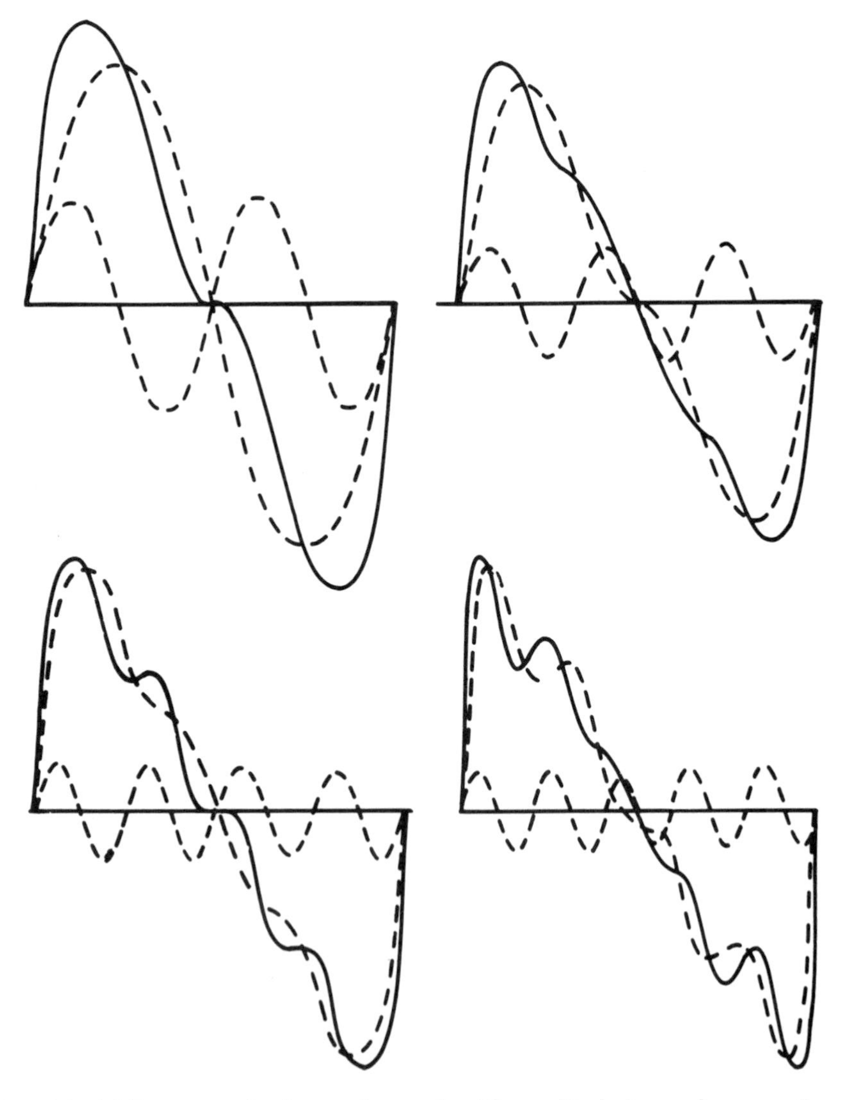

Fig. 5-23. Adding successive harmonics, each with amplitude inversely proportional to its order, causes the resultant wave to approach a sawtooth, with more and more ripples.

Adding successive harmonics, each of which is divided in amplitude by its own order (one-third of 3rd, one-fourth of 4th, one-fifth of 5th, and so on), makes the steep side get steeper and increases the number of ripples in the sloping side. They actually get shallower in depth, although they may not look it, because making more of them causes them to look sharper.

Finally, take another look at the symmetrically flattened wave, produced by adding one-ninth of the third harmonic to the fundamental. This is the maximum that can be added without double-peaking the resultant wave. But the fraction can be increased to one-sixth if as little as one-fiftieth of the fifth harmonic is also added (Fig. 5-24). Notice how this spreads the flattening effect. This process can be extended further: the third harmonic can be increased to one-fifth of the fundamental amplitude if the fifth harmonic is increased to one twenty-fifth with a little bit of seventh—the one two-hundred-forty-fifth fraction.

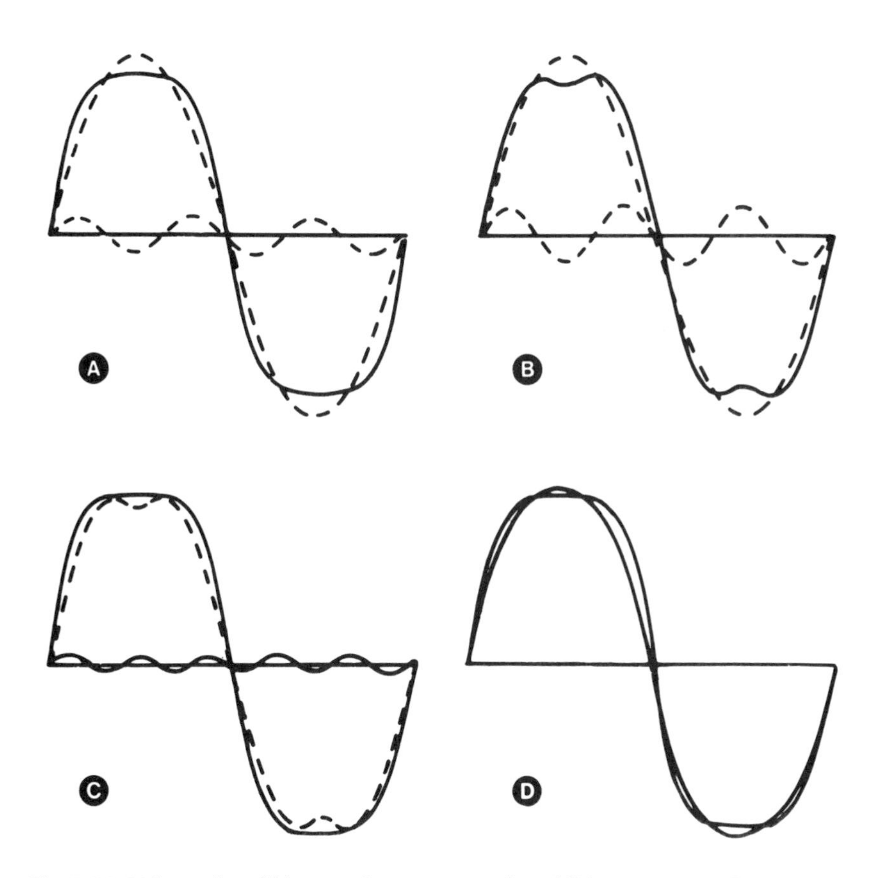

Fig. 5-24. Using only odd harmonics can cause the addition to approach a square wave: (A) fundamental with 1/9 3rd; (B) with 1/6 3rd; (C) the same with 1/50 5th added to that; (D) resultant waves at (A) and (C) compared.

These two methods of adding extra harmonics to the fundamental waveform have been pursued far enough to see where they are leading. Extended to an infinite range of frequencies, one will produce a sawtooth and the other a square wave (Fig. 5-25). In fact, a perfect sawtooth, or a perfect square wave, or a perfect triangular wave, is really only an extension of this harmonic-adding process, using the proper phase and amplitude relationship of the successive harmonics added, up to infinite frequency.

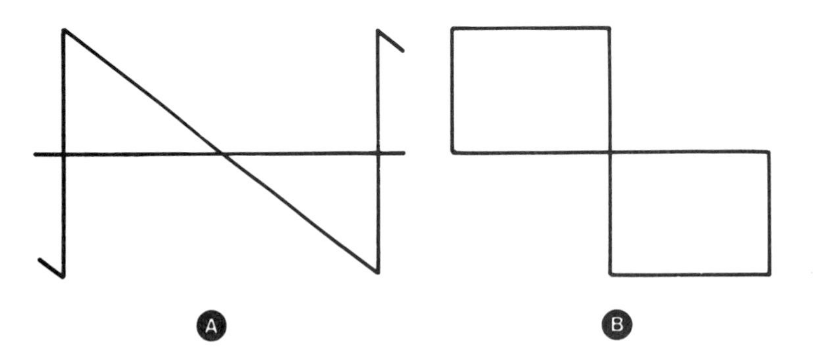

Fig. 5-25. The sawtooth (A) and square wave (B) that the syntheses of Figs. 5-23 and 5-24 approach, respectively.

If the series goes all the way to infinity (which is only a theoretical possibility), the successive amplitudes of the odd harmonics required to make a square wave ultimately are divided each by its own order: one-third of third, one-fifth of fifth, one-seventh of seventh, and so, to infinity. If this series does not go to infinity, the wave is left with as many ripples as the harmonic order at which the series fails. But by tapering the amplitudes off in the way we began, the ripples are avoided, and the departure from the square merely becomes a slight rounding of the corners a little bit.

The foregoing synthesis is largely theoretical today, although at one time parts of it may have been practical. As far as a circuit's being capable of handling the waveform, however synthesized, knowing its content, which really is analysis rather than synthesis, will tell the accuracy with which it can be handled. But there are other ways to synthesize today, as illustrated in Fig. 5-26.

Instead of adding frequencies together, the first constructs the wave form rather like an architect designs a building. At Fig. 5-26A the points numbered 1 through 7 are ordinates in height and time programmed into the system in one way or another. There are several ways of doing this. For example, the first upward slope could be set from the zero point, and the change in slope at point 1 is then initiated either by changing it after a measured time interval, or by changing it when the requisite height is reached. Then it sets the course for point 2, and so on.

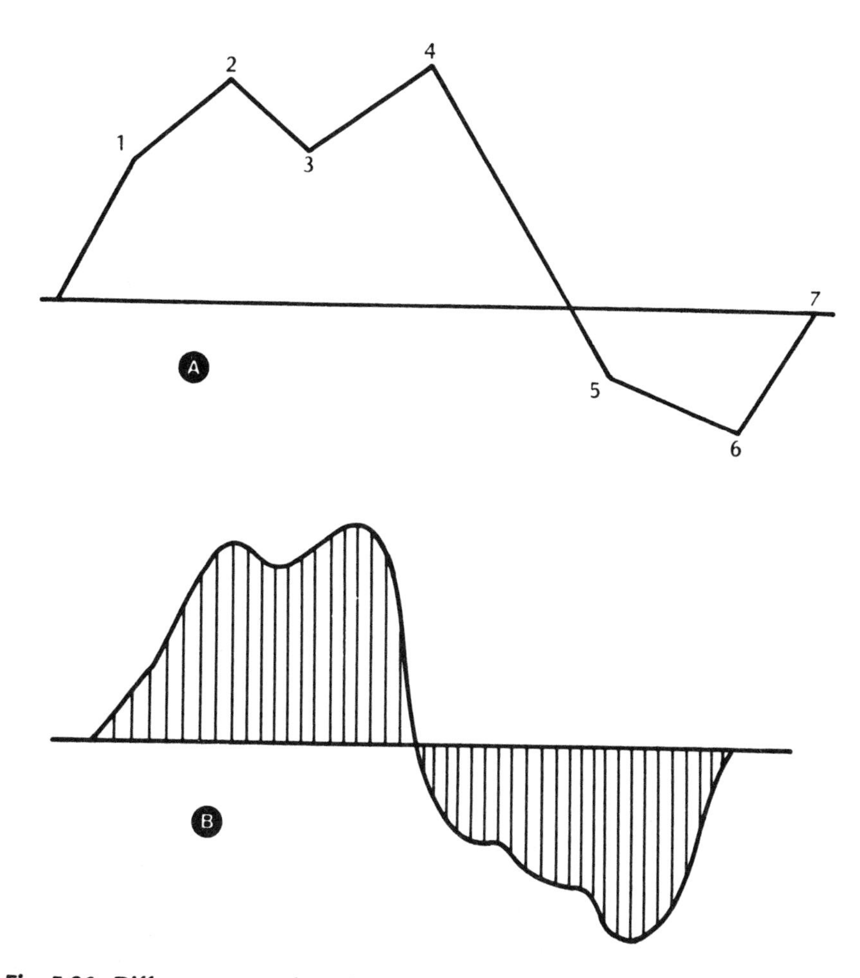

Fig. 5-26. Different ways of synthesizing a waveform today: (A) point to point; (B) successive ordinates.

Either time or amplitude (height on the waveform) could be used as reference to "trigger" the change points, except that if it is holding a steady level for part of the time, as in a square wave, amplitude does not change, so the system must have some way of measuring time between such points. However, there are still different ways of doing that. In digital technology, a clock pulse could be used to measure time. Or a slope could be set on a different waveform from the one you want, so that it will take the required time to go from one amplitude on this other waveform to another, at which second point the trigger occurs.

The other method of synthesis goes all the way to digital. At clock measured intervals, represented by the vertical lines in Fig. 5-26B, computer generated or stored digital information is used to establish the amplitude of the waveform at each successive point.

FREQUENCY RANGES

Throughout the preceding discussion, we have spoken of frequency and period, or cycle, in somewhat vague terms, purely relative: the third harmonic is three times the frequency of the fundamental; nothing said about the frequency of the fundamental.

Electronics circuits can handle a tremendous range of frequencies, depending on the application. Some earth-science and medical equipment may be concerned with very low frequencies, such that the unit Hz becomes irrelevant—the period may last several minutes, hours, or even days. At the other extreme, frequencies go up to thousands of Hz, called kilohertz (kHz), millions of Hz (called megahertz, MHz), thousands of millions of Hertz, and so on.

The lower frequencies are classified as audio, because they are predominantly used for the transmission of sound waves or their equivalent signals, although the audio frequencies serve many other applications other than sound functions. The strict audio spectrum is usually defined as from 20 Hz to 20 kHz.

Above the audio frequencies are the video and radio frequencies, those used for television signals which convey picture "information" and those used for radio transmission. The radio-video band extends from hundreds of kHz up into thousands of MHz. Above that are the microwave waveguide frequencies and laser frequencies, which involve more sophisticated (in some respect) electronics than we shall study in this book. In some respects, they are simpler—more like simple plumbing!

Examination Questions

1. To explore waveform relationships, assume in a square-cornered wave that the first one-eighth period is +5 volts, the next one-fourth period is +10 volts, the following one-eighth period, making up the first half, is +5 volts. Then the first one-eighth of the second half is −5 volts, the next one-quarter is at −10 volts and the final one-eighth is −5 volts. Calculate the rectified average and rms values of this waveform and its form factor.
2. Using the same timing as the example in Question 1, make the one-eighth periods that were 5 volts, positive and negative, 2 volts positive and negative, respectively.
3. Continuing the same exploration, assume that the first one-twelfth period is +5 volts, the next one-third period is +10 volts, the following one-twelfth period is +5 volts, and the following half period uses corresponding negative values. Calculate the average rectified and rms values of the waveform and its form factor.
4. Dividing the period into one-twelfth intervals and continuing this use of square-cornered segments, assume successive one-twelfth intervals have the following voltage values: +4, +7, +8, +8, +7, +4, −4, −7, −8, −8, −7, −4. Calculate the average rectified and rms values and the form factor of this waveform.

5. To explore the variation further, assume the waveshape is constructed by using triangular segments in addition to square-cornered ones. For this example, assume the first one-eighth period rises from zero to +10 volts in a straight line. The next one-fourth holds steady at 10 volts, then in the next one-fourth it follows a straight line from +10 to −10 volts, which it holds for another one-fourth, finally returning to zero in the final one-eighth by another straight line. Calculate the rectified average and rms values, and the form factor of this wave.
6. Consider a sloped-edge wave similar to that in Question 5 with flat tops, but change the duration of the rise and fall slopes in each half wave from one-eighth period to (a) one-twelfth period, with one-third period flat tops; (b) one-sixth period, with one-sixth period flat tops.
7. In a stepped waveform the voltage changes a step at every one-thirty-sixth period (10 degrees), with the following sequence of voltages over one complete period: +1, +2.5, +4, +6, +7, +8, +9, +10, +10, +10, +10, +9, +8, +7, +6, +4, +2.5, +1, −1, −2.5, −4, −6, −7, −8, −9, −10, −10, −10, −10, −9, −8, −7, −6, −4, −2.5, −1, Calculate the rectified average and rms values and the form factor.
8. Using a similar 10-degree interval and the following sequence of stepped values, perform the calculations called for in the previous questions: 0, +0.2, +0.6, +1.2, +2, +3, +4.5, +6.5, +8, +10, 10, +8, +6.5, +4.5, +3, +2, +1.2, +0.6, +0.2, 0, −0.2, −0.6, −1.2, −2, −3, −4.5, −6.5, −8, −10, −10, −8, −6.5, −4.5, −3, −2, −1.2, −0.6, −0.2.
9. A waveform consists of spikes in alternate directions, with zero voltage between the spikes. Thus, for one-fifth period, the voltage is zero. Then, for one-twentieth of a period, the voltage rises in a straight line to 10 volts and falls to zero in another straight line during another one-twentieth period. For the next two-fifths period, the voltage is again zero. Then for one-twentieth period, the voltage goes to −10 volts in a straight line, to return to zero in another straight line and one-twentieth period. Finally, the last one-fifth period shows zero voltage. Find the rectified average and rms voltages and the form factor of this wave.
10. Repeat the calculation in Question 9, but change the width of the base of the spikes. In Question 9, the base of each spike is one-tenth period. Calculate the same quantities when the base of each spike is (a) one-fifth period, (b) one-twentieth period.

6
Analog and Digital Bases of Operation

So far we have considered electrical circuits, as applied to electronics, in a direct current, where voltages and currents are not changing, and as alternating currents, where both flow back and forth according to some prespecified shape. That shape changes in a way specified with time as the independent variable, and amplitude, whether of voltage of current, as the dependent variable (Fig. 6-1).

Such wave shapes, whether the simple sine wave shown in Fig. 6-1, or the more complicated shapes, some of which were developed in the previous chapter and which can be synthesized from or analyzed into a complexity of such simple sine waves, are called "analog" in today's technology. This means that they vary according to a prespecified shape or waveform, with time as the independent variable.

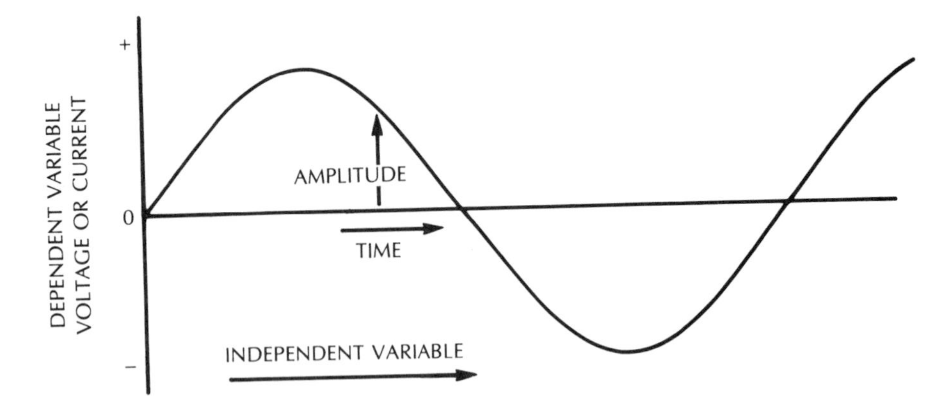

Fig. 6-1. Concept of independent and dependent variables, when independent is time.

While digital and analog can never be completely separated in the real world, they represent completely different ways in which information is stored, processed, or conveyed, whether that be in the physical world, or in the data-processing world of electronics. So before going any further, we need to grasp this distinction.

We also need to make a start on understanding the basics of information theory. Although that is not the subject of this book, it does make the understanding of what follows much easier. So let us use some real-world analogies to illustrate what we mean.

SOUND

Sounds we hear are conveyed from their source to our ears by waves, very like the alternating current waves we introduced in the last chapter. Air pressure, during the passage of such a wave, rises and falls in one direction, then rises and falls in the reverse direction. Such fluctuations in air pressure enter the auditory canal and cause the tympanic membrane (better known as your "eardrum") to move accordingly (Fig. 6-2).

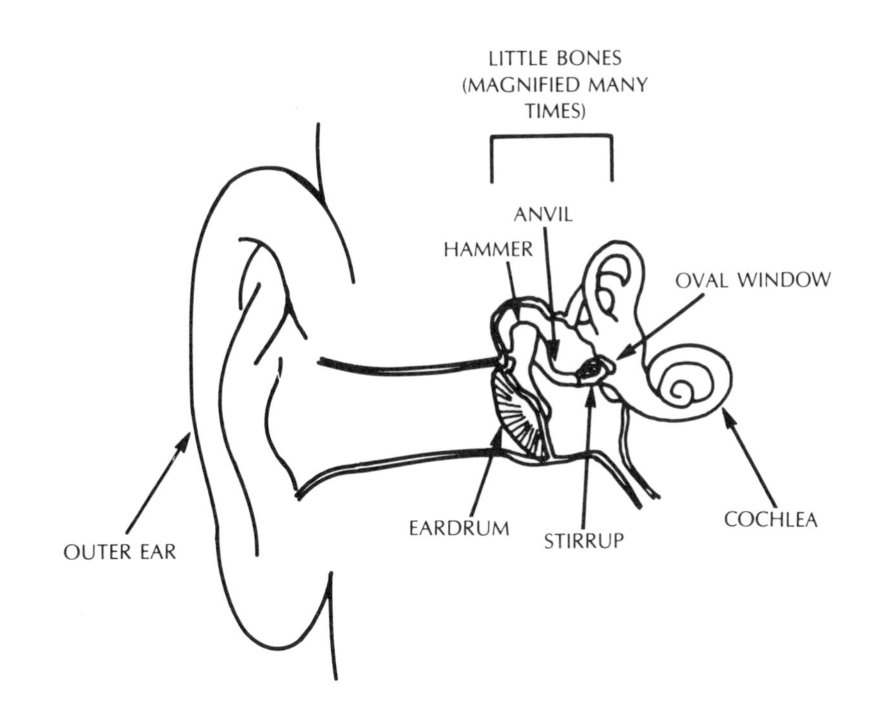

Fig. 6-2. Essential details of the human ear.

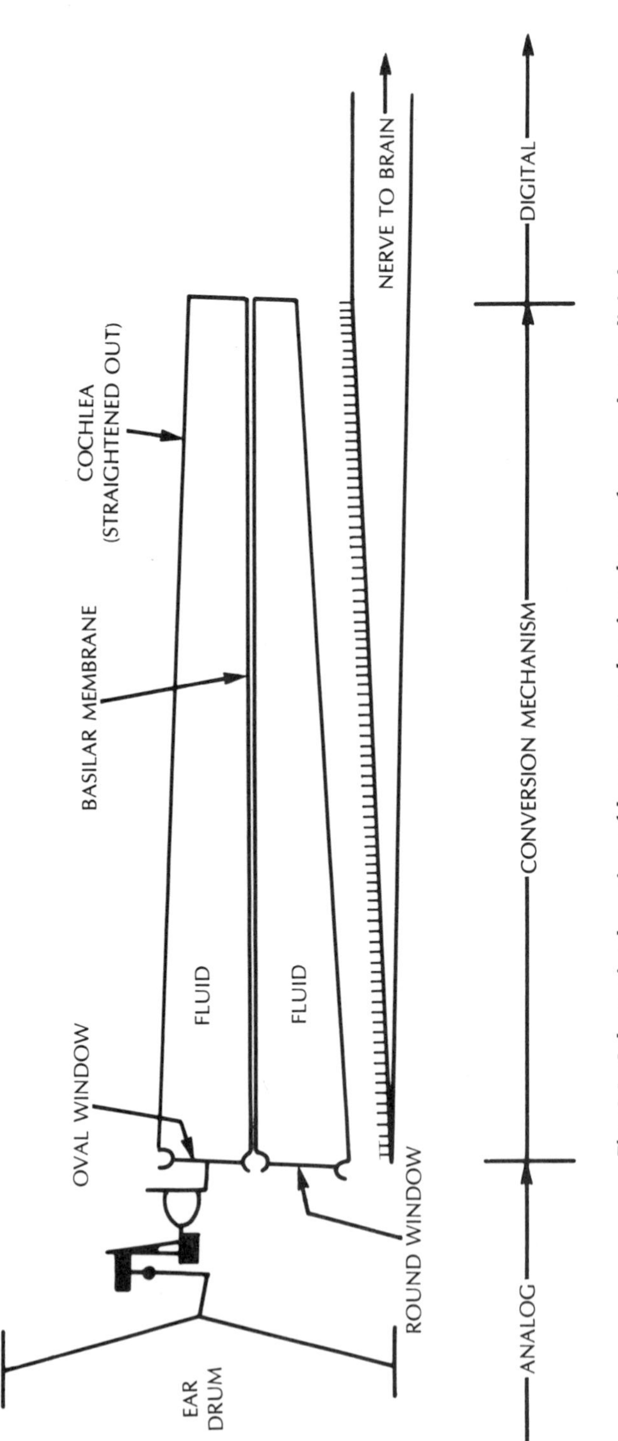

Fig. 6-3. Schematized version of human ear, showing change from analog to digital.

Such movement is analog, which means it follows the shape of the arriving wave. If it is a pure sine wave, it moves back and forth at that one frequency, following the pattern traced in the previous chapter for a sine wave. More often the waveform is much more complicated. At this point, the information arriving at and being processed by your ear is in analog form.

Next, to analyze this information, the three little bones in the middle ear transmit the same movements as a duplicated waveform to the oval window, that passes these analog movements to the fluid in the cochlea. The information representing the sound we hear is still analog. This is where it is changed into a form of digital, though quite different from the digital we shall be studying in electronic applications.

The cochlea consists of a double column of fluid in a snailshell-like structure, with a membrane called the basilar membrane (Fig. 6-3). This has the function of separating the frequency components of the vibration waveforms received by creating resonant columns of fluid, representative of each frequency detected.

Parts of the basilar membrane that vibrate in response to their individual frequencies excite hair cells that contact them, and transmit nerve impulses to the brain, which are responsible for the sensation of sound that we experience.

VIDEO

The same is true, with some differences, about human vision. But the electronic processing of signals that represent pictures we see introduces some different dimensions to the relation between analog and digital than does audio, the electronic processing of signals representing sound.

In the human eye, the crystalline lens (Fig. 6-4) focuses an image of what we see onto the retina, which is like the film or plate surface on which a camera records the picture it "sees." But the human retina holds the termination of an extremely large number of nerve fibers that convey an image of what we "see" to the brain.

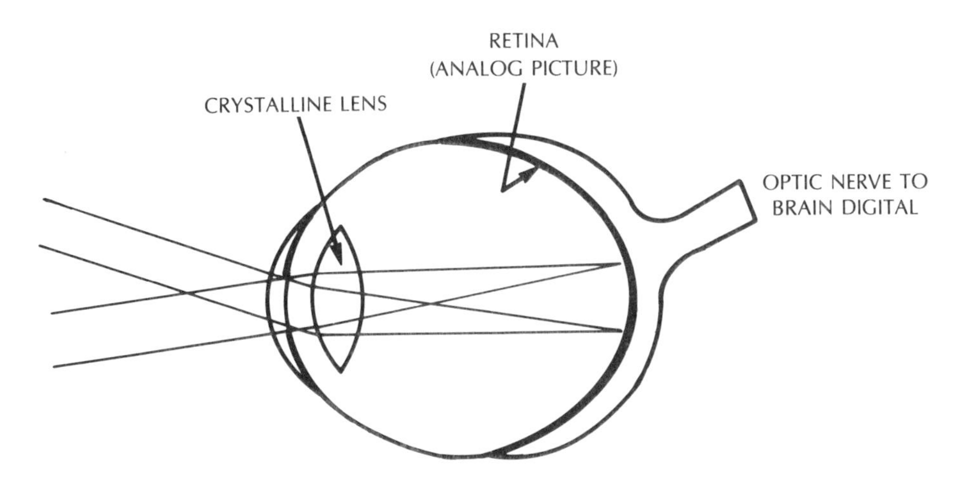

Fig. 6-4. Structure of the human eye, showing conversion from analog to digital.

Each fiber in the optic nerve conveys information about the light received by one spot on the retina, as to brilliance in a particular element of color. Some idea of the number of fibers involved can be obtained from studying your own eyesight. It conveys to you the illusion of complete, continuous objects with seemingly infinite detail. Actually, this is assembled by the brain from the information conveyed by the optic nerve about an extremely large number of points on the retina's surface.

The fact that shapes seem continuous, instead of being made up of many discrete points, shows how large the number of points is. But the fact that it is not infinite, as you might otherwise suppose, can be determined by tests of resolution.

Now, if you look at an illuminated sign consisting of thousands, perhaps millions of light bulbs, it will convey to you an impression of a complete picture. If you are far enough away from it, you may not be able to see the individual light bulbs. But each bulb is lit by an individual wire, just as your eye uses an individual nerve fiber to transmit the composite picture to your brain.

Now turn to how you view a TV picture tube, which conveys to you the illusion of a continuous color picture. If you look closely at such a tube, you will see that it has a mosaic consisting of millions of phosphorescent dots, in three different colors, each so tiny that, more than a few inches away from it, it looks like a continuous color picture.

That picture tube creates what you see, very much in the same way your brain creates an image from the information from your retina. But while your eye requires a nerve fiber to convey what each dot receives to the brain, a different method is used to illuminate that picture tube, as well as to derive the original picture from the camera tube. This method is known as *scanning*.

If the picture changes at a speed approaching $1/100$ of a second, the human eye fails to follow the change, because individual optic nerve fibers cannot accept "signals" any faster than that. So each nerve fiber must take its own signal to the brain.

But electronics can convey information at speeds of millions of "bits" a second. So scanning enables the same "channel" which is like an ultra-fast nerve fiber, to transmit all those "bits" of information. It does this by scanning the picture a line at a time.

A beam of electrons traverses upwards of 400 lines across the picture, one after the other. It does this 60 times a second, covering the complete picture. Actually it takes $1/30$ of a second to complete the picture, but during this time it covers the area twice, interlaced, to fool the eye into thinking it does it every $1/60$ of a second. As this beam traverses each line, it picks up and delivers information about every point along that line, in sequence. Now you could look at the signal thus transmitted in two ways, as either analog or digital. If you regard each line as being made up of, say 600 points just as there are 400 lines, then the signal transmitted scans those 240,000 points or more, once for every picture and sends each of them separately.

But viewed another way, each line transmits a "slice" of the picture (Fig. 6-5), which in itself appears continuous, not as so many individual points, It's true each phosphor dot, in each color, is a separate point. But it's not practical to keep them all separate in the transmission. Each color is scanned separately (but in synchronism so the beams travel together) and the information is sent at the same time, but separately, although electronics has a means of combining this on one channel so they don't get mixed up.

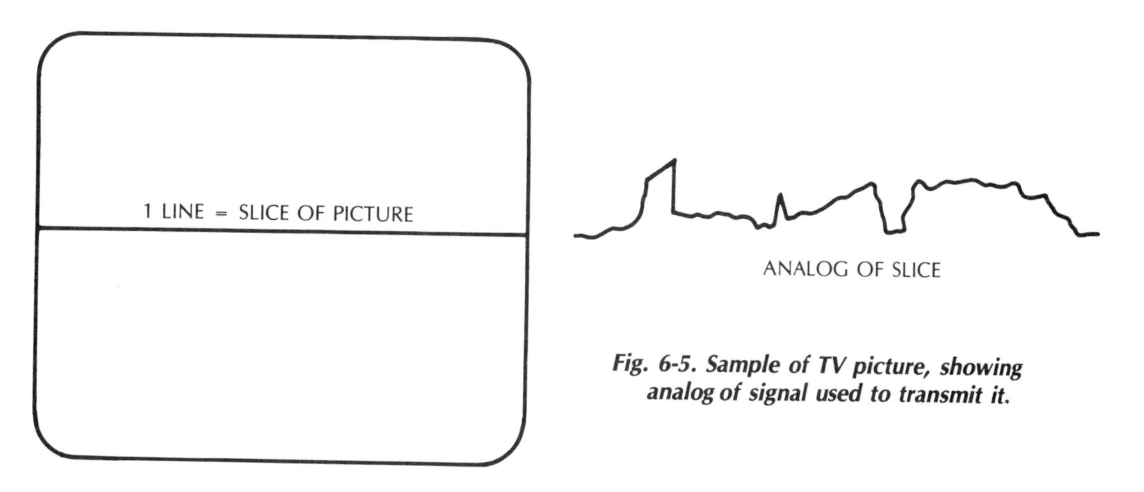

Fig. 6-5. Sample of TV picture, showing analog of signal used to transmit it.

If every dot, in their respective colors was scanned sequentially, then if one element hit the wrong colored dot, the picture would be completely ruined. So the resulting "signal" from scanning the dots, which is that applied to the picture tube at the receiving end, is not a succession of discrete dots, but a wave shape. In that sense it is analog.

But the audio waveform is a synthesis of frequencies. The same frequency content can show up as quite different shapes (Fig. 6-6). If the content is the

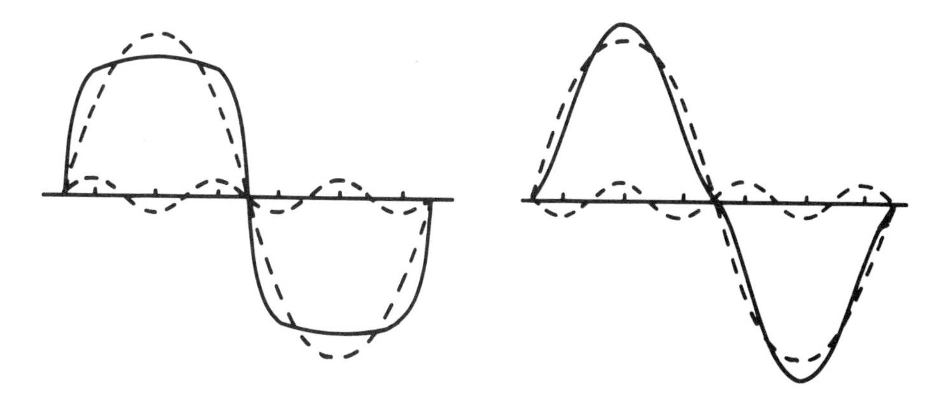

Fig. 6-6. Change in waveform for the same combination of frequencies, with their time relationship changed.

same, the sound will be the same, even though the shape is different. But in video waveform, the shape is important. So although both are, at that stage in their processing, analog in form, the critical features are quite different. This is an important factor to note when we come to conversion between analog and digital.

The human anatomy uses what, in electronics, would be a very slcw pulse rate, over a great many channels, nerve fibers. Electronic technology does the reverse. Often it uses only one channel at a very high rate, sometimes as high as trillions of elements per second.

For use in computers that use digital inputs and provide decimal readouts, the simplest system uses the same conversion for each place of decimal, either that in Table 6-1 or some other, each of which requires 4 digits of binary to represent each digit of decimal. But if the system is not concerned with numbers in the decimal system at all, it may work directly in binary.

DECIMAL	BINARY	DECIMAL	BINARY
0	0000	5	0101
1	0001	6	0110
2	0010	7	0111
3	0011	8	1000
4	0100	9	1001

Table 6-1. Conversions for Each Decimal Digit to Binary.

To show what this distinction means, Table 6-2 gives conversions for some numbers in the decimal system up to 9000, when decimal numbers are being represented in binary, and when the numbers are handled totally in binary.

DISTINCTION BETWEEN DIGITAL AND ANALOG

That should help you to see the difference between digital and analog as the two systems connect with the real world. This book is about electronics, so you want to know what difference it makes to the electronics involved, and that is the purpose of putting this chapter here.

Analog signals, whatever they represent, audio, video, or some other kind of signal, are handled as waveforms, with theoretically infinite gradations in level. For example, if a sine wave goes in, a duplicate, but probably enlarged sine wave comes out. If a more complex waveform goes in, a similar complex waveform comes out.

Other things, beside simple amplification, can be done with analog signals, but all of them involve waveform signals, more or less complex.

Digital signals, on the other hand, are always what we call "two-state." The

DECIMAL NUMBER	BINARY EQUIV.	STRAIGHT BINARY	DECIMAL NUMBER	BINARY EQUIV.	STRAIGHT BINARY	DECIMAL NUMBER	BINARY EQUIV.	STRAIGHT BINARY
10	00010000	0001010	100	000100000000	0001100100	1000	0001000000000000	00001111101000
20	00100000	0010100	200	001000000000	0011001000	2000	0010000000000000	00011111010000
30	00110000	0011110	300	001100000000	0100101100	3000	0011000000000000	00101110111000
40	01000000	0101000	400	010000000000	0110010000	4000	0100000000000000	00111110100000
50	01010000	0110010	500	010100000000	0111110100	5000	0101000000000000	01001110001000
60	01100000	0111100	600	011000000000	1001011000	6000	0110000000000000	01011101110000
70	01110000	1000110	700	011100000000	1010111100	7000	0111000000000000	01101101011000
80	10000000	1010000	800	100000000000	1100100000	8000	1000000000000000	01111101000000
90	10010000	1011010	900	100100000000	1110000100	9000	1001000000000000	10001100101000

Table 6-2. Conversions from Decimal to Binary, Showing the Difference Between Straight Binary, and a System Where Each Decimal Position is Represented by Four Binary Bits.

electronic device does not graduate the signal, as it does in handling a waveform: it is always either "on" or "off." Instead of controlling flow in gradations, to produce a waveform, it works like a switch, with no "in between" positions.

So digital equipment is made up a whole variety of elements that handle signals that are either "on" or "off," either "up" or "down" in various combinations, as predetermined by their design. Many of the devices provide for two inputs and one output, although there are more complex types, and assemblies that include a number of types in the same element.

What an element does may be specified in what is called a *truth* table, that tells the result of all possible combinations. Inputs and outputs are designated by a 1 for on, and a 0 for off. The simplest devices are called *gates*. The simplest types are the AND and the OR gates. Next to them are the NAND (not AND) and the NOR (not OR) types. Figure 6-7 shows how they are represented schematically, along with a truth table for each type.

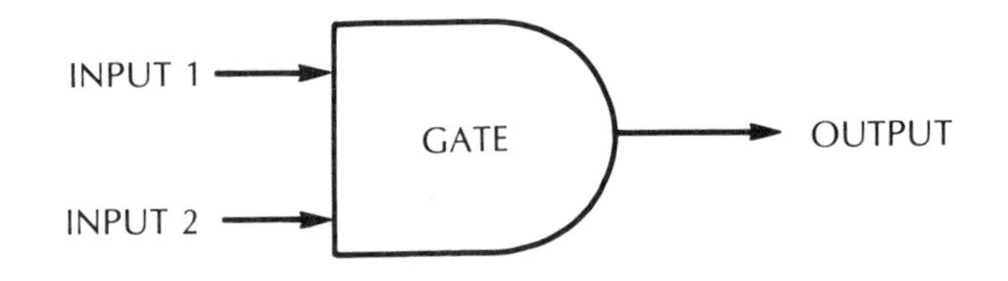

AND

INPUT 1	INPUT 2	OUTPUT
0	0	0
1	0	0
0	1	0
1	1	1

OR

INPUT 1	INPUT 2	OUTPUT
0	0	0
1	0	1
0	1	1
1	1	1

NAND

INPUT 1	INPUT 2	OUTPUT
0	0	1
1	0	1
0	1	1
1	1	0

NOR

INPUT 1	INPUT 2	OUTPUT
0	0	1
1	0	0
0	1	0
1	1	0

Fig. 6-7. Truth tables for the main types of electronic gate.

Basically, the word AND signifies that an output is only given when both inputs are present: inputs 1 AND 2 must be there to produce an output. The word OR signifies that an output is given when either input is present. But notice that an OR gate as well as an AND gate provides an output when both inputs are present. If you want to have an output only when one or the other is present, not when both are, you need a NAND gate as well, to preclude that possibility.

The NOR gate is the simple negative of the OR gate, which means an output is present only when neither input is present. NAND and NOR gates can be made by applying a simple inversion of the outputs of AND and OR gates respectively. Inversion means that 1 is converted to 0 and vice versa.

Perhaps the most important difference between analog and digital applications is that a whole different way of thinking about them is involved. Analog applications involve mathematical analysis of the waveforms handled, as we shall see in Chapter 13. Digital applications involve the logic of combinations. You have to think like a computer works, in effect.

In analog design, you are concerned with how accurately electronic devices can follow a shape, or whatever equivalent function it is they have to do. That is why Chapter 13 is titled "Linear Amplification." Precision in performance is critical. In digital design, you are only concerned with binary logic: combinations of "on" and "off" or ones and zeroes. This too can get pretty complicated, but it is never more than that: in between states, which is where analog always works, are never involved in digital.

Analog devices are designed to work in such "in between" states: neither fully off or fully on. In analog devices, the fully off or fully on state are the absolute limits of their operation, a sort of "danger point," to be avoided. In digital devices, the fully off and fully on state are the only "positions" they know: in between states are usually impossible, by design. They "trigger" from one to the other so that when they are not "on" they are "off," and vice versa.

CONVERSION BETWEEN DIGITAL AND ANALOG

Just as there appears to be a conversion between digital and analog in nature, so in electronics, a whole new family of conversions appear, different, because the parameters are so different. In biological conversions, pulse frequencies are measured in milliseconds, while in electronic digital, they are measured in microseconds, or even in one thousandth parts of a microsecond, called nanoseconds.

Where this makes a difference, as we mentioned earlier, is that biological digital makes use of a great multiplicity of nerve fibers, the electronic variety uses only one, sort of "time-sharing" it, in code. That is what we will describe now, in basic forms or possibilities. As the precise way in which these things are done is constantly changing, you should think in terms of what has to be done, rather than how it is done. What has to be done will be the same tomorrow. How it is done probably won't!

This means that in later chapters we will get into the hardware, or the kinds of hardware, to do it, here we look at what we need that hardware to do. First, the devices that generate digital codes, all of which are basically binary. Figure 6-8 shows some possibilities. Why use straight binary and some equivalent of decimal?

If your primary concern is to digitalize what are bascially analog "signals" such as an audio waveform, you don't really care about decimal numbers do you? What you are concerned with is the shape of the waveform, to a certain

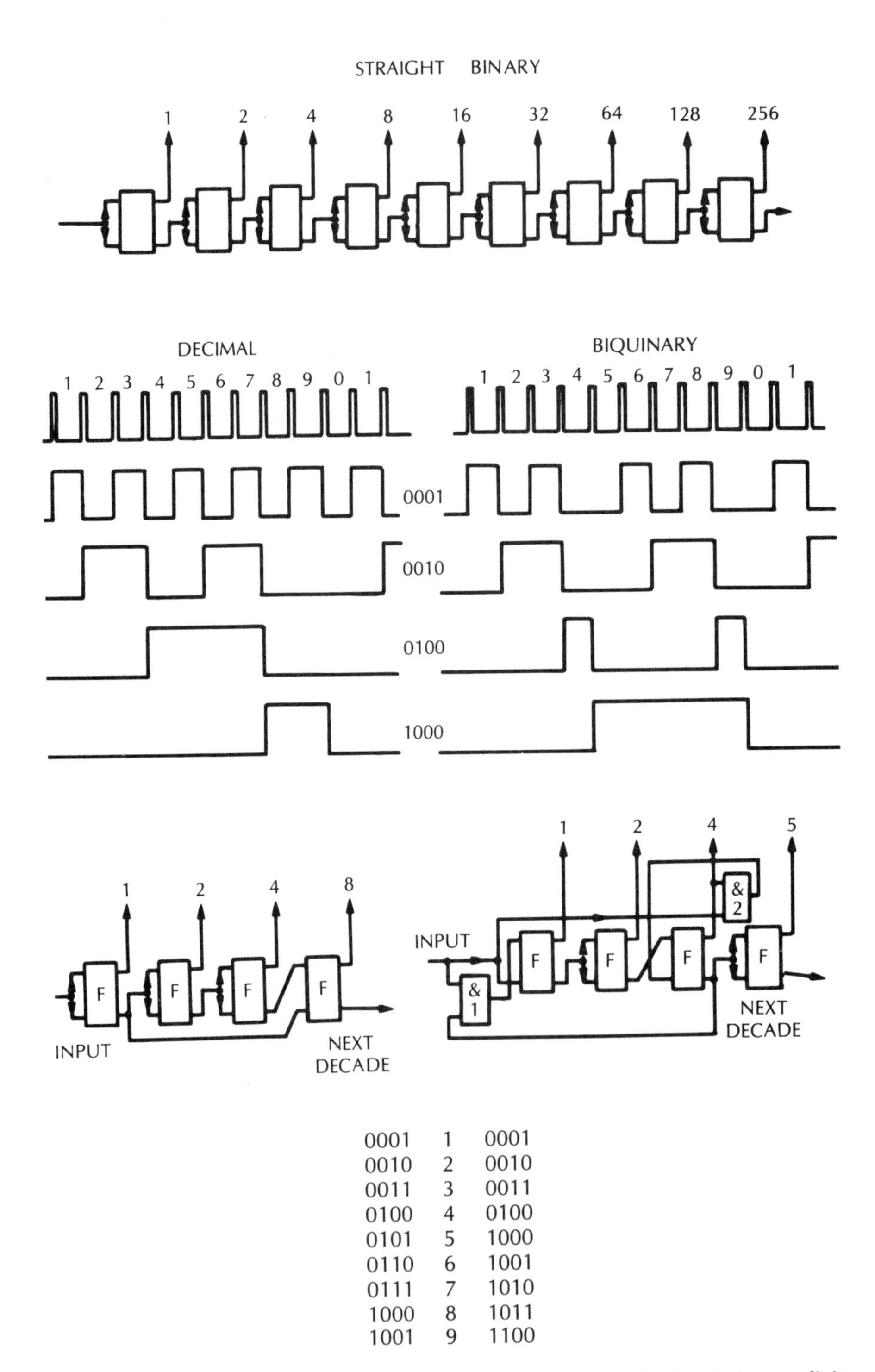

0001	1	0001
0010	2	0010
0011	3	0011
0100	4	0100
0101	5	1000
0110	6	1001
0111	7	1010
1000	8	1011
1001	9	1100

Fig. 6-8. Comparison of logic circuitry for straight binary, decimal with binary digits, or bits, and biquinary with binary bits.

degree of accuracy. Ten steps of straight binary will cover a range of amplitudes of 2^{10}, which is 1024. This means that, for example, a voltage between zero and 10 volts can be produced within 0.1%. Ten stages of straight binary will do this.

But if your aim is to read off voltages, giving a decimal readout on one of those digital meters, then you need one or the other form of decimal equivalent. At the top of Fig. 6-8, you will see the simplicity of straight binary generation. Clock pulses going in at the left, at say one every microsecond will give outputs from successive stages, until in 10 microseconds you have resolution to within 0.1%, if that is what your system is designed for.

Below that are waveform schematics within a decimal system, and a biquinary system, which is how you start to go about designing the hardware to generate them. For the decimal system, at the left, the outputs ascend as in straight binary, except that when you get to 9, for which the binary is 1001, it must recycle back to zero, and send the first pulse on to the higher decade at the same time. Only the last two digits of a decade have the greatest significant bit, representing 8. So all that is needed is to arrange for the next input that returns the 1 bit to zero, to also return the 8 bit to zero.

Note that the flip flops shown in all these schematics (Fig. 6-8) are triggered to the "high" or 1 state by the upper left input, while they are triggered to the "low" or zero state by the lower left input. The lower output, at the right of each, is high when the upper one is low.

In the straight binary, the little arrows going to each input, at the left of each flip flop, indicate that when an input is received at that point, the unit is triggered to the opposite position: if it's low, it goes high, if it's high it goes low.

In the decade circuit, this is true, except for the last one, the 8 bit. Here an input from the 4 bit, when it goes low, triggers the 8 bit high, but the next output from the 1 bit, when it goes low, which is when the input goes from 9 to ten, triggers the 8 bit low again, sending a pulse to the next higher decade.

For the biquinary circuit, the 1 and 2 bits are coupled the same way, but since they both stay low for one extra space through 5, the AND gate on the high input of the 1 bit is used so that high inputs to the 1 bit are only allowed when the 4 bit is low. The other AND gate serves to ensure that when the 4 bit is high, the next input returns it to low.

This means the first three flip flops cycle through 4, return to zero on 5, and send a trigger to the 4th input, which alternates between low and high, for every completed 5. All these circuits take unit inputs, and count them off, so that it takes 10 inputs to complete a decade. However, when converted to binary code, each decade occupies only 4 bits.

In all the arrangements of Fig. 6-8, the input is a continuous succession of pulses, all of equal value, what is called counting. Thus a number, say 3247, would have to consist of that many pulses, one after the other, at that input. But this conversion produces pulses of different value, called bits, from most significant to least significant.

Using the straight binary code, the most significant bit would be 2048, or 2^{11}. After that many have been fed in, there are $3247 - 2048 = 1199$ more.

The next significant bit will be 1024, or 2^{10}, leaving 1199 − 1024 = 175 to go. The next two bits, representing 512 and 256, are missing, will be low, instead of high, then the next is 128, or 2^7 leaving 175 − 128 = 47 to go. The next bit, representing 64, is low, but the following one, 32 or 2^5, leaves 47 − 32 = 15. 16 is low, but then 8, 4, 2 and 1 are all high to make up the remaining 15.

So the conversion to straight binary is 110010101111. That takes 12 "digits" of information. Now, if you use one of the decade equivalents, every 4 bits represents a digit in decade. Thus the decimal will be 0011, 0010, 0100, 0111, making up 0011001001000111. Or in biquinary, 0011,0010,0100,1010, making 0011001001001010.

In Fig. 6-8, each of those bits will feed out on a separate conductor, or connection. This is rather like the multicircuit systems we discussed earlier that are found in biology. It takes a lot of "circuits." In most digital work, this is narrowed down, by sending the bits in sequence along a single conductor. One circuit to make this conversion is shown at Fig. 6-9.

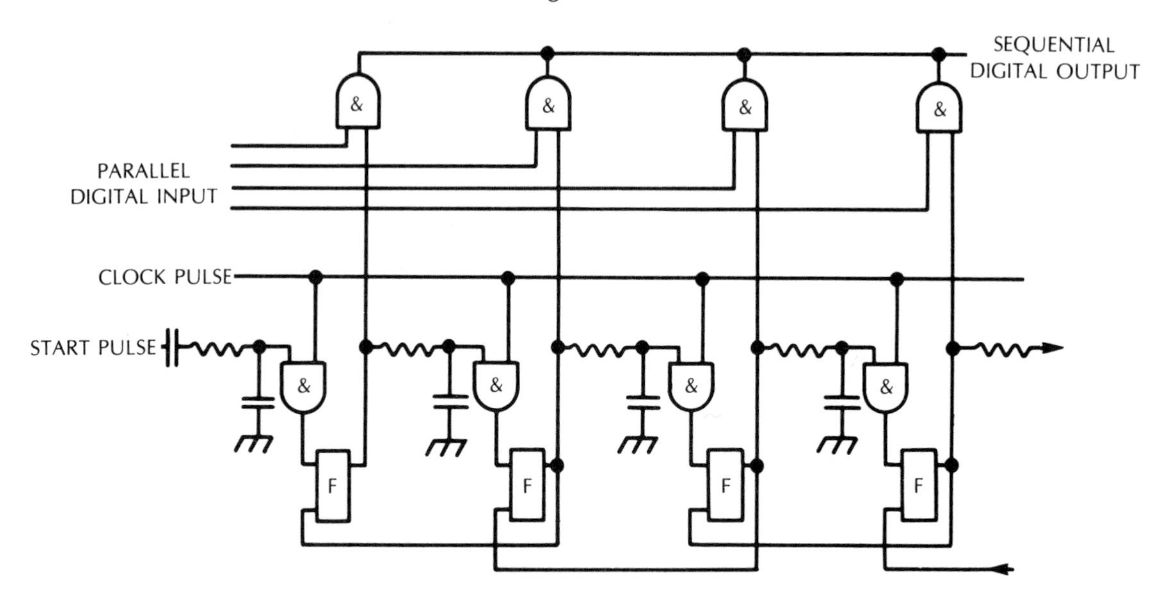

Fig. 6-9. Essential elements needed to convert parallel digital input to sequential digital output.

The parallel digital inputs come in at the left. The "and" gates along the top admit the digits that are high, as their turn comes, to the sequential output. This is controlled by a clock pulse, initiated by a start pulse. When a start pulse has made the left input of the first "and" high, which it will hold for long enough to include the next clock pulse, and no longer, as limited by the resistance and capacitors in its left input, the first flip-flop (F) will go "high."

Each flip-flop shown in Fig. 6-9 has its output (top right terminal) go high when the left top terminal goes high, and the output returns low, when the left bottom terminal goes high. Thus when the first flip-flop goes high, it starts the second AND going high so with the next clock pulse, the second flip-flop goes high. This resets the first flip-flop, back to low, and starts the input to the third AND, so that the next clock pulse sends the third flip-flop high. And so it goes, along as many flip-flops as are necessary to handle however many inputs there are. They are all sequenced in turn into the sequential digital output.

Inside computers, they can work faster if they receive parallel inputs, one circuit for each digit or bit, although when they use decade, or other "byte" systems, they may use sequencing along multi-path circuits. But this means that before feeding this totally sequenced signal into the computer, it must be decoded into separate output (Fig. 6-10). This circuit simply reverses the function of Fig. 6-9.

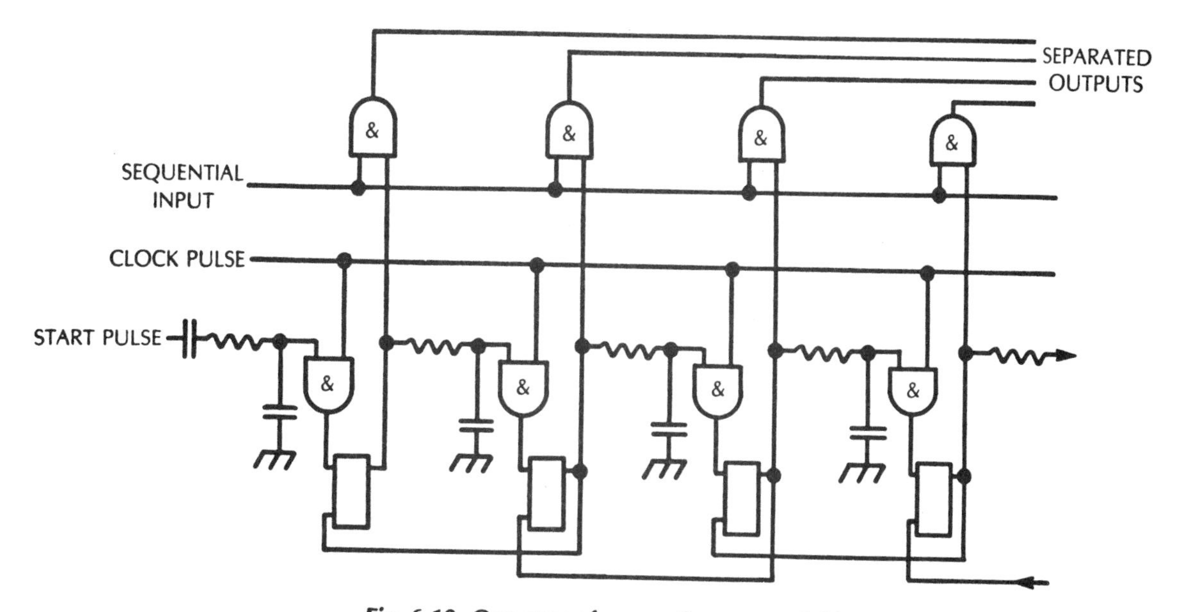

Fig. 6-10. One way of converting sequential input to parallel outputs.

Finally, to convert from digital to analog, which is what this section started to tell about, we need separate amplitudes that are proportional to the magnitudes that each bit represents. One way in which this is done is a circuit that provides 2^8 or 256 range of output magnitude. This is called "resolution."

What it means is that, if the number 256 represents, say 10 volts, then each smallest significant bit represents 10 ÷ 256 = 0.0390625 volt, or about 39 millivolts. Doubling that 8 times comes back to 10 volts. By selecting the appropriate combination of bits, any value between zero and 10 volts can be selected, to an accuracy of within 0.0390625 volt, or about 4/10 of 1%.

To convert analog to digital requires one of a variety of processes that we will consider later. Here we want to make sure you get clear the different forms that signals, or information, takes, in systems that are basically digital—that is, they handle material in a variety of ways that, collectively are called digital.

UNDERSTANDING THE SYSTEM

We have put this part, introducing digital, early in the book, because you need to learn to think about it in the right way, if you are to understand how it all works. Too many people today take the attitude that, so long as it works, that is okay, don't bother them with how it does it. If everyone did that, we'd never get beyond the present generation of systems, and when they all expired of "old age," there'd be no more. Perhaps that would be good.

But also, by doing that, we expose ourselves to quite serious misuse or abuse of the systems. For that reason also we need to understand systems more completely. So let us recapitulate the levels at which binary, or digital can operate.

First, it consists of a succession of "bits" differently arranged, in which every bit has only two states: on or off, otherwise designated high or low. There are two extremes, and lots of in betweens.

In nature, the biological binary system uses a totality of separate "conductors" or nerve fibers. Sometimes man-made binary may do the same thing. But more often the bits use the same conductor, or fiber, time-successively. However neither extreme is the best, economically. So at various points, some in between variations are used. This in turn depends on the kind of information the system is handling.

In calculators, because we humans think in decimal, the processing is done in some version of decimal, which we have introduced in this chapter. Each digit of decimal can be conveyed simultaneously on just 4 conductors. When a system is handling decimal exclusively, such as a calculator, that becomes the basis of operation. It is then common to use 4 conductors for each digit, but to run the decimal digits time sequentially.

You may have noticed how that, when you enter decimal numbers into a calculator, every extra digit you punch in, displaces those you have already entered, to the left. Thus, in entering the number we referred to earlier, 3247, you first punch in the 3. Then you punch in the 2, and the 3 moves to the left to make room for the 2. Then you punch in the 4, and the other two digits move left to accommodate it. And so on.

If you punch in a decimal, then continue, the decimal point will move to the left also, holding its place between the digits where you put it. All this is done by time-sequencing the decimal digits, over a 4-conductor circuit. But in the memory, all those digits are stored successively in space, unless the memory is running, in which case they are time-sequenced in rotation. In a sense this is always true, the difference being that a fixed memory is scanned in much the same way the clock pulse scans the individual elements as we showed in this chapter.

But microprocessors use a lot more than the 10 digits of the decimal system. Their programs are what are called alphanumeric: they use characters on a typewriter keyboard, plus some extra ones. Many of the keys, or else a separate section of the keyboard, perhaps both, carry instructions, rather simple bits.

The average typewriter keyboard has 92 characters, plus extras like space, backspace, correct, maybe, carriage return, tabulating instructions, and other things normal to the ordinary typewriter. Then when it is a microprocessor keyboard, it can have many more instruction type functions. For instance, if you are using a printer to transfer what you have entered into the system onto a printed page, one key will initiate the printing process.

Then maybe you'll need to change the type wheel, to get some words in italics. Perhaps more advanced processors will do that by themselves. But older ones need a human to change the type wheel. So an instruction is needed to stop the printer where that has to be done, and to stop it again, when the original print wheel has to be put back.

Another function most systems need, is a way to transmit data, or information, to or from a distance, for instance by telephone line. This uses a device called a "modem," which stands for "modulator-demodulator." High speed digital cannot travel over telephone lines very well. It must be conveyed by tones within the pass range of the line, usually from 300 to 3000 hertz, which is rather limited.

So the data is sent totally time sequential, using 2, and sometimes more tones, to represent "low" and "high". As it takes time for the receiver to recognize the tone, even though it is quite short in duration, this conversion takes time—it's the slowest part of the system. So another key is needed for appropriate instructions, and retrieval from memory for sending, or transfer to memory on receiving, must be in step with this much slower rate of processing.

Later on, you will want to see how some of these parts work. But to get you started, we have found the best way is for you to see what is needed to make it work.

Examination Questions

1. Assume that a waveform is to be represented digitally, with a sequence of 10 "on" or "off" type digits to represent each value. Using straight binary, the quantity represented by 1111111111 is how many times as great as the quantity 0000000001?
2. Using binary equivalent of decimal, each place of decimal requires how many places of binary?
3. If binary equivalent of decimal uses a 12-digit sequence, what is the ratio of the maximum signal that can be represented to the minimum? What would be that ratio if the same number of digits is used on straight binary?
4. NOW many digits of straight binary are needed to represent the same range of level change as 12 digits in binary equivalent of decimal?
5. The decimal system numbers that represent values of the first quarter of a sine waveform are as follows: 0, 193, 380, 556, 714, 851, 962, 1044, 1094, 1111. Using binary equivalent of decimal, work out the 16-digit codes for each value.
6. If straight binary is used, the code 1111111111 represents what value in decimal? Using this as the 10th value in Question 5, complete the table using this system.
7. Using straight binary, codes can vary from all 0s to all 1s. Using binary equivalent of decimal, the maximum proportion of 1s that can occur in a code is how much?
8. In a 4-digit binary decade, the numbers represented by a single 1 in the code are (a); the numbers represented by two 1s in the code are (b); and the number represented by three 1s is (c).
9. In the answers to Question 8, think of the possibilities for error caused by the code "slipping" one or more digit, in the case of a two-1 code. What numbers could be interchanged by such slippage?
10. In a straight binary 10-digit code, the smallest number represented by five 1s is (a); and the largest is (b).

7
Magnetism

Magnetism has properties that are at once similar to and different from current electricity. Both have the characteristics of potential and flow. Magnetic potential is known as *magnetomotive force* or *mmf,* just as its electrical counterpart is known as *electromotive force,* or *emf* (although better known as "voltage").

A magnetic field, which is regarded as a flow, is called flux. Its relative strength, measured at any given point, is equivalent with current or current density. Beyond these basic similarities, the terms of reference begin to differ.

Voltage and current are simple measures of pressure, or potential, and flow; these measures are analogous to pressure (e.g., in pounds per square inch) and quantity flow (e.g., gallons per minute) in water flow. But the magnetic equivalents have different dimensional units.

SOME BASIC QUANTITIES

The total mmf in a magnetic circuit is part of a circuit's characteristic, and the total flux (symbol ϕ) generated by that mmf is considered to depend on a quantity called *reluctance*, which is the magnetic equivalent of resistance in an electrical circuit. However, circuit calculations generally depend more on magnetomotive force per unit length than on total mmf, and on flux density per unit area, called *flux density*, than on total flux.

The reason for this important distinction is that magnetic materials have a behavior pattern throughout their volume that depends on unique relationships between mmf per unit length and flux density, or flux per unit area. For this reason, total mmf and total flux can only be calculated by considering the state of the magnetic material within a unit volume and multiplying this by length and area, respectively (Fig. 7-1).

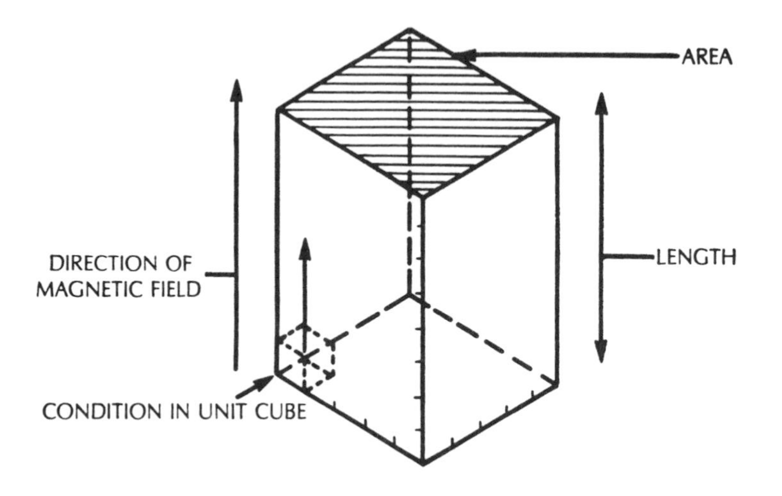

Fig. 7-1. A unit cube representation of the magnetic condition existing in a magnetic material.

In air, in regard to any non-magnetic medium, a reference to unit length or area is basically unnecessary, because the medium is linear. Non-magnetic media have no limitations or variations at different mmf gradients or flux densities. There is no more reason to relate potential and flow to gradient and density in a non-magnetic medium than there usually is to specify voltage per unit length or current per unit area, unless safe limits are being approached.

Voltage per unit length may be used in measuring drop along a line, for example, and current density may be used to determine a safe rating or possible dissipation. On the other hand, magnetism produces no dissipation in non-magnetic materials. It may induce currents that cause dissipation, about which we will learn later, but magnetism itself produces no dissipation in non-magnetic materials. So here is another difference.

Electric current does not flow through nonconductors, and hence no dissipation occurs in nonconductors due to current flow. There may be dissipation due to a fluctuating electric field, but we have not discussed that yet. Magnetic fields, on the other hand, can pass through non-magnetic media.

A useful property in magnetic circuit calculations is a quantity called *permeability*. This is the factor by which the reluctance of a portion of circuit is reduced due to the presence of magnetic material. For example, to support a magnetic field through a given length, the presence of magnetic material in that space will divide the mmf needed by the permeability of that material at the working density.

Thus, in air or some other non-magnetic medium, mmf per unit length or the magnetizing force, as this quantity is called and given the symbol H, and flux density, given the symbol B, are the same thing at any point, with the distinction that H is a quantity per unit length along the magnetic field path (from North to South or vice versa) while B is a quantity per unit area across the path (Fig. 7-2).

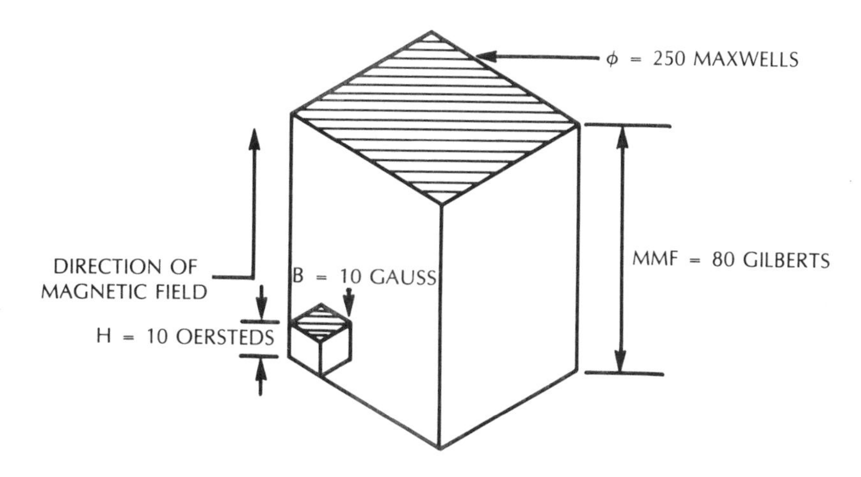

Fig. 7-2. Relationship between the quantities used in measuring magnetic properties.

Now, if a space occupied by non-magnetic material, such as air, has a B of 10 gauss, which is 10 *maxwells* (otherwise known as "lines") per *square centimeter*, the accompanying H is also 10 *oersteds*. But if the space has a length of 8 centimeters (the units for both *oersteds* and *gauss* are metric, centimeters and square centimeters, respectively) and a cross-sectional area of 25 square centimeters, the mmf for the whole length will be 8 × 10 or 80 *gilberts*, and the total flux in the whole area will be 25 × 10 or 250 *maxwells*. Otherwise expressed, an *oersted* is a *gilbert per centimeter* length, and a *gauss* is a *maxwell per square centimeter* of cross-sectional area. If this same space is filled with magnetic material in which permeability is 3000, the relationship changes. If the mmf does not change but remains at 80 gilberts, the flux will be multiplied by 3000 to yield 750,000 maxwells (also commonly called "lines").

On the other hand, if the flux is controlled to 250 maxwells or some other figure, then the mmf accompanying this field strength will be reduced. For 250 maxwells, the mmf would be 80 divided by 3000 or 0.00267 gilbert. If the space were of different proportions (such as one-three-thousandths as long) but still air, the ratio would hold, whatever B and H (in which case, both would always be the same).

That calculation seems simple, but it is complicated by the fact that, in any practical magnetic material, permeability is not a constant. That value of 3000 would be correct only at one particular value of B and H. However, such a figure of permeability may be quoted as a minimum figure, which means B could be higher or H lower than the 3000:1 relationship would indicate, which would generally be regarded as "better."

As well as varying in permeability, and the way that permeability varies with B or H in any one material, magnetic materials vary in other ways, more complicated than anything analogous in the simpler electrical circuits. The first of these properties was originally given an evaluative or qualitative designation of "hardness" or "softness."

There is a reason for using this kind of distinction: the materials are generally physically hard or soft to correspond with their magnetic properties of the same name, although physical hardness is not definitely related to magnetic hardness. The simple notion, which does not have any simple index of hardness or softness, like the physical counterpart, is that a "hard" magnetic material requires much magnetic effort to magnetize it, and then much magnetic effort to demagnetize it. "Soft" magnetic materials, on the other hand, magnetize and demagnetize quite easily. They take magnetism easily, but do not hold it after the source is removed. The reason why no simple numeric index is given is because the relationship is too complicated for that. The best way to understand the relationship is to consider what happens throughout magnetization and demagnetization within a unit cube of a particular magnetic material.

A way to generate a magnetizing force under control is with an electric current in a coil, wound around the magnetic material (Fig. 7-3).

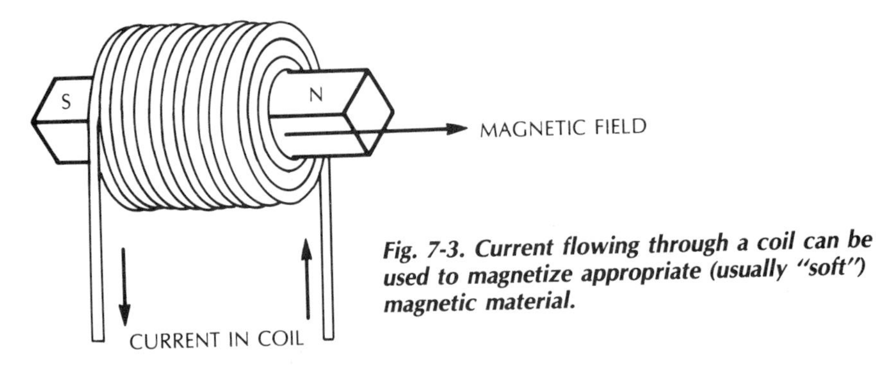

Fig. 7-3. Current flowing through a coil can be used to magnetize appropriate (usually "soft") magnetic material.

So that the behavior can be related to a unit cube of the material, the path for the magnetism generated by the current to take should be wholly within this material (no air gaps or other material in the magnetic circuit). Then the ampere turns (current multiplied by number of turns carrying it) is divided by the magnetic path length to get a figure that is strictly proportional to H in the magnetic material: ampere-turns per unit length (which is different from oersteds and may be stated per inch or per centimeter, but be sure which).

An ampere-turn per centimeter is 1.257 oersteds. An ampere-turn per inch is 0.495 oersteds. Thus, if the total length of the magnetic circuit is 20 inches (measured along a mean path, Fig. 7-4) and the current is 1 amp flowing in 2000 turns, the magnetizing force is 100 ampere-turns per inch or 49.5 oersteds.

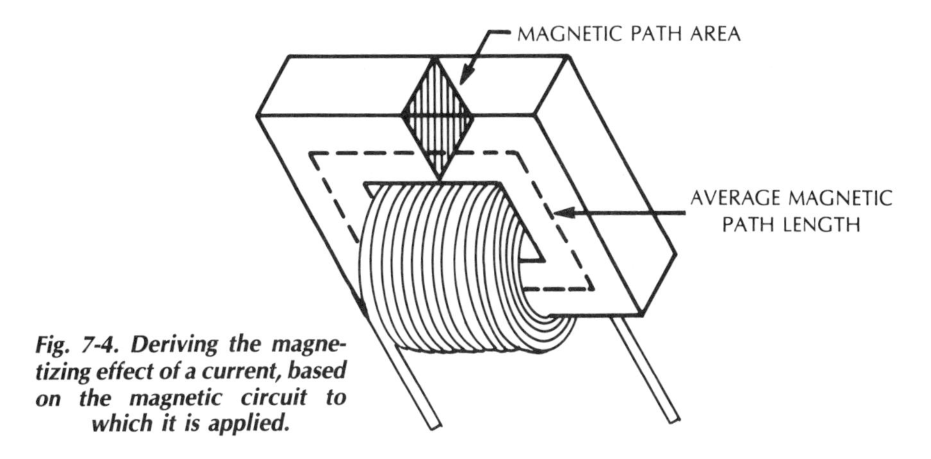

Fig. 7-4. Deriving the magnetizing effect of a current, based on the magnetic circuit to which it is applied.

As a small H is applied to a demagnetized piece of material, B (flux density) begins to grow, but not linearly. Field density is in lines per square inch or per square centimeter (the latter is gauss)—again be sure which, and convert if necessary. A square inch is 6.45 square centimeters. As more H is applied, the rate of growth in B increases to a maximum and then falls off, until further H eventually adds very little extra B. This is called saturation (Fig. 7-5).

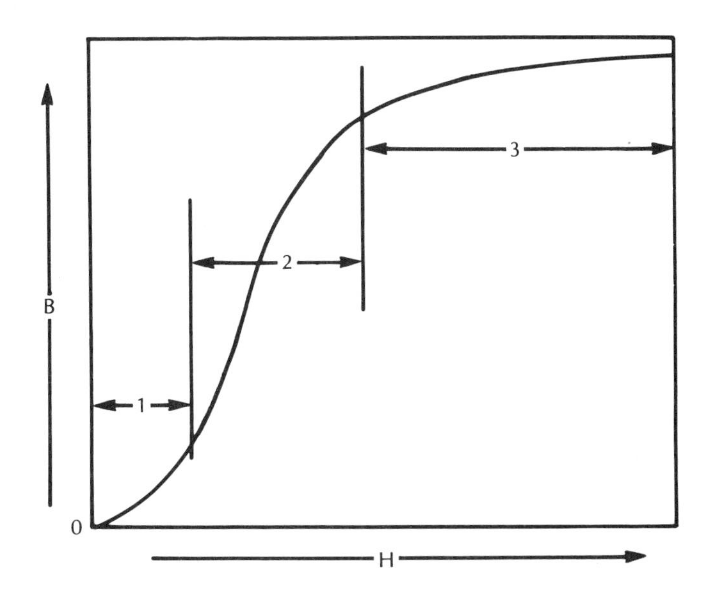

Fig. 7-5. The three general domains in magnetizing a material: (1) initial; (2) high permeability region; (3) saturation.

As the magnetizing current is removed again (and notice that the current in the coil must continue to flow to maintain the H in the material) some of the B falls off. In a hard material, a large proportion of that saturation density will remain, even when the H is totally removed by switching the current off. In a soft material, magnetization begins to fall off rapidly when the current is totally removed (Fig. 7-6). Very little reverse current would remove the residual. The amount of magnetism remaining, when the magnetizing force is removed, measured in appropriate B units, is called "retentivity."

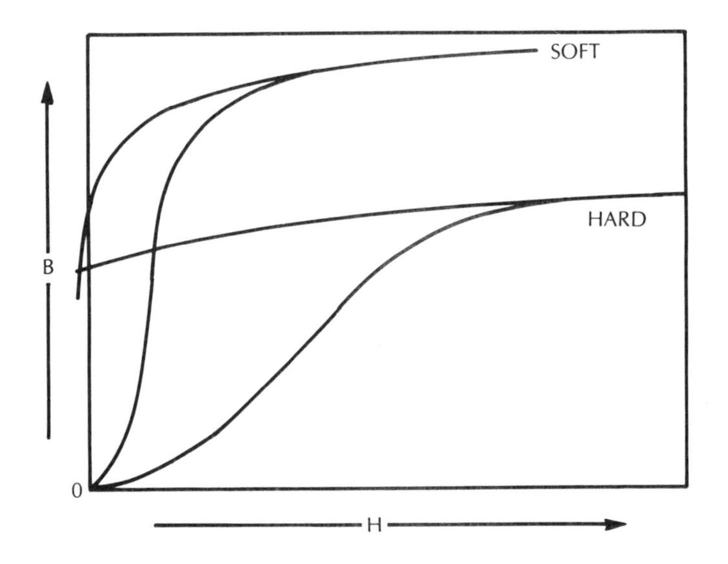

Fig. 7-6. General distinction between hard and soft magnetic materials, shown by their magnetizing characteristics.

Now, suppose the current in the coil is reversed and again increased gradually. Early reverse H will begin to cancel the remaining B and then magnetize the material in the opposite direction. The point at which the original magnetization is neutralized is designated by the reverse H needed to do this, which is called the "coercivity" of the material. Taking the reverse-direction current up to the saturation H level will result in a reverse of the magnetism in the material, which it will then hold after this H is removed, due to the retentivity in that direction (Fig. 7-7).

Soft materials have an indeterminate retentivity and very low coercivity. Hard materials have high figures of both. But even in that, there is no consistent relationship. But to see the value of different properties of individual materials, let us see how they are used.

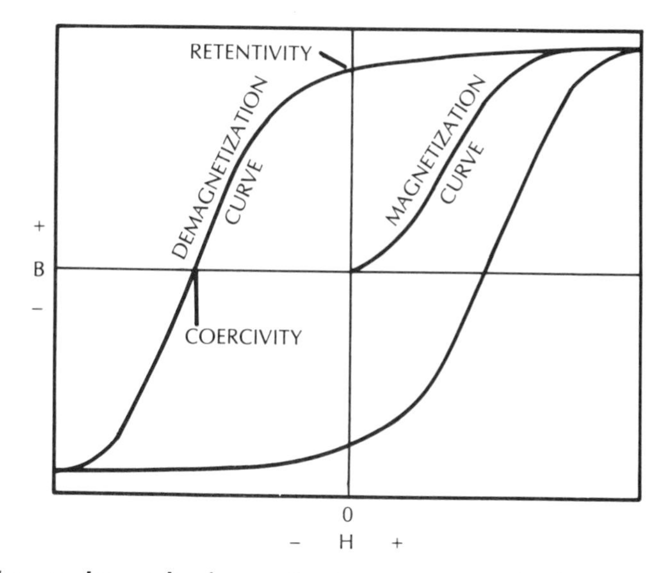

Fig. 7-7. The complete cycle of magnetization and demagnetization in a typical hard magnetic material.

PERMANENT MAGNETS

A permanent magnet is made of a magnetically hard material that retains its magnetism against endeavors to demagnetize it. It is put into a magnetic circuit, which may consist of soft material to convey the magnetic field to the point where it must be used (Fig. 7-8). Now the permanent magnet is "charged."

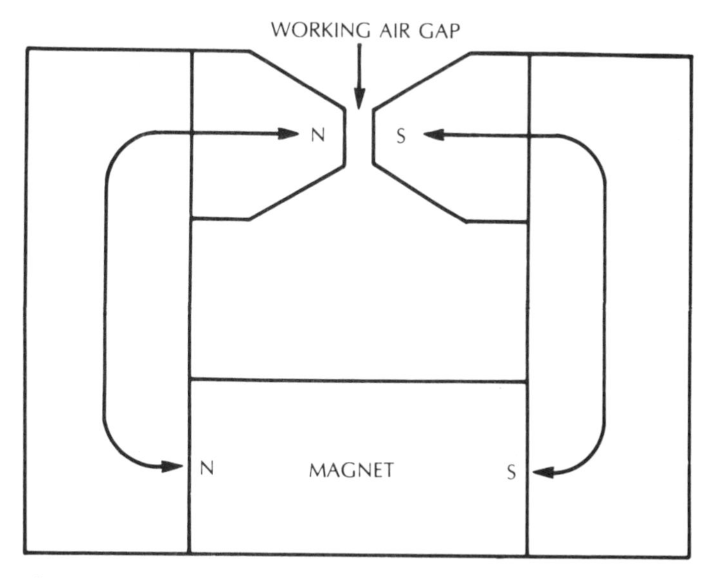

Fig. 7-8. How a permanent magnet supplies a work area air gap.

During charging, everything magnetizable is magnetized by a powerful magnetic field (Fig. 7-9). This field must be strong enough to saturate the permanent magnet part of the magnetic circuit. Then when the magnetic circuit is removed from the charger, the permanent magnet retains some of that magnetism to keep a field going in the rest of the magnetic circuit. How much magnetism is retained depends on the circuit. Notice that charging magnetizes the return path or work circuit the reverse way from that provided by the residual from the permanent magnet after charging.

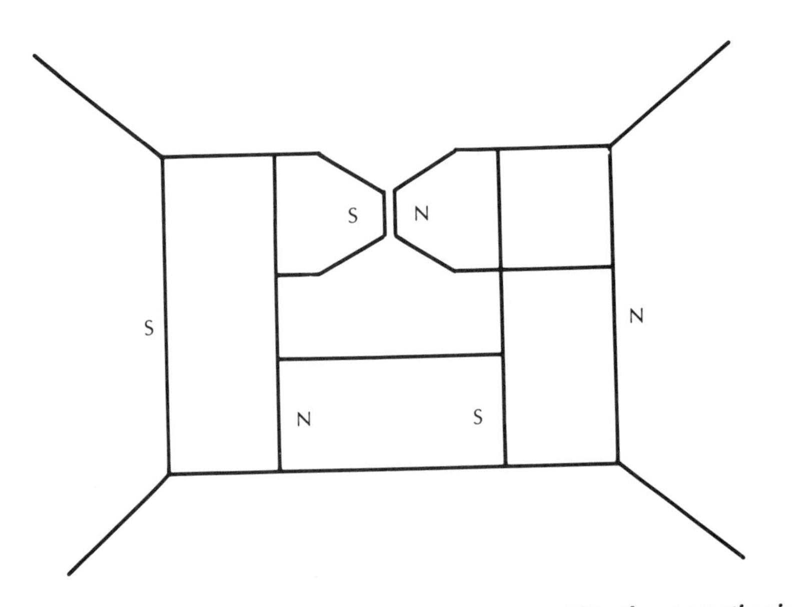

Fig. 7-9. A method of magnetizing a permanent magnet within the magnetic circuit in which it is to serve.

Optimizing the magnetism that a given sized magnet will deliver to a working magnetic circuit is a process of matching up mmf generated by the residual magnetism or remanence in the permanent magnet, with the mmf needed to drive that same residual through the circuit. To the permanent magnet material, this mmf is a demagnetizing force, and thus moves the operating point round the loop of magnetizing sequence for the magnetic material of which the permanent magnet is made; top left quadrant in Fig. 7-10.

A line representing the reluctance of the passive or soft part of the magnetic circuit will find where the material demagnetizes to, and how much field it can deliver at that point. This needs "translating" by the relative dimensions of the magnet material.

Suppose the reluctance figures to 8 for the magnetic circuit, and the magnet itself has an area of 25 square centimeters with a length of 12 centimeters. Applied to a centimeter cube of the magnet, that actual reluctance of 8 for the whole magnet loading reduces to 8×25 over 12 for each centimeter cube, which yields a value of 16.7, and determines the line to be drawn to represent it.

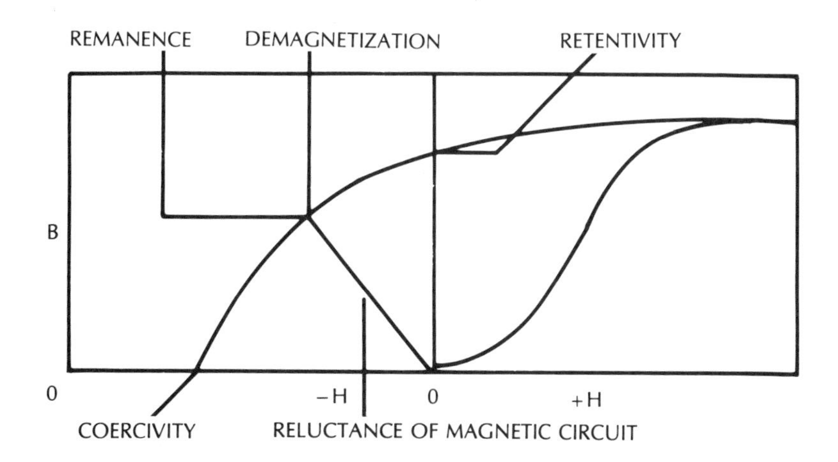

Fig. 7-10. Practical operating condition of a permanent magnet coupled to its magnetic circuit.

If the magnetic circuit is relatively closed (has no air gaps) and is of a magnetic material with a high permeability throughout (Fig. 7-11), little demagnetization occurs. Some of the earlier permanent magnet materials had high retentivity, but not much coercivity (at least as compared with more modern ones).

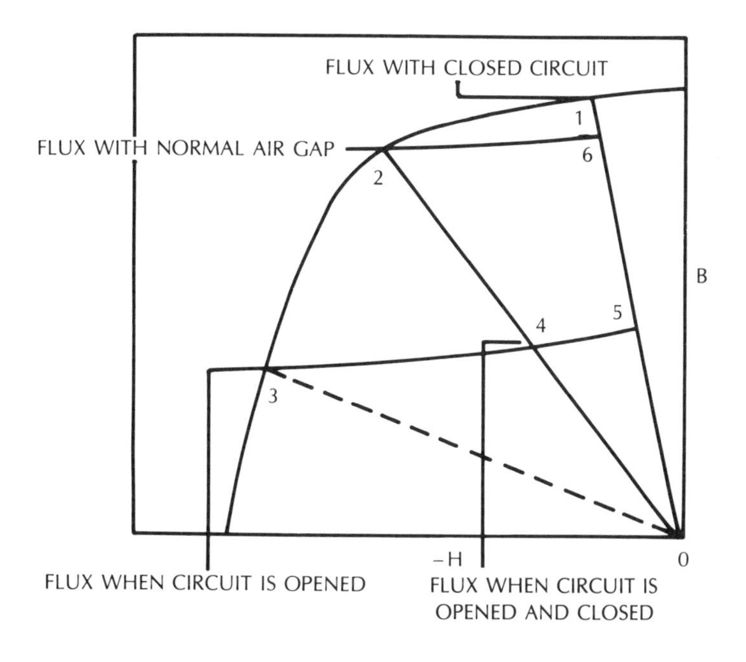

Fig. 7-11. Sequence of events caused by removing a magnetic circuit from a typical long type permanent magnet: (1) magnetized with air gap closed; (2) normal condition with a working air gap; (3) condition if magnetic circuit is removed altogether; (4) condition when magnetic circuit is replaced, following removal; (5) condition when air gap is closed following complete removal.

DEMAGNETIZATION

Such materials would demagnetize themselves if a high-reluctance external path were provided. For this reason, materials of this type require a "keeper" to complete the magnetic circuit when the magnet is not attached to its regular circuit. If the keeper is removed without placing the magnet in a work circuit immediately, the magnet is at least partially demagnetized (Fig. 7-12). To avoid opening the circuit in this way, you must slide the magnetic assembly into place as the keeper is removed and vice versa.

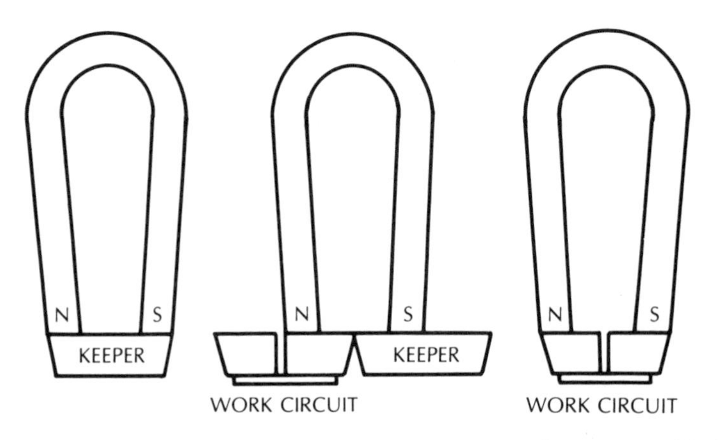

Fig. 7-12. The method of replacing a "keeper" (which closes the air gap) with the work circuit, to avoid the loss indicated in Fig. 7-11.

Modern magnetic materials have a much higher coercivity. This means they do not as readily lose their magnetism when presented to a high-reluctance circuit. However, they still lose some. The magnetic sequence loop is not reversible. If a high-reluctance path or circuit is presented to a permanent magnet, the operating point of the material moves around the top left quadrant of the curve in a counterclockwise direction.

In doing so, as well as approaching the coercivity point by going to the left, it loses some of its remanence. Now, if the circuit is again closed with a high-permeability, low-reluctance circuit, the material starts to take a different loop: it will move back toward the vertical zero line (as far as the reluctance load line) but not along the demagnetization curve by which it came down. Rather, it tends to move parallel to the negative part of the curve where negative magnetization is removed (not shown in Fig. 7-11, but it can be surmised from Fig. 7-7).

If the circuit is again opened, it will start a new loop again, lower than the first. Looking at this sequence shows why it is never good to dismantle and reassemble the circuit of a permanent magnet, unless you have a charger handy with which to remagnetize it.

The next distinguishing characteristic among permanent magnets is the two principal groups of materials used and their naturally related magnet shapes, which are most readily recognized by the basic difference in shape—long or short. This can best be understood by tracing this development.

EARLY PERMANENT MAGNET CONCEPTS AND CALCULATIONS

All older permanent-magnet materials were best suited for a long "slug" of the material, due at least partially to the fact that this was the only way to generate adequate mmf for use in an external magnetic circuit and partly because a piece of magnetic material magnetized "long-ways" (longitudinally, for those who prefer a more scientific designation) can be regarded as producing two quite discrete point-source poles (Fig. 7-13). The old horseshoe magnet is an example of this type.

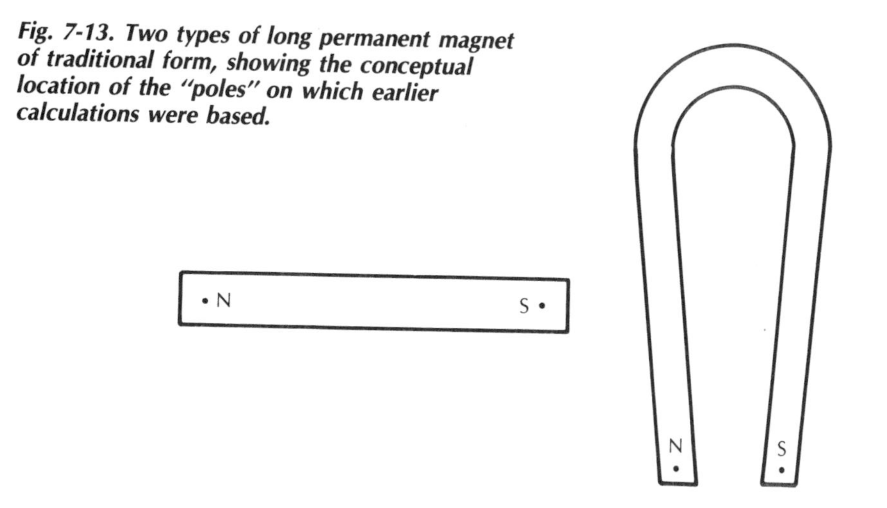

Fig. 7-13. Two types of long permanent magnet of traditional form, showing the conceptual location of the "poles" on which earlier calculations were based.

The poles result from the cumulative effect produced by each element of the total magnet. If a center of action for each pole is identified, the magnetic behavior can be described by viewing each such pole as a point source of magnetism, one North and one South, by association with earth's natural magnetism.

If we calculate the total reluctance of all paths from one pole to the other, we must take into account the distribution of the field between poles. The field is strongest along the shortest path, which is the direct one from one pole back to the other—a straight line. The field thins out in inverse proportion to the distance a particular route takes to get from one pole to the other.

At the same time, the flow of magnetic force or field (which can be identified by putting a magnetic compass needle at such a point within the field) maintains equipotential ("Northness" to "Southness") between adjacent flow lines (Fig. 7-14).

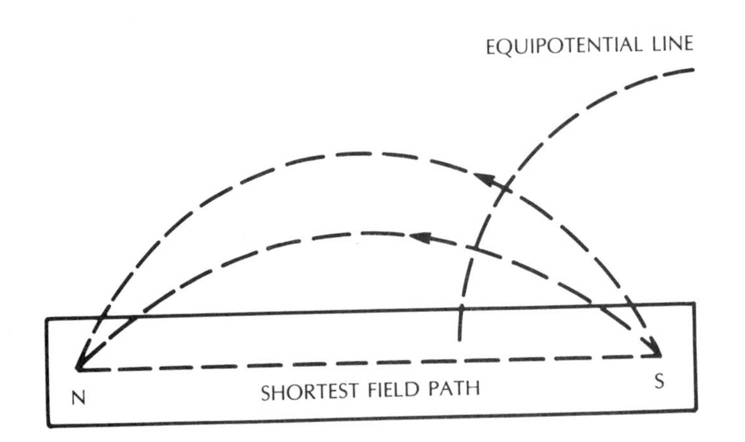

Fig. 7-14. Relationship in the field of an open permanent magnet on which earlier calculations were based.

It is this distribution density, coupled with the equipotential law, that determines the shape of the magnetic field produced. A magnetic field can be mapped, very approximately in a plane over or under the magnet, by sprinkling iron filings on a sheet of paper (Fig. 7-15).

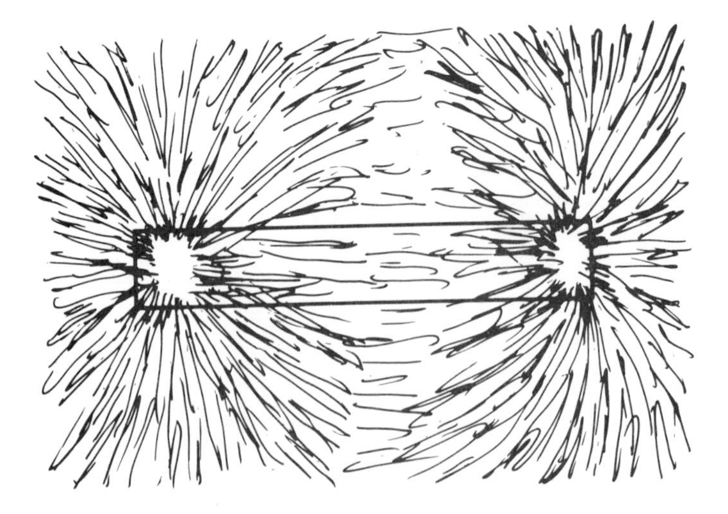

Fig. 7-15. Typical iron filing "map" of the magnetic field surrounding an open bar magnet.

Because the density is inversely proportional to the length of the path, the total field passing from one pole to the other is the same as if all possible paths leading from one pole to the other took the shortest possible path at uniform density. The total possible directions by which magnetic paths can leave one pole to go to the other is the same as the surface area of a sphere, because they distribute uniformly in all directions at the point of the pole. As magnetic paths become longer, they spread further apart as they leave the pole, resulting in the proper distribution density.

So, if the distance between the poles is given in some unit of length, such as inches or centimeters, and the mmf of the magnet is H multiplied by the distance in these units between equivalent poles (which will usually be a short distance from the end surfaces of the physical magnet), then the total field passing between poles can be determined by the remanent B at 1 centimeter from the poles, along the path between them (in the appropriate units, per square inch or per square centimeter) multiplied by the area of the sphere's surface, which is 4-pi square centimeters.

This statement may bother you, because you immediately think about this specific sphere at a one centimeter radius, at which point the field distribution is already uneven. But the surface area of a sphere is always 4-pi times the radius squared, *whatever radius is used.* This is true, right down to zero radius, which is the point source where the density (in theory, at least) is uniform. And density along the straight path remains the maximum all the way from one pole to the other. Only as you leave the straight path does the density fan out to reduced values.

That bit of theory used to be the basis of calculations for magnets in open space. But as soon as such a magnet is harnessed into a magnetic circuit, the 4-pi formula no longer applies. And if the magnet is remagnetized while coupled to a magnetic circuit of lower reluctance than the air around it, it will retain more, provided it is not uncoupled from the circuit. Usually, the reluctance of any magnetic parts of the circuit is low, compared to any air gaps in the circuit, which is where the magnet is put to work in most electronic applications such as a moving-coil instrument, transducer, motor, or some other device using a permanent magnet.

Long magnets have high retentivity with low coercivity (relative to other materials. Such figures can only be relative; what was once high may no longer be regarded so, as better materials are developed). This relationship explains why a magnet needs length to build up an adequate total mmf. Made in short, fat shapes, they readily demagnetize themselves and do not match the work areas for which they are needed.

OPTIMIZING A MAGNET FOR ITS APPLICATION

For any air gap to be supplied with a magnetic field, there is an optimum or best shape of magnet, depending on the material of which the magnet is made, to supply that field. The best shape is determined by the *BH Product* of of the material's demagnetization curve (Fig. 7-16).

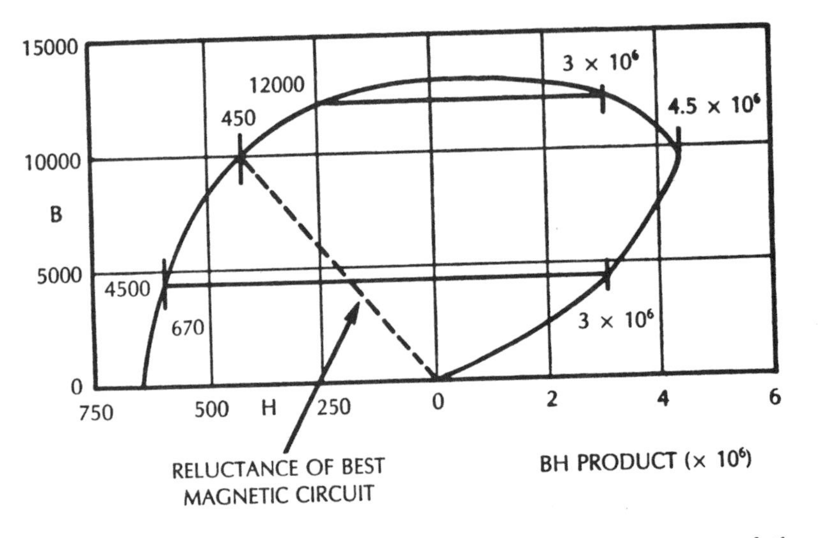

Fig. 7-16. Curve showing the significance of BH product and BH-max, relative to permanent magnet material.

If you take this curve and multiply together the field density and demagnetizing force at each point along the curve, you will find a point along this product curve at which it reaches a maximum. This is the value described as the *BH-max* of that material. This value determines the maximum magnetic energy that this magnet can deliver to an external magnetic circuit. This can be delivered only by arranging the air gap or the reluctance of the external circuit so it matches this BH-max point on the curve.

The earlier high-efficiency (high BH-max) materials developed were alloys of iron, cobalt, nickel, aluminum and copper, carefully produced with a large crystalline structure, and with their crystals aligned during manufacture so they would magnetize most rapidly (and more efficiently) in the direction for which they were designed. These materials greatly increased retentivity, without materially increasing coercivity, over the earlier materials developed. This was the first breakthrough from the more primitive magnetic materials.

Enter Ceramic Magnets

Later improvements took a different trend. At somewhat lower retentivity, coercivity was multiplied several times (Fig. 7-17). Such magnets may not have a larger BH-max than previous materials, but they do have a more useful shape for utilizing what they produce: one which is less susceptible to demagnetization.

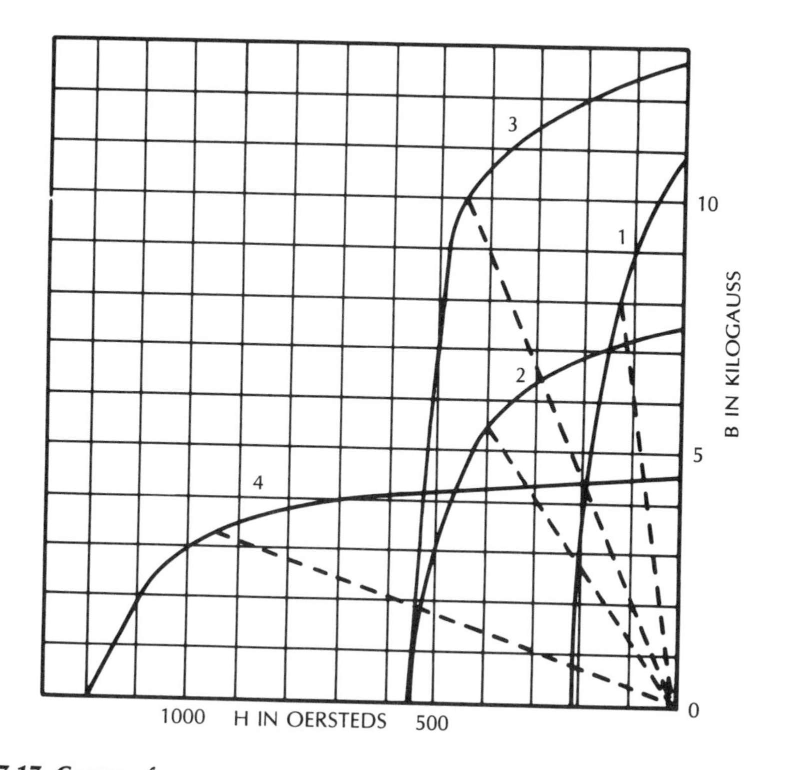

Fig. 7-17. Curves of some permanent magnet materials: 1. Cobalt steel (a very early material); 2. An early anisotropic material; 3. An improved anisotropic material; 4. A typical newer ceramic material.

Because of the greater coercivity, a short magnet length would maintain a greater mmf. And because of the lower retentivity, a larger magnet area is needed to get the amount of field needed. So these materials naturally lead to a short, fat magnet shape. The newer ceramic magnets are in this class.

Ceramic magnets are invariably magnetized "along" one of the shorter dimensions, so they have a North pole on one face, distributed all over the face, and a South pole on the other (Fig. 7-18). It is impossible to apply the old 4-pi formula to these magnets, because they cannot possibly be viewed as having point poles from which an expanding sphere of lines originate.

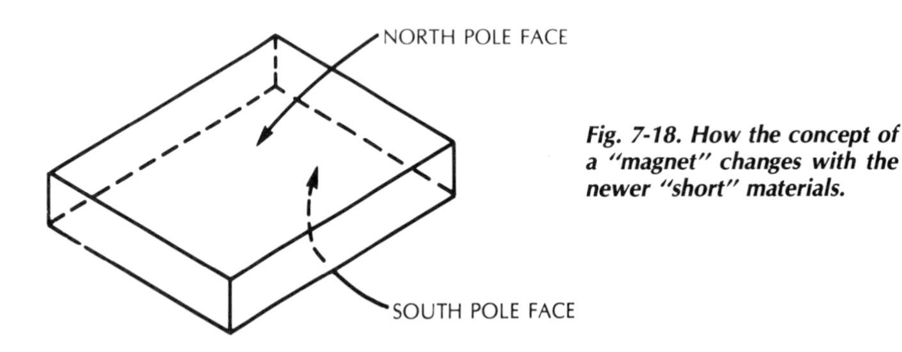

Fig. 7-18. How the concept of a "magnet" changes with the newer "short" materials.

Over most of the surface area the only effective path from one pole to the other is the direct one—through the magnet itself. Viewing this as an air path to return the generated mmf of the magnet, it does not go far beyond the point on the demagnetization curve representing the BH-max. Thus, removing such a magnet from its magnetic circuit, while it will deteriorate it from a maximum performance standpoint as magnetized with the circuit attached, will not do so to the same extent as was common with earlier shaped magnets. In fact, this type of magnet is quite effective for refrigerator door closures and other places where it may spend an appreciable portion of its life having its magnetic circuit opened and closed.

SOFT MAGNETIC MATERIALS

So much for permanent magnets. Before moving into magnetic calculations, we should look at the soft magnetic materials. As we said earlier, distinction between soft and hard arose because the earlier material forms were physically hard and soft. Soft iron was once the best soft magnet, while hardened steel made a good permanent (hard) magnet, by comparison. Nowadays, some of the soft magnetic materials are quite brutal to tools used to work them, which challenges the notion that they are really "soft." Early departures from the simple soft iron saw silicon added, which softened the magnetic properties while hardening the physical properties. Later, some of the nickel alloys yielded magnetically soft materials. Still newer, and widely in use, are the ceramic varieties called "ferrites."

It's doubtful that many readers will become metallurgists. Most will accept materials available and be concerned with utilizing whatever properties the metallurgists are able to deliver. Just as hard magnetic materials have a variety of properties, so do their soft counterparts. From the magnetic softness viewpoint, the important thing in regard to all of them is a hysteresis loop, as the total magnetization sequence graph is called, with a small area. If the magnetization and demagnetization curve followed the same line, retracing every step, even if that line is very curved (nonlinear), the material would take no power or energy in its magnetization-demagnetization cycle.

The up part of the curve represents energy put into making the magnetic material behave as a magnet. The down part represents that energy being returned or withdrawn. To the extent that the down curve fails to follow the up curve, the difference—the area between the curves—represents energy not retrieved and, thus, spent or dissipated.

But beyond this simple relationship between the magnetization cycle and energy loss, precise behavior gets complicated, as it did with the hard variety of material. There is no simple Ohm's law of B and H, relating to energy dissipated, as there is in an electrical circuit with resistance, current and voltage in a direct inter-relationship. The area contained within the hysteresis loop, which is a direct measure of energy dissipated per cycle, changes in a complicated way with the values of B and H applied, and the complications vary with different magnetic materials.

For example, some soft materials have a relatively "curvy" loop (Fig. 7-19), while others run up pretty steeply to a certain point and then saturate suddenly.

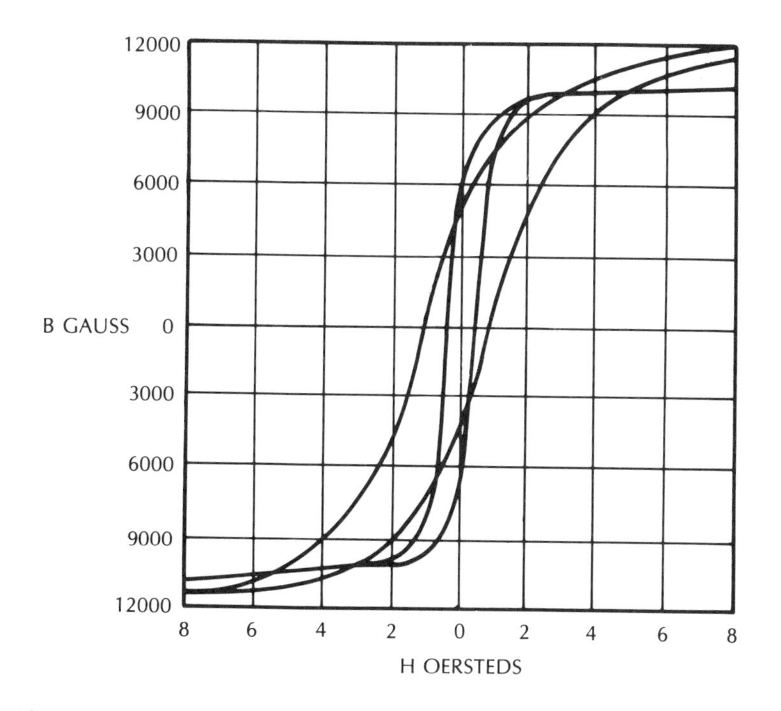

Fig. 7-19. Hysteresis loops typical of different low-loss ("soft") materials.

To some extent, what they look like on a graph depends on the scale to which the graph is drawn and the degree to which the curve runs into saturation, but there are quite definite differences between different materials (Fig. 7-20).

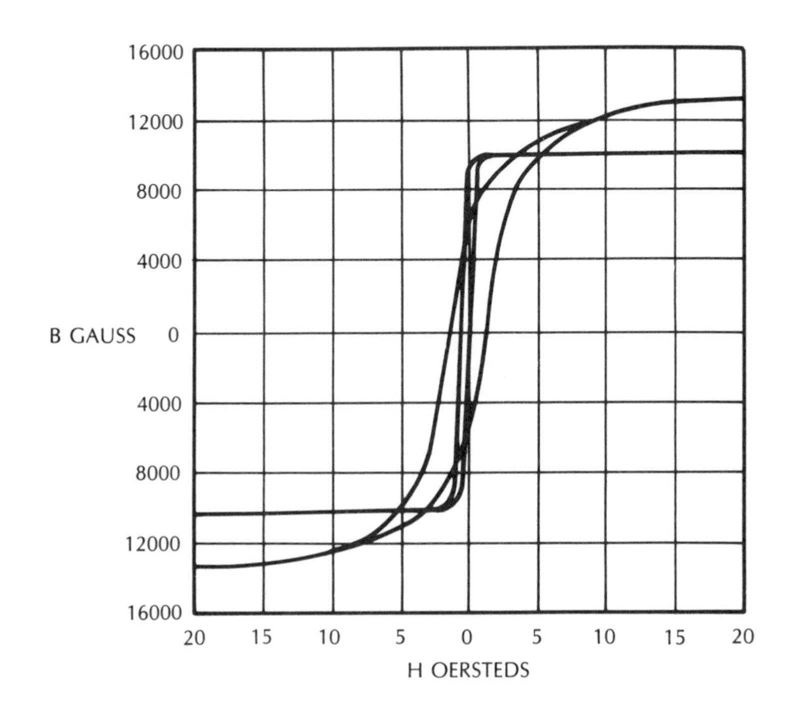

Fig. 7-20. The same loops, plotted to a different scale, to show how scale can make them look different.

If an alloy is made for absolute minimum energy loss, particularly for use in applications where there is not much energy, which means the "signal"—maximum values of B and H—is small, such a material is usually limited in how much signal it can handle, since it saturates at a lower density than other materials. If another alloy is made for handling a larger signal, with a high saturation density, it will not have the lowest possible loss, although it may be a low-loss type compared to almost anything else; the earlier, more common materials (Fig. 7-21), for example.

Another difference shows in materials that are designed to achieve a more linear magnetization curve—less kinky, and thus less likely to affect the electrical circuit, to which the windings are connected, by causing distortion or other disturbance—while others may be deliberately designed to saturate very suddenly. The latter are used for saturable reactors in so-called magnetic amplifiers. But before even inductors using magnetic cores became that sophisticated in design, there were more common variations that had to be accommodated. These were used for circuits in which ac (usually signal) and dc (usually supply) flows in the same windings.

Going back to the raw magnetic material, of which cores of such inductors are made, when ac magnetization is superimposed on dc magnetization, the result is a hysteresis loop that is off-center (Fig. 7-22).

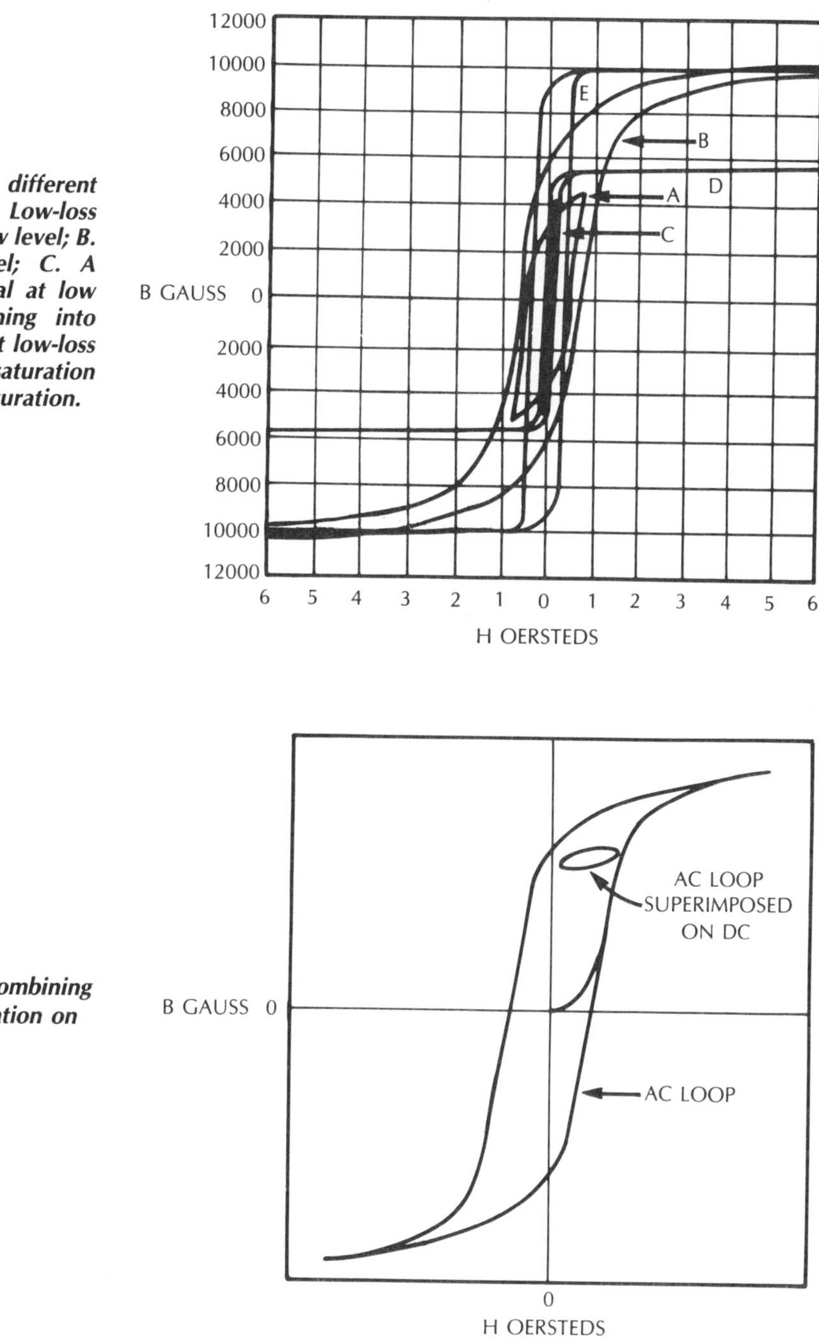

Fig. 7-21. Samples of different "soft" materials: A. Low-loss silicon-iron alloy at low level; B. Same at higher level; C. A minimum loss material at low level; D. Same running into saturation; E. Different low-loss alloy, allowing higher saturation density, shown at saturation.

Fig. 7-22. Effect of combining ac and dc magnetization on magnetic material.

The degree by which it off-centers will depend on a number of factors that are considered more fully in Chapter 10. The important thing to notice here is that the off-centering seriously impairs the B-H ratio (permeability) within the loop, which reflects adversely on the inductance such a component can provide in its electrical circuit.

Where such an inductor is used on pure ac, or with a kind of combined ac and dc (known as Class B) which appears the same as pure ac to the magnetic core, the relationship between B and H is far from constant, and the area enclosed in the loop does not hold a convenient fixed relationship, such as the square or some exponent of the magnitude of ac B or H swing.

MAGNETIC CALCULATIONS

In all magnetic circuits, for design if not for use reasons, magnetic calculations are necessary. In the permanent-magnet category, the magnet must be designed to deliver the desired field where it is wanted, which is usually across an air gap. This involves calculating the reluctance of the magnetic path external to the permanent magnet material, and projecting back the required magnetic flux to produce the specified field into that reluctance. Then the magnet is designed to deliver that amount and a means must be found to magnetize it. An example will show how it is done.

Figure 7-23 represents a possible magnet assembly for a moving-coil instrument. The part around the coil has already been designed, as follows: the gap in which the coil will rotate has a radius of 1 centimeter and a width of 1 millimeter; the length of the curve is 2 centimeters, and the width of the whole assembly (depth into the page) is 1 centimeter.

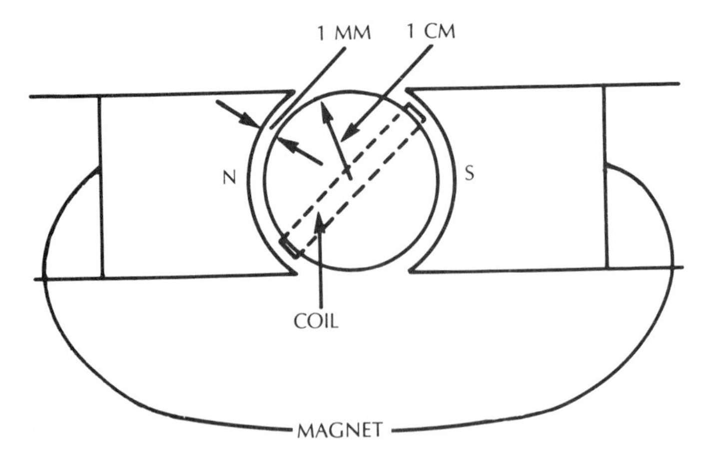

Fig. 7-23. Design requirements for a magnet to energize a working air gap.

Thus, the area of the gap that has to be supplied with a magnetic field is 2 square centimeters each side of the cylindrical center piece. There are two such gaps, each 1 millimeter wide (or "long" from the viewpoint of the magnetic field that crosses them). They are in series in the magnetic circuit; one is from the left-hand pole piece to the central cylinder and the other from the central cylinder to the right-hand pole piece. Thus, the total air gap to be supplied with a field is 2 square centimeters in area by 2 millimeters total magnetic length. From these quantities, the air-gap reluctance is 0.2 divided by 2 or 0.1 magnetic ohm. Notice that all dimensions must be in centimeters or square centimeters.

In considering how to supply this gap with a magnetic field, of 10,000 gauss (also referred to as 10 kilogauss), we have to consider the part of the circuit made of magnetic material, not so much from the viewpoint of calculating its reluctance, as of ensuring that it does not saturate. The area of the magnetic path at the gap is 2 square centimeters, so, assuming the material of which the pole pieces are made does not saturate at 10 kilogauss, the sectional area of the pole pieces also should not be less than 2 square centimeters.

Assume the pole pieces extend back to the end blocks, and a block of magnet material is inserted at the back, as shown. For the moment, we fix all dimensions except the length of the pole pieces, because these must be in accord with the magnet length needed.

In calculating the rest of the circuit, two factors have to be considered: the additional gilberts needed to drive the field through the reluctance of the magnetic material between the magnet and the air gaps, and the additional maxwells needed to make up for losses due to leakage.

Any magnetic assembly of this type shows magnetic effects by attracting steel objects, such as screwdrivers, in its vicinity. This attraction is evidence of field leakage that is not going to the working air gap where it is put to use.

The amount of leakage is difficult to calculate and even expert magnet designers do not attempt to calculate leakage. Rather, they estimate it, based on previous experience with magnets of various shapes. In a magnet of this shape, it would be fair to estimate 50 percent more flux for leakage, which means one third of the total flux will take leakage paths, rather than reaching the working air gap. If you also allow 50 percent of the reluctance of the air gap, as additional reluctance due to the magnetic circuit between the magnet and the poles of the gap, both maxwells and gilberts needed from the magnet will be increased by 50 percent, so the reluctance figure stands at 0.1 magnetic ohm.

Using a material with the demagnetization curve shown by Curve 3 in Fig. 7-17 for a first calculation (because this was the best material available a couple of decades or so ago), it yields a BH-max of 4.5×10^6, which is obtained at a remanence of 10 kilogauss with a demagnetizing force of 450 oersteds.

First, since the required density is 10 kilogauss, allowing for the one-third loss requires the area of the magnet to be 3 square centimeters. The total flux delivered by the magnet will be 30,000 maxwells. With a reluctance of 0.1 magnetic ohms, this will require a magnetizing force of $0.1 \times 30{,}000$ or 3000 gilberts.

Now for the length. The magnet will work at 450 oersteds, so the length needs to be 3000 divided by 450 or 6.7 centimeters. The magnet is drawn to scale, assuming the depth dimension to be 1 centimeter throughout (Fig. 7-24).

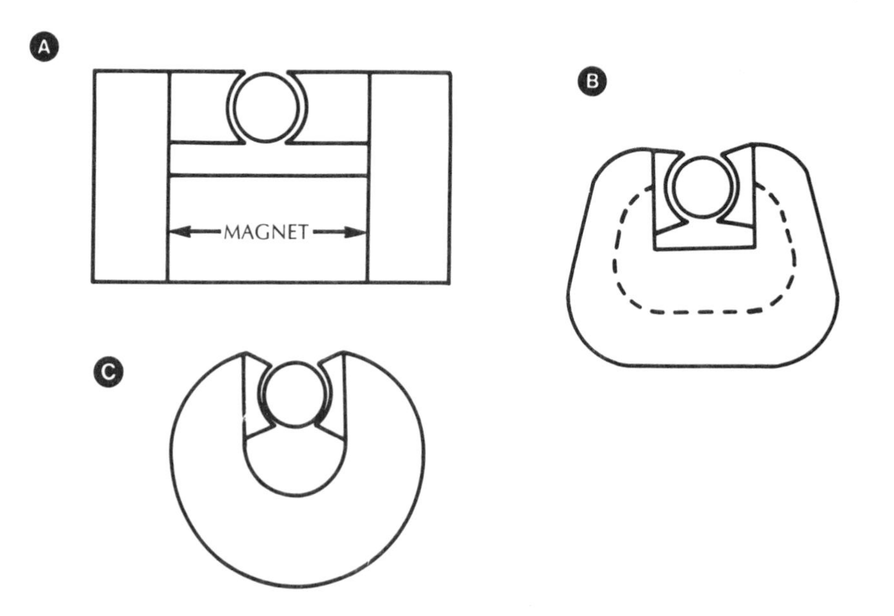

Fig. 7-24. (A) The magnet designed in the text; (B) and (C) are two more practical shapes from the viewpoint of economy which do, however, pose problems with magnetization.

The assembly has an unwieldy shape, which would result in more leakage than estimated, an estimate based on visualizing a more compact design.

Other possible solutions, in general outline and using the same material, are shown in Fig. 7-24B and C, each of which utilizes more the magnetic circuit as the actual magnet. However, such magnet shapes involve difficulties in both casting and magnetizing the magnet material that we will not discuss in detail here.

Considering the possibilities of using a short magnet type, curve 4 in Fig. 7-17 represents the demagnetizing curve for a ceramic material that has a retentivity of 4.5 kilogauss, a coercivity of 1200 oersteds, and a BH-max of 3.2 $\times$ 10^6, achieved at 960 oersteds and 3.33 kilogauss. If the same maximum density is required (B equals 10 kg), and using the same assumption for the loss due to leakage as before, resulting in a required 30,000 maxwells, the area of this magnetic material needs to be 30,000 divided by 3330 or 9 square centimeters. To achieve the required 3000 gilberts magnetizing force, the length needs to be 3000 divided by 960 or 3.125 centimeters. Such a magnet could be made in two parts, each 4.5 square centimeters in cross-section and 3.125 centimeters long (Fig. 7-25).

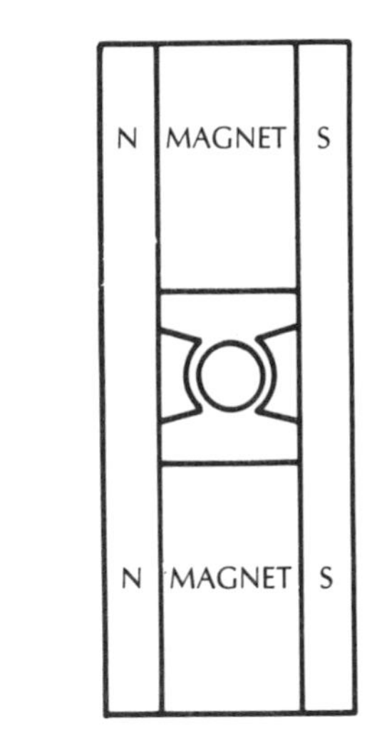

Fig. 7-25. Two "short" magnets may be used to supply the same work gap as that in Fig. 7-23.

One more possibility is now worth considering (and has, in fact, been used). A good way to avoid leakage is to turn the magnetic circuit inside out. Leakage occurs because the magnet is on the outside of the working area (being much bigger). With the short design, what would happen if the magnet is made a slug inside the coil, appropriately shaped (Fig. 7-26)?

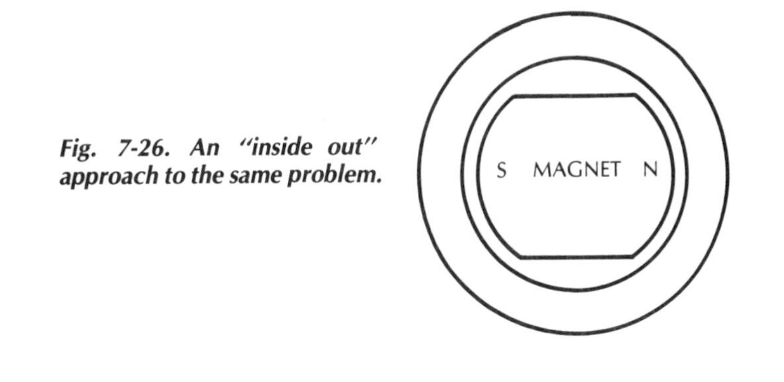

Fig. 7-26. An "inside out" approach to the same problem.

Using the same operating point, the maximum density would be 3.33 kg, which is one-third of that obtainable at the gap the other way. But you do utilize all of it, because it has nowhere else to go—all leakage goes through the working gaps. If the outer ring that provides the return path allows two air gaps of the same dimensions discussed so far, the reluctance of the working space is unchanged at 0.1 magnetic ohm.

Taking the shortest magnet length of about 1.2 centimeters, with an area of about 1.8 square centimeters, the air gap serves as a load to a unit cube of the magnet with a specific reluctance of 1.8 divided by 1.2 $\times$ 0.1 or 0.15 magnetic ohm. Drawing this line on the graph in Fig. 7-17, by using values of

H equals 600 and B equals 4000, Fig. 7-27, and drawing a straight line from this point to zero, the working point on the curve yields a remanence of just over 4000 gauss, which is the possible working density, rather than the 3330 figure at BH-max.

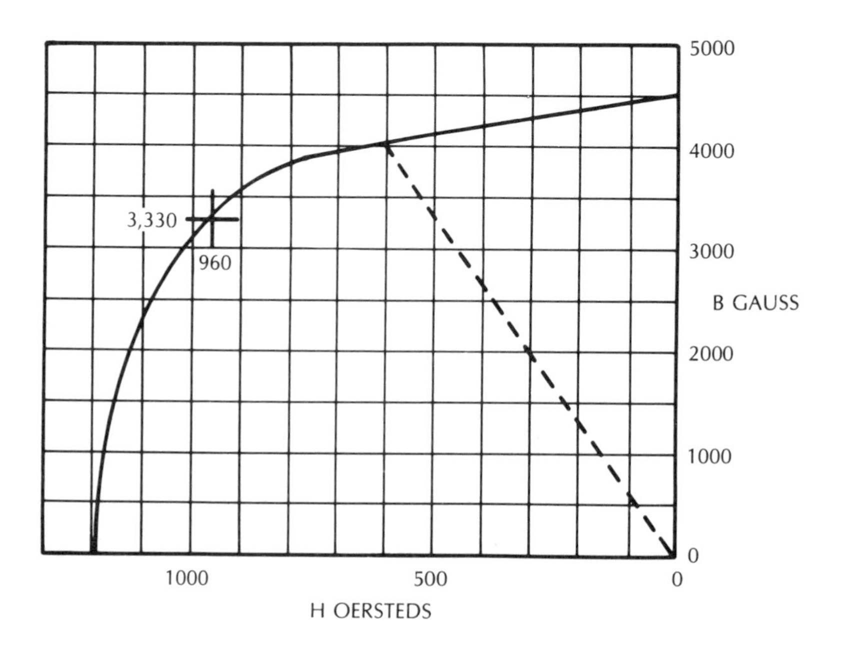

Fig. 7-27. The short (ceramic) magnet curve redrawn to show the BH-max point, and an improved operating point resulting from this approach.

Finding how to use magnetic material is only half way to actually using it, which requires the thing to be magnetized, with the magnet in position in its working magnetic circuit. This means the whole assembly must be placed in a magnetic field that will saturate not only the whole magnetic circuit but especially the magnet material. The soft magnetic materials are saturated in a direction opposite to that remaining after the job is done, due to the permanent magnet.

The important thing is that the magnetizing force must be great enough to saturate the permanent magnet material. The saturation force needed will usually be more than twice the coercive force necessary to demagnetize the magnet in the opposite direction. Here, perhaps, is a disadvantage with the shorter type magnet materials. Because their coercive force is so high, the magnetization necessary to saturate them is even higher, even though the remanent flux density (or retentivity) is not as high as for the long materials.

Most of the better materials today, both hard and soft, are anisotropic, which means they exhibit different magnetic properties in different directions through the material. This allows utilization of the better properties in the direction in which they will be magnetized at the expense of those same properties in directions where they will not be used. One way this is achieved, although each material requires different and sometimes secret processes to achieve the optimum, is to cool the material from a high temperature in a powerful magnetic field. Some materials completely lose their magnetic properties at higher (red-hot) temperatures. If the material is cooled from this point with a strong magnetic field maintained through it, the molecules will be oriented by it, making the material easier to magnetize in that direction than any other. Some of the soft magnetic materials possess a similar property; however, with soft material the molecular orientation may be achieved by cold rolling. In such materials, the permeability and saturation density will be higher along the direction of rolling than across it.

In yet another kind of magnetic material, the magnetic particles are suspended within a non-magnetic medium, and the spacing and arrangement is controlled during manufacture. The final magnetic material has a great many magnetic particles embedded in the non-magnetic material. The resulting magnetic properties of the material depend on the spacing and orientation of the magnetic particles. In this case, a magnetic field can be used in the material's production to orient and space the magnetic particles optimally for the purpose.

The presence of the many tiny gaps between the particles of magnetic material controls the effective permeability of the total material, which is like some percentage of air from the magnetic viewpoint; the space between the tiny air gaps is filled with high permeability magnetic material. For example, if one-two-hundredths of the average magnetic path, measured along the path of magnetization, consists of non-magnetic material, the overall permeability is controlled to something a little less than 200. If the minimum permeability of the magnetic material embedded in the substance is 3000 (as a solid), the overall permeability will vary between 200 $\times$ 3000 divided by 3200 or 188 and the full 200.

Examination Questions

1. The following columns of information give the operating conditions in a magnetic material, B in gauss, H in oersteds, area of material in square centimeters, and length in centimeters. For each, calculate the total flux in maxwells and the magnetomotive force in gilberts.

B	H	AREA	LENGTH
5000	1.5	12.5	60
7500	2.4	25.4	45
10,000	3.6	6.45	25.4
3300	0.7	64.5	12.7

2. Work the following additional examples, as required in Question 1. The column headings are the same as in Question 1.

B	H	AREA	LENGTH
1250	1250	6.45	2.54
2500	2500	12.9	0.254
5000	5000	25.8	0.1
6300	0.1	6.45	25.4

3. In the following examples, area is in square inches and length is in inches, while B and H are in gauss and oersteds, respectively. For conversion, take 1 square inch as 6.45 square centimeters and 1 inch as 2.54 centimeters.

B	H	AREA	LENGTH
5000	1.25	2.5	10
10,000	3.5	1.25	5
12,500	6.5	1.0	1
2500	0.45	5.0	3.5

4. Work the following additional examples, with the same column headings as Question 3.

B	H	AREA	LENGTH
2300	1.03	2.25	10.5
3450	1.73	1.5	5.25
6300	3.5	0.65	2.0
4750	0.75	0.25	6.0

5. A section of a magnetic circuit has a cross-sectional area of 6.5 square centimeters and a length of 27.5 centimeters. The total flux is determined as 55,000 maxwells and the magnetomotive force is 35 gilberts. Find the B and H in this section of circuit.
6. Another section of the same circuit has a cross-sectional area of 3.75 square centimeters and a length of 17.5 centimeters. Assuming no leakage and that the material is working at the same permeability as the section in Question 5, calculate the B, H and magnetomotive force for this section of magnetic circuit.

7. Find the H generated by 10 amps flowing through 1000 turns, if the length of the magnetic circuit path that passes through the coil is 20 inches.
8. Find the current needed, in the same situation as Question 7, to produce an H of 500 oersteds; of 1200 oersteds.
9. An air gap area of 1 square inch and a width of 0.05 inch requires a flux density (B) of 10,000 (maxwells per square centimeter, or gauss). The magnetic material produces its BH-max at 10,000 gauss, 450 oersteds. Assuming 50 percent additional flux for leakage and additional reluctance for the magnetic circuit, calculate the magnet shape (area and length) needed.
10. If the air gap in Question 9 is doubled in area, with the same length, and the same magnet is used (as designed in Question 9), find the working flux density (B), assuming the magnet is recharged and the curve in Fig. 7-17, Curve 3 applies.
11. Using Curve 4 in Fig. 7-17, calculate the magnetic proportions needed to replace the magnet in Question 9, using the same leakage assumptions (although a new magnetic circuit will be needed).
12. Recalculate the charge brought about by a change in the gap, as in Question 10, using similar assumptions.

8
Electromagnetism

The previous chapter shows that current in a coil produces a field that is capable of magnetizing a magnetic material. If the coil is not wound around magnetic material, it still has a magnetic effect, with a permeability of one. Without performing any more experiments, the fact that an electric current produces a magnetic field might suggest to us that magnetism could also produce an electric current. It can, in fact, but not in the most obvious way. The reciprocity or two-way action works like this:

Effect 1. Current is electricity in motion, and it is the motion of the charge that makes up current and thus causes magnetomotive force to be generated.

Effect 2. Similarly, a moving or changing magnetic field results in the generation of an electromotive force, just as electric charge in motion generates magnetomotive force.

In each effect, the movement of something that may be considered as a quantity (coulombs of electricity, or maxwells of magnetic field) results in the generation of potential or motive force—volts of electricity or gilberts of magnetizing force. Those are the basic relationships. Now, let's see how they relate to one another in a little more detail.

In the previous chapter, we related ampere-turns to the gilberts generated. In this one we shall relate maxwells per second to volts generated, and see how the changes have to occur to produce gilberts or volts respectively.

SPATIAL RELATIONSHIPS

The first step is to see the spatial relationship between movement and generation each way. An electric current flowing in a coil produces its effect because the wire of which the coil is wound, and through which the current flows, is formed into a circular, repetitive shape. The coil may be square or rectangular, rather than circular, but it always encloses or encircles the area in which mmf is generated (Fig. 8-1).

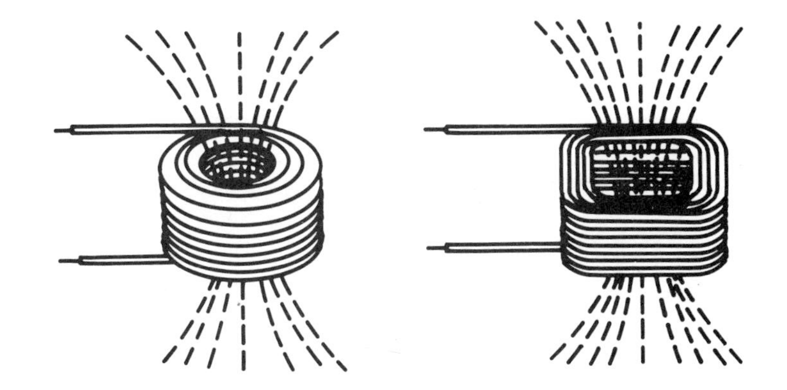

Fig. 8-1. Magnetic field generated by a current in a coil.

Coiling the conductor multiplies the effect of the electricity moving (current flowing) in the individual turns, so that 1 ampere in 1000 turns has the same magnetizing effect as 1000 amperes in 1 turn.

Deriving the Rules

But the conductor, or wire, does not have to be shaped into a loop or circular turn to produce a magnetic field when electricity flows through it. A magnetic field is produced in the proximity of a straight conductor (Fig. 8-2).

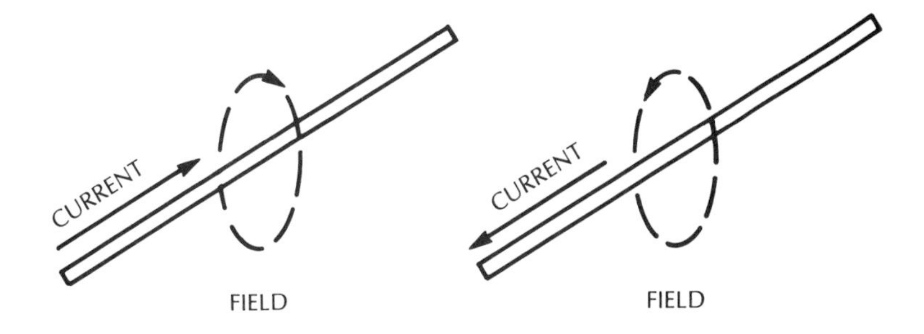

Fig. 8-2. A magnetic field is produced around a straight conductor carrying a current.

The field rotates about the wire in a fixed relationship to the current flowing in it. Reverse the current and the field rotates the opposite way. Now, wrap the wire to make a turn, and the field becomes concentrated in the center of the turn (Fig. 8-3) as observed earlier.

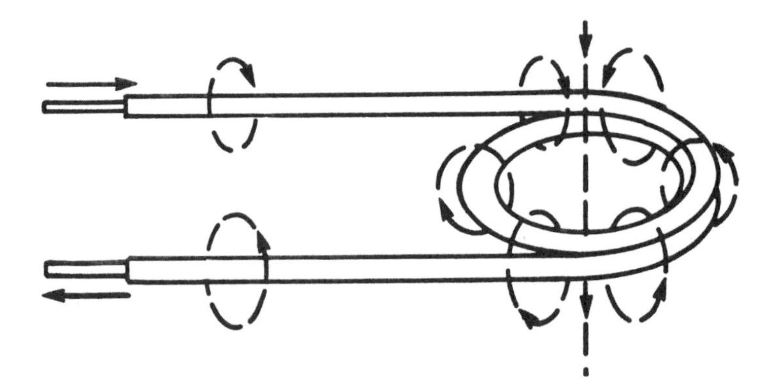

Fig. 8-3. Winding the conductor into a coil concentrates the field along the axis in the center of the coil.

There are two rules that aid in remembering which way magnetic fields relates to the current producing it.

1. If you look in the direction in which current is flowing along a wire, the magnetic field around the wire moves in a clockwise direction; this is the direction in which a compass needle (its North end) will point.
2. If you look at a turn of wire or a coil from a direction such that the current appears to rotate in a clockwise direction, you are looking at the South pole end of the coil, toward the North pole end.

When a magnet is moved in relation to a wire or a coil, an electromotive force (voltage) is generated. If the circuit is closed, a current will flow. There are several ways to relate the relative directions of a magnetic field, its movement and the generated emf, but the one that makes the relationship easiest to reconstruct and understand is to derive it from the two rules already given.

FORCE DUE TO CURRENT IN A MAGNETIC FIELD

First combine a magnetic field produced by current in a wire with a magnetic field present due to some outside source of magnetism. Suppose a wire is going into the page in Fig. 8-4 and that a magnetic field consists of flux going from left to right across the page. According to Rule 1 above, the field around the wire moves clockwise. This means it will add to the external field above the wire and subtract from it, or neutralize the same field below the wire.

In the study of magnetism, we showed that a magnetic field tends to equalize its energy distribution in space. To do this in the situation shown in Fig. 8-4, the field above the wire needs to expand, while that below the wire needs to contract. What we're saying is that the combination of the field and current shown produces a mechanical force which tends to move the wire down the page.

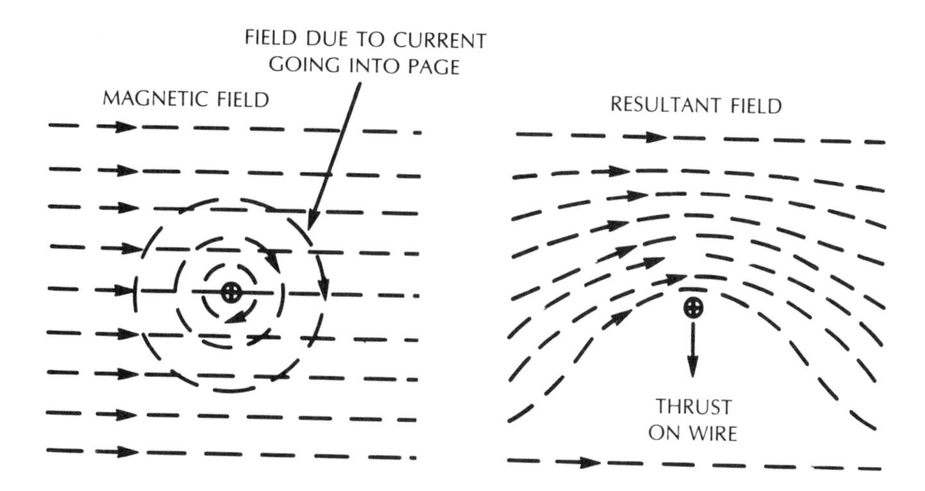

Fig. 8-4. Field due to current in a conductor, combines with an external magnetic field, to produce a mechanical thrust on the conductor.

Notice that the three quantities: (1) electric current (or voltage), (2) magnetic field and (3) movement (or mechanical force) are always mutually at right angles. This is what makes comprehension of the relationship a little difficult to retain without some rules of help.

MOVING A CONDUCTOR IN A MAGNETIC FIELD

Another situation, a variation of the one just mentioned where a field is moved in relation to a wire or coil, is that of moving the wire in the field. It should be fairly obvious that these are different versions of the same situation, using a different element viewed as stationary. In Fig. 8-4, the effect would be the same whether the wire moved down the page, or the source of magnetism providing the field moved up the page.

The other variable that can confuse the situation is the respective roles of cause and effect. If the magnetic field combined with the current flowing causes movement, that is one thing. If movement of a conductor through a field causes emf that tends to produce current (if the circuit is completed so the current has somewhere to go) that is another thing. The easiest way to see this is to think in terms of the conservation of energy—the fact that perpetual motion "won't work."

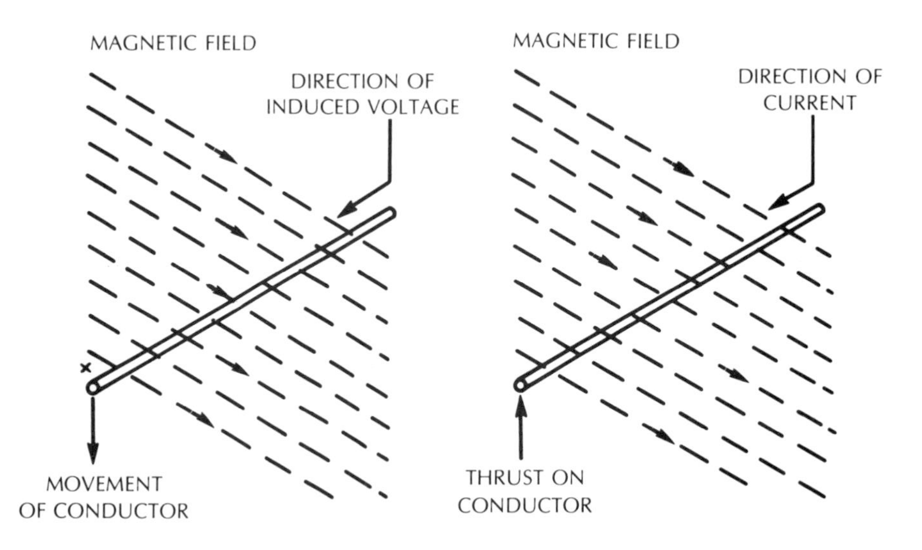

Fig. 8-5. Relationship between the generation of a voltage due to movement of a conductor in a magnetic field, and the force produced on the same conductor when current is allowed to flow due to that generated voltage.

This means that if movement generates an emf (voltage) that tends to produce current, then allowing that current to flow (by completing the circuit), while the conductor is moving in the same field that generated the emf producing it, will cause a force that tends to stop that movement (Fig. 8-5).

Left-Hand and Right-Hand Rules

These facts are combined in the left-hand and right-hand rules attributed to Fleming. In each case, the thumb, first finger (index finger) and second (middle) finger are extended in positions mutually at right angles (Fig. 8-6). In each case, too, the three extremities have the same basic significance:

- The thu**M**b signifies the direction of **M**otion.
- The **F**irst finger signifies the direction of the magnetic **F**ield or **F**lux.
- The se**C**ond finger signifies the direction of **C**urrent.

With those significances, here is how the two hands apply: the right-hand rule gives the relationship for current generation; it is called the generator rule. The left-hand rule gives the relationship for motor action, where the current flowing produces movement or a force tending to produce that movement.

Apply it to the situation portrayed in Fig. 8-4. Your second finger is pointing into the page, while your first finger points from left to right; this leaves your thumb (left hand, remember) pointing down the page. If you want to reverse this, you will have to put the book over to your right.

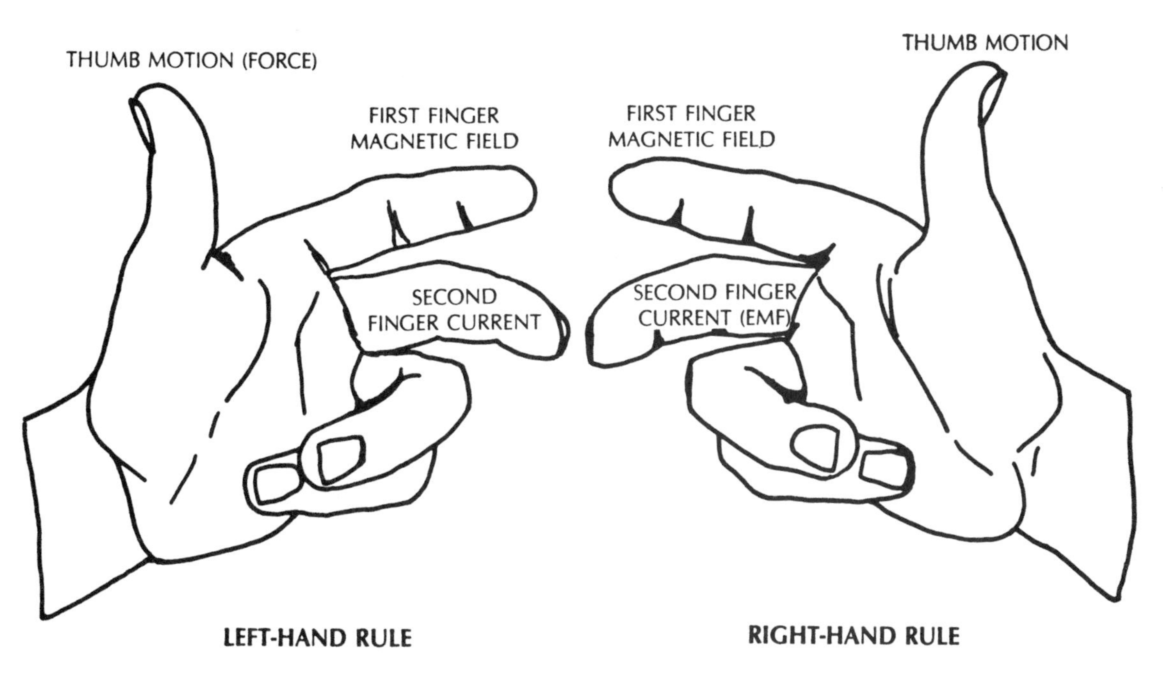

Fig. 8-6. The Left-hand (motor) and Right-hand (generation) rules attributed to Fleming.

Now, with your second finger pointing into the page and your first finger pointing across it (to the right), your thumb points up the page. This means that upward movement will be needed to generate an emf tending to drive current into the page. Now turn your right hand over, so the first finger still points to the right. In this position, the second points away from the page and your thumb now points down the page. This means that moving the conductor downwards will produce an emf tending to cause a current flow out of the page.

ELECTROMOTIVE FORCE GENERATED

So much for directions and inter-relationships. Now we come to values. Ampere turns were related to oersteds in the previous chapter. Electromotive force is defined, in practical units, as follows:

- If 10^8 (100,000,000) maxwells per second is the rate of change in which a magnetic field cuts a conductor, the emf generated will be 1 volt.

Notice that this is the total rate of change of the field cutting the total conductor. If 50,000 maxwells cut a 100-turn coil that passes through that field 100 times, the total coil will cut through 100 × 50,000 or 5,000,000 maxwells. Then, if the rate at which this field rises from zero or falls to zero takes one-sixtieth second, this represents a rate of change of 60 × 5,000,000 or 300,000,000 maxwells per second, which will generate 3 volts.

Figuring through examples like the above carefully will help you realize that the relationship is due to change—not to a steady field. This is not always easy to grasp. It is not the field that generates the voltage, but the change in field. If the same change occurs in one-one-hundred twentieth second instead of one-sixtieth, the voltage generated during the shorter period will be 6 volts.

Of course, as we said in Chapter 5, changes are not always steady. In fact, they are usually not in alternating systems. The field does not rise from zero to some stated value or fall from such a value to zero at a steady rate, except in very special cases. Generally, it follows some waveform of change, such as sinusoidal. This means that at the instant when the field is changing at a rate of 600,000,000 maxwells per second, the voltage generated will be 6 volts.

If waveforms are sinusoidal, it means that the maximum change occurs when the field is actually passing through zero (or a mid value), so the 6 volts will occur at that moment. If 10,000,000 maxwells is the maximum field (for example), when it reaches that value it will momentarily not be changing. This means the voltage generated at peak field will always be zero (Fig. 8-7).

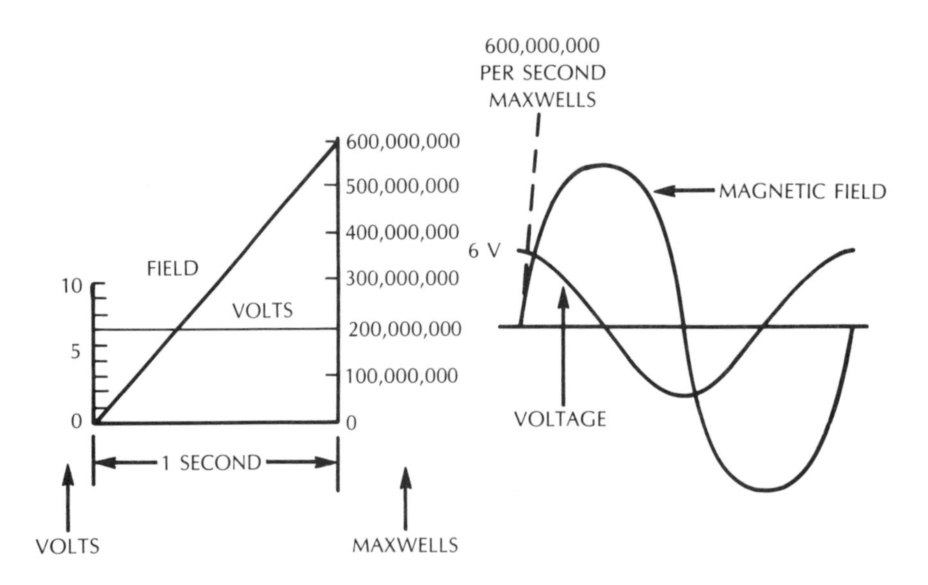

Fig. 8-7. Relationship between a changing magnetic field and the voltage it generates: left, the simple relationship; right, as applied to a sinusoidal waveform.

MECHANICAL FORCE GENERATED

The remaining relationship to consider puts to work the force produced by a current flowing in a conductor that passes through a magnetic field. This force is based on the field intensity and the length of the conductor in that field. It is not dependent on the total field, because the conductor only cuts across the field at one point, so the relevant density is maxwells per centimeter, rather than maxwells per square centimeter.

This is not the usual way of specifying density, so the way to avoid another set of units is to specify density by square centimeter, and then multiply by the length of the conductor in centimeters. The force produced, measured against current in a conductor, is specified in centimeters of conductor when the field is specified in square centimeters.

A current of 1 ampere flowing through a field of 1 gauss for a distance (conductor length) of 10 centimeters will produce a force of 1 dyne. This is a very small force—that needed to accelerate 1 gram (the weight of a cubic centimeter of water) at 1 centimeter per second per second, which is a little more than one thousandth of the gravity weight of 1 gram (gravity averaging about 981 dynes per gram).

So a current of 1 ampere flowing through a field of 1000 gauss for 10 centimeters will produce a force of 1000 dynes, or a force equal to the weight of 1000 divided by 981 or 1.02 grams. Make the current 100 amps and the field 10,000 gauss, and the force reaches 1.02 kilograms (about 2¼ pounds) for a length of 10 centimeters of conductor.

ENERGY TRANSFER AND BALANCE

So far, we have considered the basic relationships mentioned earlier under Effects 1 and 2. In practice, voltages and current occur together, as do movements and forces. For example, an electric motor uses current in a field to produce a field that makes the motor turn. But as the motor turns, the same coils that produce the force to make it turn also generates a voltage to oppose the current in the coils (Fig. 8-8).

A balance is achieved when the applied voltage is just a little bigger than the voltage generated due to motor rotation. The difference in voltage then drives the current through the coil that produces a force just adequate to maintain the rate of rotation. If the motor is not turning fast enough to achieve this balance, the voltage difference is greater. As a result, the current is higher, producing more force, which speeds up movement until balance is restored. This is how energy balance is restored when a mechanical load is applied to a motor. The mechanical load momentarily slows the motor down, so more current begins to flow, due to the greater voltage difference, until the force produced is equal to the extra mechanical load applied.

The same kind of energy balance is maintained during generation. Movement of the conductor in the magnetic field causes an electrical voltage to be generated.

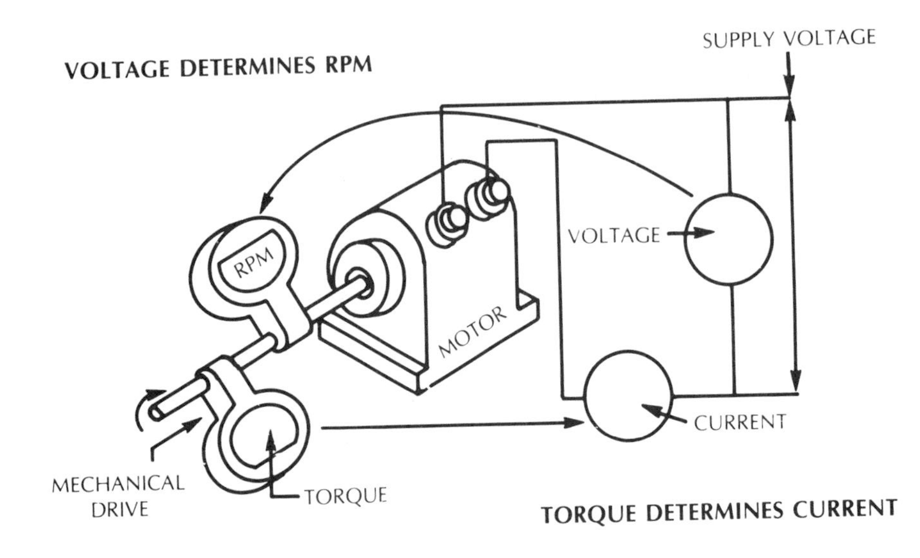

Fig. 8-8. Relationship between voltage, rpm, current and torque in a basic electric motor.

Provided no current flows, no force is applied to the moving conductor. The only energy needed to keep the conductor (coil) moving is a sufficient amount to overcome friction in the generator mechanism. But as soon as electrical current is taken from the voltage generated, this reflects back as force opposing the movement (Fig. 8-9). If the generator is being driven by some kind of engine, the increased force necessary to produce the current in the conductor results in a greater demand on engine power.

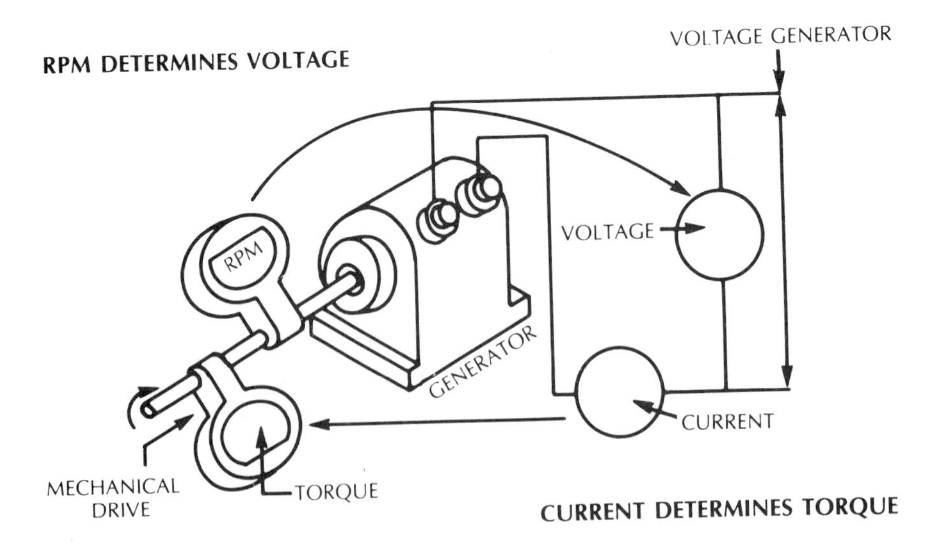

Fig. 8-9. Relationship between voltage, rpm, current and torque in a basic electrical generator.

REVERSIBILITY

We will not go into details about various types of motors and generators in this book, because this is a big subject in itself and scarcely within the scope of basic electronics. But to understand the principles involved consider a permanent magnet motor-generator. The speed of rotation determines the voltage generated by the coils as they pass through the magnetic field. This is a direct transformation—voltage into speed and vice versa; the two are inseparably locked together. At the same time, the current flowing in the coils, since they are in the magnetic field, produces torque or the turning force, and vice versa; these two are also inseparable.

If the rotating part is driven faster than the speed required to produce a voltage equal to that of the electrical circuit connected to its terminals, the generated voltage will be higher than the external circuit voltage. Consequently, current and electrical energy will flow from the generator to the external circuit. The electrical energy delivered will equal the excess mechanical energy that drives it faster than the balance speed.

If the rotating part runs more slowly than the balance speed, the voltage it generates will be less than that of the external circuit, so current will reverse (as compared with that discussed in the previous paragraph) and energy will flow from the external circuit into the machine (motor-generator). Now the electrical energy absorbed from the external circuit will equal the excess mechanical energy that is dragging the speed below that of balance.

EFFICIENCY

In a practical situation, friction and other losses—both mechanical and electrical—will cause the machine to run slightly slower (as a motor) or faster (as a generator) than the exact balance speed, so just enough energy is taken from the electrical circuit (as a motor) or the mechanical drive (as a generator) to keep it running, even when no energy is taken from the other side.

Applying a mechanical load to the machine to make it work as a motor increases the electrical input by the amount of current needed to supply the required torque. Not all the energy will be converted into mechanical energy, due to electrical losses in the coils and magnet material of the machine.

Similarly, if the machine is used as a generator, it will take a small amount of mechanical energy just to spin it at the balance speed, with no electrical energy flowing either way. Then, when electrical power is taken from it, extra mechanical torque must be supplied, since the machine runs a little slower, to deliver the necessary energy to the electrical circuit.

Here again, not all the mechanical power put in comes out as electrical power, because of both mechanical and electrical losses. If the conversion were perfect, every mechanical horsepower would produce 746 electrical watts, or every 746 electrical watts would produce 1 horsepower. Because of inefficiency—the fact that no machine is 100 percent efficient—this never quite happens. For example, if a motor is 80 percent efficient and delivers 2 horsepower, this represents 80 percent of the input power. For each horsepower delivered, it requires 746 watts, making 1492 watts. Since this is 80 percent of the input, the electrical input required is 100 divided by 80 × 1492 or 1865 watts.

If the same machine is operated as a generator to deliver 2000 electrical watts, it represents only 80 percent of the mechanical energy put in, which would thus be equivalent to 100 divided by 80 × 2000 or 2500 watts. Converting this to horsepower, 2500 divided by 746 or 3.35 horsepower as the required mechanical input.

TRANSDUCERS

Motors and generators were the first transducers—devices for converting electrical energy into mechanical energy and vice versa. Nowadays, many more varieties exist. A device that converts minute vibrations in the air into mechanical movement, which, in turn, generates minute electrical voltages or currents, is called a microphone. A device that reverses the process, accepting electrical energy at the input to produce a mechanical or acoustic energy output, is another kind of transducer.

Phonograph pickups, cutting heads, loudspeakers, as well as vibration pickups and generators, each use these principles. Also, there are instruments for measuring voltages and currents. Figures 7-23, 7-24, 7-25 and 7-26 show the magnet part of such an instrument and Fig. 7-23 shows the position for the coil. When current flows through this coil, it passes, say, up under the North pole and down under the South pole. This makes the left side of the coil move up and the right side move down, according to the Left-Hand Rule.

By leading the current into the coil through light springs that also control movement (Fig. 8-10), the coil moves until the restoring force of the springs balances the force of the current trying to turn the coil. The position at which the instrument needle comes to rest then indicates the value of the current.

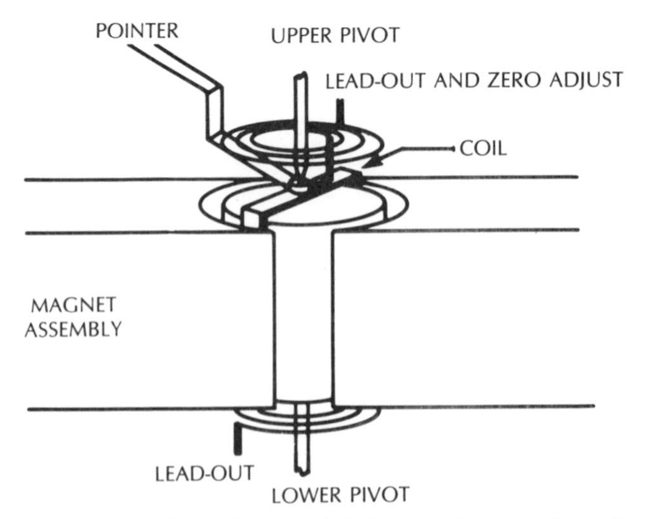

Fig. 8-10. Details of the essential elements in a moving coil meter.

INDUCTANCE

All electric circuits possess inductance, but coils most especially. This is an effect caused by changes in the current flowing in the circuit or coil. When current in a coil starts to increase (from zero, say) it generates a growing magnetic field around the coil. But this same magnetic field, because it is changing, generates an emf in the same coil. By the rules we gave earlier, the emf tends to oppose the growing current producing it (Fig. 8-11).

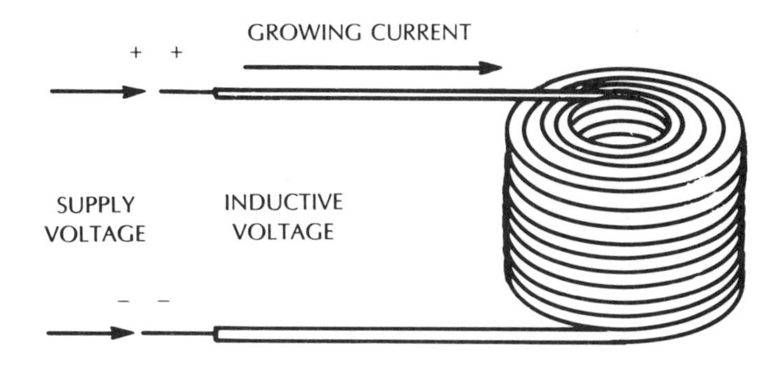

Fig. 8-11. Inductive effect of a coil when current is increasing due to an external applied voltage.

When the current in the coil starts to decrease from some value it has previously attained, the magnetic field decreases to match. And this, too, being a changing magnetic field, generates a voltage that tends to oppose the *change* in current, which means it tends to maintain the current already flowing. So a growing current is opposed by what is called an *inductive* or *back emf,* while a falling current is similarly opposed by a voltage that tends to keep the original current flowing (Fig. 8-12).

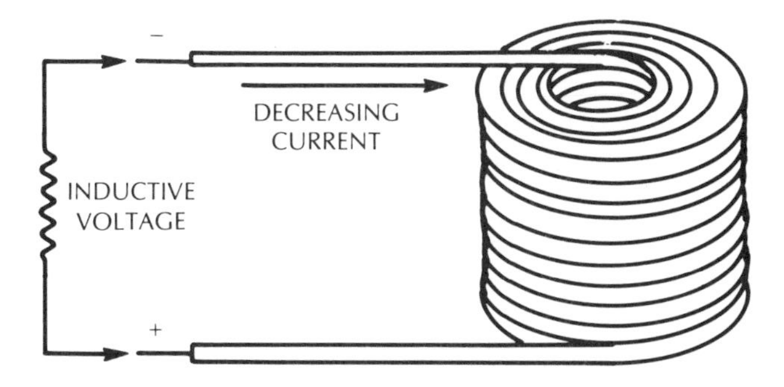

Fig. 8-12. Inductive effect of a coil when current is decreasing due to a voltage in an external resistance.

If two coils are in proximity, the changing field from one will produce an effect in the other. The same changing field will produce an emf in both coils (Fig. 8-13). This effect is called *mutual inductance.*

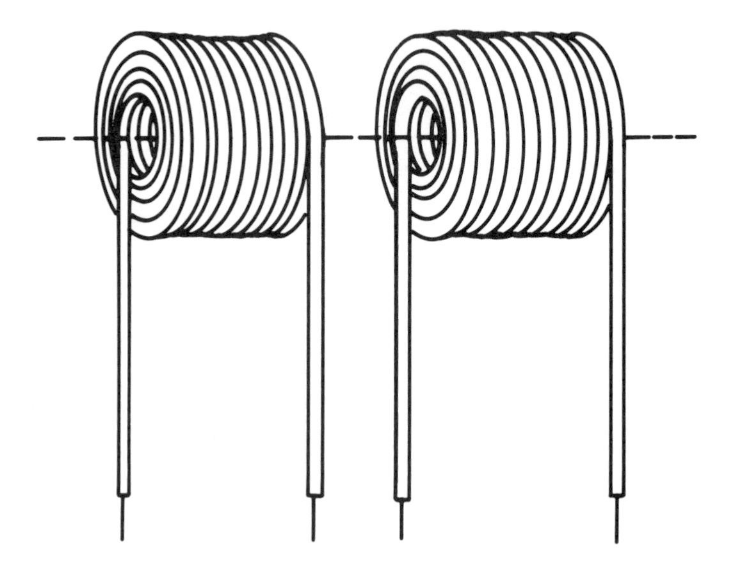

Fig. 8-13. Two coils having a common axis, in proximity, exhibit the property known as mutual inductance.

If both coils occupy the same space (which is, of course, a physical impossibility), then the voltages induced in them by a changing current would be exactly proportional to the numbers of turns in each coil, because the same changing field would be identical in each.

LEAKAGE

If the second coil does not carry any current due to the emf induced, and if the coils are reasonably close, the voltages induced are still very close to being proportional to the turns in the respective windings. However, if current is taken from the second winding due to the voltage induced in it, the secondary current will produce its own magnetic field, distinct from that due to the current in the first coil, and voltages will no longer be strictly proportional to the turns.

When this last condition occurs, the field that passes between the windings, due to the fact that current is flowing in both of them (and thus the combination acts as if the two were one coil, in this sense) is called "*leakage flux*" (Fig. 8-14). If two sets of coils were perfectly mixed so that no magnetic field could get between them, this would be an ideal transformer. Then, whatever current flowed, the induced voltage would be exactly proportional to the turns in each winding. But in any practical transformer, the coils have a physical separation, even if they are sandwiched many times, to reduce the leakage effects to a minimum.

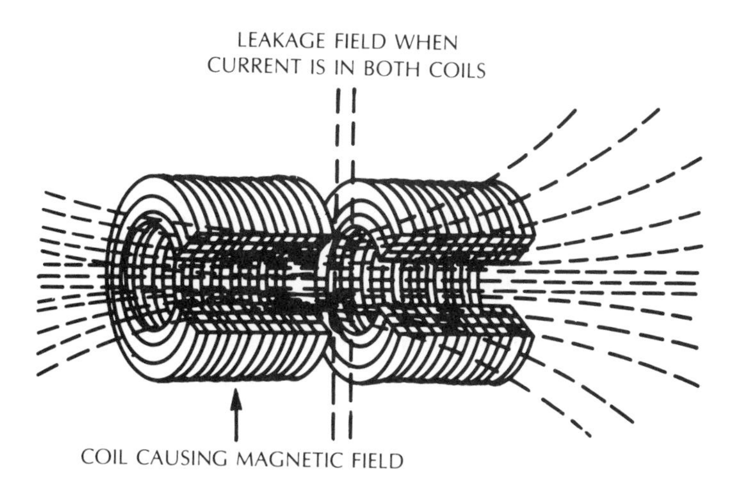

Fig. 8-14. When one coil carries current, the field links both coils. When the current changes a voltage is induced. When the second coil also carries current due to this induced voltage, a leakage field appears between the coils, changing the inductance in both.

And because of this, the transformer can be regarded as *equivalent* to an ideal transformer having the same ratio between the turns in its windings, with the addition of inductances, called *leakage inductances,* between the windings (Fig. 8-15).

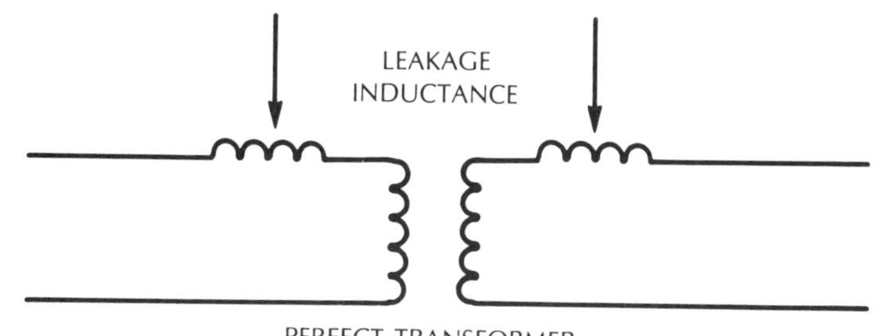

Fig. 8-15. Equivalent circuit of two coils in a transformer with a magnetic core.

Most practical transformers employ cores of soft magnetic material, as described in the previous chapter. This core goes through both coils, which increases the main inductance of the windings themselves (exclusive of leakage) by thousands of times, due to the permeability of the core material. Thus, the main inductance of the windings, as measured individually externally, is usually at least a thousand times the leakage inductance component.

ELECTROMAGNETIC RADIATION

The relationships described at the beginning of this chapter show that a change in the position of an electrical charge establishes a magnetic field, and that change in a magnetic field produces electrical induction, which is equivalent to an electrical charge. The pioneers of radio spotted this double effect and speculated that if the right conditions could be set up, these two effects would be mutually self-maintaining: the changing electric field would maintain the changing magnetic field and vice versa.

There is just one difficulty in demonstrating this in a relatively small space. It involves a very high velocity of movement or change—the velocity of light, which is approximately 300,000,000 meters per second, or 186,000 miles per second. How do you "chase" such an effect?

In normal electrical circuits, one effect predominates over the other. Currents radiate magnetic fields in which the rate of change is sufficient to generate only minute electrical voltages; or electrical voltages (e.g., in capacitance circuits) generate electrical fields, but the associated currents are insufficient to generate an appreciable magnetic field.

In certain types of circuits, such as radio antennas, conditions are more nearly balanced, so that energy is passed out from the antenna into space (Fig. 8-16).

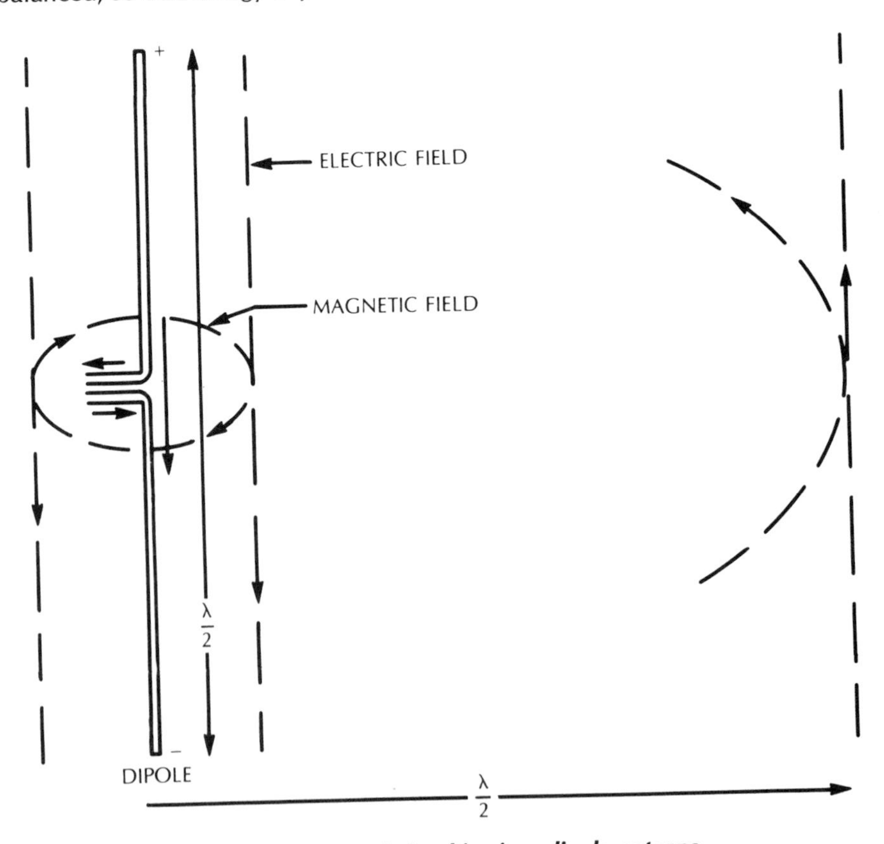

Fig. 8-16. Basic relationships in a dipole antenna.

Assume that such an antenna, called a "*dipole*," has current and voltage fed in at its center. The antenna will be resonant to the frequency at which its length is half a wavelength at the propagation velocity (speed of light). Thus, if the overall length is 1.5 meters, the wavelength is 3 meters, which means the resonant frequency is 300,000,000 divided by 3 or 100 MHz (100,000,000 hertz).

Current builds up due to the charge passed out into the air, so a maximum current exists at the point adjoining the feed-in point at the middle. The progressive charge builds up along the length, so a maximum voltage appears at the ends. Both current and voltage are alternating at a frequency of 100 MHz.

The current also generates a magnetic field precisely in phase with the current. So at this precise frequency, the antenna absorbs energy fed into it because of the maintaining action. The current produces a field which induces the voltage to maintain the changing current. This maintaining action builds up both magnetic and electric fields which continue to maintain one another, outward from the antenna at the speed of light. The magnetic field expands in radiating circles or rings about the antenna as the axis, while the electric field remains parallel to the antenna. Successive alternations follow one another out into space (Fig. 8-17).

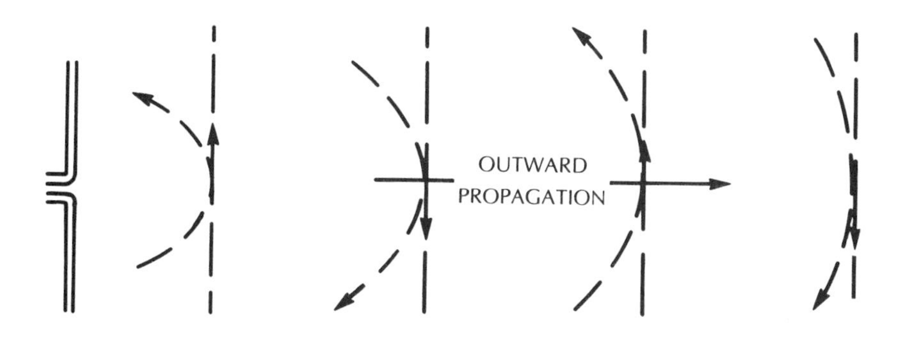

Fig. 8-17. How such a dipole radiates electromagnetic waves.

Notice an important relationship about these electromagnetic waves, as they are called. At all points in their propagation, there is a mutual right-angle relationship: the direction of the electric field, the direction of the magnetic field, and the direction in which the combined field or wave is instantly moving, are always at right angles at a given point in space and a given instant in time.

Close to a dipole antenna, the electric field runs parallel to the antenna and the magnetic field consists of circles with the dipole as center or axis. The direction of propagation is away from the antenna, radially. As the propagation expands, reducing in intensity because the same energy is spread over a larger space, a portion of the wave gets to look like three intersecting directions, all virtually straight lines and mutually at right angles (Fig. 8-18).

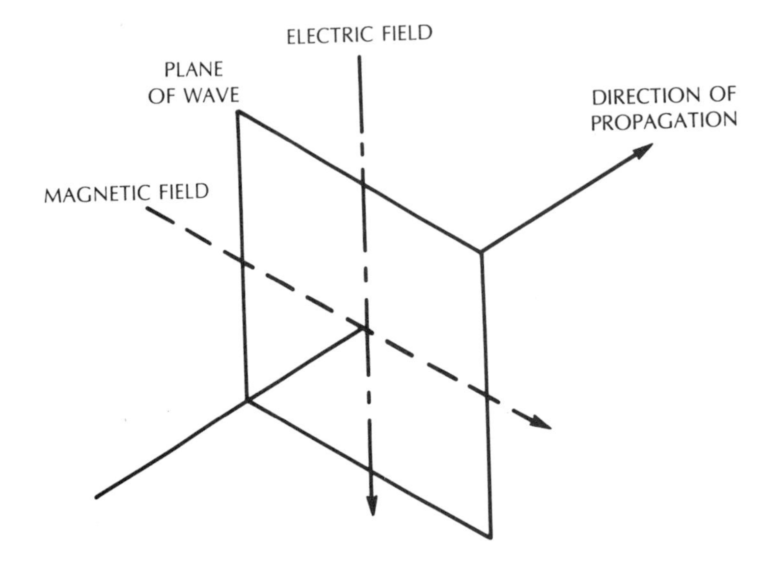

Fig. 8-18. Relationships between electric and magnetic fields, and the direction of propagation, all mutually at right angles.

Another kind of radiator, sometimes also called an antenna, produces a straight, central magnetic field by using a coil to change the energy from the electric circuit to electromagnetic radiation (Fig. 8-19).

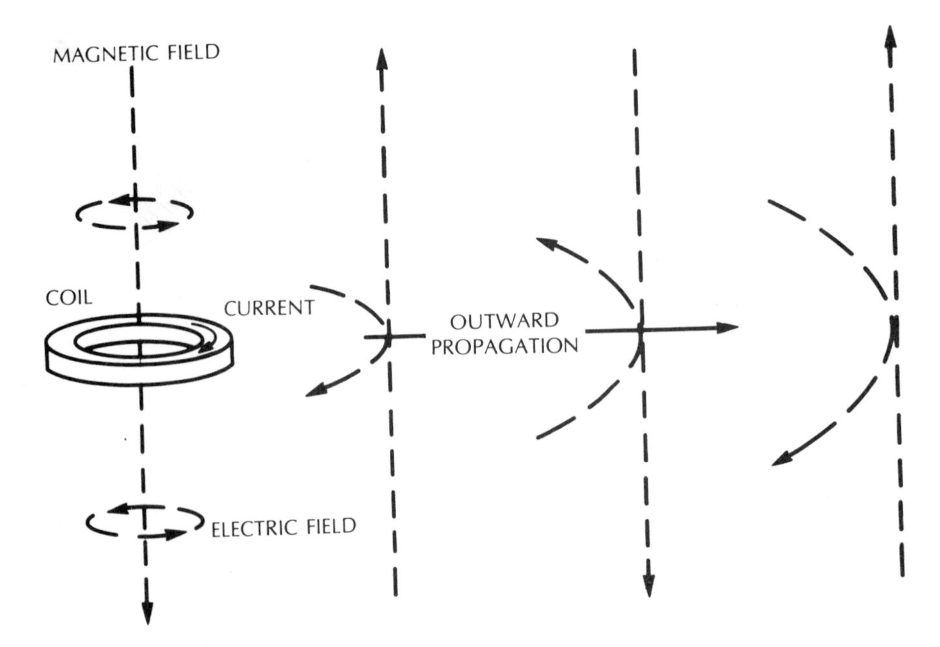

Fig. 8-19. A coil can also generate electromagnetic waves, like a dipole with the roles of electric and magnetic fields interchanged.

This has the same overall effect, except in a region close to the radiating antenna where the positions of the magnetic and electric components of the wave are virtually reversed; the magnetic field is straight and the electric field is circular.

Both types of dipole have a similar radiation pattern, which can also serve to pick up waves coming from the same directions as the transmitted waves are radiated. If the dipole has a vertical axis, the maximum radiation or pick-up is in the horizontal plane, with zero radiation or pick-up vertically. The distribution pattern, in three dimensions, is something like a doughnut (Fig. 8-20).

By using more than one dipole at critical spacing, equivalent to half a wavelength of the frequency radiated or received, which is written as $\lambda/2$, λ (the Greek letter lambda) being the symbol for wavelength, the radiation or reception pattern can be changed. Still there is zero radiation or pick-up vertically. Now the maximum radiation occurs along the line between the dipoles, in either direction, when they are connected anti-phase, that is, so when the top of one is positive the top of the other is negative. And there is zero radiation or pick-up at right angles to this direction, because in that direction, the two dipoles neutralize each other (Fig. 8-21).

By connecting them in phase, so that both are positive or negative together, the reverse applies: maximum radiation or pick-up occurs where the anti-phase produces zero, and vice versa. Use of more than 2 dipoles, consistently connected, can sharpen this directionality pattern.

Another way to change patterns uses reflectors. A single reflector, spaced a quarter wavelength away from the dipole ($\lambda/4$), acts like a mirror behind the dipole. A reflector is a conducting rod, half a wavelength long, with no connection in its middle. Thus you could think of it as a dummy dipole. A dipole has a break in the middle, where current is injected, or detected, while the reflector just connects straight through, and thus "short circuits" its action as a dipole (Fig. 8-22).

This means that the polarized wave virtually disappears right at the position occupied by the reflector. This can be regarded as equivalent to introducing an exactly equal and opposite radiation at the point occupied by the reflector, which radiates outward, just like a dipole radiates outward.

So if the reflector is a quarter wave away from the dipole at an instant when the dipole is positive at its top end, the reflector is receiving a signal that is at the zero point, going from positive to negative. But because this is short-circuited, it does the equivalent of radiating a signal that, at the same instant, is at the zero point, going from negative to positive. So, in another quarter wave, this reflected wave will be positive again.

The reflector, by itself, will stop the radiation "behind" it, and double the intensity of radiation "in front of" it, whatever direction the wave arrives from. So, combined with the dipole, it doubles the intensity of waves arriving from a direction opposite to the reflector, and neutralizes waves coming from "behind" the reflector.

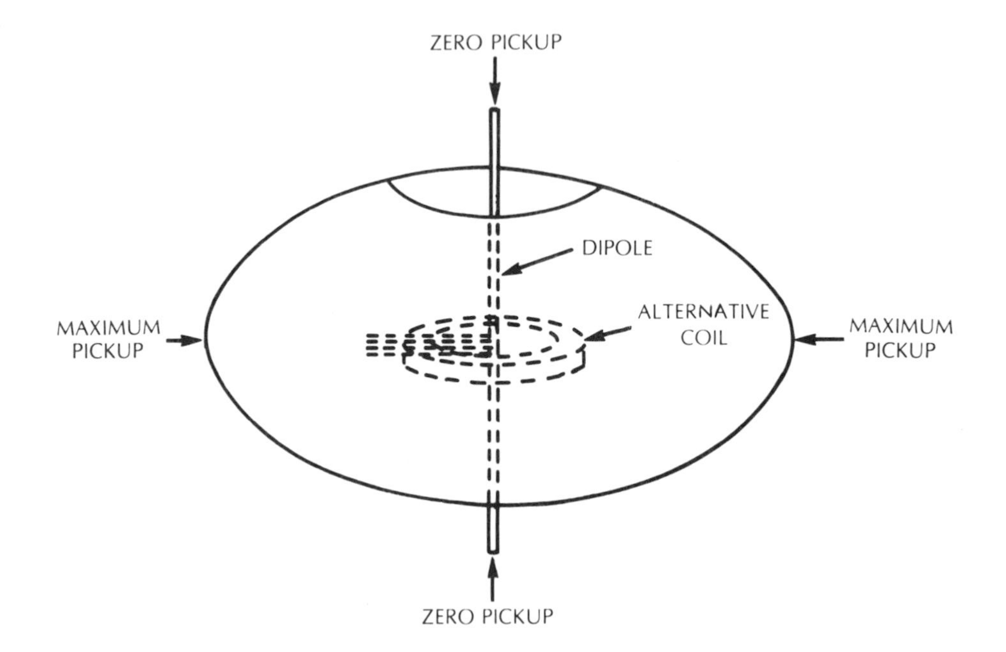

Fig. 8-20. Radiation and/or pickup pattern of either kind of dipole is like a doughnut.

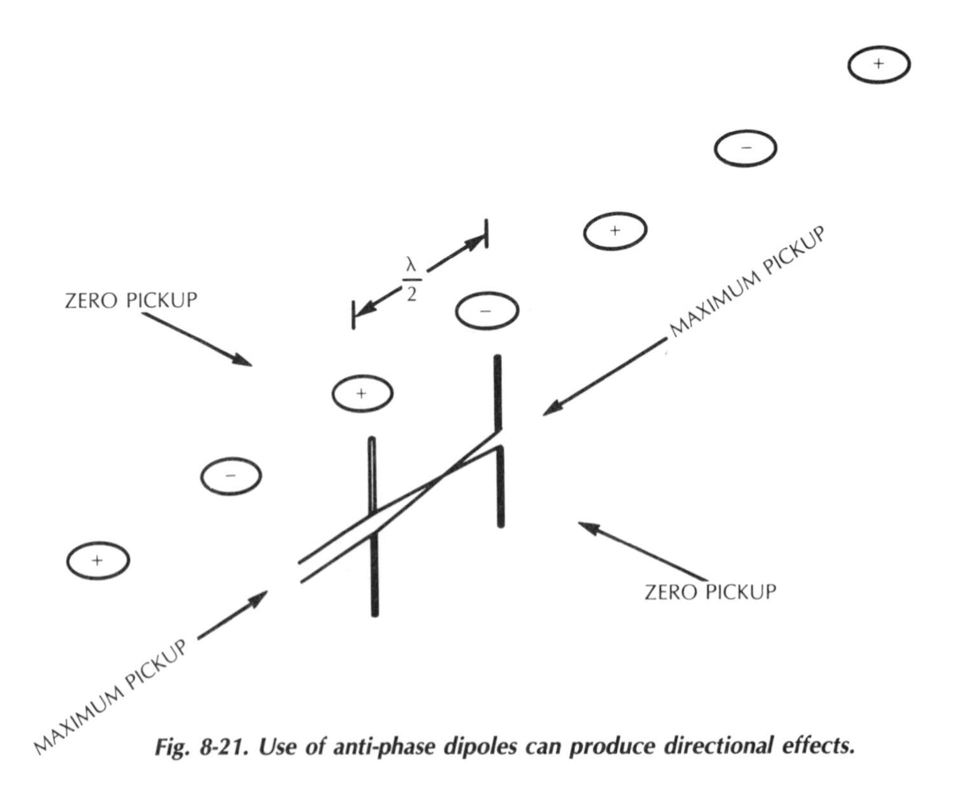

Fig. 8-21. Use of anti-phase dipoles can produce directional effects.

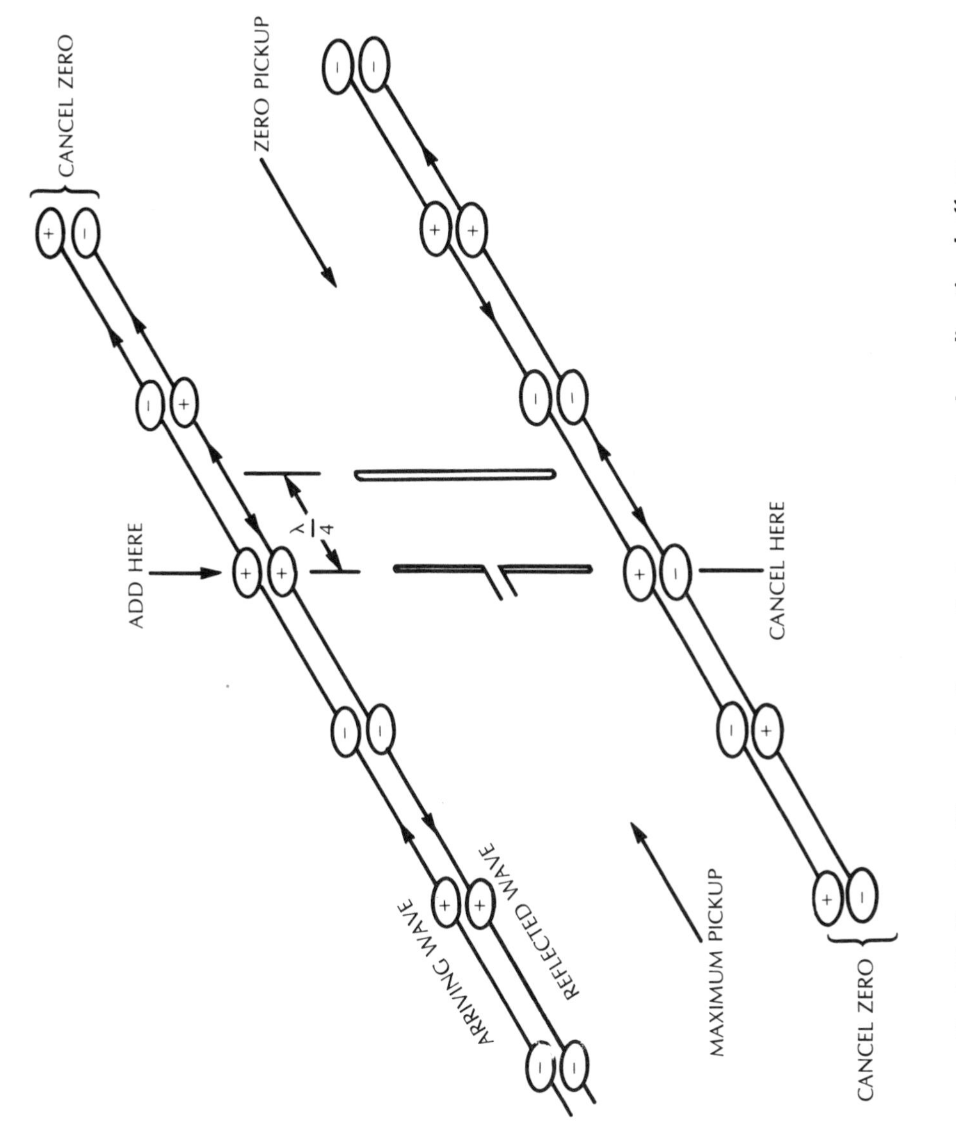

Fig. 8-22. Use of a "dummy" reflector dipole can also produce directional effects.

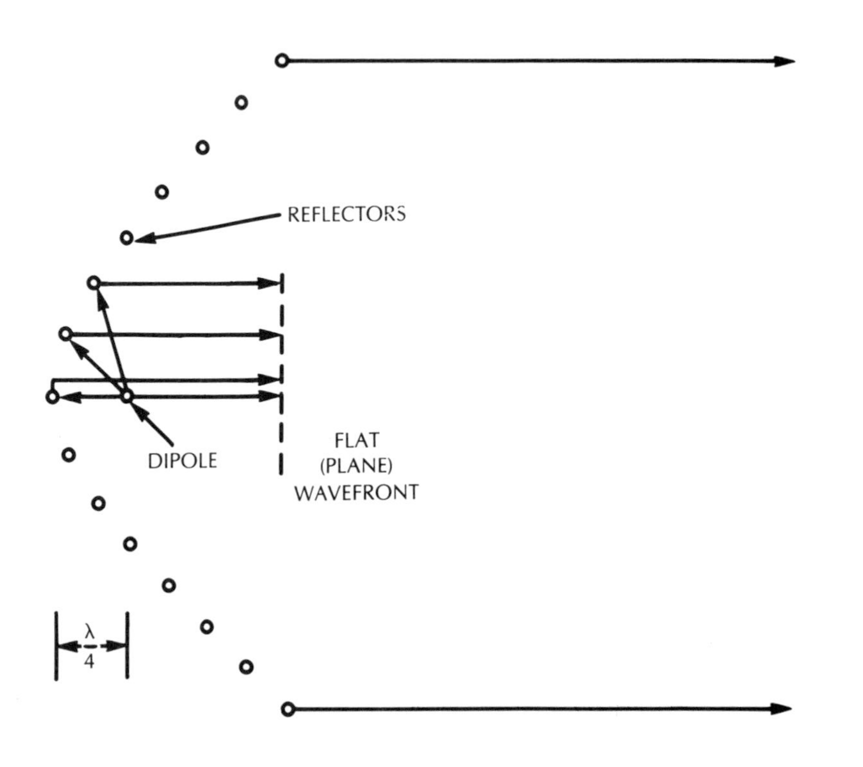

Fig. 8-23. A parabolic reflector can beam a wave, or receive a beamed wave.

This same principle can be used to build up a parabolic reflector (Fig. 8-23). As shown, it represents transmission, but the same effect is obtained on reception. The reflector immediately behind the dipole is a quarter wave from it, so the reflected wave proceeds in the opposite direction in phase with the direct wave. The reflectors at 90° to the dipole are spaced a half wave from it, so again the waves traveling in the desired direction are in phase. And all the other reflectors follow the contour of a parabola.

One more essential ingredient to modern electromagnetic radiation, particularly for the shorter wavelengths (in the centimeter range) is the waveguide, shown in its simplest form at Fig. 8-24. This can best be viewed as a sort of "pipe" that contains electromagnetic waves, by surrounding them with almost perfect reflection. The waves are polarized as shown in Fig. 8-24, which causes circulating currents as shown in the top and bottom inner surfaces of the rectangular "pipe." This gives rise to alternating potentials in the end or side faces of the tube.

Waveguides handle frequencies of shorter than a critical wavelength that depends on the dimensions of the guide, and thus of a higher frequency.

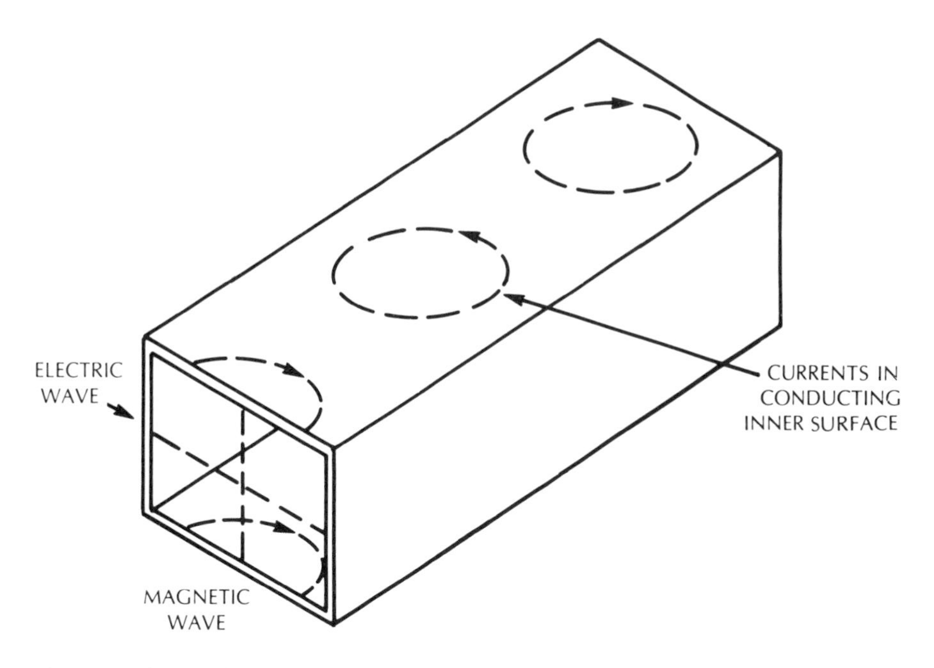

Fig. 8-24. The conducting inner surface of a waveguide serves to contain waves propagated along it.

Wavelengths that are much shorter than the dimensions of the tube travel in the same way they do in open space, but are just confined to the tube. However, where the wavelength approaches the dimensions of the tube, the waves start to "ricochet" along the tube (Fig. 8-25).

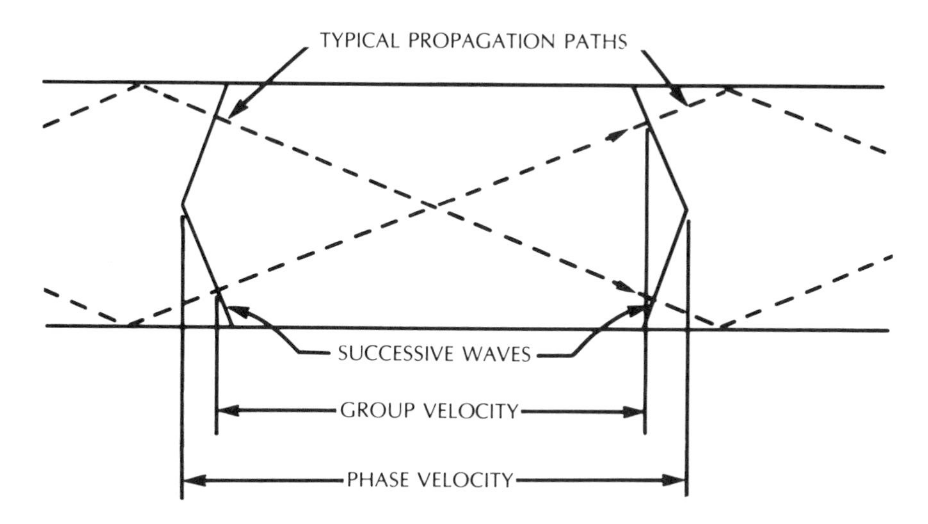

Fig. 8-25. Relationship between different velocities inside a waveguide, which diverge in value as the cut-off frequency is approached.

They can be regarded as traveling in complementary slanting directions that reflect off the walls at an angle depending on the relation between the waveguide dimensions and the wavelength. This gives rise to two different apparent velocities at which the wave proceeds along the tube.

Along the slanting direction, the wave travels at its propagation velocity, the speed of light. But measured along the direction of the tube, this is less than the speed of light, shown in Fig. 8-25 as "group velocity," or the speed at which a given packet of energy at this frequency proceeds along the tube.

At the same time, the combined effect of successive waves, with the complementary criss-crossing, produces a succession of positives and negatives that are further apart along the tube than they are along the direction of propagation, presenting an apparent speed of propagation along the tube that is greater than the speed of light, shown in Fig. 8-25 as "phase velocity."

By using plates spaced apart, using the critical spacing of a waveguide, a waveguide lens can be built, in which the effective speed of propagation through the lens is greater than the speed of light, using the phase velocity (Fig. 8-26).

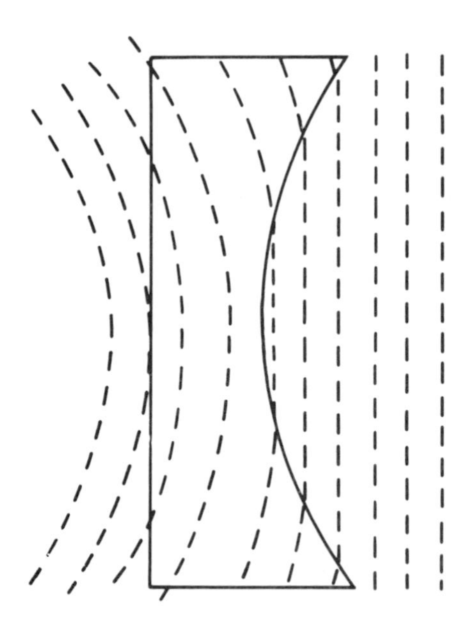

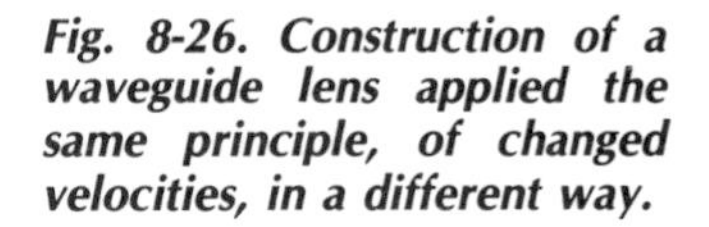

Fig. 8-26. Construction of a waveguide lens applied the same principle, of changed velocities, in a different way.

With this configuration, which is very frequency critical, the shape of the lens, which is made up of many plates shaped as shown and critically spaced apart, the contour of the lens is the opposite of optical lenses. In optical lenses, the light travels more slowly in the glass than in air, so a convex lens is needed to converge rays, slowing down the middle part of the beam more than its edges. With a waveguide lens, the wave is speeded up in effect, as it passes through the lens, so a concave lens, as shown, converges the beam. This gives you an introduction to electromagnetic radiation that will get you started on all the different things that can be done with them. Lasers are an even later development in the use of polarized electromagnetic waves.

Examination Questions

1. A 1000-turn coil has a 120-Hz 20-volt peak (40-volt peak-to-peak) square-wave across it. The up and down portions of the wave are equal. What is the shape of the magnetic flux waveform and what is the peak flux through the coil?
2. The important factor produced in Question 1 is the magnetic field generated. If the following changes in the magnetic field are required, what change in voltage will be needed: (a) same peak flux, but at a frequency of 60 Hz; (b) same peak flux in each direction, but changes one way in 90 percent of the 60-Hz period, changing back again in the remaining 10 percent? (Note: part (b) is an idealized waveform used in television reception.)
3. A coil of 200 conductors (turns) passes through a field of 12,000 gauss and carries a current 5 amperes. If the length of each conductor (turn) within the active field is 150 centimeters, find the total force on the coil.
4. The dimensions of a coil similar to that in Question 3 are given in inches and the flux density in lines per square inch. It has 200 turns and is in a field of 55,000 lines per square inch. The coil carries a current of 2 amperes and has a length per turn of 25 inches. Find the total force produced.
5. A permanent magnet motor-generator, when run at 1000 revolutions per minute, generates 200 volts dc. When it is connected to a dc voltage of 200 V, it runs at 950 revolutions per minute and draws a current of 0.5 amp with no mechanical load coupled to it. How much power does it take to run the machine idle at 950 revolutions per minute?
6. Of the power taken in Question 5, how much is lost in the resistance of the machine, and how much is mechanical or of some form other than resistance loss?
7. A mechanical job requires precisely 5 horsepower of effort, which is supplied by an electric motor. The machine runs 8 hours a day for a 5-day week. The machine doing the job is old and works at 65 percent efficiency. It is proposed to change it for one of 85 percent efficiency. Find the saving in cost resulting from installing the new machine, if the factory buys electricity at 1¢ a kilowatt hour (the energy drawn by 1 kilowatt during 1 hour).
8. For another job, using the same choice of motors as in Question 7, the motor is only working for approximately 20 percent of the time. The rest of the time it is idling, ready for use when needed. When idling, the older machine takes 1.5 kilowatts of power, the new one only 500 watts. Find the cost saving by changing motors for this job.
9. What will be the total length of a dipole for use with a transmission frequency of 300 MHz?
10. From the information given about how a reflector is used with a dipole, deduce the effect of using an antenna that is (a) precisely half, (b) precisely double the frequency for which it is designed. Consider the effect on both efficiency and directionality.

9
Reactance

Two basic components—coils and capacitors—provide a property known as reactance. While such reactances are never quite perfect, an understanding of the idealized property helps visualize how these components function. Actually, coils and capacitors can be regarded as providing pure reactances, plus other effects that are unavoidable in practical components. Primarily, in the basic sense, a reactance is an energy-storing device, and the use of this stored energy in various ways produces effects described as "reacting," which differs from "resisting" (the property of resistance).

INDUCTANCE

A coil provides what is commonly known as a positive reactance in regard to the relationship between voltage and current in an ac circuit. The inductance stores energy in the form of a magnetic field, which is maintained by an electric current associated with it.

In a perfect inductance, the coil would have zero resistance, so current is not limited by the inductance itself. With a steady dc flowing and with zero resistance, no voltage drop would appear across the terminals of a pure inductance—voltage zero, current steady at some value. This is the condition of constant stored energy in the inductance.

With no current flowing, the stored energy is zero. To get current flowing, a positive voltage is applied, which causes the current to rise as long as the voltage is applied. When the voltage drops to zero, the current ceases to rise and settles to a steady value (Fig. 9-1). Therefore, the rate of current rise depends on the voltage applied.

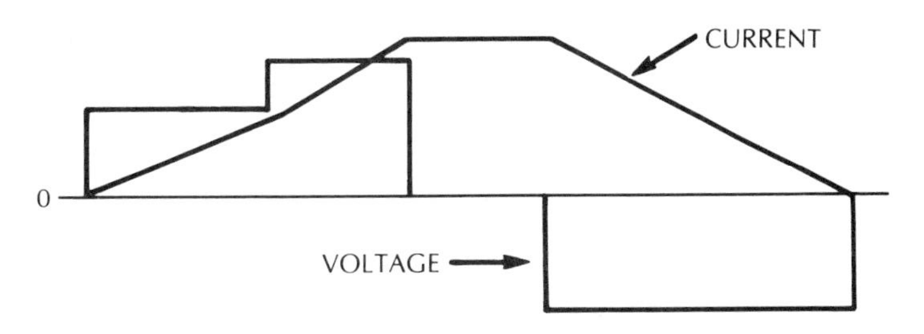

Fig. 9-1. Sequence showing that change of current (slope of the graph) is proportional to voltage, in an inductance.

Energy in an electrical circuit is a product of voltage, current and time. So as long as voltage is applied and current is rising, more energy continues to be added to the store in the inductor, measured in watt-seconds or whatever unit is appropriate. When the voltage drops to zero and the current becomes steady, the energy input ceases and the inductor stores the precise amount it has until the current changes again.

The unit of inductance is the henry, the value where applying 1 volt will cause the current to rise at the rate of 1 ampere per second. So, if 1 volt is applied to a 1-henry inductance for 10 seconds, the current will rise to 10 amperes. The average current during that 10 seconds will be 5 amperes, so the energy stored into the inductor will be 5 × 10 or 50 watt-seconds.

If 10 volts is applied for 1 second, the final current will also be 10 amperes. The average current during that 1 second will also be 5 amperes, but the voltage applied is 10 instead of 1, so the energy stored is 5 × 10 × 1 or 50 watt-seconds again. It is fairly evident that a given current in an inductance of specified value always represents the same stored energy, however that current is reached (Fig. 9-2).

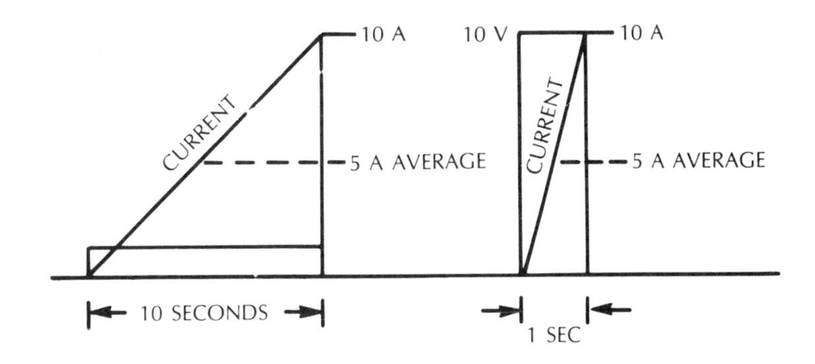

Fig. 9-2. The energy stored in an inductance is the same, whether 10 volts produces 10 amps in 1 second, or 1 volt produces 10 amps in 10 seconds.

The energy stored in the inductance is returned when the current is reduced to zero again. Suppose, for example, that some kind of load, like a resistance, but a load designed to produce a constant voltage drop (such as a zener diode, described in Chapter 11), is connected across the inductance at the same time as the source of current is removed (Fig. 9-3). Now, the current that has been flowing through the inductor continues flowing through the diode until the voltage drop reduces the current to zero again. If the zener voltage is 5, this will also be the voltage across the inductance as it works to reduce the current. It will take 2 seconds for the current to fall to zero. The average current will be 5 amperes, the voltage 5 volts and the time 2 seconds; the result is 5 × 5 × 2 or 50 watt-seconds, which will be dissipated in the zener diode.

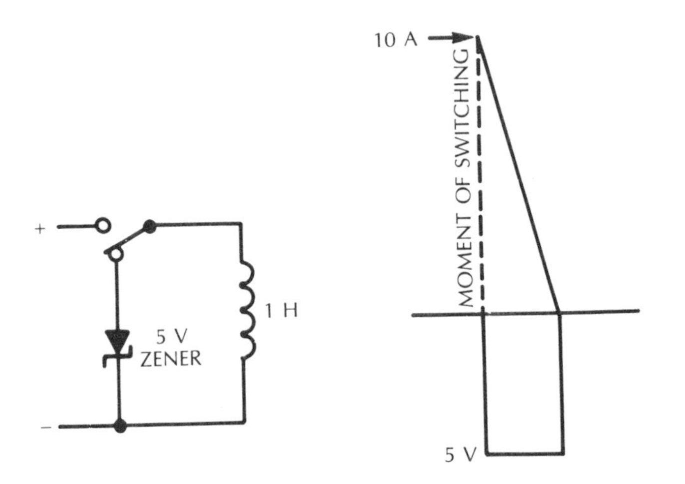

Fig. 9-3. A method of dissipating energy in a manner that allows a constant voltage to exist until the energy is dissipated, resulting in a straight-line rate of current decrease.

An alternative provision for dropping the current in the inductor could be a silicon diode, so its forward conduction direction becomes operative when the source of current is removed. When the voltage is applied to build up the current (Fig. 9-4), the diode is not conducting at all.

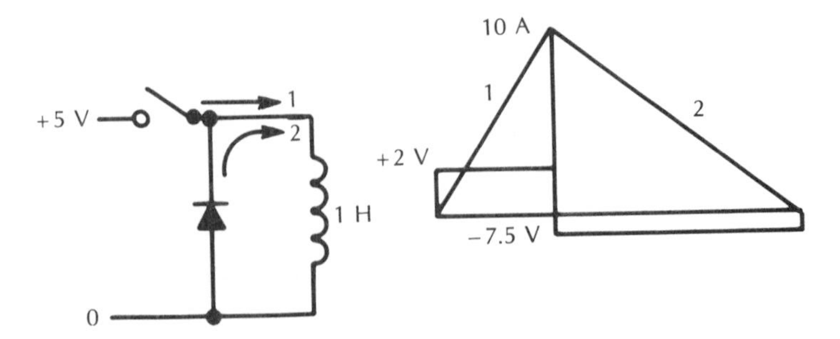

Fig. 9-4. An alternate dissipation method to that illustrated in Fig. 9-3, using a silicon diode.

But when the current supply is removed, the current already established in the inductance has to continue flowing in the diode through its conducting direction. A silicon diode has a forward voltage drop of about 0.75 volt. So a 10-amp current in a 1-henry inductance would take 10 divided by 0.75 or 13.3 seconds to drop to zero. The average current will be 5 amps, voltage 0.75, time 13.3 seconds, which produces 5 × 0.75 × 13.3 or 50 watt-seconds, as before.

CAPACITANCE

Where an inductance stores energy in the form of a magnetic field, a capacitance stores energy in the form of voltage. The basic measure of capacitance is the farad, which says a coulomb stored (1 ampere flowing for 1 second makes a coulomb) will produce 1 volt across 1 farad. The practical unit of capacitance is the microfarad (mfd), which is one-millionth of a farad. One milliamp flowing for 1 second into a capacitance of 1 mfd will produce a voltage of 1000 volts.

Energy is stored by these voltages in a way similar to that which an inductance stores energy as current. If a capacitor of 1 mfd is charged to 1000 volts by feeding 1 milliamp into it for 1 second, the average voltage across the capacitor at which the 1 milliamp has to be supplied is 500 volts. So the energy stored is 500 milliwatt seconds or 0.5 watt-seconds.

Suppose a larger capacitor, say, 500 mfd with a 200-volt rating, is charged to that voltage by applying a current of 10 milliamps for 10 seconds. The charge is 100 millicoulombs, or 0.1 coulomb, which agrees with multiplying 500 by 200 and dividing by a million (because the 500 is in microfarads). Now, the average voltage will be 100 volts, the current 10 milliamps and the time 10 seconds, yielding an energy storage of 100 × .01 × 10 or 10 watt-seconds.

If the same capacitor were charged only to 100 volts and the 10 milliamp current was applied for 5 seconds, the average voltage would be 50 volts, resulting in an energy storage of 50 × .01 × 5 or 2.5 watt-seconds. Just as the energy stored in an inductance is proportional to the square of the current storing it, so the energy stored in a capacitor is proportional to the square of the storing voltage for that value of capacitor.

REACTANCE

From these basic energy-storage concepts of inductance and capacitance as pure elements, we can develop a concept of the quantity known as reactance as it relates to each component. In the foregoing, each is defined in terms of steady rates of change: the rate of change of current in an inductance, due to a steady voltage across it, and the rate of change of voltage across a capacitor, due to a steady current into it. While some circuits can be described, explained or calculated in such terms, these components are much more often used in circuits where combinations of ac and dc are present, and where the ac tends to be of a sinusoidal waveform or of a more complicated waveform that can be analyzed in terms of multiple sine waves (of different frequencies).

Viewed in these terms, an inductance (pure) cannot sustain a dc voltage across it. If the average (dc) voltage is other than zero, the associated current continues to rise indefinitely. Similarly a pure capacitance cannot pass a dc current. If the average current is other than zero, the voltage across the capacitance must continue to acquire an ever-accumulating potential (voltage).

Looking at the same properties in a different way, either component can, under different circumstances, separate ac from dc. A dc current through an inductance does not produce any corresponding voltage drop across it, so an inductance produces voltages corresponding to ac components of current, while ignoring dc. A capacitor will pass ac currents due to ac voltage components, while ignoring dc voltages, which merely produce a constant charge across the capacitor. These properties are useful and constitute the principal reason, very often, why such components are found in a circuit.

Having eliminated the dc component from among those affected by a reactance, the next step is to examine what happens to ac components a little more closely. First take inductive reactance. By definition, with an inductance of 1 henry, a current that is changing 1 ampere per second will produce a reactive voltage of 1 volt. The 1 volt is present at any instant when the rate of current change is 1 ampere per second. Now assume that the current is 1 ampere (peak value), alternating at 100 Hz. This means it goes through a sinusoidal fluctuation from zero to 1 ampere in one direction, back to zero, to 1 ampere in the opposite direction, and back to zero again, all within one one-hundredth second (Fig. 9-5).

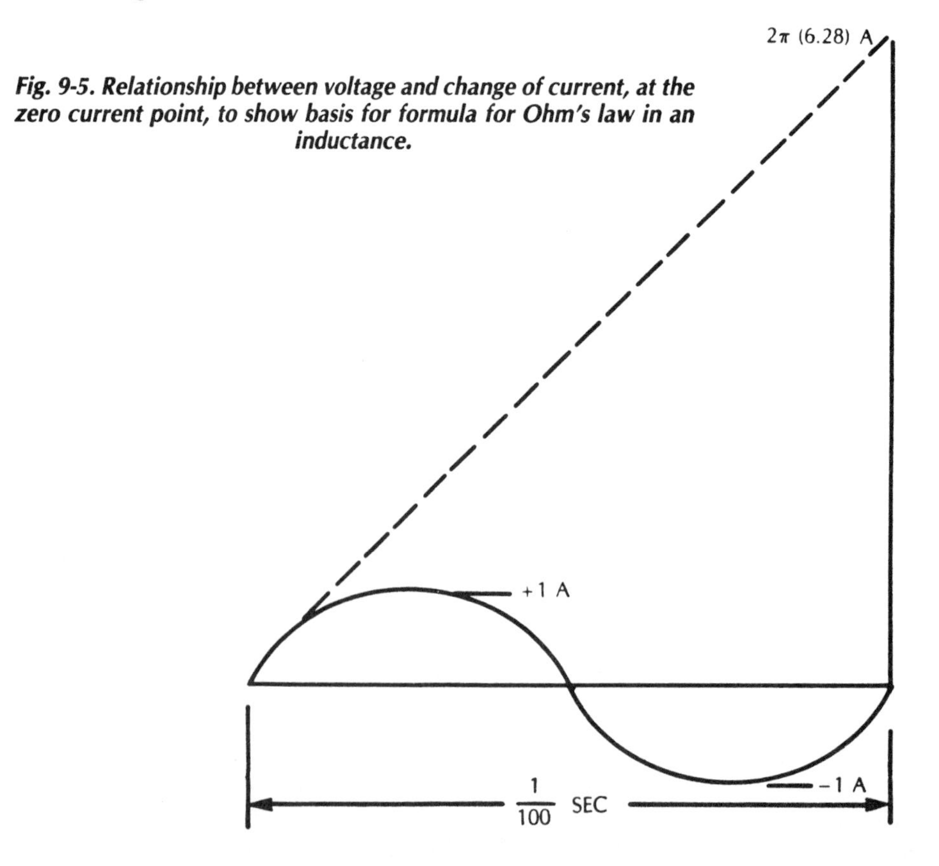

Fig. 9-5. Relationship between voltage and change of current, at the zero current point, to show basis for formula for Ohm's law in an inductance.

The voltage across a reactance is proportional to the rate at which current changes at every instant. When current is at its full 1 amp value in either direction, it is momentarily not changing. The maximum rate of change occurs as the current value passes through its zero value. At this moment it is changing at a rate that, if the rate persisted for an entire one one-hundredth second, the current would grow to 2 π times the actual maximum reached in either direction from zero (Fig. 9-6). So the peak voltage is 2 π times the frequency, times the inductance in henries, times the peak current, in this case 2 π $\times$ 100 $\times$ 1 or 628 V.

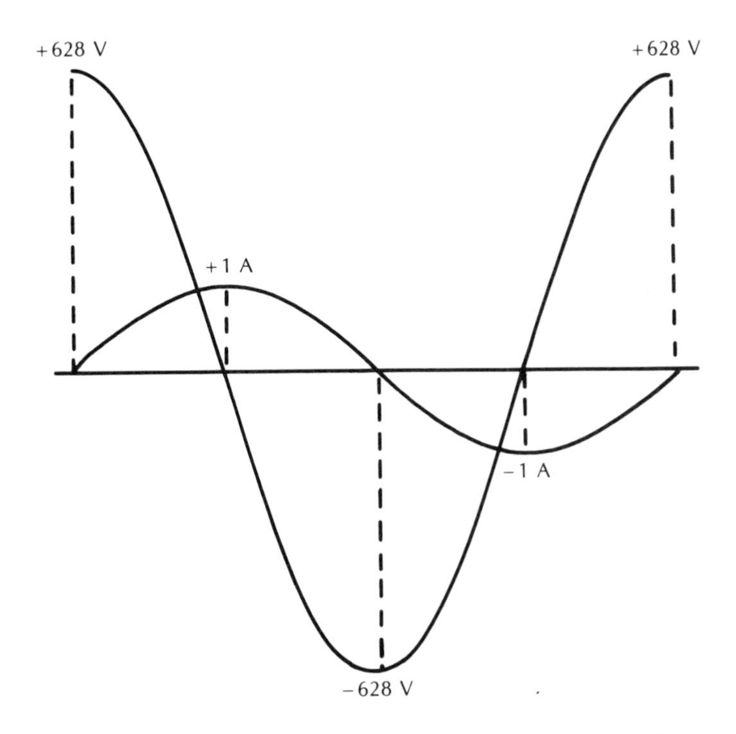

Fig. 9-6. Relationship between voltage and current in an inductance.

As the waveform is taken to be sinusoidal, it makes no difference to the ratio between the voltage and current waveform values if different references are used, such as the rms values instead of the peak values. Where reactance differs from a resistance is that it reveals a ratio between voltage and current that is never instantaneous—it gives the ratio of the sizes of the waveforms. But each waveform is zero at the other's maximum values, so this ratio does not exist at any instant in time.

A similar relationship happens with a capacitive reactance. Current is proportional to the change of voltage. With a steady voltage, no current flows. As we calculated, using 1 mfd as the value, a current of 1 milliamp results in a voltage change rate of 1000 volts per second. Suppose again that the current and voltage have a frequency of 100 Hz. At the moment when the current is

1 milliamp peak in either direction, the voltage change rate will be 1000 volts per second. If this change rate continued for an entire period of one one-hundredth second, the voltage would reach 10 volts (1000 × one one-hundredth).

But during that period, this current falls to zero, rises to 1 milliamp in the opposite direction, falls to zero again, and rises to its original 1 milliamp value—all within that one one-hundredth second. So the maximum voltage reached will be 10 volts divided by 2 π, or about 1.59 volts (Fig. 9-7). Thus, we have a starting point for calculating the reactance of inductances and capacitances. An inductance of 1 henry, with a current of 1 amp at 100 Hz, will produce 628 volts. A capacitance of 1 mfd with 1 milliamp flowing at 100 Hz will produce 1.59 volts.

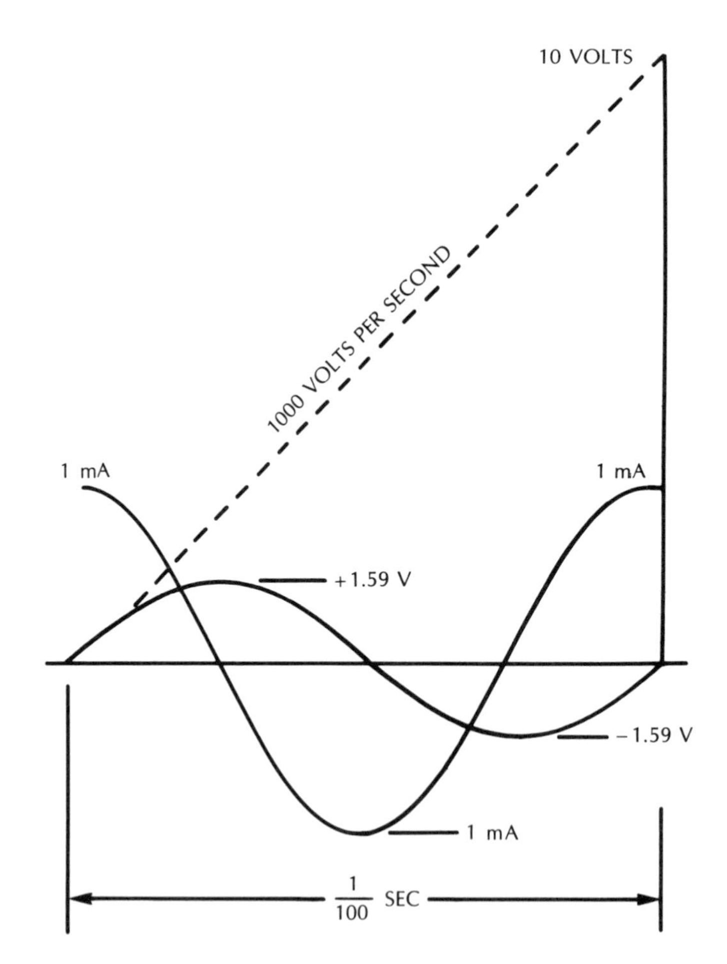

Fig. 9-7. Relationship between current and change of voltage, at the zero voltage point, to show basis for Ohm's law in a capacitance.

In resistances, a ratio of 1 amp and 628 volts represents a value of 628 ohms. Using a similar ratio, an inductance of 1 henry has a reactance of 688 ohms at 100 Hz and a capacitor of 1 mfd has a reactance of 1590 ohms (multiplied by 1000, because the current was 1 milliamp, instead of 1 amp) at 100 Hz.

Now, how does frequency affect reactance? With inductance, the higher the frequency, the more rapidly current changes; thus, the higher is the voltage generated by the rate of change. Therefore, reactance is proportional to frequency. With capacitance, the higher the frequency, the shorter the period during which the voltage has to change (for a given current value, which determines the rate, in volts per second, of voltage change) so reactance is inversely proportional to frequency. The formulas are:

$$X_L = 2\pi fL \tag{13}$$

$$X_C = \frac{10^6}{2\pi fC} \tag{14}$$

In these formulas, L and C are the symbols for inductance in henrys and capacitance in microfarads, respectively, and f is for the frequency in hertz. X_L stands for the reactance of an inductance and X_C for the reactance of a capacitance, both in ohms.

Compare the relationships represented in Figs. 9-6 and 9-7. In each, reactance defines the relationship or ratio between voltage and current, even though they do not coincide in time, just as resistance defines a like relationship or ratio where coincidence in time is absolute; the time relationship is the basis for designating reactance as positive or negative.

This may be explained several ways. The important thing is that they be given opposite signs, so that formulas will work mathematically. Here is one way to view the matter.

In inductive reactance, the current waveform follows the voltage waveform by a quarter-wave delay (90 degrees). In a capacitive reactance, the current waveform *precedes* the voltage waveform by a quarter wave. Viewing voltage as the *cause* and current as the *effect* (which is arbitrary, admittedly), inductance has a positive time delay, while capacitance has a negative delay, which is an advance. Viewed this way, current *anticipates* voltage in a capacitive reactance.

Since the quantities for both resistance and reactance (in ohms) are a measure of voltage caused by current, the above explanation may give some people difficulties. If current is viewed as the cause and voltage as the effect, then an advance, a sort of positive "kick" as given by an inductance, is positive, and delay, introduced by a capacitance, is negative. The important thing is to be consistent—either explanation yields the same result.

TIME CONSTANTS

The foregoing considers an inductance and capacitance under two conditions:

- An assumed constant rate of change of current or voltage, respectively, which results in a constant voltage or current;
- An assumed alternating wave of sinusoidal form and a known frequency.

There is another and, in some senses more basic set of relationships derived from the use of reactances—the relative time constants.

Both inductances and capacitances are normally used in circuits that also contain resistance, whether that resistance is part of the same component that provides the inductance or capacitance or whether it is part of the external circuit into which the component connects and into which it operates.

A constant rate of current change in an inductance to yield a constant voltage and a constant current in a capacitance to yield a constant rate of voltage change are really unreal assumptions, except in very special circuits designed to have such characteristics. By far, the more usual kind of circuit includes inductance and-or capacitance in conjunction with ordinary resistances.

If a resistance and inductance in series are applied across a dc voltage (Fig. 9-8), we can use the symbol E for the total dc voltage applied throughout, with E_R and E_L representing the voltages across the resistance and inductance at any instant and I the current at the same instant. When the dc is first connected, no current will flow. For this reason, no voltage will be dropped across the resistance because voltage is proportional to current in resistance. Because of the full dc voltage across the inductance element at this moment, current will start to rise in it, as calculated by the relationship given earlier.

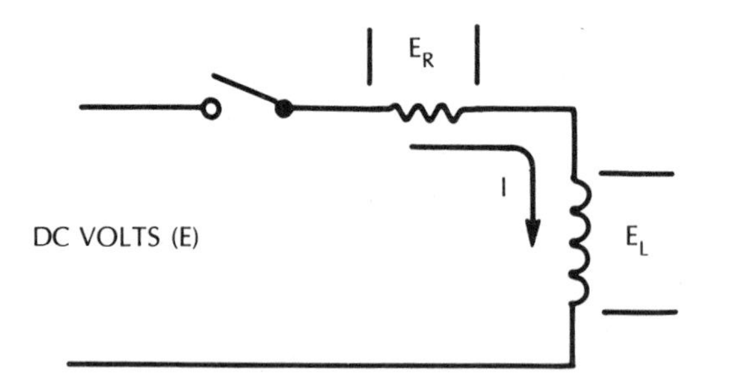

Fig. 9-8. Simple arrangement for figuring the time constant of an inductance-resistance combination.

But as soon as current starts to build so will a voltage drop across the resistance to correspond (Fig. 9-9). Since the total voltage (E) remains at the same dc value, this means the voltage across the inductive element will decrease by a like amount from the full initial value. As the voltage across the inductance decreases, so does the rate of current rise, which means that the initial rise in voltage across the resistance does not continue at a uniform rate; instead it curves over.

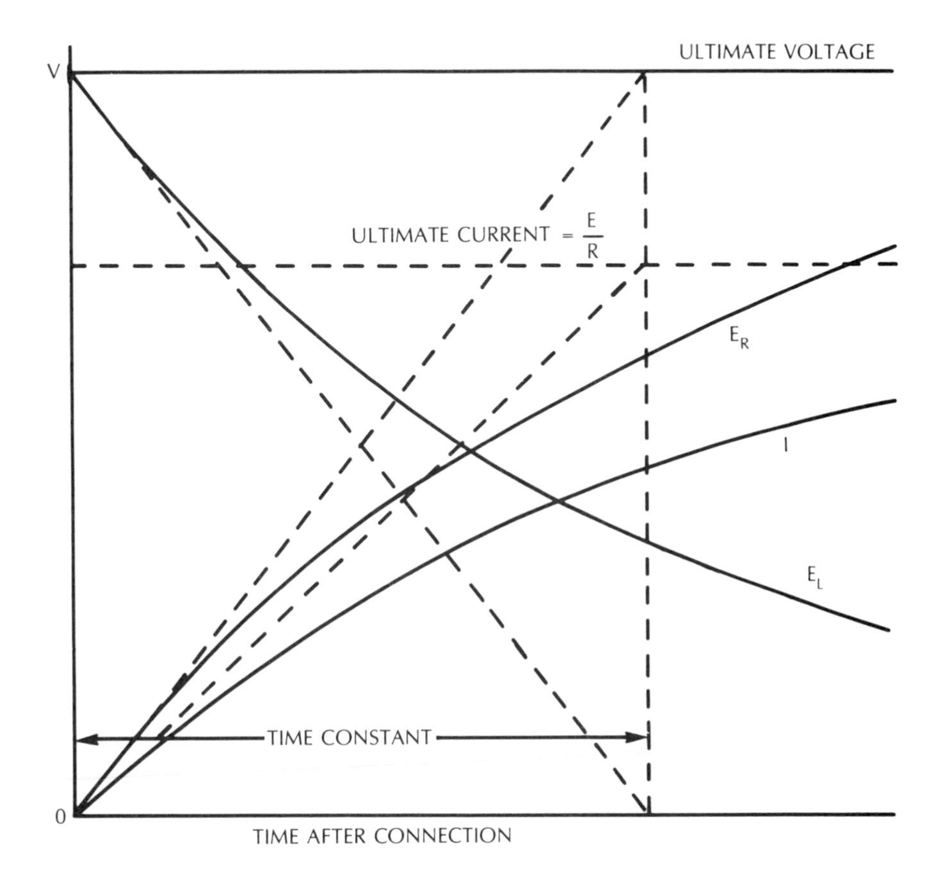

Fig. 9-9. Relationship between voltage and current in an inductance resistance combination.

Eventually, the circuit comes to a steady state where the inductance no longer plays a part. The current is controlled by the resistance only, and, being constant, no voltage appears across the inductance; all of it appears across the resistance. Only if the applied voltage changes or is disconnected does this change. Then the inductance suddenly produces a voltage that tries to keep the same current flowing, as discussed earlier.

The exact shape of the current and voltage curves before the steady state is reached is called an exponential curve, which is somewhat complicated to calculate (unless you use a computer that is designed for that calculation). But there is a dimension about the curve that is of some importance in specifying circuit behavior—its time constant.

Suppose, by some means, the rate of current change when the voltage is first applied is maintained consistently for a long enough period to bring the current up to the point where the total supply voltage is dropped across the resistance. Then the time constant is the interval required for that current to be reached at this steady rate (Fig. 9-10).

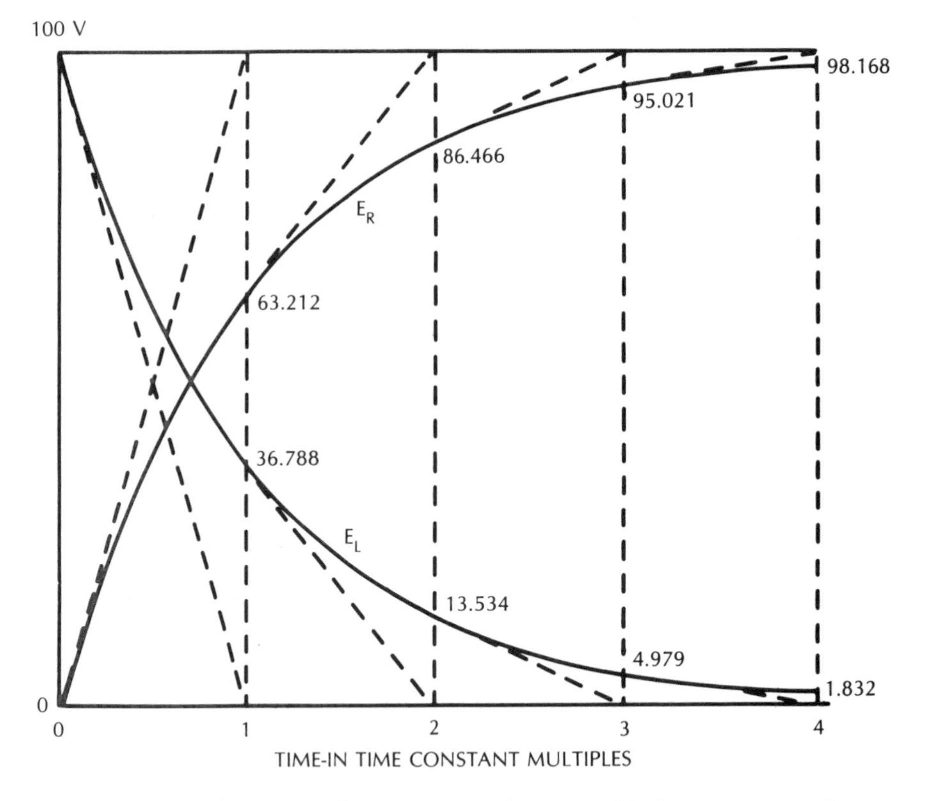

Fig. 9-10. Relation between voltages across resistance and inductance over a longer period of time.

But because the rate of current rise decreases during this interval, the value of the current actually attained will be only 0.63212 of the final value. The easiest way to understand this figure is by its relationship to the "other part," the portion of the original voltage appearing across the inductance as time goes by. This voltage starts at the full dc value and, if the rate of current growth were constant for the time constant period, would stay at that voltage; then, it would suddenly drop to zero, a situation difficult to create in a practical circuit. Instead, it drops steadily right from the start, as indicated by the curved graph line. It reaches a value 1 over ϵ of its original in the time constant period. The quantity ϵ is the base of natural logarithms and all exponential functions. It has a value calculated by adding together the terms of an infinite series of the form:

$$\epsilon = 1 + \frac{1}{1} + \frac{1}{1 \times 2} + \frac{1}{1 \times 2 \times 3} + \frac{1}{1 \times 2 \times 3 \times 4}$$

$$+ \frac{1}{1 \times 2 \times 3 \times 4 \times 5} + \frac{1}{1 \times 2 \times 3 \times 4 \times 5 \times 6} + \dots \quad (15)$$

$$= 1 + 1 + \frac{1}{2} + \frac{1}{6} + \frac{1}{24} + \frac{1}{120} + \frac{1}{720} + \dots$$

which to five decimal figures approximates:

$$1 + 1 + 0.5 + 0.16667 + 0.04167 + 0.00833 + 0.00139 + \dots$$

Adding these figures successively shows the successive approximation:

1

2

2.5

2.66667

2.70834

2.71667

2.71806

The further the series is taken, the more closely it approaches the true value of ϵ which, to 5 decimal places, is 2.71828.

The reciprocal of ϵ, also to 5 decimal places, is 0.36788. This means that if the dc voltage, which initially appears across the inductance, is 100 volts, then after precisely one time constant period it drops to 36.788 volts. After a second, equal period of time, it will drop to a value obtained by dividing by ϵ (or multiplying by 1 over ϵ) again, which gives 13.534 volts. After the third time constant interval, it is divided by 2.71828 again to yield 4.979 volts. The fourth period drops it to 1.832 volts, the fifth to 0.674 volt, the sixth to 0.248 volt, and so on. Ten times the time constant interval reduces the original 100 volts to 0.005 volt (nearest third decimal figure).

Figures for various powers of ϵ can be obtained from exponential or natural logarithm tables, which is where the above figures were taken. If much of this kind of work is needed, a computer with such calculations built in can be used.

But this gets a little beyond the basics of electronics. The figures given above, plotted in Fig. 9-10, show how the time constant thing works for inductance.

While in theory the reduction goes on forever, you can see that it gets smaller so fast that a few of them reduce the residue to quite a small piece of the original value. The voltage appearing across the resistance and the current through both can be found by subtracting these values from the total 100 volts dc. Thus, successive voltages across the resistor at time constant intervals are zero (at the start), 63.212, 86.466, 95.021, 98.168, 99.326, 99.752; the tenth is 99.995, which differs from 100 by only 0.005 percent.

The value of the time constant is easy to calculate, given the operative values of inductance and resistance. The formula is:

$$T = \frac{L}{R} \tag{16}$$

where T is in seconds, L in henries, and R in ohms. As with other calculations, different units can be used. Thus, an inductance of 1 henry in series with a resistance of 10 ohms will produce a time constant of one tenth second or 100 milliseconds.

A similar thing happens with a resistance and capacitance. If the two in series are connected across a voltage (Fig. 9-11), with the capacitor initially uncharged (having no voltage across it), maximum current starts to flow when the connection is made (unlike the inductance combination where there was zero current at the start).

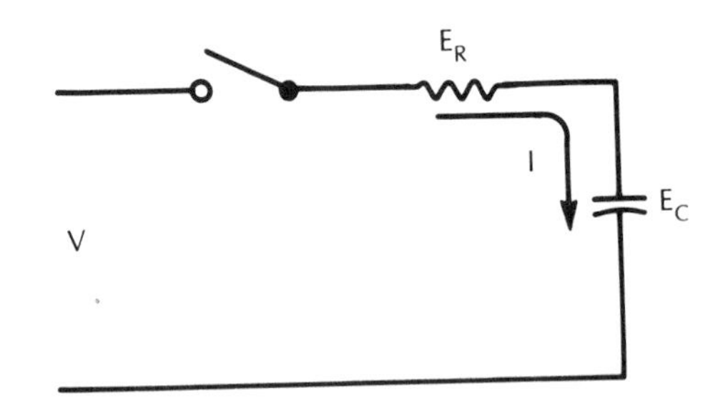

Fig. 9-11. Arrangement for considering resistance and capacitance combination.

Voltage starts to rise across the capacitor at a maximum rate and falls off as the charge current drops, because the voltage drop across the resistance is decreasing (Fig. 9-12).

The time constant for a capacitance and resistance combination is found by the formula:

$$T = RC \tag{17}$$

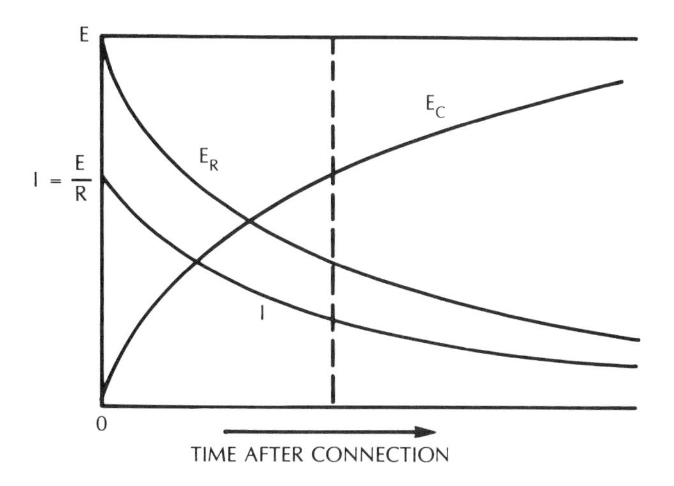

Fig. 9-12. Voltage and current curves associated with resistance capacitance combination.

where T is in seconds if R is in ohms and C is in farads. If R is in ohms and C is in microfarads, then T is in microseconds. Thus, a resistance of 1 megohm (1,000,000 ohms) with a capacitance of 1 mfd yields a time constant of 1 second.

INDUCTOR AND CAPACITOR TYPES

So far, we have discussed ideal inductances and capacitances. To be able to use this information in actual electronic circuits, we need to know something about the actual inductors and capacitors that can be bought (or made) and put into circuits.

The simplest form of inductor is a coil of wire. It is not a pure inductance, because it possesses resistance as well as inductance. The resistance of an inductor can be regarded as resistance in series with its inductance. The simple inductor is either too big for modern circuits or it has far too much resistance to be practical, so most practical inductors have a magnetic core of some kind to accentuate the inductance, so the relative resistance is lower.

If a coil without a core has an inductance of 0.1 henry and a resistance of 100 ohms, putting a core into it raises its inductance to 10 henries—an improvement of 1000 times. By changing the coil so it has about one thirtieth of the number of turns wound into the same space, and by using a correspondingly heavier wire gauge (almost 6 times the diameter), the inductance of the new coil would be 0.1 millihenry (thousandths of a henry) without the core. Putting the core in brings the inductance up to 0.1 henry and reduces the resistance to 0.1 ohm, due to the heavier gauge wire and fewer turns.

With an air-core inductor (one without a magnetic core) the presence of dc in the winding, in addition to ac and inductive voltage due to the ac, has no influence on the ac effects. But where a magnetic core is used, the properties of the magnetic material may be changed by the presence of dc magnetization, where a dc component is present.

In inductors designed to function with dc present in the windings, the core is usually designed to control the dc magnetization so that the performance on ac is optimized for that level of dc. This is a process involving more advanced techniques than are discussed in this book. For the purposes of this discussion, an inductor with a dc rating should be used at that rating for utilization of its designed properties.

If an inductor is used at a current value lower than its rated dc, it will have a somewhat higher inductance than the value that would exist at the rated dc, but it will not be as high a value as could be built into the same component if it were optimized for the lower dc value. If a coil is used at more than its rated dc, inductance very rapidly diminishes.

Cored inductors may also have a maximum ac rating, which means that the inductor will saturate if used in a circuit where the ac is beyond that rating (usually given in volts at a specified frequency). Again, the desired performance will not be realized. Saturation density corresponds with a voltage that rises proportionally to a given frequency. This means that a maximum voltage of, say, 200 at 60 Hz will correspond with a maximum of 400 volts at 120 Hz. We're assuming that the maximum voltage is set by the inductance properties of the component. Sometimes it may be a flashover limitation, due to close spacing of terminals or something. In this case, changing the frequency will not increase the flashover voltage. Usually, this situation applies with higher rated voltages, in the thousands of volts, for example.

Capacitors also come in various forms. Historically, the best capacitors used natural mica with or without an oil filling as a dielectric (insulation between plates). But these were very bulky by modern standards for any given capacitance value and voltage limits. A 1-mfd capacitor occupies the better part of a cubic foot of space!

Modern materials (synthetic plastics, and electrolytic films), formed by electrical processing, allow very large capacitances to be produced by comparatively small capacitors. Many are polarized, which means they must be operated with a dc polarizing voltage to maintain the film inside the capacitor. This voltage must not exceed a specified limit, or the capacitor will start to leak seriously and probably will destroy itself. On the other hand, an adequate voltage must be provided or the capacitor will deteriorate.

With many electrolytic capacitors, the polarity of the applied dc is important: they must be connected so the positive dc goes to a specified terminal and the negative to the other. Other electrolytics are designed to be reversible. But reversible does not necessarily mean they will operate on ac only; they may not be capable of "reversing" quite that rapidly!

A related factor here is the voltage that accompanies the ac, which depends on frequency and reactance combinations. In most electrolytics, the dc voltage is fairly large and the ac that passes through the capacitor produces a small ac voltage difference across the capacitor. If the ac voltage across a capacitor is the major voltage of an order similar to the normal dc voltage for that capacitor, the rate of change of storage is heavy and rapid. As a result, the capacitor may overheat due to internal losses. So a capacitor to withstand large ac voltages must be designed and specified for that purpose.

TEMPERATURE COEFFICIENTS

All inductors and capacitors, like other components, possess temperature coefficients which relate to a value change with temperature. Sometimes this may be important and sometimes not. Failure to specify a temperature coefficient usually means the component is intended for use in a circuit or situation where the variation likely to occur with temperature change must be unimportant.

Temperature coefficients are usually important in tuned circuits (discussed in Chapter 10) because the precise value of inductance and capacitance has a bearing on the operating frequency, which may have to be held within close tolerances, such as a very small fraction of one percent. Where this is necessary, two approaches are possible:

- Make the circuit so that temperature does not affect its operation, either by using components with zero temperature coefficient or by combining positive and negative changes to get that effect;
- Put the entire circuit containing the critical components into a temperature-controlled oven that maintains a temperature within close limits of a specified reading.

In many cases where extremely close frequency tolerance is needed, both methods are combined. The components used are selected to minimize a change of frequency with temperature, and then the entire package is housed in a temperature-controlled oven, so that the temperature in which the circuit operates does not vary more than a degree or so.

Changes in the inductance of an air-core coil occur due to the fact that the wire of which it is wound is subject to a change in physical size (by linear dimension) with temperature. Since the dimensions of a self-supporting coil (one without a permanent "former" on which the coil is wound) all change in the same proportion (length of coil as well as its diameter, for example), such an inductor always has a positive temperature coefficient of the same order as the expansion coefficient of the wire of which it is wound.

However, a coil wound on a former can be made to change its shape, due to a difference in the temperature coefficients of the wire and the material of which the former is made. By careful attention to design (using methods beyond the scope of this book), a coil can be made in which the inductance does not vary appreciably with temperature, because the shape change offsets the linear dimensional change.

Inductors with magnetic cores are more complicated. The core may have some kind of temperature coefficient, which is not usually constant over an appreciable range of temperature. Thus, it may be possible to design an inductor in which the value remains very close to constant over a limited temperature range.

Here, the word "limited" can mean almost anything. It may mean 10 degrees or it may mean 100 degrees, depending on the causes of variation. It also may mean a dimensional change in the core, or a change in the core's magnetic properties due to temperature variation.

In capacitors, a change in value with temperature can be positive or negative, again depending partly on the materials of which the capacitor is made and partly on the physical dimensioning within the capacitor. This, too, is a highly specialized art that is not discussed in this book.

Usually, the important thing in equipment where temperature coefficients matter is that the combined effect does not change the working frequency generated by the inductance and capacitance combination. This means that if an inductor has a positive coefficient of 2 percent per degree Centigrade and the capacitor with which it works has a negative coefficient of 2 percent per degree Centigrade, the product of the two components which control frequency will have a combined coefficient of close to zero.

Alternatively, the inductor may be designed to have a zero temperature coefficient by the method described above, which means that the associated capacitance should also have a zero temperature coefficient. There are such components. Or the same effect may be achieved by combining parts of the capacitance with positive and negative coefficients. Suppose the total capacitance is 1000 micromicrofarads (pF), made up of 600 pF with a negative temperature coefficient of 1 percent per degree C and 400 pF with a positive temperature coefficient of 1.5 percent per degree C. The combined effect of a change in temperature will then be zero.

SKIN EFFECT

As currents flow through a conductor, the same characteristics responsible for electromagnetic radiation and reactance also cause another effect—skin effect. As the frequency of the current increases, the current has a tendency to crowd at the surface of a conductor, skin effect can be understood by considering the field generated by each element of current.

When the frequency is relatively low, so the magnitude of the field generated by an alternating current is big, the current in a relatively small conductor is uniformly distributed throughout the cross-section of the conductor, and can be considered as a lump current concentrated at the center of the conductor. Actually, because of the relative dimensions involved, whether the current is all at the center of the conductor or whether it is distributed throughout its cross-sectional area makes no difference to calculations.

But when the frequency gets higher, so that field size becomes commensurate with conductor dimensions (of the same order, say, ten times), we have to consider the effect of fields generated by the current within as well as outside the conductor (Fig. 9-13).

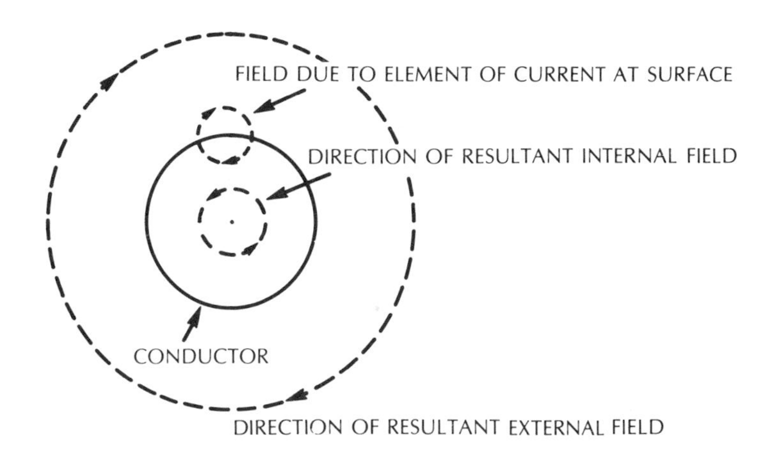

Fig. 9-13. How an electromagnetic field at very high frequencies induces a "skin" effect.

The field encircling any element of current near the surface of the conductor produces a field direction inside the conductor that is opposite to that outside the conductor. The effect of that field is to concentrate the current, more and more as frequency increases, into a thin skin at the surface of the conductor. From the viewpoint of inductance and radiation, surface crowding has very little effect on the current. The field inside the conductor does not materially affect the field outside. But since all the current is concentrated in only part of the conductor, the result is an increase in the effective resistance of the conductor. From the viewpoint of resistance, a concentration of current at the surface is like taking away the inside, so the cross-sectional area is considerably reduced. In fact, hollow pipe is used in circuits that carry high-frequency energy. To prevent overheating, a coolant is circulated through the pipe.

Examination Questions

1. Find the energy stored in the following inductances when the currents listed are flowing through them: 1 henry, 10 amps; 5 henries, 10 amps; 10 henries, 5 amps; 1 henry, 5 amps.
2. Write the formula for energy stored in an inductance (in watt-seconds), given its inductance in henries and the current flowing in amps.
3. Find the energy stored in the following capacitances when the voltages stated appear across them: 1000 mfd, 100 volts; 1000 mfd, 50 volts; 10 mfd, 50 volts; 50 mfd, 10 volts.
4. State the formula for energy stored in a capacitance (state units used).
5. When an alternating current flows in an inductance, maximum energy is stored at: maximum voltage, maximum current, or both? Indicate which answer is correct.
6. When a capacitance is subjected to alternating current and voltage, maximum energy stored in the capacitance occurs at: maximum voltage, maximum current, or both?
7. Find the reactance value of the following components at the stated frequencies: 5 henries, 60 Hz; 250 mh, 1000 Hz; 1 mfd, 120 Hz; 3.3 mfd, 512 Hz.
8. Find the frequencies at which the following components have the reactance stated: 3.3 henries, 1000 ohms; 2.5 henries, 314 ohms; 7.5 mfd, 3200 ohms; 13.2 mfd, 1250 ohms.
9. Find the time constants of the following component combinations: 5 henries, 300 ohms; 16 mh, 220 ohms; 25 mfd, 1000 ohms; 0.1 mfd, 220,000 ohms.
10. Find the resistance value needed with each component to produce the time constant named: 3 henries, 2 seconds; 5 millihenries, 4 microseconds; 0.1 mfd, 75 microseconds; 16 mfd, 3 seconds.
11. The inductor in a tuned circuit is designed to have a zero temperature coefficient over the working temperature range. The required capacitance value is 1500 pF, which is made up of values having positive and negative temperature coefficients. Part of this capacitance has a value of 550 pF and a temperature coefficient of +4 percent per degree C. What must be the capacitance and temperature coefficient of the other part?
12. In the circuit in Question 11, suppose the inductor has a positive temperature coefficient of 1 percent per degree C, using the same 550 pF as part of the capacitance. What must be the tolerance of the remainder's capacitance now?

10
Reactance In Circuits

In practical circuits and components, pure reactances never exist alone. Reactance is invariably associated with resistance. At every instant in an ac circuit, Kirchhoff's laws apply. That is to say the sum of the voltages across a resistance and reactance in series is always equal to the voltage across the combination (Fig. 10-1) at every instant.

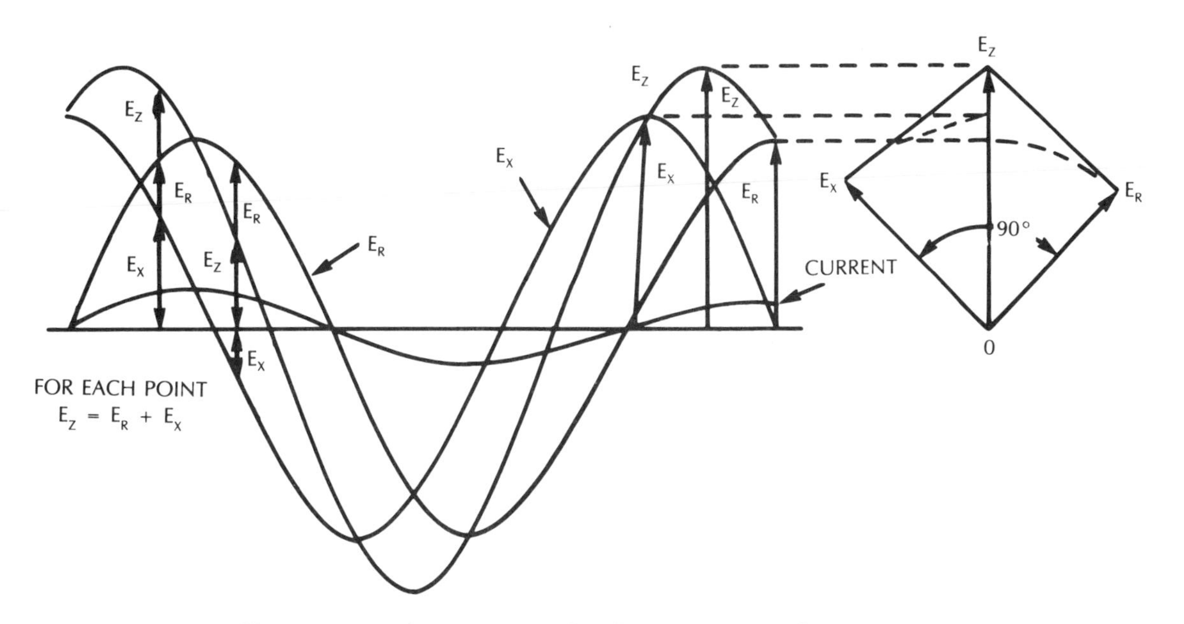

Fig. 10-1. Waveforms associated with a resistance and reactance in series.

But this does not mean that the voltage across them as a nominal ac value (which is usually rms or root-mean-square) is the sum of the voltages across the components separately. Across the resistance the maximum voltage coincides with the maximum current, but at the same time the voltage across the reactance is zero. The maximum voltage across the reactance occurs at zero current, at which time the voltage across the resistance is zero. Maximum voltage across the combination will occur at some time between the points where it is maximum across the components, but it will not be the sum of the maxima across them individually because of the time difference.

Here is where the concept of the rotating vector, first shown in Fig. 5-9, can prove useful. If the vectors for the voltage across the resistance and reactance are drawn separately, the vector for the voltage across the combination is found by drawing the diagonal to the rectangle of which the first two vectors are two sides. So the Pythogorean Theorem comes in useful:

The square of the hypotenuse (longest side) of a right-angled triangle is equal to the sum of the squares on the other two sides (Fig. 10-2).

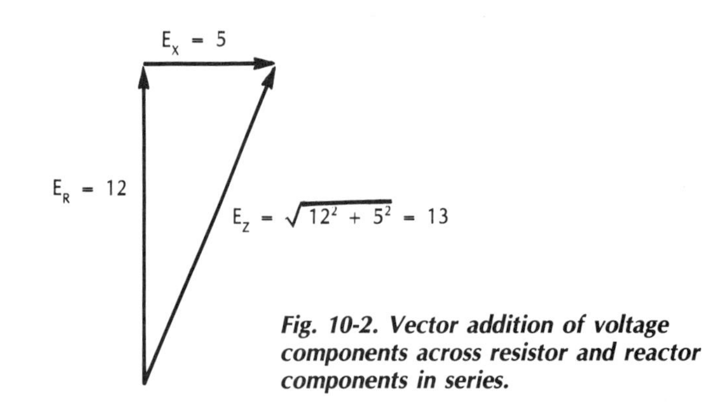

Fig. 10-2. Vector addition of voltage components across resistor and reactor components in series.

Thus, if the peak voltage across the reactance is 12 volts and the peak voltage across the resistance is 5 volts, the peak voltage across the combination is found by squaring those two voltages and extracting the square root:

$$\sqrt{12^2 + 5^2} = \sqrt{144 + 25} = \sqrt{169} = 13 \text{ volts}$$

That could be the rms voltage as well as the peak voltage, provided each measure corresponds, but using the peak voltage is more obviously linked with the application of vectors, as seen in Fig. 10-1. Figure 10-3 shows the relationship in a circuit consisting of resistance and reactance (of either kind, for a particular frequency) in series.

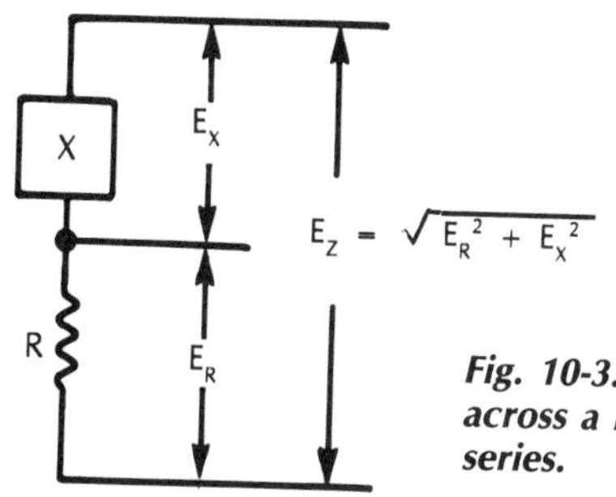

Fig. 10-3. Method of calculating the voltage across a resistor and reactor combination in series.

The same thing can happen with the combination of currents in parallel circuits. If a reactance is in parallel with a resistance, with the same voltage across both, peak currents occur at different times. Peak current in the resistance coincides with the peak voltage, while peak current in the reactance occurs at zero voltage. Now, vectors are used to add currents instead of voltage (Fig. 10-4).

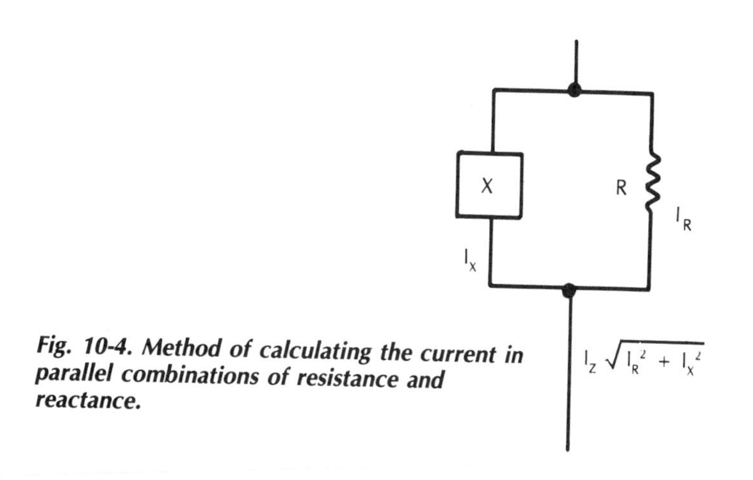

Fig. 10-4. Method of calculating the current in parallel combinations of resistance and reactance.

For example, if the current through the reactance is 3 amps and that through the resistance is 4 amps, the current through the parallel combination is found by squaring these two current values and extracting the square root:

$$\sqrt{3^2 + 4^2} = \sqrt{9 + 16} = \sqrt{25} = 5 \text{ amps}$$

POWER FLOW IN CIRCUITS WITH REACTANCE

These calculations do not tell all about what happens in ac circuits containing both reactance and resistance. In pure resistance circuits, power can flow only from the source to the resistance, the resistance still absorbs power, but the reactance absorbs power only for a part of the cycle, delivering the same power back to the circuit at other times in the cycle.

In a dc circuit, power is strictly a product of voltage and current. If a voltage is 10 and the current is 5 amps, the product 10 × 5 equals 50 gives the power in watts, without question. In an ac circuit, this is no longer true if the circuit includes reactance.

IMPEDANCE

Suppose the current in the series circuit is 2 amps; therefore, the resistance is 2.5 ohms (5 divided by 2) and the reactance is 6 ohms (12 divided by 2). The impedance, as the combination is called, is 6.5 ohms (13 divided by 2). But the power in this case is dissipated only by the resistance, 5 volts at 2 amps, making 5 × 2 or 10 watts. That could be the peak power or the average power, according to whether the voltage and current are both peak or rms, respectively. For a sinusoidal waveform, peak is always root 2 (1.414 approximately) times the rms value. Since this applies to both voltage and current waveforms, the peak power is root 2 squared, or twice the average power. Conversely, the average power is half the peak power. This can be shown graphically by plotting curves of voltage, current and instantaneous power (Fig. 10-5).

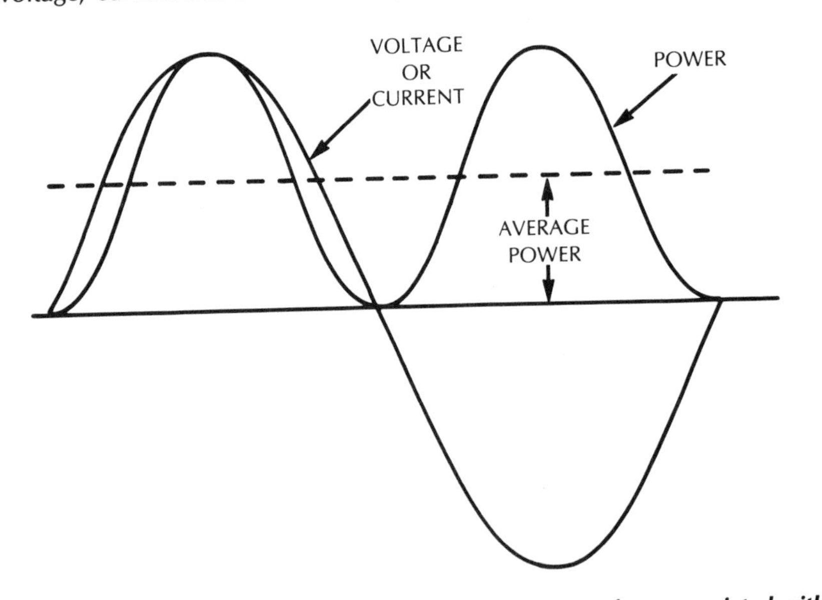

Fig. 10-5. Voltage or current waveforms, and the power waveform associated with a resistance connected to a sinusoidal supply.

In a circuit that contains reactance, plotting the voltage and current in the combination produces a different picture. In Fig. 10-5, because current always flows in the same direction as the applied voltage "pushes" it, power is always flowing the same way, from the power source to load, which is considered as a positive power flow. In a circuit with reactance, sometimes the voltage and current are on the same side of the zero line (Fig. 10-6) and sometimes on opposite sides.

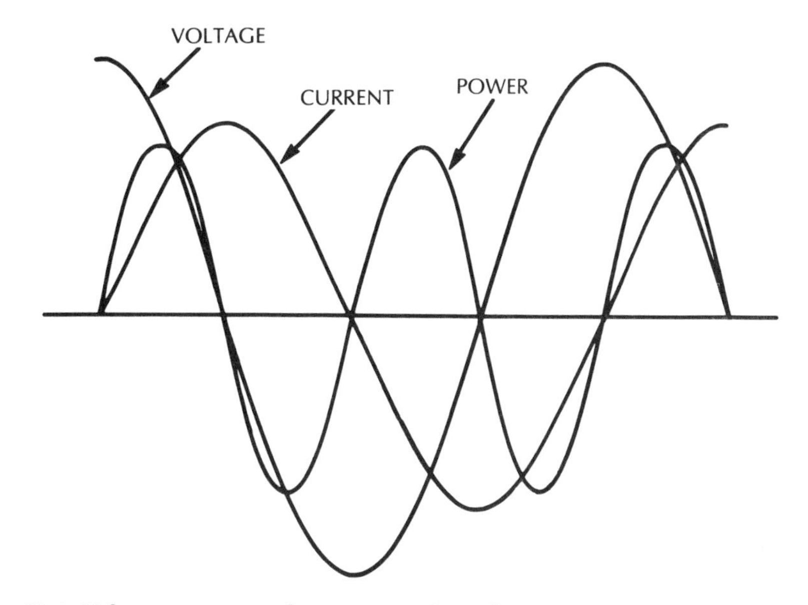

Fig. 10-6. Voltage, current and power waveforms in a pure reactance connected to a sinusoidal supply.

When current and voltage are both on the same side, power is going from the source into the load, as with a resistance. But when they are on opposite sides, the reactance component is delivering its stored power back into the circuit. In a pure reactance circuit, power alternately flows into and out of the reactance, so the net power, averaged over a cycle, is zero flow either way.

These facts give rise to different qualities used in ac circuits to describe what is quite simple in dc circuits. The product of voltage and current is called *volt-amps*, or *VA* because it is not watts unless there is no reactance in circuit. However, the average power delivered to the circuit is still average watts. If the average actual power is divided by average volt-amps, the fraction thus obtained (usually given as a decimal) is called the *power factor* of the circuit. Thus, in a circuit consisting of 2.5 ohms resistance in series with 6 ohms reactance, making an impedance of 6.5 ohms, with an rms current of 2 amps flowing, the average volt-amps is 13 × 2 or 26 VA, while the average power is 5 × 2 or 10 watts (Fig. 10-7). The power factor is 10 divided by 26 or 0.385.

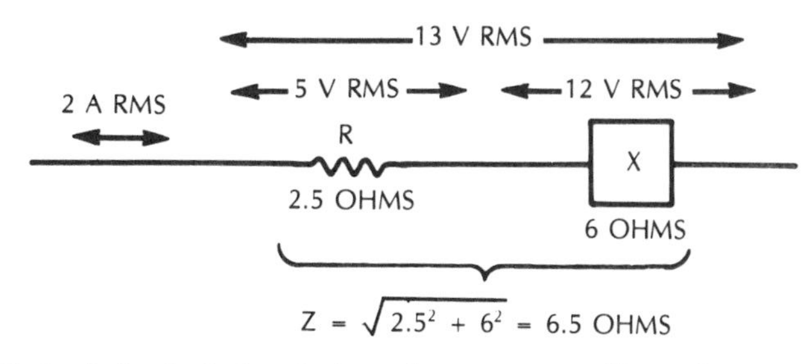

Fig. 10-7. Method of calculating the impedance presented by resistance and reactance in series.

In a circuit that carries 3 amps reactive and 4 amps resistive, assume the voltage applied is 12 volts. This means that the reactance is 12 divided by 3 or 4 ohms, while the resistance is 12 divided by 4 or 3 ohms. The two in parallel pass 5 amps, which at 12 volts represents an impedance of 12 divided by 5 or 2.4 ohms. In this case, the VA (again assuming those figures are rms) is 12 × 5 or 60, while the power is 12 × 4 or 48. The power factor is now 48 divided by 60 or 0.8. Exploring this example a little further shows how the power flows during the cycle (Fig. 10-8). If the average VA is 60, the peak value, obtained by multiplying together the peak voltage and peak current (which has no real or physical significance and thus cannot be shown on the waveform diagram), is 120.

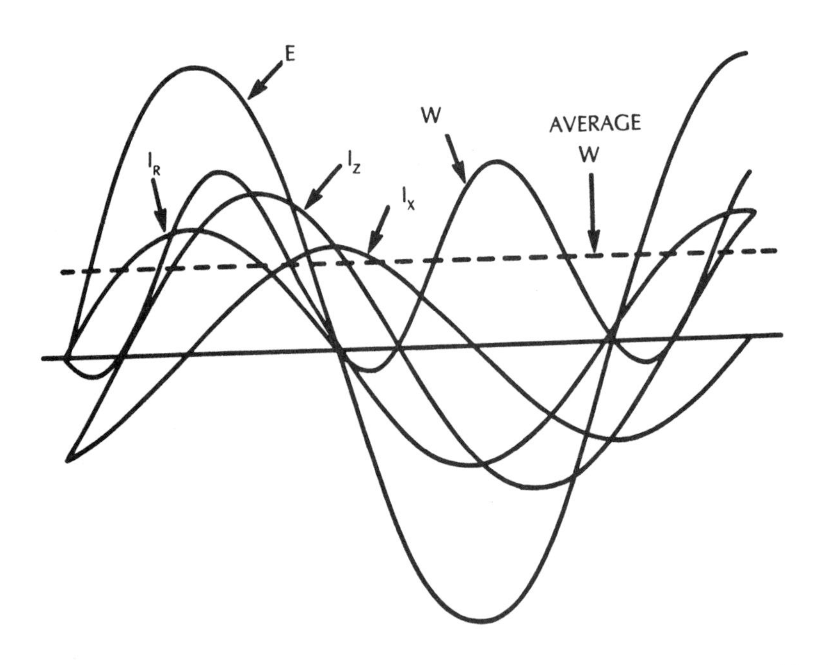

Fig. 10-8. Voltage, current and power waveforms associated with a parallel combination of reactance and resistance.

Because of the phase difference between the voltage and current, the peak positive power occurs half-way between the voltage and current peaks in the same direction, where it is 0.9 times the peak produced by multiplying together the non-coincident peaks (that resulted in 120), or 0.9 × 120 or 108 watts (instantaneous). At a point midway between where the voltage and current waveforms cross the zero line, the maximum reverse wattage appears, and this is 0.1 × 120 or 12 watts (instantaneous). With the wattage fluctuating cyclically (twice each voltage or current cycle) between +108 and −12 watts, the average is ½ (108 − 12) or 48 watts, as calculated earlier.

In the above discussion, resistance and reactance have been treated as separate entities. In the physical world of components, they may not always be so readily separable. Inductances in particular can consist of series resistance (due to resistance of the wire in the winding) and effective parallel resistance, which is discussed later.

For purposes of calculation, it makes no difference whether the resistance is part of the reactance component or whether they exist in two separate components, unless you want to measure the voltage and current associated with the pure reactance when it is part of a component that is not pure.

For calculation purposes, we need to combine resistance and reactance to form impedance, for which the general formula is:

$$Z = R + jX \tag{18}$$

where "j" signifies that the two quantities are not simply added, but combined by adding the squares and taking the square root. The "j" is put in front of a quantity in which the voltage is at right angles to the current in phase (by vectors).

J AS AN OPERATOR

Actually, the symbol "j" can stand for the "square root of minus (1)." It is usually called "root minus one," but there is really no reason to specify a number. If you think of "j" as another sign, similar in its effect to the minus sign, it will be easier to understand.

You may have been taught that minus quantities do not have square roots, in which case that definition may be mystifying. Really, this concept provides a convenient short cut in calculating. If you view a minus quantity as a reverse in direction from a plus quantity (Fig. 10-9) then the square root of minus can represent a point half-way to reversal, which is 90 degrees. Using the other half of the square root completes the reversal to minus. Just as multiplying minus by minus makes plus, so multiplying root minus by root minus makes minus.

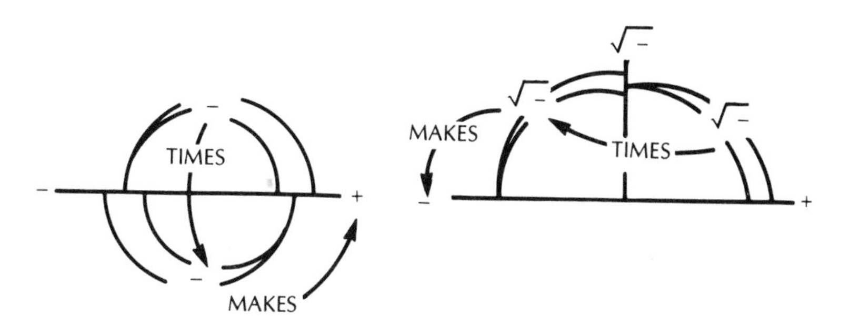

Fig. 10-9. Comparing minus times minus making a plus, with root-minus times root-minus making minus, graphically.

Completing this rotational symbology, if multiplying minus by root minus makes minus root minus (−j), then multiplying by root minus (j) one more time brings us back to plus. We have created a number "field" in which there are more than just positive and negative numbers. There are also positive and negative root minus numbers, generally called imaginary numbers (Fig. 10-10).

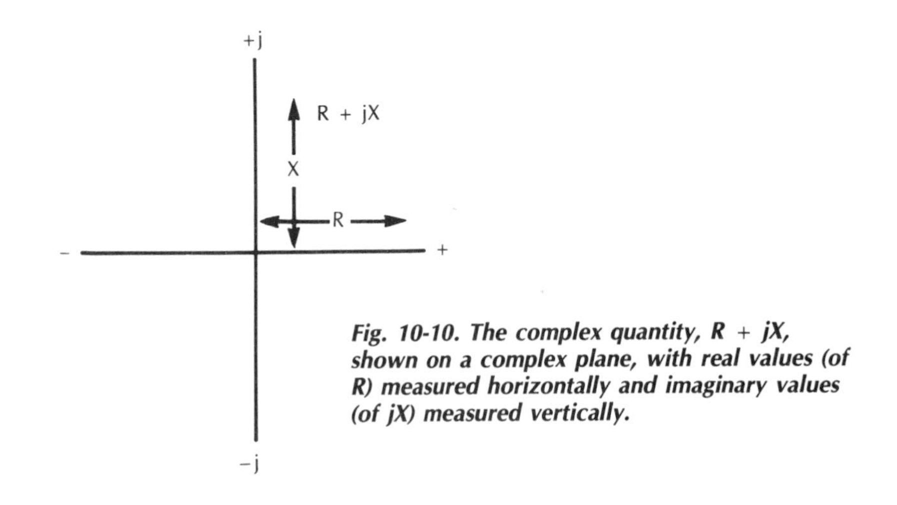

Fig. 10-10. The complex quantity, R + jX, shown on a complex plane, with real values (of R) measured horizontally and imaginary values (of jX) measured vertically.

These two sets of numbers (known as real and imaginary) have the properties of quantities measured at right angles. Measuring left or right from a vertical reference line in a horizontal direction does not change the imaginary component of the new value to the left or right, because imaginary components are measured up and down. Measuring up or down from a horizontal reference line in a vertical direction does not change the real component of the new value, because real values are measured left and right.

Thus, the number R + jX represents a position in this field, in the top right quadrant, to be precise. Changing the value of R moves the position representing the complex value to the left or right, without altering its vertical location. Changing the value of X moves the position up or down, without altering its horizontal location.

EQUIVALENT SERIES AND PARALLEL VALUES

Now we can consider calculating the equivalent reciprocal quantities. The impedance family (resistance, reactance and impedance) all measure volts when unit current flows through them. The reciprocal, or admittance family (conductance, susceptance and admittance), measures current when unit voltage is applied to them. To take this by easy steps:

- **Admittance** is the reciprocal of impedance.
- **Conductance** is the reciprocal of resistance.
- **Susceptance** is the reciprocal of reactance.

The symbols are Y for admittance, G for conductance and B for susceptance. Thus:

$$Y = 1/Z \quad G = 1/R \quad B = 1/X \tag{19}$$

Suppose we are dealing with quantities actually in series. They are impedances, representing voltages across resistance and reactance that add vectorially at 90 degrees when the same current flows through each. Then, using a reciprocal value enables us to find an equivalent parallel circuit that will result in equivalent currents that add vectorially when the same voltage is across each (Fig. 10-11).

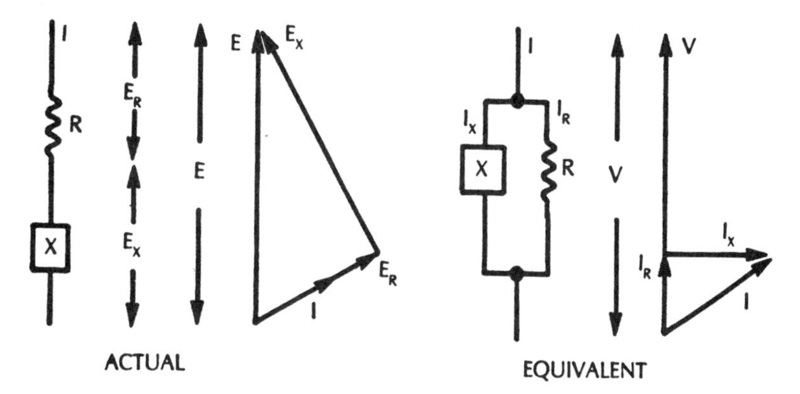

Fig. 10-11. Equivalent parallel network and vector diagrams, showing how the same resultant voltage and current can be achieved with a series or parallel combination.

Each combination has the same E and I. So, proceeding from equation (19),

$$Y = \frac{1}{Z} = \frac{1}{R + jX} \tag{20}$$

Now, if you multiply the top and bottom of this expression by $R - jX$, the product in the denominator is:

$$(R + jX)\,(R - jX) = R^2 - j^2X^2$$

But because j^2 is -1 (or just $-$) the denominator becomes:

$$R^2 - (-X^2) = R^2 + X^2$$

and the whole expression becomes:

$$Y = \frac{R - jX}{R^2 + X^2} \tag{21}$$

From this, the separation of the "real part," which has no "j," from the "imaginary part," which has a "j" in front of it, enables separation of the in-phase and reactive parts of the admittance.

$$G = \frac{R}{R^2 + X^2} \qquad (22)$$

and

$$B = \frac{X}{R^2 + X^2} \qquad (23)$$

Those are the equivalent conductance and susceptance elements of the actual series combination. But to make this more familiar, let us see what equivalent values of resistance and reactance this means in a parallel equivalent. Resistance is the reciprocal of conductance and reactance the reciprocal of susceptance, so:

$$R_e = \frac{1}{G} = \frac{R^2 + X^2}{R} = R \times \frac{R^2 + X^2}{R^2} \qquad (24)$$

and

$$X_e = \frac{1}{B} = \frac{R^2 + X^2}{X} = X \times \frac{R^2 + X^2}{X^2} \qquad (25)$$

Notice that both values are larger than the series values given in formula 18 by the "top-heavy" expression by which each quantity is multiplied in its last expressed form. Let us put in some values to illustrate what these formulas mean.

Suppose, at the frequency in question R is 10 ohms and X is 100 ohms. Then R^2 equals 100 and X^2 equals 10,000. Therefore,

$$R_e = 10 \times \frac{10{,}100}{100} = 1010 \text{ ohms}$$

and

$$X_e = 100 \times \frac{10{,}100}{10{,}000} = 101 \text{ ohms}$$

Reactance, in that example, has increased by 1 percent, from 100 to 101 ohms, while the resistance in parallel has changed from one tenth the reactance (in series) to over 10 times the new value of reactance, from 10 to 1010, which is a change of 101 times.

Suppose the frequency is doubled, which will make it 200 instead of 100 if the reactance is inductive. Working the same calculation will show that X_e becomes 200.5 ohms and R_e becomes 40,100 ohms. The higher the frequency, the less the resistance affects the equivalent reactance, while the ratio by which the resistance is multiplied becomes greater.

But go the other way, reducing frequency and reactance to one tenth the original, so X equals 10, the same as R. Now,

$$X_e = 10 \times \frac{200}{100} = 20 \text{ ohms}$$

and R_e is the same; both values are double, when considered as equivalent parallel elements. Reducing the frequency and reactance by another factor of 10, to one hundredth of the original, and there's a reversal in the way that reactance and resistance affect one another. Now X is 1 in series with 10 ohms (R). The new value of resistance becomes:

$$R_e = 10 \times \frac{101}{100} = 10.1 \text{ ohms}$$

while the reactance becomes:

$$X_e = 1 \times \frac{101}{1} = 101 \text{ ohms}$$

Now the resistance hardly changes value while the reactance is multiplied by approximately the square of the ratio between resistance and reactance.

RESONANCE

If both inductive and capacitive reactances are combined in the same series or parallel combination, they tend to cancel one another's effect (Fig. 10-12).

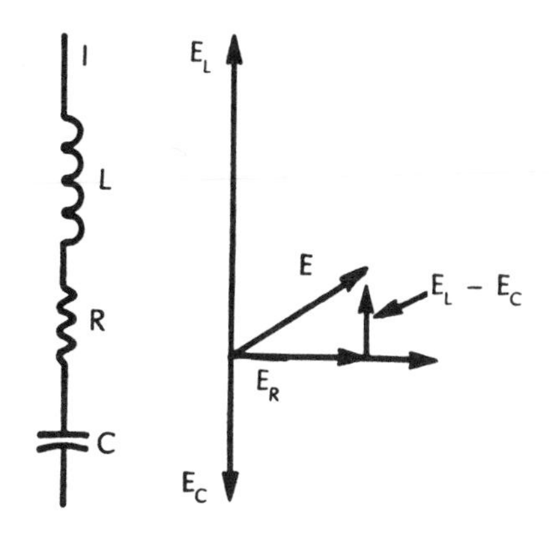

Fig. 10-12. Vectors for a series resonant circuit, slightly off resonance.

In series, the voltages tend to cancel, because they are in opposite phase. And in parallel, the currents tend to cancel, because they are in opposite phase (Fig. 10-13). In either case, when they exactly cancel or neutralize, because both reactances are equal, the condition is known as *resonance*. This is usually defined as the frequency at which, because reactances are equal, the resultant impedance is pure resistance. This definition is not always strictly true, but the exceptions are more involved than we will consider here.

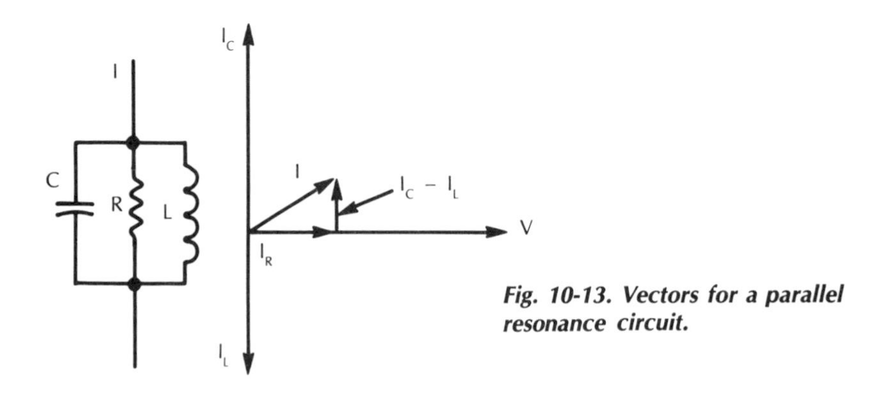

Fig. 10-13. Vectors for a parallel resonance circuit.

To explore the possibilities of resonance, take a 0.1-mfd capacitor, which has a reactance of 1590 ohms at 1000 Hz, and an inductance of 0.254 henry, which also has a reactance of 1590 ohms at 1000 Hz. If this inductance has a resistance of 10 ohms, and the two components are connected in series, the combination will have an impedance that is pure resistance of 10 ohms at 1000 Hz.

But at 1200 Hz the inductance has a reactance of 1.2 times 1590, or 1910 ohms, while the capacitor has a reactance of 1590 divided by 1.2, or 1330 ohms. The combined reactance, instead of being zero, will now be 1910 - 1330 or 580 ohms, and because the inductance is bigger it will be an inductive reactance. At 800 Hz, on the other hand, the inductive reactance is 0.8 times 1590, or 1275 ohms, while the capacitive reactance is 1590 divided by 0.8, or 1990 ohms, yielding a difference or resultant reactance of 1990 - 1275 or 715 ohms, this time capacitive.

If the combined reactance is plotted against frequency, the curve shown in Fig. 10-14 results in a series resonance curve. The minimum is zero reactance, but when the resultant reactance gets down to the region of the series resistance, very close to 1000 Hz, the resistance sets the limiting value. Figure 10-14 also shows a curve with a resistance value of 100 ohms.

Now, put the same combination together in parallel (Fig. 10-15). Using formulas (24) and (25) the inductance can be converted into an equivalent of 1590 ohms reactance with 254,000 ohms resistance in parallel, instead of 1590 ohms in series with 10 ohms. Actually, transferring the circuit from series values to the equivalent parallel will change the equivalent reactance by a fraction that is the square of 10 divided by 1590, or (1 over 159)^2 equals 1 over 25,400, or about 0.0039 percent, which is a negligible change for most purposes.

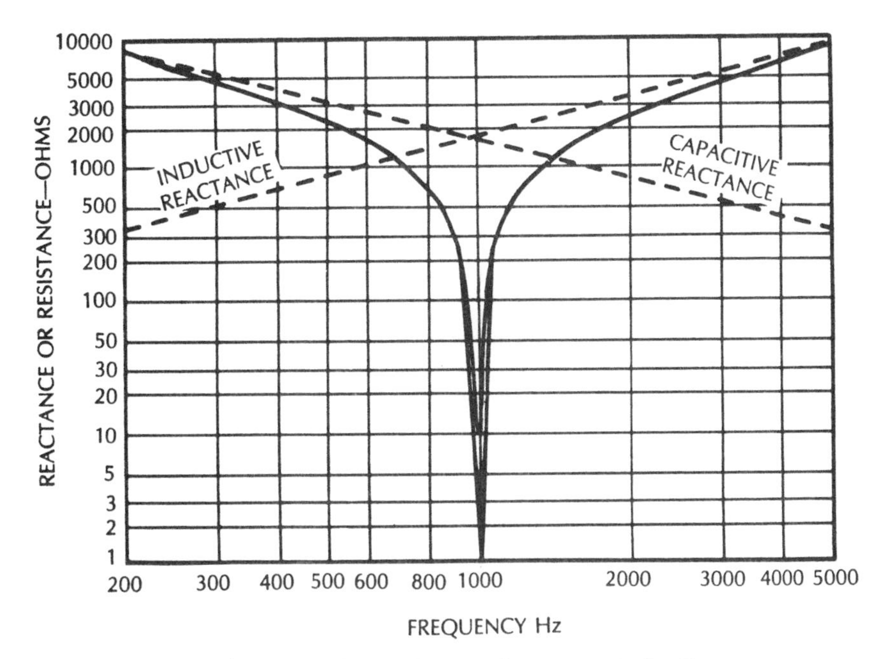

Fig. 10-14. Curves for a series resonant circuit.

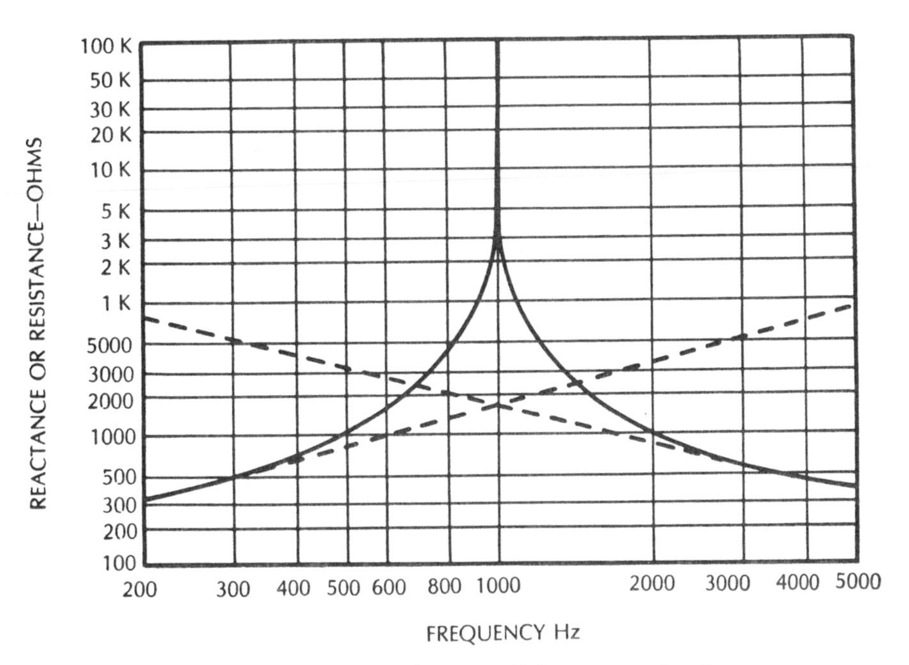

Fig. 10-15. Curves for a parallel resonant circuit.

But suppose the actual resistance is 100 ohms instead of 10 ohms, then the equivalent parallel value will be 25,400 instead of 254,000 and the shift in reactance will be 0.39 percent instead of 0.0039 percent. Alternatively, using the original inductance with a resistance of 10 ohms at a frequency of 100 Hz instead of 1000 Hz, which requires a 10-mfd capacitor instead of 0.1 mfd, the reactances are both 159 ohms instead of 1590 ohms, which gives a similar relationship.

BRIDGE CIRCUITS

Another way of combining reactance into the same circuit is shown in the bridge circuit introduced in Chapter 3. And reactances can be put into such a circuit in an almost limitless variety of combinations. Two major varieties of such ac bridge circuits can be distinguished: those that are frequency-dependent and those that are independent of frequency.

The second variety may never be an absolute distinction, in the sense that no bridge will ever be completely balanced at all frequencies. But in theory the latter type hold their balance over a wide range of frequencies, while the first group is quite critical, since changing the frequency will change the condition of balance.

FREQUENCY-INDEPENDENT BRIDGES

If two arms of the bridge are capacitances and two are resistances (Fig. 10-16), balance is achieved at any frequency if the capacitances have the same ratio as the resistances.

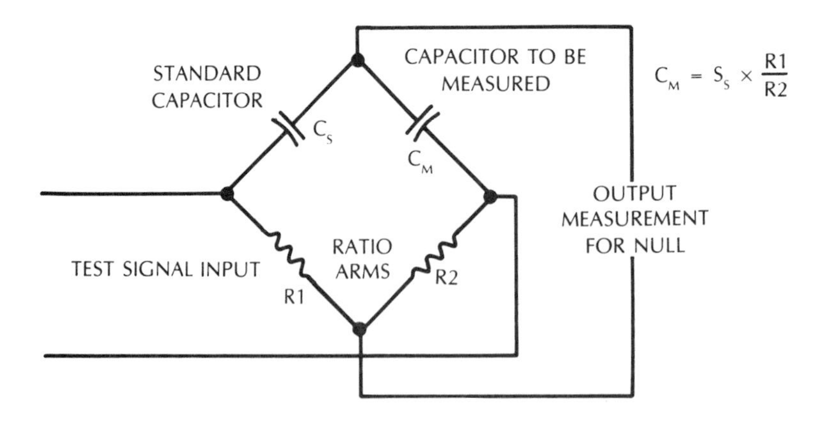

Fig. 10-16. Circuit of a simple capacitance bridge.

This principle is used in a simple capacitor checker in which a standard capacitor (inside the checker) is switched into one of the capacitor arms, while the one to be measured is connected into the other. Adjusting the resistance for balance, a calibrated scale on the resistance knob indicates the capacitance of the unknown, based on the ratio of the resistance arms at that position of the knob (Fig. 10-17).

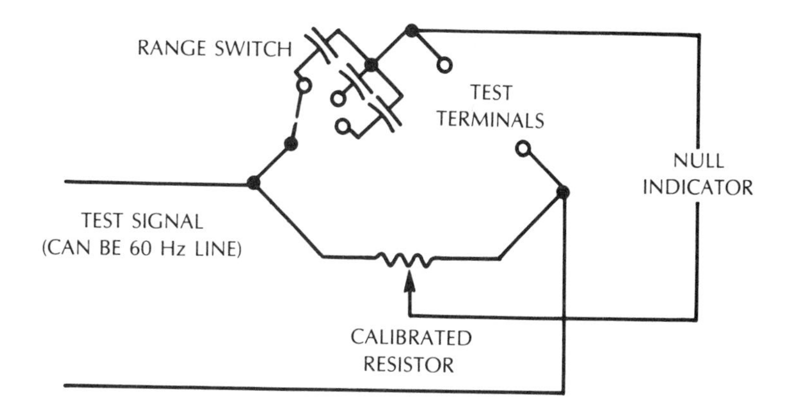

Fig. 10-17. Revised capacitance bridge circuit as used in a type of capacitance checker widely used for many years.

In theory, the same thing could be done with inductances, but it is not quite so simple, because inductances are not so close to perfect reactances as capacitors are. To get balance, both inductances would need to have exactly the same ratio of reactance to resistance at the test frequency. For this reason, to measure inductance, the usual bridge circuit uses a capacitor-resistor combination to balance an inductance-resistance combination in the opposite arm (Fig. 10-18).

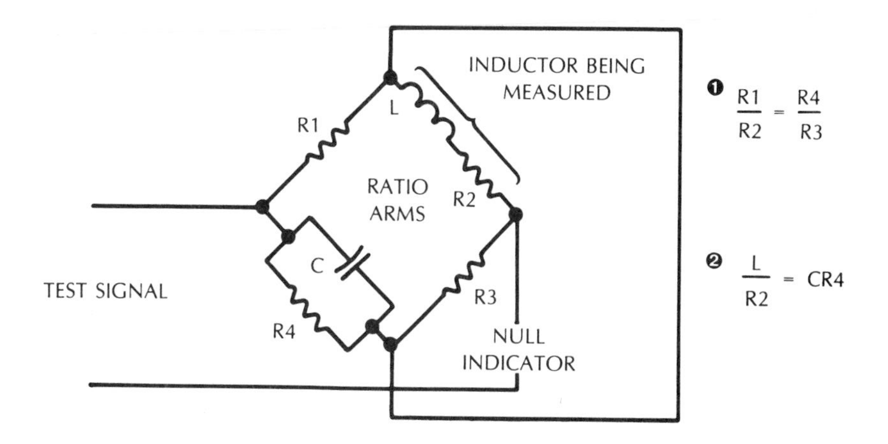

Fig. 10-18. Circuit of the most-used bridge for measuring inductance.

In this circuit, when the resistance in the inductive arm is in series, that in the capacitive arm should be in parallel. Then, if the reactances behave as pure inductive and capacitive values with respective fixed resistance values associated with each, the bridge is balanced irrespective of frequency.

Balance requires two conditions: that the amplitudes of voltages across pairs of arms is identical, and that their phase relationships are also identical. If any of the total of six values in the circuit is different from that required for balance, it will unbalance the bridge.

The condition for balance in both respects can be calculated with more complicated reactance-impedance formulas for series and parallel combinations. But a simpler way is to use time-constant figures. The time constant of the inductance with its associated resistance should be the same as that of the capacitor with its associated resistance. Then, for the remaining two arms, the time constant of the inductance and the resistance in either adjacent arm must be the same as the time constant of the capacitor with the resistance in the other adjacent arm. Thus, in a sense, these opposite arms are "ratio arms" like the adjacent ones in the dc bridge or the resistance arm in the capacitor bridge.

EFFECTS OF SPURIOUS ELEMENTS

Each of these bridges is basically independent of frequency. But the independence may not hold perfectly. Inductances possess capacitance between the turns of the winding, which makes the component behave as a parallel resonant circuit at the top end of its usable frequency range. Below this, the effect of the capacitance begins to shunt the inductive reactance.

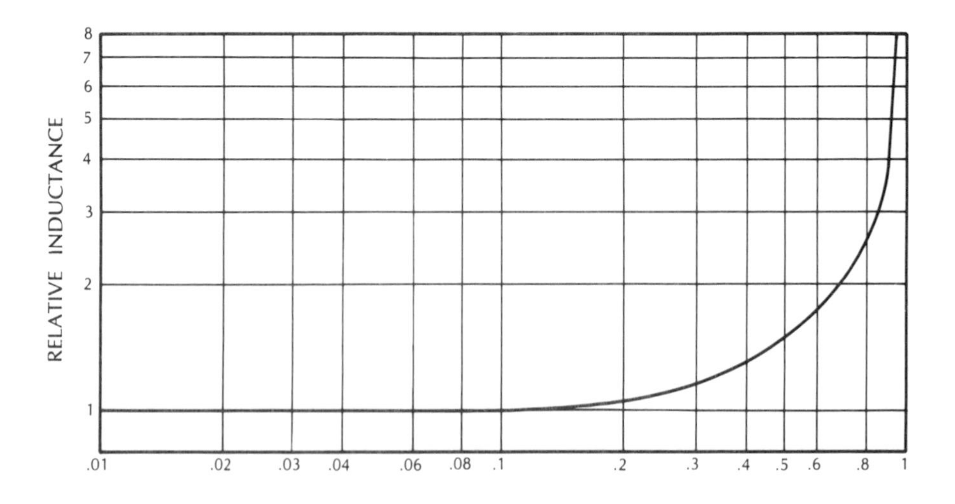

Fig. 10-19. Curve showing the effective inductance value changes as its self-resonant frequency is approached.

But remember that shunting a reactance of one type with one of the opposite type tends to neutralize or cancel currents. So the same inductance will pass less current than it should (as a perfect inductance) as frequency increases and thus will present an effective inductance value that appears to be increasing with frequency as its natural resonance is approached (Fig. 10-19).

Capacitors also possess inductance, due to the connecting leads, but this usually takes effect at a much higher frequency (values considered) than the point where capacitance begins to affect inductance.

COMPLICATED NATURE OF INDUCTORS

Inductances with magnetic cores also are subject to other losses that are more like parallel resistances, although such an equivalent resistance does not have a constant value, either with a varying frequency or with a varying applied voltage at the same frequency (Fig. 10-20). Study of this characteristic is more involved than we can discuss here. It is mentioned so that you are aware that an inductance may have an effective resistance different from its measured value.

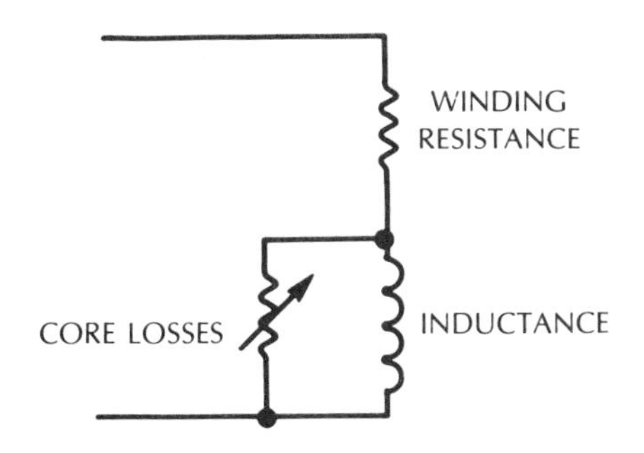

Fig. 10-20. Equivalent circuit for an inductance with an iron (or some form of magnetic material) core.

Just as the series resistance can be treated as an equivalent parallel value, so the effective parallel value due to core losses can be treated as an additional series value. Thus, if the reactance of an inductance is 1000 ohms at a particular frequency, its winding resistance is 7 ohms and the losses are equivalent to a parallel resistance of 100,000 ohms at the same frequency. The shunt resistance is 100 times the reactance at that frequency and measurement level, so is equivalent to a series value one hundredth of the reactance, or 10 ohms. With the actual 7-ohm winding resistance, the effective resistance of this inductance at this frequency and level will be 17 ohms (10 + 7).

Assuming, as sometimes comes close to the fact, that the equivalent parallel resistance due to losses is constant, changing the frequency will change only the reactance. Halving the frequency would make the reactance 500 ohms, so losses are now a resistance which is 200 times the reactance. The equivalent series value will be 500 divided by 200 or 2.5 ohms, making an equivalent resistance at this frequency of 7 + 2.5 ohms equals 9.5 ohms. Doubling the frequency raises the reactance to 2000 ohms. Now, the losses are equivalent to a resistance of only 50 times the reactance, so the equivalent series resistance is 2000 divided by 50 or 40 ohms, making a total equivalent resistance of 47 ohms at this frequency.

Q VALUES

Because of the number of variables, inductances are often given a "Q" value at a specific frequency or frequencies. This is the ratio of the reactance to the equivalent series resistance. For the three frequencies named, the particular inductor above would have a Q of 1000 divided by 17 or 59 at the first frequency, 500 divided by 9.5 or 52.7 at the lower frequency, and 2000 divided by 47 or 42.5 at the higher frequency. Under the above circumstances, the highest Q value exists when the two components of equivalent resistance are equal. To drop the 10-ohm equivalent parallel resistance at the first frequency to a value of 7 ohms requires that the frequency be dropped by the square root of 10 over 7, and that the reactance be reduced by the same ratio to 836 ohms. The equivalent resistance is now 14 ohms, yielding a Q value of 836 divided by 14 or 59.75.

The usual Q curve for an inductance is shown in Fig. 10-21 which shows that Q stays nearly constant over a restricted range of frequencies near its maximum value before dropping off quite rapidly at higher or lower frequencies beyond that.

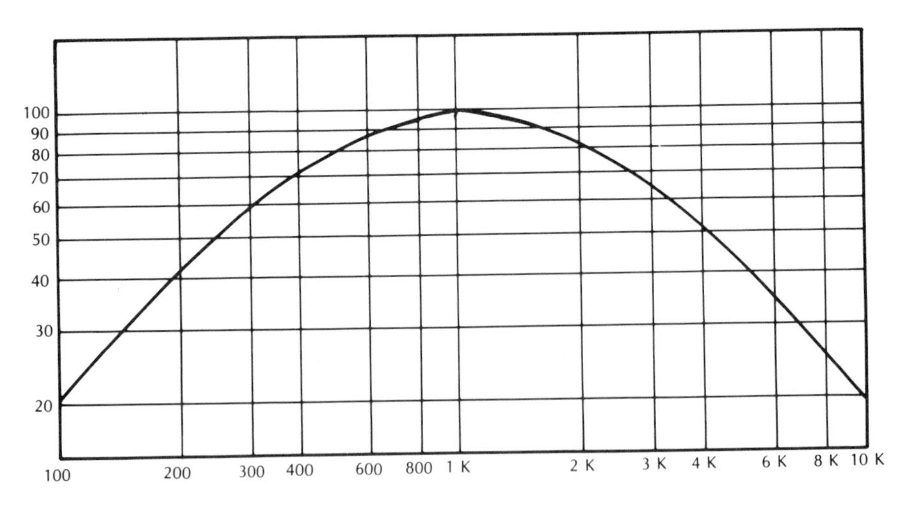

Fig. 10-21. Typical Q curve for an inductance.

Reverting to the discussion of bridge circuits, opposite arms contain elements related by a reciprocal relationship (Fig. 10-22). If one value is decreased, the other must be increased by the same ratio to retain balance.

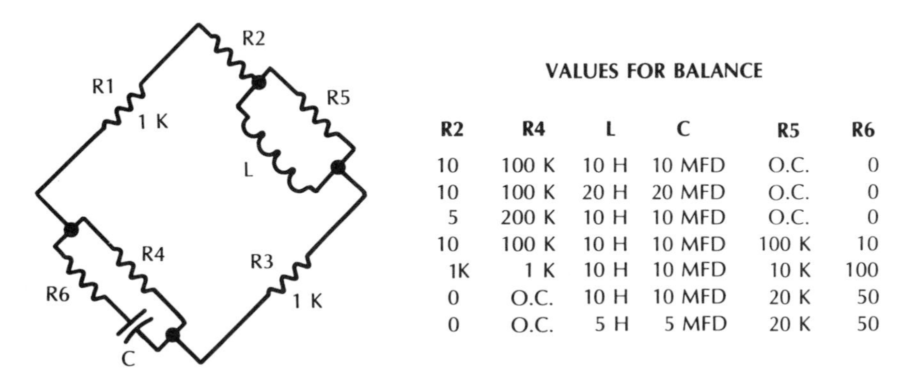

VALUES FOR BALANCE

R2	R4	L	C	R5	R6
10	100 K	10 H	10 MFD	O.C.	0
10	100 K	20 H	20 MFD	O.C.	0
5	200 K	10 H	10 MFD	O.C.	0
10	100 K	10 H	10 MFD	100 K	10
1K	1 K	10 H	10 MFD	10 K	100
0	O.C.	10 H	10 MFD	20 K	50
0	O.C.	5 H	5 MFD	20 K	50

Fig. 10-22. Values that produce balance in a more complete inductance bridge, that allows for shunt as well as series losses.

If one value is decreased, the other must be increased by the same ratio to retain balance. If one value is a pure inductance, the other must be a pure capacitance. If the inductance value is increased, the capacitance value must be increased also (remember that reactance increases with inductance, but varies inversely with capacitance). If the inductance has an equivalent series resistance, the capacitance must have a parallel resistance to balance it. If the inductance has a parallel resistance, it will be counterbalanced by a resistance in series with the capacitor.

FREQUENCY-DEPENDENT BRIDGES

The above conditions, except for the pure resistance case, apply to a bridge that is frequency-independent. But balance can occur at one frequency for other configurations and sets of values. For example, at a particular frequency, the series resistance can be expressed as an equivalent parallel resistance. So it is possible to use a series resistance with both inductance and capacitance and achieve balance *at one frequency only*.

This introduces the other type of bridge. The best way to view it, as related to a family of electronic circuits, is as a way of augmenting the discrimination that a resonant circuit can produce. In a resonant circuit, whether series or parallel, the impedance at resonance approaches either zero or infinity, respectively. It never gets to zero or infinity because of the residual resistance.

If it could reach zero or infinity, a simple resonant circuit could block that one frequency by completely short-circuiting it (series circuit) or by open-circuiting it (parallel circuit). The frequency-discriminating bridge can produce a complete null—zero transmission—at one frequency. But to achieve this null, both amplitude and phase must be correct. If either the resistance value or the reactance ratio is incorrect, the bridge will not null at any frequency.

The condition for achieving a true null can be approached in several ways. Perhaps the simplest is to consider the in-phase condition, because if the two points of the null network are not in phase with each other, the presence of the correct amplitude will not achieve a null. The in-phase condition occurs when the ratio of reactance to resistance in each reactive arm is the same.

So, if the series resistance in the inductive arm is twice the series resistance in the capacitive arm, the in-phase condition will occur when the inductive reactance is twice the capacitive reactance also. To calculate the remaining resistance arms, a knowledge of the reactance values when this occurs (Fig. 10-23) is required.

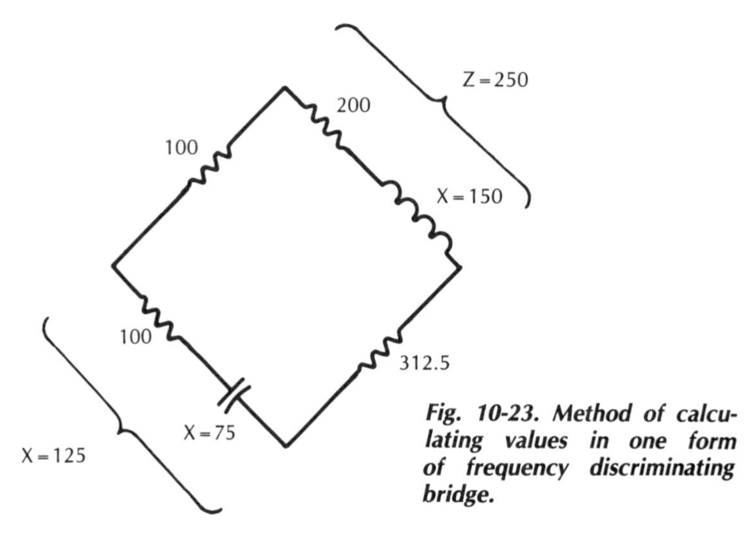

Fig. 10-23. Method of calculating values in one form of frequency discriminating bridge.

Since the inductance-resistance combination produces an impedance of 250 ohms at an in-phase condition, where the impedance of the capacitor-resistance combination is 125 ohms, the remaining arm must be 250 divided by 100 or 2½ times 125 ohms, which is 312.5 ohms. The frequency must be correct to produce the reactances named.

REACTANCE FILTERS

Both kinds of reactance find various uses in filter design, which we will only touch on in this book, because it is a very specialized subject. However, proper use of filters requires a little knowledge of what affects their performance, so we introduce that much here.

Filters come with combinations of elements to cut off high frequencies and pass low frequencies, to cut off low frequencies and pass the high ones, to pass a band of frequencies, and reject frequencies beyond that band, or to reject a band of frequencies and pass those on either side of it. But filters are generally designated by what they pass, rather than what they reject, with one exception.

Thus low pass filters cut off frequencies above a certain point. High pass filters cut off frequencies below a certain point. Band pass filters reject frequencies both below and above the band passed. Band reject filters do not pass the band specified as rejected. To illustrate their properties, we will use the most basic of all four types, which are those that pass lower frequencies and cut off higher ones. These are illustrated at Fig. 10-24.

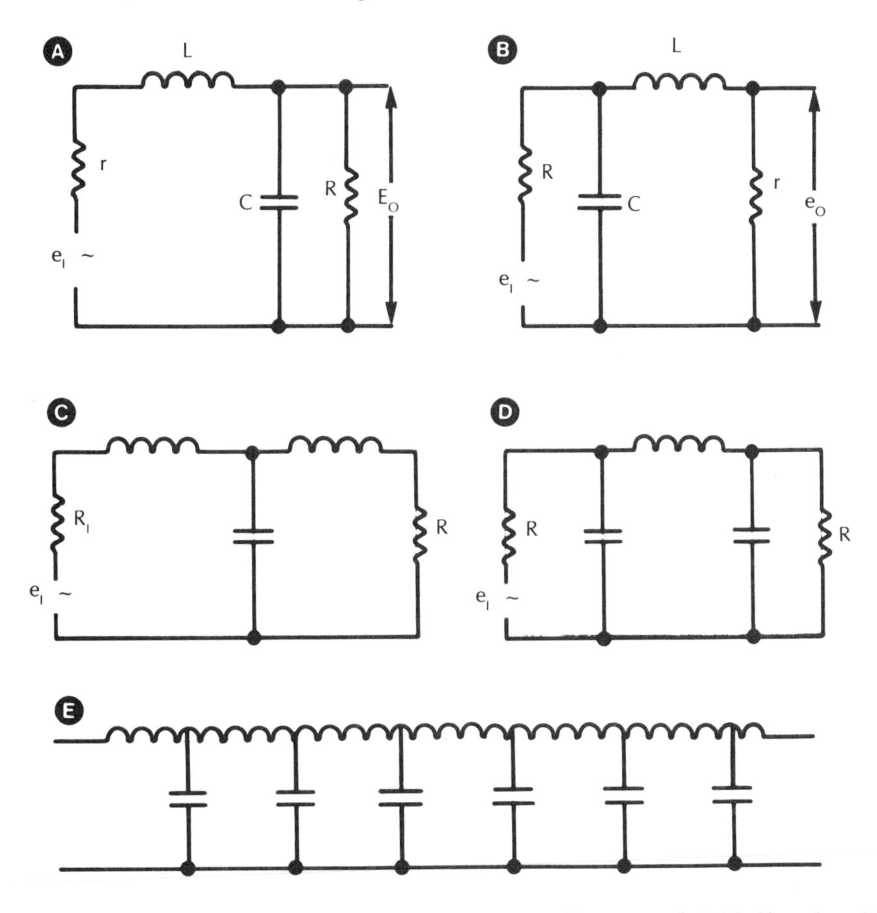

Fig. 10-24. Some basic forms of inductance-capacitance filter: (A) and (B) "half-sections;" (C) and (D) "full sections;" (E) the concept of distributed elements.

In (A) and (B) of Fig. 10-24, there are two "half-section" filters, using an inductance L and a capacitance C. Their response is characterized in terms of signal in, e_i~ and signal out, e_o. Correctly used, the configuration at (A) requires an input source impedance (or resistance) of r and an output load resistance of R. Conversely the configuration at (B) reverses the positions of R and r.

If you will compare these schematics with those at Figs. 10-12 and 10-13, you will see that in each case the resistance designated "r" relates to L and C in the same way the resistance does in the series resonant circuit of Fig. 10-12. And in each case the resistance designated "R" relates to L and C in the same way the resistance does in the parallel resonant circuit of Fig. 10-13.

This means that if smaller than the correct values of "r" are used, the resonance effect will appear, or be increased, while larger values of "r" will make the filter "droop" instead of peaking, as shown in Fig. 10-25. Conversely, larger than correct values of "R" will result in peaking, while smaller values will cause droop.

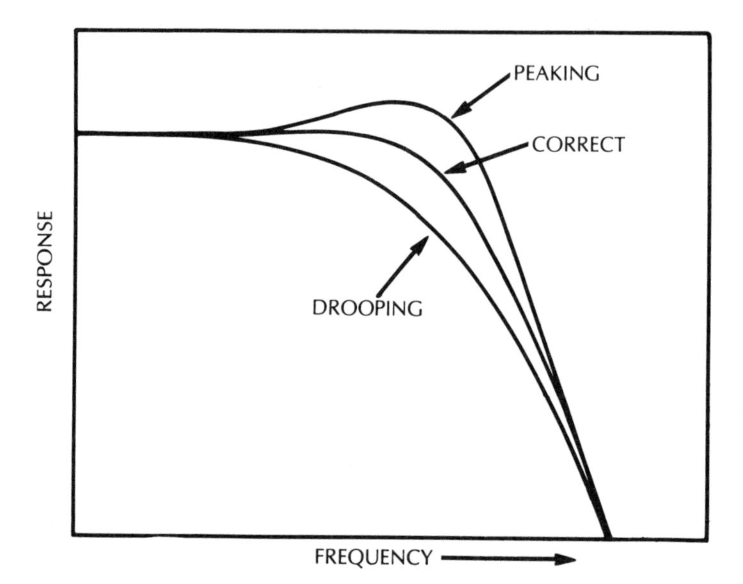

Fig. 10-25. How response changes with varying the values shown in Fig. 10-24 (see text).

As we said, the types at (A) and (B) are termed half-section filters. Those at (C) and (D) are full section filters, but again, of opposite types. Keeping the same nomenclature for the terminating impedances or resistance, smaller values at either end will result in peaking, with the circuit of (C), while larger values cause droop. Conversely, larger values at either end cause peaking with the (D) configuration, while smaller values cause drooping.

For comparison, (E) represents a "transmission line," in which, instead of lumped values of L and C, both inductance and capacitance are uniformly distributed alone the line. The inductance is due to the inductance of the actual conductor of which the line is made, unless some special material is used to artificially increase it, called "loading." The capacitance, is the actual capacitance between the line and adjoining conductors, also distributed along its actual length.

A schematic such as (E) can only approximate the elements, because they are actually infinitely divided, so that neither L nor C predominates at the ends. However, like the lumped circuits, there is a characteristic impedance that suits the combination of L and C. The line will work correctly when terminated at each end with that characteristic resistive impedance. Other values will make it behave differently.

Correctly terminated, if such a line has no losses (in either conductor or insulation) its only effect will be to produce a time delay proportional to its length: it will not have a cut-off frequency. In practice this is never realized, because there are always losses, which are also greater at higher frequencies. However, at frequencies before this happens, using wrong impedances can have different effects.

When correctly terminated, neither inductance nor capacitance effects predominate. If the line is opened at any point, then measuring either way, its impedance will "look like" the terminating impedance. But this is only true if it *is* correctly terminated. Using higher resistance values makes the capacitance predominate over the inductance, and using lower resistance values makes the inductance predominate over the capacitance.

Some filter networks, more complicated than those in Fig. 10-24 A to D are designed to approximate the line properties of E, but they never do quite, because that works only when the elements are both infinitely divided throughout.

SOME PRACTICAL FEATURES

The thing to watch for in practical circuits is that inductances seldom have either pure series or parallel resistance components. More sophisticated bridges are used to measure complicated external impedances. For example, a combination of a capacitor with resistors in one arm be used to analyze the behavior of an inductance over a small frequency range.

One more thing about balancing bridges with practical inductors in the circuit: We assume that a pure sinusoidal waveform of a single frequency is used for the test. If the waveform has any impurities that make it contain other frequencies, it may null for the main frequency but not for those in the impurities, so a complete null is not possible. When the fundamental or main frequency is nulled, the other frequencies will remain, practically undiminished. Nulling consists of eliminating the main frequency, so only residue frequencies remain. Figure 10-26 shows successive waveforms through null for two types of harmonic content.

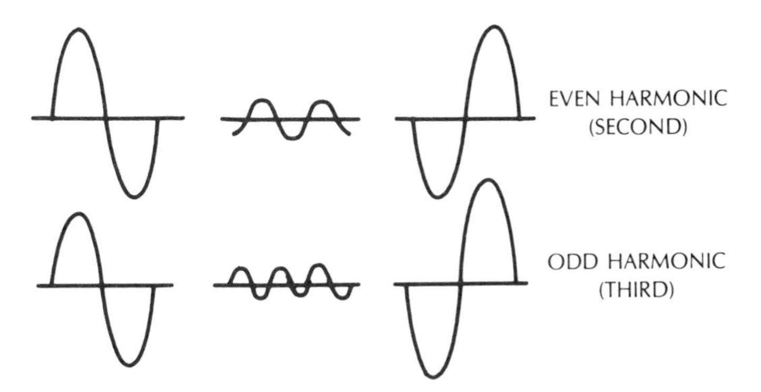

Fig. 10-26. Waveforms deviate from a true sine wave when either more than one frequency is present in the test signal, or when a component, such as inductance, is non-linear.

Also, the test waveform may be completely pure, but the inductance may not behave linearly. When a sinusoidal voltage is applied to it, the current taken may not be perfectly sinusoidal, or vice versa. This means that the pair of arms including the inductance will not produce a voltage containing only the input frequency at the junction. Because of the spurious frequency(ies) contained in the inductance voltage-current relationship (it is not pure reactance, because it causes distortion as well as additional resistance) the arms with the capacitance will have only the input frequency, while the inductance has an additional frequency. Null is again impossible.

Examination Questions

1. A series of reactance components have the following combinations of reactance and resistance, respectively. Calculate the impedance of each. 4.8 ohms, 1.4 ohms; 1680 ohms, 260 ohms; 160 ohms, 300 ohms; 75 ohms, 100 ohms; 90 ohms, 400 ohms.
2. In the following combinations, the first figure is the measured dc resistance of an inductor, the second figure is the measured impedance. Assuming the measured resistance is the only resistive component, calculate the reactance in each case. 22 ohms, 122 ohms; 26 ohms, 170 ohms; 3.2 ohms, 13 ohms; 40 ohms, 85 ohms; 13 ohms, 85 ohms.
3. The following sets of figures relate to series combinations of resistance and reactance. Find the voltage across the combination and the currents in the equivalent parallel elements that will produce the same combined voltage and current in both relative magnitude and phase.

CURRENT	VOLTS ACROSS RESISTANCE	VOLTS ACROSS REACTANCE
25 mA	1.12 V	3.84 V
1.7 amp	2.6 V	16.8 V

4. Find the series and parallel values represented in the voltages and currents in Question 3.
5. At what frequency will a combination of 1 mh and 1 mfd resonate? What will be the reactance, in series and in parallel, when this frequency is either halved or doubled?
6. A winding self-resonances at 100 kHz. Putting a capacitor of 0.0001 mfd in parallel with it reduces the resonance frequency to 50 kHz. Find the value of the inductance and its effective self-capacitance.
7. A bridge of the form shown in Fig. 10-18 is balanced when R1 is 1000 ohms, R3 is 10,000 ohms, C is 0.05 mfd and R4 is 1 megohm. Find the values of L and R2.
8. When the frequency is changed in the previous question, the balance does not hold. By changing the value of R4 and adding a resistor in series with C, a balance can be obtained that will hold for an octave. To achieve this, R4 needs changing to 2 megohms and the resistor in series with C has to be 50 ohms. What are the effective series and parallel resistance elements with the inductor?
9. A special tuned circuit consists of an inductance of 0.1 henry in series with an 0.05-mfd capacitor; across the series combination, an 0.04-mfd capacitor is connected in parallel. This combination has a minimum point where the series elements resonate and a maximum point where the series combination resonates with the parallel capacitor. Find the two frequencies at which these happen. (Hint: Use an adaptation of the Thevenin principle and verify your result).
10. In a configuration similar to Question 9, the series elements are in a sealed box. The external capacitor is 0.01 mfd. The points of minimum and maximum impedance occur, respectively, at 49 kHz and 83 kHz. Find the values of inductance and capacitance in the sealed box.

11

Diodes and Simple Semiconductors

The first diodes date back quite a long way, although semiconductors have only recently "come into their own." The old crystal set used a rudimentary diode, called a cat-whisker (actually a coiled, pointed metal prod) critically poised against a carborundum crystal to "detect" or rectify radio-frequency signals. Later, more precise diodes were made of germanium, and later of silicon, using a controlled amount of impurity in part of the structure to produce the needed characteristics.

BASIC CHARACTERISTICS

Junctions between pieces of metal, germanium or silicon, that are pure or have various kinds of impurities in controlled quantities as low as a few parts in a million, have the property of conducting current freely in one direction while behaving virtually as insulators to voltages that would drive current through them the opposite way.

While this is the basic property of any semiconductor, which means it "half conducts" or conducts current only one way, any such device has a number of other properties that may be pertinent to its use. Typical current-voltage characteristics for a germanium diode are shown in Fig. 11-1. The material does not suddenly switch from a conducting to a nonconducting state. It goes through a transition, in the region of zero voltage and current, where its resistance changes quite rapidly, but not instantaneously. The forward resistance, as measured by taking the voltage drop when 100 milliamps of current is flowing, which may be 0.25 volt, would thus be 0.25 divided by 0.1 is 2.5 ohms. The reverse resistance, determined by measuring the reverse current when 10 volts reverse voltage is applied, which may be 5 microamps, is thus 10 divided by 0.000005 or 2 megohms. Based on these figures, the ratio of reverse to forward resistance is 2,000,000 divided by 2.5 or 800,000:1.

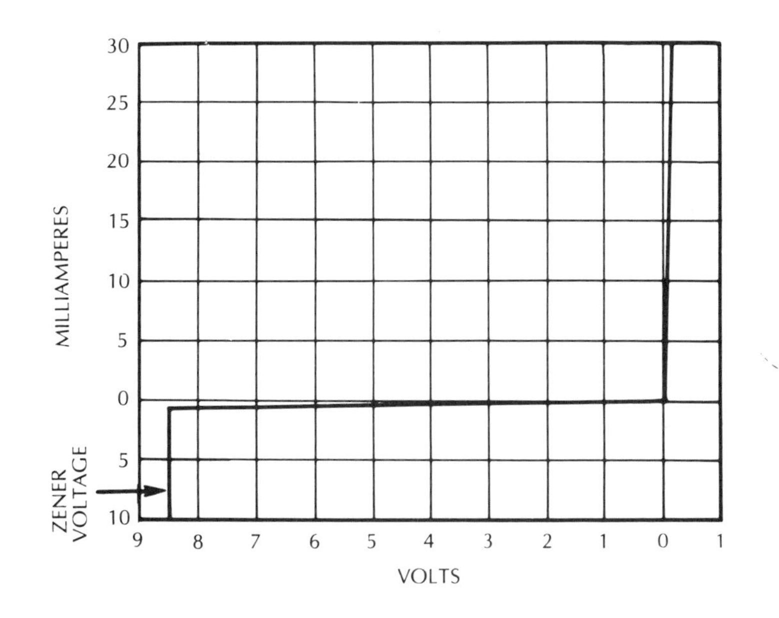

Fig. 11-1. Typical characteristics for a germanium diode.

This sounds like a very high ratio, as indeed it is. But such a statement can be misleading, because it merely relates equivalent resistances at two quite specific pairs of voltage and current values. For example, if the reverse leakage current is measured at 5 volts instead of 10, the current may be 4 microamps instead of 5, which represents 1.25 megohms instead of 2 megohms. Similarly, if the forward measurement is made at 50 milliamps instead of 100, the voltage drop may be 0.22 instead of 0.25, resulting in a forward resistance of 4.4 ohms instead of 2.5.

Thus, with this new set of values, the ratio is 1,250,000 divided by 4.4 which makes 284,000:1, a value much lower than 800,000:1. Come down to 1-volt reverse voltage, and the reverse current may be 3 microamps, making a reverse resistance of 0.33 megohm. And at 10 milliamps forward current, the voltage drop may still be as much as 0.15 volt, representing 15 ohms forward resistance at this current. With these figures, the ratio drops to 330,000 divided by 15 or 22,000:1.

At zero voltage and current, there is no basis for determining a resistance value. At this point the slope of the voltage-current curve is changing most rapidly. At 0.1 volt reverse, the current may still be 1 microamp, while at 1 milliamp forward, the volt drop may be 0.05 volt. This would represent a reverse resistance of 100,000 ohms and a forward resistance of 50 ohms, a ratio of only 2000:1.

What these figures tell is that forward-reverse ratios depend on the magnitude of signal being handled. If four such diodes are connected in a bridge (Fig. 11-2) to convert ac to dc, for each cycle or direction of ac, two diodes conduct and two do not. The two that conduct produce the specified forward voltage drop for the current being rectified, while the two that do not conduct produce the specified leakage current for the reverse voltage they sustain. Thus, if four such diodes rectify 10 volts at 100 milliamps, the forward voltage drop is 0.25 volt each, or 0.5 volt total for the two that are conducting at the same time. The leakage current is 5 microamps for each reverse diode, a total of 10 microamps, which will not be noticeable against the load current of 100 milliamps, being only one ten-thousandth of the value.

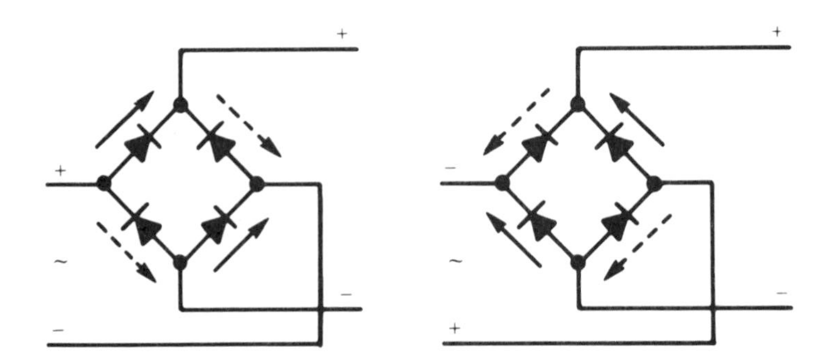

Fig. 11-2. The solid arrows show the direction of current flow for each phase of the ac. The dashed arrows indicate nonconduction.

But if the current comes down to 1 milliamp, still at 10 volts, the leakage is still 10 microamps, which is now a 1 percent loss. Also, if the voltage drops to 1 volt, the 0.5-volt forward drop (at 100 milliamps) means the input must be 50 percent higher to get the 1 volt out. All the losses are relative to the working voltage and current.

LOGARITHMIC PROPERTY

Over a certain range, the relationship between voltage and current is almost perfectly logarithmic in a good diode. For example, if the forward current at 100 milliamps produces a 0.25-volt drop and at 10 milliamps the drop is 0.15 volt, then, at 1 milliamp it will be 0.05 volt: each 0.1 volt represents a 10:1 ratio of current (Fig. 11-3).

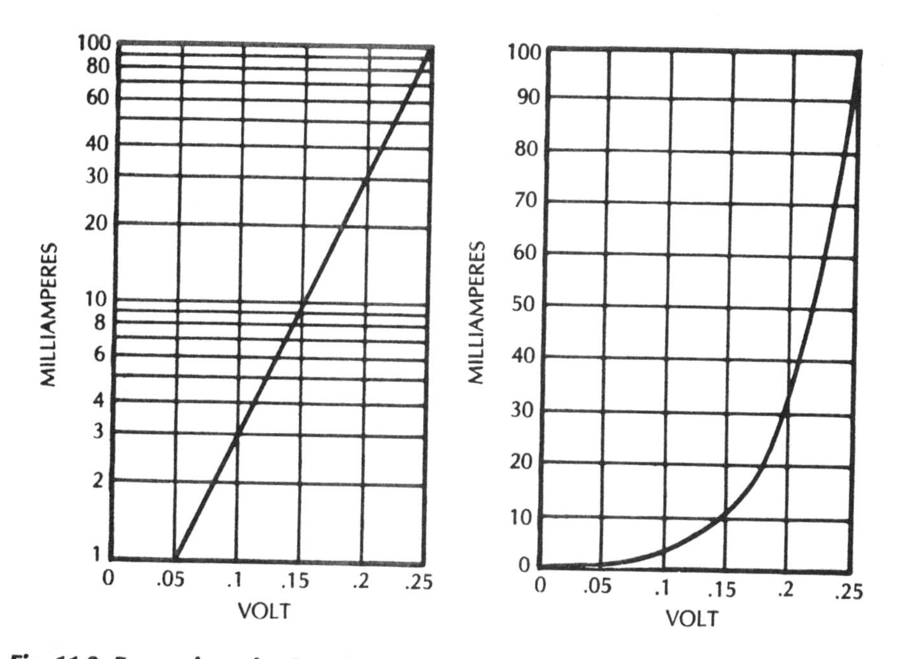

Fig. 11-3. Forward conduction characteristic plotted to a logarithmic scale (left) and to a linear scale (right).

Following this log law, 50 milliamps would show a 0.22-volt drop, 22 milliamps, 0.18 volt; 5 milliamps, 0.12 volt; and 2 milliamps, 0.08 volt. Such diode characteristics have been used to provide a logarithmic scale reading for a milliameter. If the meter takes 0.25 volt at 100 microamps to read full scale, then putting such a diode in parallel with it (Fig. 11-4) will permit the meter to read 100 milliamps at full scale.

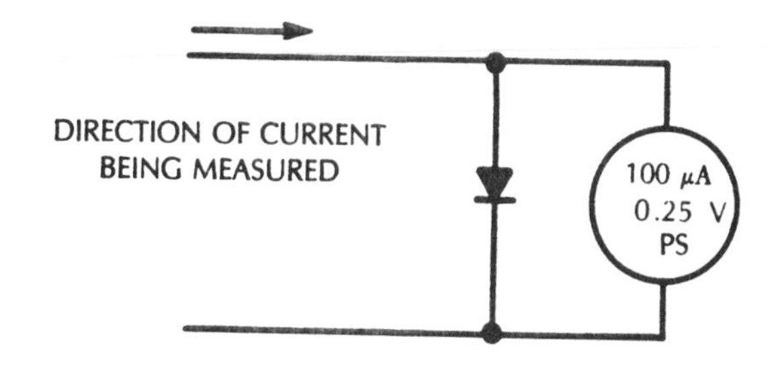

Fig. 11-4. Diode connected as shown can provide an approximately logarithmic scale for a moving coil meter movement.

At 0.6 scale reading, the voltage across the meter will be 0.15 volt, so the diode will pass 10 milliamps. At the 0.2 scale reading, the voltage will be 0.05 volt, so the diode will pass 1 milliamp, while the meter will pass only 20 microamps (one fifth of full scale). The scale of such a meter is shown in Fig. 11-5.

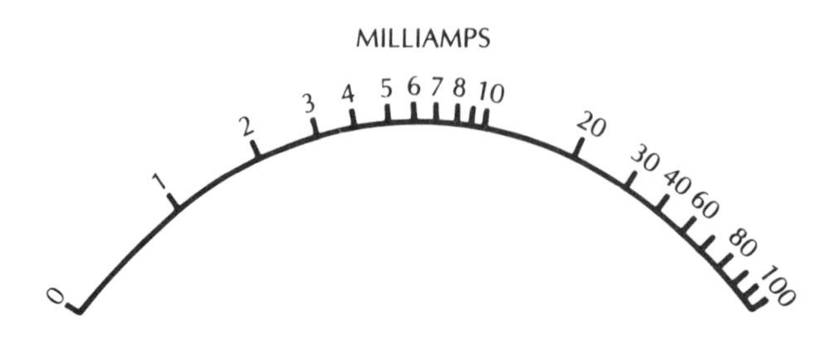

Fig. 11-5. Typical scale achieved with the circuit of Fig. 11-4.

One limitation of such use is that the forward resistance of a semiconductor diode, as well as its reverse resistance (which is not used for this application), can be quite dependent on temperature. Thus, the accuracy of calibration will depend on the diode being kept at a constant operating temperature. This is usually achieved by putting it in a temperature-controlled oven, where the temperature is held constant by a thermostat.

ZENER VOLTAGE

A diode has other properties besides the change that occurs between forward to reverse voltage and current. The high reverse resistance is not maintained to an indefinitely high voltage. A diode has what was once known as a "breakdown" voltage. If reverse voltage is applied until the diode starts to conduct at a higher rate than its leakage current (which accounts for the high reverse resistance), the current suddenly increases and, unless the current is artificially limited, the diode is quickly destroyed by this high current at a relatively high voltage.

For years, the only thing about this feature was that it dictated a safety reverse-voltage rating for the diode. But later it was discovered that the so-called "breakdown voltage" had a very useful feature, if the current flowing is controlled to avoid destruction of the diode. Like any other destructive action caused by electricity, the damage to a diode under these circumstances is due to heat, caused by the electrical energy dissipated.

If the current starts to rise at, say 50 volts, when it has been not more than, say, 10 microamps until that time, then it suddenly rises to, say, 100 milliamps, the dissipation would suddenly rise from 500 microwatts, which is safe, to 5 watts (or higher, if nothing limits the current to 100 milliamps) and destroy the tiny diode almost instantly.

The destructive action occurs because the diode has no internal means of limiting this dangerous rapid current rise. With no increase in voltage drop, current can rise almost indefinitely, unless a control device external to the diode limits it. Putting a resistance in series with the diode achieves this external control.

Maybe the forward rating of the diode is 100 milliamps at a voltage drop of 0.25. If this 25 milliwatt limit is exceeded, the delicate conduction zone probably would overheat and destroy the diode. If 25 milliwatts is the safe limit one way, it will probably be the same value the other way. At 50 volts, 25 milliwatts represents 0.5 milliamp or 500 microamps.

This is still a big step up from the leakage current of 10 microamps. So, with a supply of 60 volts (Fig. 11-6), a drop of 10 volts at 500 microamps will require a series resistor of 20 K. Then the diode will pass the same 500 microamps with a 50-volt drop, which equals its maximum rated dissipation of 25 milliwatts.

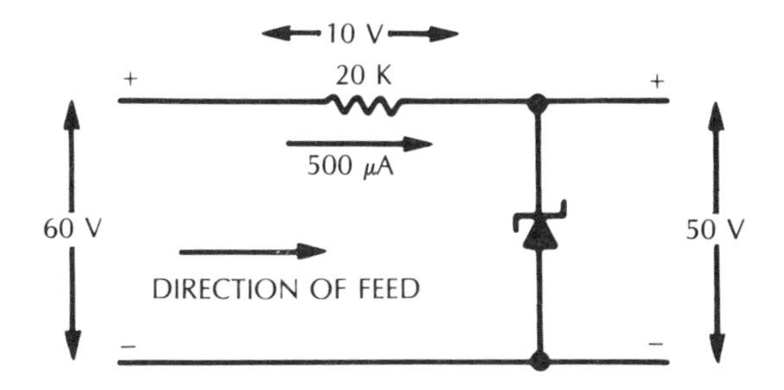

Fig. 11-6. Simplest possible circuit using a zener diode as a regulator.

Larger zener diodes, as components built specifically for this use are called (named for the man who discovered the characteristic) may be capable of dissipating as much as several watts. Zener diodes are manufactured specifically to utilize a stabilized value of this voltage. The important property is that the voltage remains very close to constant over a considerable current range above the leakage current. The important thing, if such a diode is designed to maintain a constant voltage in spite of current variations, is that the current remains well above the leakage current value and never exceeds the maximum established by permissible dissipation.

For example, if the applied voltage may vary from say 51 to 60 volts (Fig. 11-7), the current through the 20 K resistor will vary from 50 to 500 microamps, while the voltage will stay very close to 50 across the zener diode, because the lowest current is still 5 times as much as the leakage current.

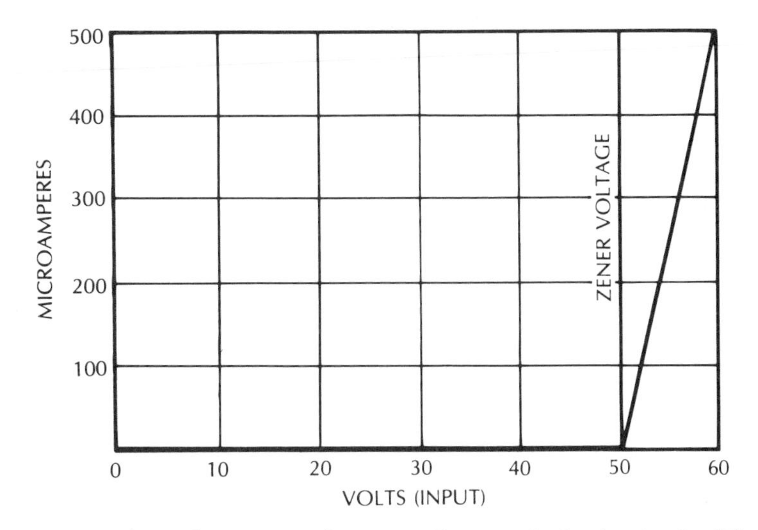

Fig. 11-7. Input volts against current, drawn on a linear scale, for the circuit of Fig. 11-6.

But obviously, if the applied voltage drops much below 51 volts, this is no longer true. In fact, below 50 volts the current becomes the leakage current, and there will be very little voltage drop across the 20 K resistor.

While most diodes have something approaching a logarithmic change of current with applied voltage in the forward direction for smaller changes in the region near maximum current, say, above one-tenth maximum, the voltage drop does not change very much. With a germanium diode, this maximum drop may vary from about one-tenth to nearly half a volt, depending on the type and intended application.

Contact Potential

With silicon diodes, the forward voltage drop is somewhat higher, but it reaches nearly the maximum drop more rapidly when viewed against a linear current scale (Fig. 11-8). Such a variation may still be close to a logarithmic rate at low currents, which means that at lower forward voltages, the forward current is extremely low compared with normal working currents. It is at a level that might be considered leakage current in comparison to circuit operation. Because of this effect, the voltage at which the current reaches the working region is called "contact potential."

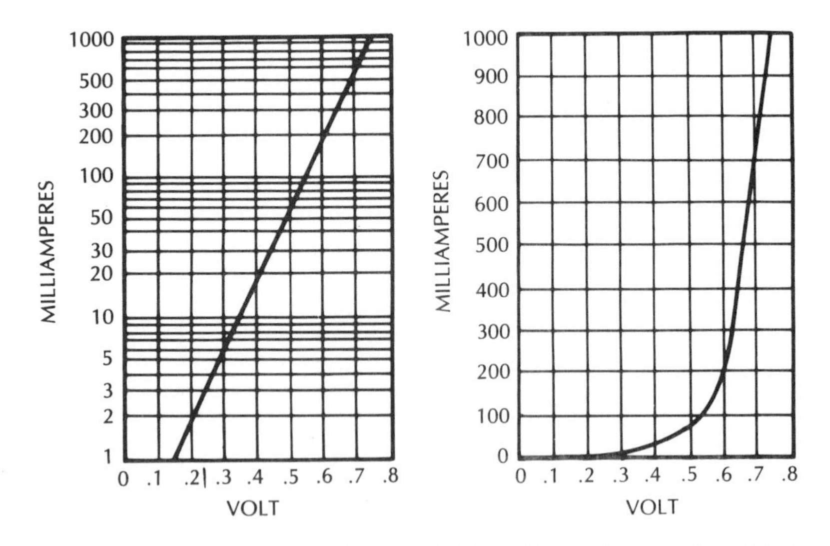

Fig. 11-8. Forward characteristics of a typical silicon diode, plotted to logarithmic and linear scales.

Silicon diodes have a much higher reverse-to-forward ratio than do germanium; therefore, they may be regarded as almost perfect insulators to reverse voltages. The reverse current of some silicon diodes before the zener voltage is reached may be in the nanoamp (one-thousandth of a microamp) region.

Of course, these and practically all semiconductor parameters are affected by temperature changes.

By far, most diodes are used for the simple process of "rectifying" current, which they do very well since they permit current to flow one way but not the other. And in electronic instruments, rectifiers are used quite frequently for measurement purposes.

Before the days of digital instrumentation, when moving coil instruments were used as indicators, the bridge arrangement was very common. But more recently, silicon diodes, and transistors in which two of their terminals behave as diodes, are much more commonly used in a different way. With the older germanium diodes, they behaved as a sort of "on-off" switch with the changeover occurring at close to zero voltage.

But with silicon diodes and transistors, the voltage at which this change takes place is away from zero, which makes a whole new set of technology available for its use. Silicon diodes do not begin to conduct, to any detectable extent, until an appreciable fraction of a volt is applied in the "forward" direction. It is true that the precise voltage at which this may be regarded as occurring depends on other values in the circuit, as well as temperature and other variables, but the effect is so much more noticeable than with germanium, that the voltage at which this "switching" takes place is given the name of *contact potential.*

Reverse current is infinitesimal, until *Zener voltage* is reached, so it can be regarded as zero. But forward current is also very small, until *contact potential* is reached, so it too can be regarded as zero. However, putting the same diode in a circuit with much higher resistance values will make the contact potential apparently lower.

INSTRUMENT RECTIFIERS

In the past, when electrical and electronic meters and instruments more commonly used a moving-coil movement (Fig. 8-10), which has the property of measuring average current in only one direction. The average value of an ac current flowing for an equal time and at an equal amplitude in alternate directions is zero, which is what such a meter will read if applied to ac. To measure ac, a moving-coil instrument needs a rectifier to make all the current flowing through the coil move in the same direction.

This can be achieved by one of two forms of rectification: full-wave or half-wave. Full-wave utilizes both halves of the ac waveform, but each flows through the movement in the same direction. Instrument rectifiers have four diode elements (Fig. 11-2). Two permit the current to flow in one direction and prevent it flowing in the other direction, and vice versa, according to the momentary alternating cycle.

In a power supply rectifier, a diode must handle the full applied voltage as a reverse voltage (unless several diodes are used in series to achieve an adequate reverse voltage capacity), and the leakage current is relatively unimportant. If an instrument rectifier handled the full voltage being measured, the leakage current would be a substantial portion of that applied to the meter, with its multiplier; thus, accuracy would be impaired. To an instrument rectifier, the diodes are connected across only the moving coil, so that the reverse voltage is as low as possible, thus keeping the leakage current to a minimum (Fig. 11-9).

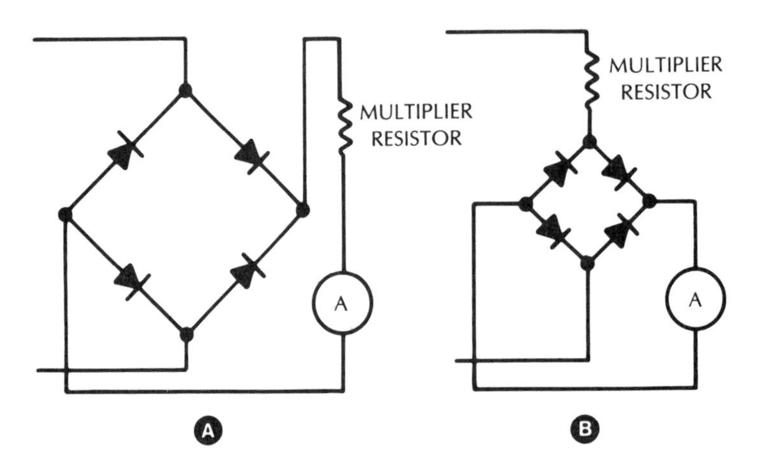

Fig. 11-9. Comparing the use of a rectifier at the meter input voltage (A), with using it directly across the meter coil (B).

In substance, looking from the ac input terminals, the two diodes connected to each terminal serve as a switch to connect the current from that terminal to the correct terminal of the movement. In whichever direction the current is momentarily flowing, it must pass through two diodes in series with the movement, while a nonconducting diode is across each forward conducting diode and the meter movement. Thus, leakage current through the two nominally nonconducting diodes does not pass through the movement, and the forward voltage drop across the conducting diodes is in series with the voltage drop across the instrument coil. Provided reverse current is less than one-thousandth of the forward current, which any good modern rectifier will ensure, the loss of current through the movement due to leakage will cause an error of less than 0.2 percent.

CALIBRATION

If the meter is used to measure voltage, it is calibrated on current, and the series resistor used with it is chosen to enable it to indicate voltage. Usually, this calibration is based on an assumed sinusoidal rms current and voltage. This involves the form factor discussed in Chapter 5. Thus, if the full-scale reading is to be 100 volts rms, this is 141.4 volts peak on a sinusoidal waveform, which corresponds to 141.4×0.637 or 90.07 volts (90 is well within 1 percent) rectified average.

This means that the meter must be arranged so that a rectified 90-volt average, which is what it measures, gives an indication of 100 volts rms, which it represents. If the meter takes 100 microamps for a full-scale reading, and the rectifier is better than 99 percent efficient (because less than 1 percent of the current is lost as leakage current), then the meter's total resistance in ohms at the ac terminals must be 90 divided by 0.0001 or 0.9 megohm.

Actually, for a full-scale reading, the meter will drop its 0.25 volt full scale (if this is the figure in the movement specification). This is average, which is 0.35 volt peak, plus about 0.35 volt peak across each rectifier diode, making a total peak voltage drop of 1.05 volts for a full-scale reading. This corresponds with 1.05 × 0.637 or 0.67-volt rectified average. So the series resistor (Fig. 11-9) must drop 90 − 0.67 volt or 89.4 volts. A 0.9 megohm 1 percent resistor should produce a reading within 1 percent. Because of the high-value series resistance, the scale will be close to linear. At the very bottom end, there may be some loss of linearity, due to the fact that the rectifier loses its very high efficiency at very low currents and voltages.

LOW-VOLTAGE SCALES

Now, assume that the meter is required to read 2 volts (rms) full scale. This changes the picture. Peak voltage is now 2.828 and the rectified average is 2.828 × 0.637 or 1.801 volts. The rectified average dropped across the meter at this point on the scale will be 0.67 volt, as before, leaving 1.801 − 0.67 or 1.134 volts to be dropped across the series resistor, which makes the required value 1.134 divided by 0.0001 or 11.34 K.

But now look at the center-scale reading, which would represent 50 volts on the 100-volt scale, because this yields so close to half meter current as to show indiscernible error. At center-scale reading, the peak voltage across each rectifier diode will be 0.3, instead of 0.35 for full-scale, while the meter drop will be half of 0.35, or 0.175 volts, yielding a total peak drop of 0.775 volts for half scale, as against 1.05 volts for full scale. This represents an average of 0.775 × 0.637 or 0.504 volt.

The current will be half the full-scale current and, therefore, will produce half of the 1.134 volts across the series resistor or 0.567 volt. Thus, the total rectified average voltage to produce a half-scale reading will be 0.504 + 0.567 or 1.071 volts. This would represent an rms value of 1.071 × 1.11 or 1.189 volts. So half scale on the 2-volt range will not be 1 volt, as 50 volts is half of 100 volts, but 1.189 volts, which is quite different—about 19 percent different to be precise.

Going down to one-fifth scale, the peak voltage across the meter will be 0.07 volt, while that across each diode will be 0.23 volt, making a total of 0.53 volt peak, which is 0.53 × 0.637 or 0.338 volt rectified average. One fifth of full-scale current will produce one-fifth of 1.134 volts across the series resistor, or 0.227 volts, making a total of 0.338 + 0.227 or 0.565 rectified average, or 0.565 × 1.11 which is 0.627 volt rms. So a one-fifth scale reading, on the 2-volt full-scale range, will represent 0.629 volt, rather than 0.4 volt, which is one-fifth of 2. This is more than a 50 percent difference at this point on the scale (Fig. 11-10).

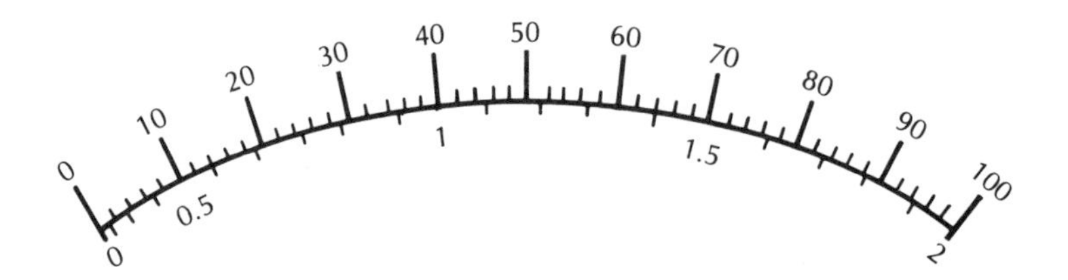

Fig. 11-10. Comparison of scales drawn to measure 100 volts, and 2 volts, as full-scale reading on ac.

Actually, these calculated values may not be strictly accurate, because the meter integrates the effect over the entire waveform. Those calculations, in one respect (use of the form factor), assume a sine-wave form. The voltage may have that form, but the current will not have when the resistance across the sinusoidal voltage is changing throughout the cycle.

INPUT RESISTANCE

At the full-scale reading, the peak voltage is 1.05 at a peak current of 141.4 microamps, so the input resistance of the meter and rectifier at full-scale peaks will be 1.05 divided by 0.0001414 or 7.4 K. At half-way to peak current, which is 70.7 microamps, 0.95 volt makes the resistance at this point on the waveform 0.95 divided by 0.0000707 or 13.4 K. At one-fifth peak current, which is 28.28 microamps, the voltage is 0.53, making the resistance at the point 0.53 divided by 0.00002828 or 18.7 K.

So the instrument consists of a fixed resistance of 11.34 K in series with one that, when giving a full-scale reading, varies from 7.4 K at peaks to 13.4 K halfway to peaks and 18.7 K at one-fifth of the peak amplitude. For a half-scale reading, the resistance of the rectifier and meter is 13.4 K at peaks and 18.7 K at two-fifths of peak amplitude.

The above assumes that the current is sinusoidal. But if the voltage is, which is what the instrument is calibrated for, the increased resistance at points below peak will reduce the current. So the 11.34 K value will be too high to allow a full-scale reading with 2 volts rms of a sine waveform.

The only real way to calibrate such a meter is to use carefully measured voltages and mark the scale accordingly, after first finding the resistance that makes 2 volts rms read full scale. Calculation does not provide a direct, reliable way. One thing is evident: The scales for 100 volts full-scale and for 2 volts full-scale require a very different spacing; you cannot use a simple 50-factor multiplier (Fig. 11-10).

HALF-WAVE RECTIFIER

Many meters use a half-wave rectifier (Fig. 11-11), which halves the drop through the diodes because current each way passes through only one diode. It also saves a couple of diodes, since only two are needed instead of four. But current flows through the instrument only during one half wave. During the other, it bypasses the movement.

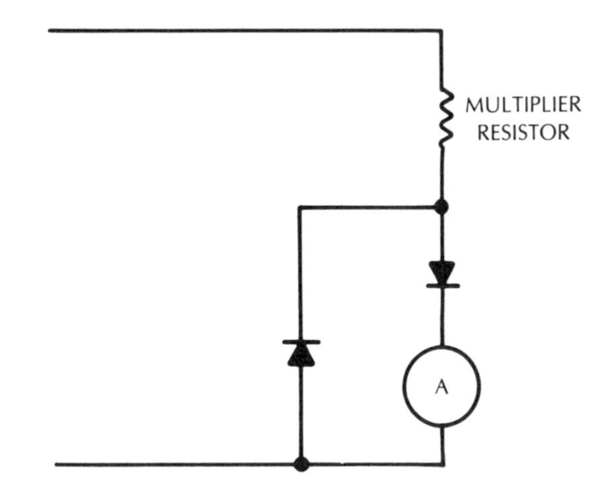

Fig. 11-11. Half wave meter rectifier circuit, at one time used on some inexpensive meters.

This means that the average rectified current is half of 0.637 or 0.3183 of the peak, again assuming a sine waveform. To get the same full-scale reading, the series resistor needs to be about half the value necessary with a full-wave rectifier. For 100 volts full-scale, using a 100 microamp movement, the resistance needs to be 450 K, instead of 900 K.

MULTIMETERS

Many meters are designed to measure either ac or dc, with different connections. For example, putting a capacitor in series with an ac meter will keep any dc out of it (Fig. 11-12). Disconnecting the rectifier altogether, so the coil responds only to the dc part of whatever current flows through it, will keep the ac out (Fig. 11-13).

Such changes of connection become complicated, and the sets of series resistors for measuring ac and dc need to be different. For example, if the meter uses 100 microamps for a full-scale reading, it needs 1 megohm in series with it to make it read 100 volts dc full scale, 900 K to make it read 100 volts ac full scale with a full-wave rectifier, or 450 K if the rectifier happens to be half wave.

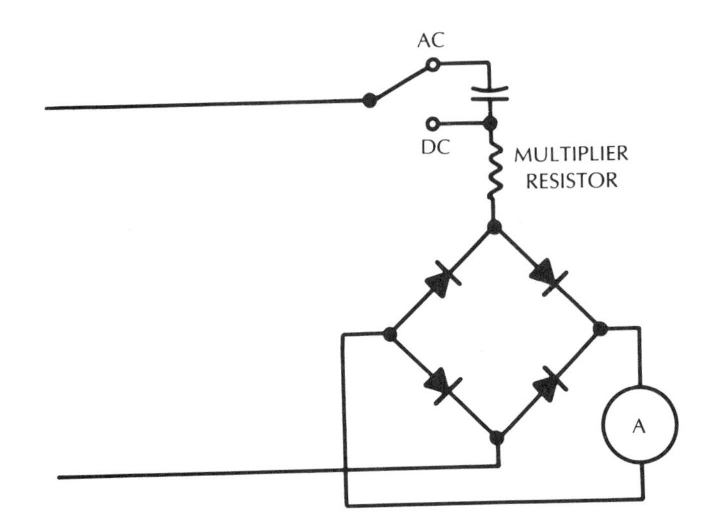

Fig. 11-12. The capacitor eliminates any dc component when the meter is set to read ac.

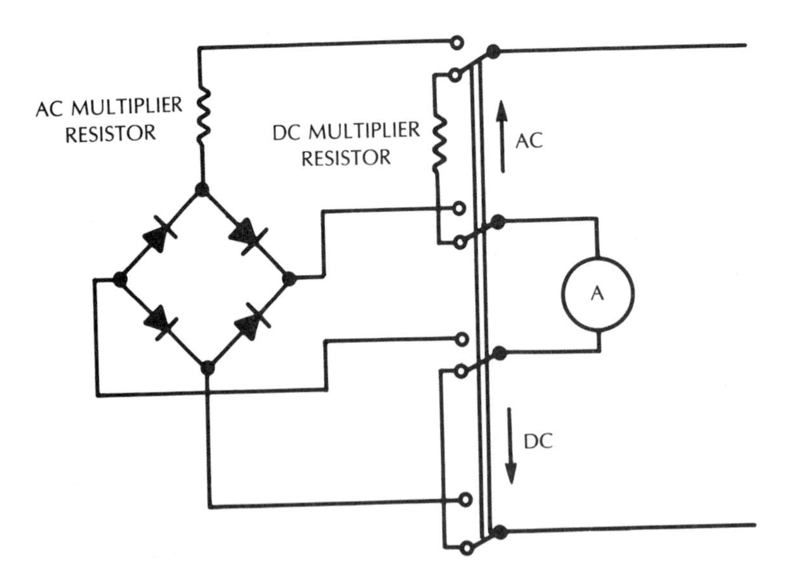

Fig. 11-13. This switching arrangement eliminates any possibility of reading ac when the meter is set to read dc.

This can be simplified in multi-range instruments by using the same set of series resistors for both ac and dc, and then shunting the coil on dc so the total current through the coil and its parallel resistor is the same as for the full-scale ac reading (Fig. 11-14). Suppose the coil resistance is 750 ohms (75 millivolts at 100 microamps for full scale). Then, with a full-wave rectifier the resistor needs to be 750 divided by 0.111 or 6750 ohms, while for a half-wave rectifier the resistor needs to be 750 × 0.45 divided by 0.55 or 535 ohms.

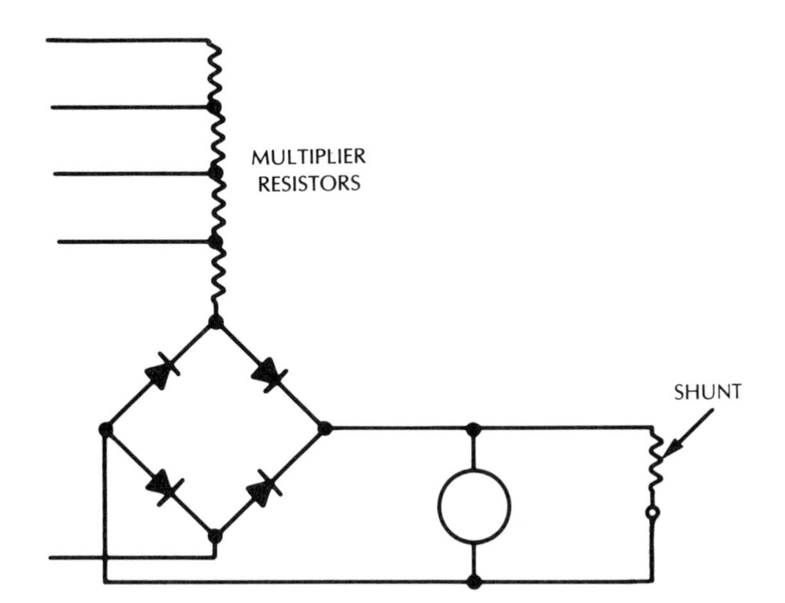

Fig. 11-14. One method of modifying a universal meter to read ac or dc, where the shunt resistor provides ac rms and dc readings, provided the ac is sinusoidal in form.

Some circuits merely connect the parallel resistor, without changing anything else. This will correct the reading, assuming the type of waveform is as specified. But if an ac waveform is connected when the switch is set for dc, or vice versa, the reading will be incorrect. For example, if the voltage is dc and the meter is set for ac, then with a full-wave rectifier the reading will be 11 percent high. An actual voltage of 90 will read as 100 volts.

With a half-wave rectifier, it depends on which way the dc voltage is connected to the meter. One way the reading will be zero, because the current will bypass the meter through the shunt diode. The other way it will read twice that of a full-wave rectifier meter, so that 45 volts will read as 100. Of course, switching in a capacitor for ac readings will prevent this error, because a dc voltage will then be blocked altogether.

Now, suppose the voltage is ac and the meter is set for dc readings. With the full-wave meter, the reading will be 10 percent low or 90 percent of the actual. With a half-wave meter, it will be 45 percent of the actual voltage, which is very low. As stated before, if the rectifier is removed from the circuit so the current passed by the voltage multiplier resistor is passed directly to the coil, the meter will indicate only dc, ignoring any ac that passes with it.

For this reason, notice that the rectifier must be removed from the circuit (as in Fig. 11-13), not merely bypassed (Fig. 11-15). When the rectifier is simply bypassed, the other diode remains across the instrument in the direction in which the meter normally reads, so current passes through either diode. But ac currents flow in both directions. This means that the diode will bypass current in the unwanted direction, but not in the normal-reading direction, so that some reading will still be obtained on ac, although the relationship of the reading to the actual voltage is not easily determined because the meter-shunting effect is a very variable quantity.

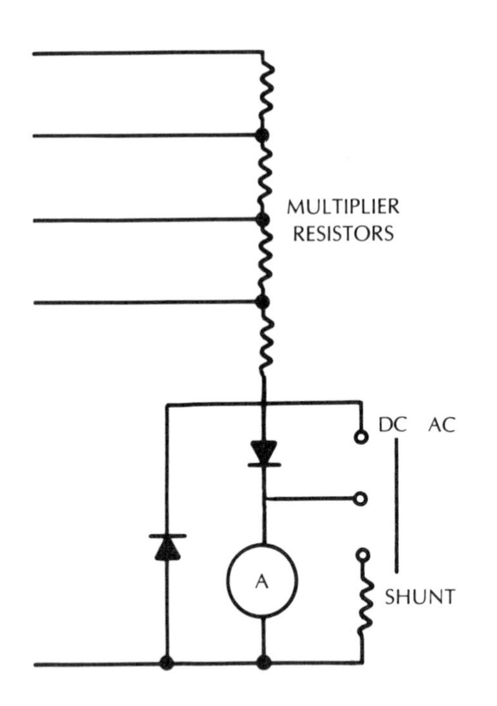

Fig. 11-15. An alternative method of changing from ac to dc with a half-wave instrument rectifier.

Another cause of error arises when the waveform is not sinusoidal, or when the voltage being measured is a combination of ac and dc (or both). Two examples of regular nonsinusoidal waveforms are square and triangular. There are many other shapes that are periodic and truly ac (equally balanced in both directions). A square wave, rms, peak and rectified average all have the same value. So the meter will read 11 percent high on this waveform, just as it does on dc.

If the waveform is balanced (equal up and down times) the half-wave rectifier will also register 11 percent high, because the negative-going halves will not be measured and both are identical. On a triangular waveform, the average square is one-third of the peak square, so the rms is the square root of ⅓, or 0.577 times peak. The rectified average is a simple 0.5 times peak. The meter will read 11 percent higher than the true rectified average because that is how it is calibrated. This will be 0.555 times peak, which is a little lower than the true rms of 0.577 times peak (3.75 percent lower to be precise).

An unbalanced square wave is something else again. Take an example where the up time is half the duration of the down time (Fig. 11-16). Now, peak voltage is a term with some ambiguity. If the up part is twice the amplitude of the down (which would be required for total flow each way to be equal or balanced) then, the peak-to-peak voltage would be three times the down amplitude. If the peak is derived from the standard form, where it is half of peak to peak, then the peak will register as 1.5 times the down amplitude. But if the peak indicates the highest deviation—in either direction—from center (the level about which the flow is balanced), the maximum deviation is in the up direction, where it is two times the down amplitude.

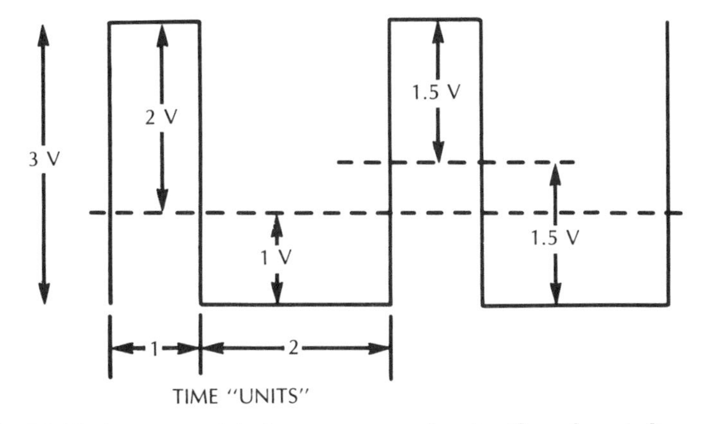

Fig. 11-16. An asymmetrical square wave showing the relevant dimensions.

The quantity that is of interest here is the relationship between the rectified average and the rms. The rectified average is found by multiplying each amplitude by its duration and then dividing by the total duration. If the up half wave has an amplitude of 2 units for 1 unit duration, the product is 2; if the down part has an amplitude of 1 unit for 2 units duration, the product is again 2; the total is 2 + 4 or 4. The total duration is 1 + 2 or 3 units, so the rectified average is four-thirds or 1.333.

The rms is found by squaring individual values, multiplying by the duration, summing, dividing by the total duration, and taking the square root of this result. In this case, the up has an amplitude of 2 units, the square of which is 4; if the duration is 1 unit, the product, therefore, is 4. If the down amplitude is 1 unit, of which the square is 1, and the duration is 2, the product is 2. The total of squares is 4 + 2 or 6. Dividing by the total duration of 3 yields an average square six thirds or 2. So the rms is the root of 2, or 1.414. This yields a form factor of 1.414 divided by 1.333, which is approximately 1.061. So this waveform will read approximately 5 percent high of the true rms, because the meter is designed for a waveform with a form factor of 1.11.

Take as one more example, to see the trend, a wave in which the up half wave is one-third the duration of the down and the amplitude of the up is three times the amplitude of the down. The rectified average is now six fourths or 1.5 times the down amplitude. The rms is found by squaring 3, which is 9, adding 3 times the down amplitude squared, which is 1, making 9 + 3 or 12, and dividing by 4, to get an average square of 12 over 4 or 3. Rms is the square root of this, or the root of 3 is 1.732. This yields a form factor of 1.732 divided by 1.5 or 1.15, which is now more than that for a sine wave. This waveform will read approximately 4 percent low of the true rms.

With mixed voltages, made by combining ac and dc, different things can happen. First, provided the ac peak is not higher than the dc value, the presence of the ac will not affect the dc reading, because no rectification is taking place. So if the meter is switched to dc, it will read the dc directly and correctly and ignore the ac components.

In this case, the ac can be measured separately by switching to ac, if this feature is included within the meter (Fig. 11-12), or by adding an external capacitor, which need be large enough only to ensure that the reactance of the capacitor is less than one tenth of the total meter resistance on the range used, which will ensure that the reading error is less than 1 percent. Be sure to use the ac position, too, to get a correct ac reading.

If the ac peak exceeds the dc value, different things happen. If the rectifier is still in the circuit for dc readings (which is sometimes done so that the meter will read voltages either way without having to reverse connections) the reading will be higher on a full-wave rectifier than the dc component because part of the ac riding it is "turned around" (Fig. 11-17).

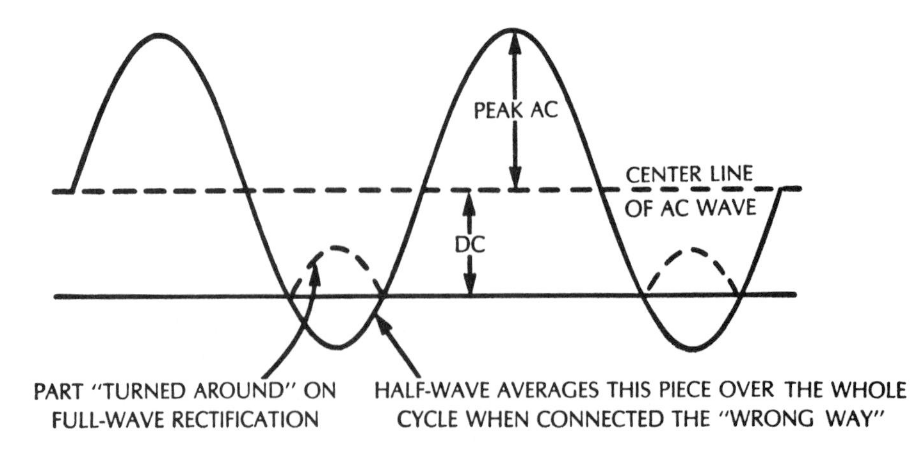

Fig. 11-17. Some of the things that happen when a combination of ac and dc is being measured, where the peak ac is greater than the dc.

With a half-wave rectifier, when the ac peak is less than the dc value, the meter will read zero one way and the correct dc the other. If the ac peak is more than the dc value, a reading will be obtained, even when the dc connection allows current to flow the wrong way, because the ac peaks exceed the dc part for part of the reverse half-cycle.

Connecting the meter the correct way for the dc part will cause the meter to read higher than the dc component, because part of the ac cycle is chopped off and only when that chopped off part is retained does average over the whole cycle yield a true dc indication, ignoring the presence of the ac component.

TEMPERATURE COEFFICIENTS

Diodes present a variety of additional characteristics that may be specified in different ways. Temperature affects both forward conduction and reverse insulation characteristics. But it is not normally practical to give these properties a "temperature coefficient" rating, because too many things are varying at the same time. The main thing to recognize is that a rise in temperature is dangerous.

Perhaps more important to diode operation, particularly in circuits where power rectification is involved, is the range of safe values that can or must be considered. The actual safe value varies according to how much of the time, or for how long at a time, the potentially dangerous current or voltage is applied.

Where a rectifier feeds into a resistance load, the current waveform follows the voltage waveform, both of which look like a sine wave with the down half waves turned upward. The important figure in this mode of operation is probably the average current flowing. Thus, a diode or rectifier operating under this condition may be rated in terms of the dc output current, which will be the average of the ac rectified value.

But ac may not be rectified in such a way that its current waveform matches the voltage output to produce a dc value that averages the rectified ac. This happens only when the rectifier feeds a resistance load directly, so true dc is not obtained; instead, a raw rectified ac (Fig. 11-18) results.

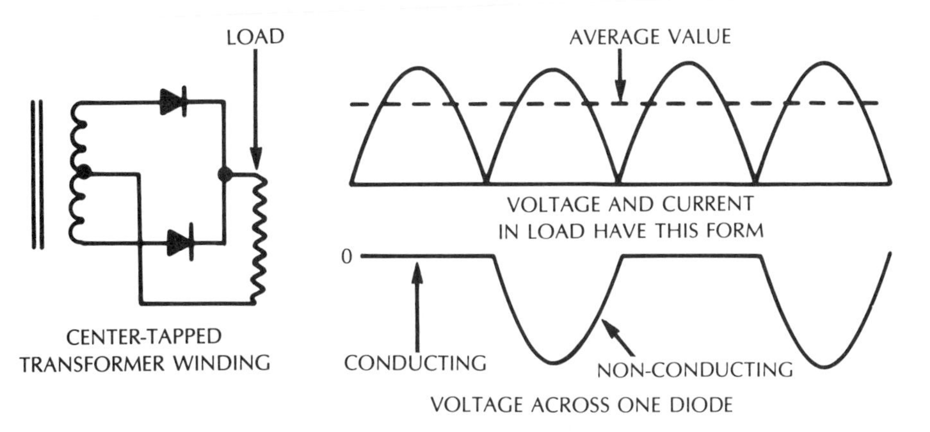

Fig. 11-18. Waveforms associated with full-wave rectification when no smoothing is employed. The load resistor receives raw rectified ac.

A current that really is the average of what the ac would be can be achieved as a steady DC output by using an inductance (commonly called a choke in this application) to smooth out the voltage variation at the rectifier output (Fig. 11-19).

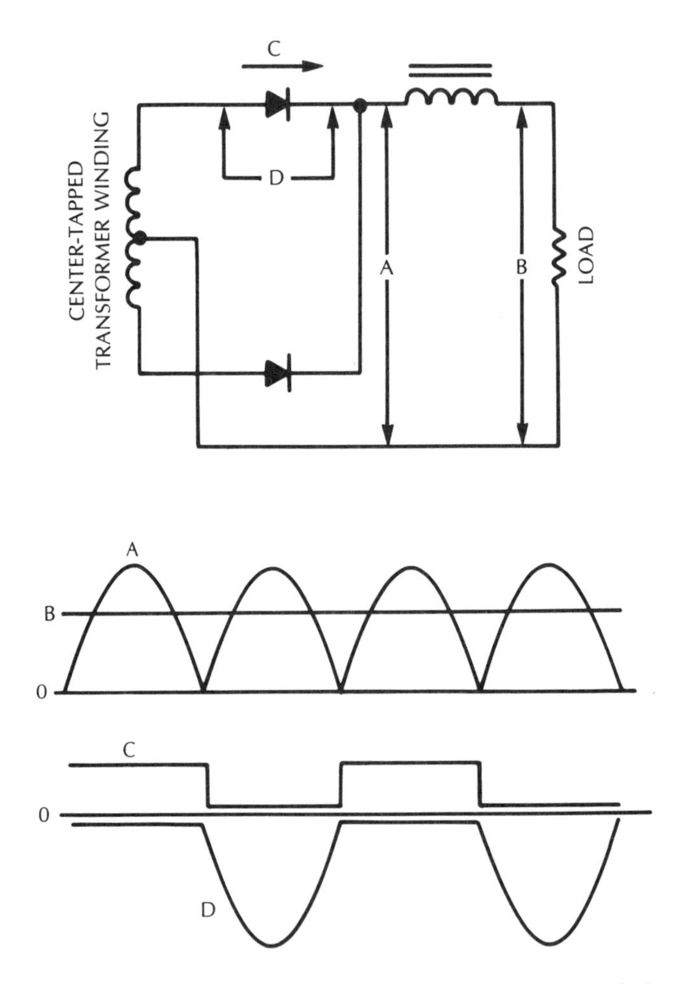

Fig. 11-19. Waveforms associated with full-wave rectification when choke smoothing is employed to produce a good approximate dc.

Here, as in Fig. 11-18, each diode conducts for all of its own half-cycle of ac, but it conducts at a practically steady current through that half cycle. The choke smooths the difference in voltage between the raw rectified ac at the diode output and the smooth dc value delivered to the load. The reverse voltage applied to each diode is the same as for (Fig. 11-18).

With another method, the rectifier charges a capacitor which feeds the dc load, providing a considerably smoother dc voltage (Fig. 11-20). But this method reduces the current-delivering capacity of the rectifier diode. True, it passes only the same average current (for the same dc output) whichever circuit is used. But this current is concentrated into brief pulses at the peak of each ac wave or half wave.

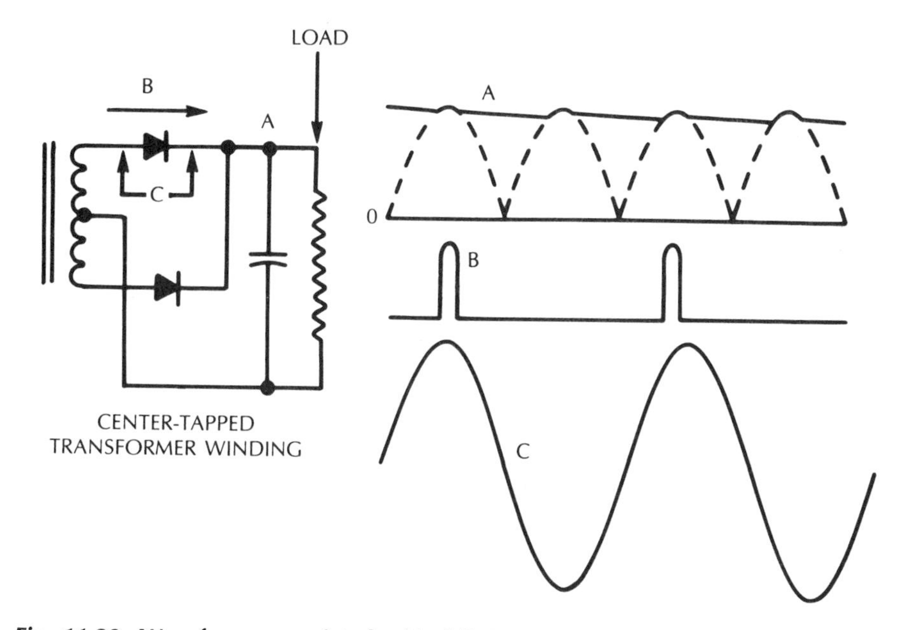

Fig. 11-20. Waveforms associated with full-wave rectification when a storage (or reservoir) capacitor is used to provide a good approximate dc.

Suppose the average rectified current (dc output) is 1 amp and the rectifier keeps the capacitor charged during one tenth of each half cycle. This means that the average current during the one-tenth of the time that current is passing through the diode (one or other) must be 10 times the average current delivered by the capacitor to the load.

If the voltage drop across the diode is constant, whatever the current, this will not increase its dissipation. It will be 10 times the current at the same voltage drop, but for only one-tenth of the time; averaged out, the dissipation would be the same. But the voltage drop is not constant as the current increases. Making the current 10 times as high may increase the voltage drop to twice the voltage (although not to 10 times), which means the dissipation is doubled by concentrating the current into pulses in this way.

RATINGS

To meet this situation, diodes or rectifiers sometimes have more than one maximum rating. There is a maximum dc rating, based on a fairly uniform averaged current in which the peak is not more than about 1.57 (which is pi over 2) times the average. Also, there is an instantaneous peak rating, which is the limit of current to which short-duration pulses can safely rise, provided the average rating is also met.

Sometimes the pulse rating is based on a repetitive pulse in a service cycle; e.g., the pulse may be presumed to be repeated at intervals 50 times the duration of the pulse, which means the charging period is one-fiftieth of the cycle or half cycle, as the case may be (Fig. 11-21). In this case, a third rating may sometimes be added—the absolute maximum current. In contrast with a maximum pulse, based on a prescribed repetition interval, this specifies the maximum current that can be safely handled non-repetitively; that is, as a pulse isolated in time from any other pulses.

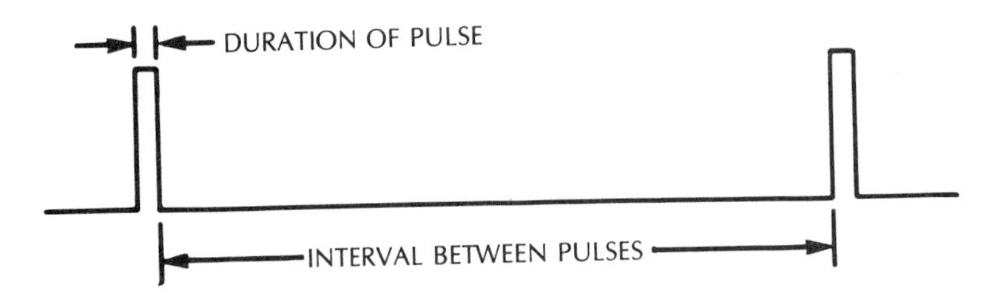

Fig. 11-21. An illustration of the service interval between pulses, used for specifying the pulse rating of rectifier diodes.

Exceeding this absolute maximum current value, even for a virtually instantaneous pulse, runs the danger of blowing the diode instantly. Exceeding the average current rating, as an average current, risks burning out the diode after a few cycles, and the greater the excess, the more quickly will it blow. Exceeding the instantaneous pulse rating in a circuit using a storage (sometimes called a "reservoir") capacitor will also burn out the diode after a few cycles. To be safe, the average current, over the entire cycle, should not exceed the average rating, nor should the peak current exceed that rating. But if a single pulse exceeds the absolute maximum pulse current rating, the diode will probably burn out without waiting for a second pulse to arrive.

Reverse voltage is also a variable. For one thing, calculating its precise value in a rectifier circuit may pose problems because the reverse voltage (ac component) is usually added to the dc to produce the total reverse voltage applied to the rectifier (see Fig. 11-20). If the dc is close to peak ac, total reverse voltage is close to twice peak ac.

The reverse voltage rating is usually associated with a temperature coefficient, too. But there is a further danger here, because the zone where the current flows, and across which the reverse voltage appears in the opposite direction, is quite small physically, and the temperature that matters is the instantaneous temperature of this small zone.

The average temperature of the diode may be safe, or the reverse voltage may be safe at that temperature. But if a peak current momentarily raises the temperature of the conduction zone, to be followed immediately by a reverse voltage that reaches peak so the conduction zone does not have time to cool to the diode's average temperature (which will be its average throughout the cycle), the diode may blow on the reverse voltage, because the safe value is momentarily reduced.

This chapter has dealt with diodes, as they have developed, and as they are still used, as separate entities, discrete components. However, like the more complicated semiconductor elements introduced in the next chapter, far more of them are used in the microminiaturized circuitry of today.

The type of instruments developed in this chapter are seldom used today, because the digital variety are far more accurate. But the principles developed through them are still valid: basics never change. The discussion of dissipation is still valid, perhaps more so, when diodes are so tiny that thousands of them are crowded onto a chip less than a quarter of an inch square.

The same comments are true with respect to the more complicated semiconductor elements introduced in the next chapter. Often thousands of them are crowded into similarly compressed spaces. The question of dissipation then acquires a two-ended significance. Not only is it important to minimize dissipation in such extremely tiny parts, but providing a power supply to feed them also becomes a major factor.

Examination Questions

1. A zener diode is rated at 6.3 volts and has a dissipation of 500 milliwatts. Find the permissible current.
2. If the diode in Question 1 is to be used with a maximum supply voltage of 10 volts, determine the value and rating of the resistor to put in series with the diode.
3. An 8-volt zener diode has a maximum dissipation of 400 milliwatts and is to be used with a 12.6-volt supply. Find the resistor that will be needed in series to prevent the diode from exceeding this dissipation. What will be the current if the supply drops to 9 volts?
4. An input to a supply may vary from 9 to 12 volts. The output must be regulated to 7.5 volts, and it must supply a load that takes 50 milliamps at that voltage. Find the smallest rated zener diode that can be used if its dissipation is not to be exceeded when the input goes as high as 12 volts.

5. A certain multimeter is used to measure a sequence of unknown voltages, each of which may be ac, dc, or a combination of both. Below is a tabulation of each set of readings, with the ac-dc switch set each way; the terminals are connected to the voltage one way (1) and then the other (2). Find what kind of circuit the meter uses, and what kind of voltages are being measured.

VOLTAGE	dc		ac	
	1	2	1	2
1	10	0	22.2	0
2	4.5	4.5	10	10
3	11.09	1.09	24.6	2.42

6. A different type of multimeter is used to measure a different set of voltages from those in Question 5. In this instance, reversing the meter connections makes no difference to any of the readings, but changing the setting of the ac-dc switch does. Find what the voltages are and the type of circuit the meter uses.

VOLTAGE	dc	ac
1	10	0
2	9	10
3	10	4.5
4	14.5	14

7. A meter is designed to measure ac waveforms in three different quantities: (a) peak-to-peak; (b) peak (indicating whichever peak is higher, if the wave is asymmetrical) and (c) nominal rms (which is the rectified average multiplied by the standard sine-wave form factor of 1.11). This meter is used to measure the following waveforms, each of which is adjusted to read 10 volts peak-to-peak. Calculate the readings obtained in the other two positions in each case: a square waveform, a triangular waveform, two asymmetrical waveforms, one with the down half wave 2.5 times the up time, the other with the down time 3.5 times the up time.
8. The same meter used in Question 7 is applied to a symmetrical waveform that measures 10 volts peak-to-peak, 5-volts peak, and 1.11 volts rms. Assuming the waveform consists of a step form which is 5 volts positive for a unit duration, zero volts for another duration, 5 volts negative for unit duration, and zero volts for a duration equal to the other zero-volt duration. Find the ratio of the zero-volt durations to the 5-volt durations, and the rms reading obtained, as well as the true rms value.

12
Compound Semiconductors

Diodes were in use for many years before the effects of putting two (or more) of the diode materials in close proximity within the same unit was discovered. The result is the transistor. Nowadays, the transistor is commonplace, although, like its predecessor the vacuum tube, the way it functions remains a mystery to the uninitiated.

In use, transistors are simpler to understand, as well as being more common in today's world, than was the vacuum tube, which is why the discussion on transistors appears ahead of thermionic tubes. Somewhat complicating the understanding of these solid-state devices, as they are collectively called, is the fact that transistors effectively use quite a variety of different characteristics associated with the elements of which they are made. This gives greater variety than was ever possible in vacuum tubes.

THE CURRENT-AMPLIFYING TRANSISTOR

The first transistor developed contained two diodes placed close together with one electrode common and the others separate (Fig. 12-1). Apart from any common property between the two of them, which is invariably used for the application to which they are put (otherwise separate diodes would be used), each diode functions, when separated from the other, as a normal diode. In other words, it conducts current one way but not the other.

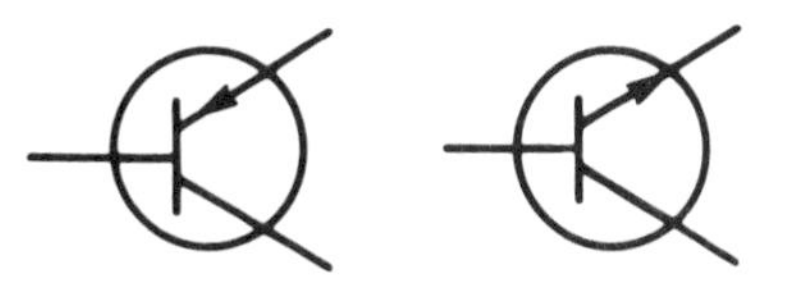

Fig. 12-1. Either type of transistor (pnp, left, or npn, right) acts primarily (in the absence of any interaction between its parts) as two diodes, with one electrode (the base) common to both.

The most common application involves using one diode element, in which the electrodes or terminals are called the emitter and the base, predominantly in its conducting direction. It may occasionally be reversed, but the reverse voltage has no further effect on the action of the other diode.

When the emitter-base is conducting a current which is controlled by the circuit to which it is connected, and the collector-base diode has a reverse voltage applied to it, the latter will conduct a current almost exactly equal to the control current between the emitter and base (Fig 12-2). The benefit that accrues from this arrangement is that current changes in the circuit containing the emitter and base are achieved at a low voltage or impedance, while a corresponding change in current is produced in the circuit containing the collector and base at considerably higher voltage or impedance. This constitutes amplification, although as an active device, the transistor can be used for purposes other than simple amplification.

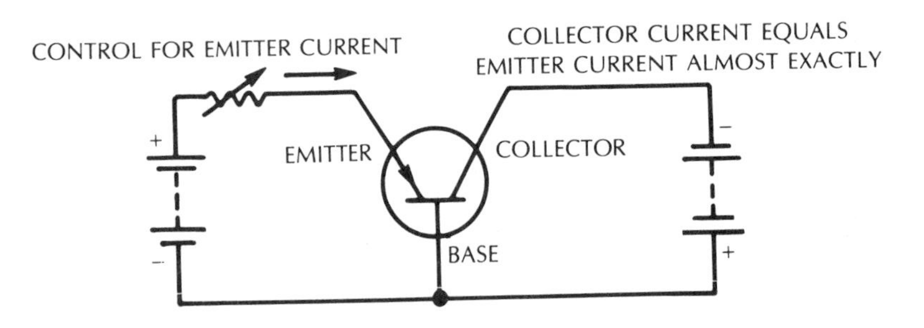

Fig. 12-2. Basic circuit for current "amplifying" transistor.

The basic quantity, from the viewpoint of the transistor, is the relationship between the current change in the emitter-base circuit, and the corresponding change in the collector-base circuit. This ratio is given the greek letter alpha. In most modern transistors, this is very close to unity (1).

GROUNDED (COMMON) EMITTER

For many practical circuits, the most useful arrangement makes the emitter electrode common to the input and output circuits, rather than the base. The input electrode is usually the base. The current at the base is, by Kirchhoff's first law, the difference between the current at the emitter and that at the collector (or the algebraic sum, to use Kirchhoff's definition). Thus, if alpha is 0.992, the collector current is 0.992 of the emitter current, and the base current must be 0.008, which is (1 − 0.992) of the emitter current (Fig. 12-3).

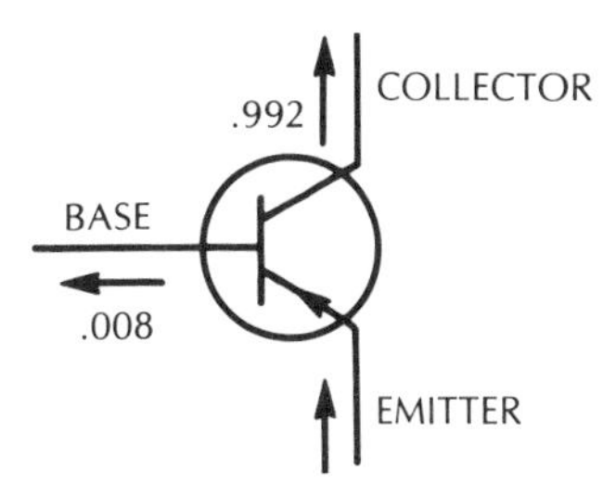

Fig. 12-3. The readings indicate the significance of the (ac) values of alpha (α) and beta (β) in a grounded emitter configuration.

The relationship between the base current and the collector current (in this case, 0.992 divided by 0.008 or 124) is called the *beta* of the transistor. The formula for beta is:

$$\beta = \frac{\alpha}{1 - \alpha} \tag{26}$$

The above assumes that the alpha is constant as the current varies, and that the beta, therefore, also has a fixed value. In fact, alpha varies at different current values. Therefore, beta also has more than one value. For example, suppose, at an emitter current of 1 milliamp, alpha is 0.992, while at 2 milliamps it is 0.990. Then at 1 milliamp, beta is 0.992 divided by 0.008 or 124 and at 2 milliamps it is 0.99 divided by 0.01 or 99.

Both those are dc parameters, values of beta, relationships between dc values of current. But over the change from 1 to 2 milliamps, which could be a signal swing, an ac relationship can be derived. For a change in emitter current from 1 milliamp to 2 milliamps, the collector current changes from 0.992 to 1.98 (2 × 0.99) milliamps, a difference of 0.988 milliamps. This results in an ac beta between these current values of 0.988 divided by 0.012 or 82.3.

Another variation might make the dc beta at 2 milliamps change to 0.9925, so the dc beta at that current is 0.9925 divided by 0.0075 or 132.3. With this change, the ac alpha is 1.985 − 0.992 or 0.993, and the ac beta calculates to 0.993 divided by 0.007 or 141.9. Thus, we can see that the ac beta can differ from the dc beta in various ways, according to the way the dc beta changes, but dependent on the change in dc alpha, which is the primary quantity of a transistor.

To measure ac beta, the collector voltage should be held constant, so that only the current changes in the collector circuit. If the collector voltage also changes, as it most often does in a practical circuit, the theoretical value of the ac beta is not realized. If a series resistance is used in the collector supply circuit, the voltage will rise when the current drops. This rise in voltage will tend to reduce the drop in current. Conversely, the voltage will drop when the current rises, which will tend to reduce the current rise.

This effect can be viewed as a collector-source resistance or conductance (also called admittance). It is as if the change in voltage due to the external resistance is reduced because the change in current is shared with this internal collector resistance or admittance. Thus, if the collector resistance is equal to the external resistance, the two will share equally, and the voltage developed in the external circuit will be only half that predicted by beta times the base input current change times the collector resistance (Fig. 12-4). If the internal effective collector resistance is twice the actual external resistor value, only one-third the voltage will appear at the collector, because the transistor will "soak up" one third of the current change.

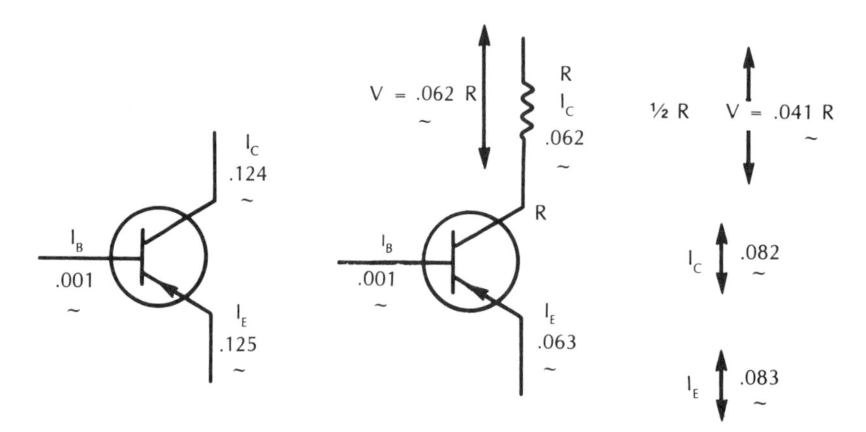

Fig. 12-4. The distribution of signal currents (assuming an ac beta of 124) and the effect of using a collector resistance to obtain a signal voltage output from the effective current gain, when the external value is equal to the internal value (R); also, when the external value is half the internal value.

The input (emitter-to-base) diode behaves, basically, just like a simple diode, when looked at in the grounded-base configuration (Fig. 12-5). But when the emitter becomes the common or grounded terminal, the much smaller current in the input circuit makes the input impedance appear to be that of the simple diode, effectively multiplied by the working current gain (beta).

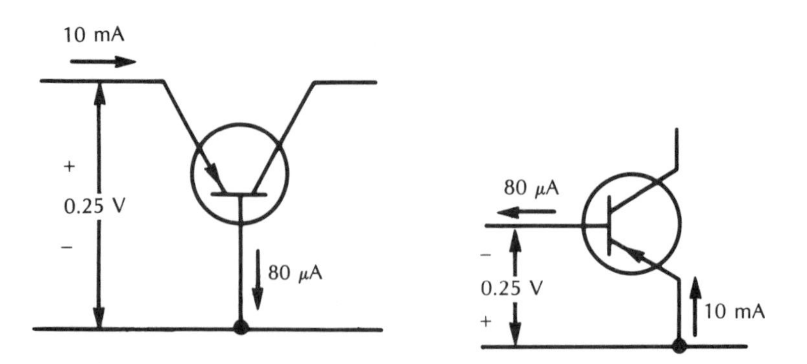

Fig. 12-5. Input dc conditions for common-base (left) and common emitter (right) circuits.

If the collector voltage does not change so the full theoretical beta is realized, the input impedance duplicates that of the simple diode over the current range in the emitter circuit, *multiplied by beta*. This is quite nonlinear (resistance changes with instantaneous current or voltage). However, for small changes, a value can be taken that does not vary as much, being the average at the operating point chosen.

If the operating point is at an emitter current of 10 milliamps and the dc beta is 124, this means that the base current is 10 milliamps divided by 125, or 80 microamps. The dc voltage between the base and emitter may be 0.25 volts, which makes the dc emitter-to-base resistance 25 ohms. For a small change in current, say, by 1 milliamp to 9 or 11 milliamps, the same voltage may *change* by 0.01 volt (which is 10 millivolts). This means that ac emitter-to-base resistance is 10 ohms at this operating point. Measured the other way, with the emitter grounded so the current measured is that in the base rather than the emitter, the ac input resistance is 125 × 10 or 1.25 K. This value is achieved only if the working beta of 124 is realized.

Now, suppose that the collector resistance is about 1.5 K and has a resistor, also 1.5 K, connected between the supply and the collector. This will reduce the working current gain to half the beta value, or 124 divided by 2 or 62. Now, the ac input resistance will be the input resistance of the diode as measured at the emitter, multiplied by only 63, which makes 630 ohms instead of 1.25 K.

In the collector resistor is raised to 3 K, the working current gain will be only one-third of beta or 124 divided by 3 or 41 (to the nearest whole number). The input impedance will now be about 320 ohms. Notice that increasing the collector load (as the series resistor is called, except that the value may be modified in ways described in the next chapter) reduces the effective base input impedance (Fig. 12-6).

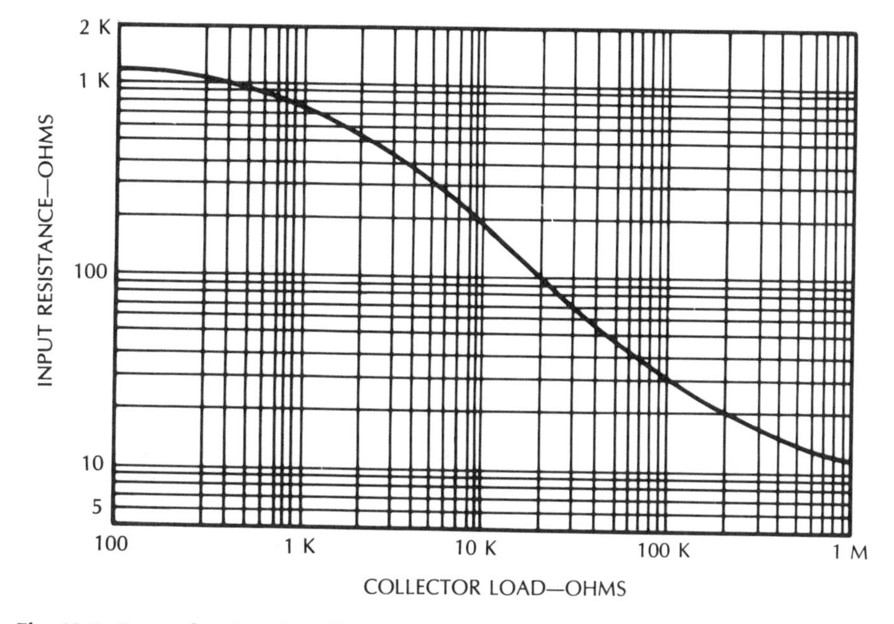

Fig. 12-6. Curve showing the effect of collector load resistance on the input resistance in a common emitter stage.

GROUNDED (COMMON) BASE

If the transistor base is grounded as is common to both input and output circuits, a different set of relationships prevails. The base input impedance is now virtually unaffected by the collector circuit impedance, and will be 10 ohms to ac (or current changes). Part of the reason for this is that the transistor's internal collector resistance is different when measured from collector to base than it is when measured from collector to emitter. The change has the effect of multiplying the value by the working gain of the transistor.

Thus, the same transistor will have a collector resistance of 125 × 1.5 K or 187.5 K in this configuration. But here again, if a very high collector resistance produces correspondingly high (relative) voltage swings, it can influence the current gain, and the effect that the failure to produce the predicted current gain at the collector has on emitter current will change the input impedance.

For example, if the collector resistor is 187.5 K (which is a high and not usually practical value), the current gain will drop to 62. This means that the current change that should be produced in the collector circuit influences, instead, the emitter current, so that the input resistance will now rise to about 20 ohms (ac) instead of the original 10 when the collector voltage does not affect the emitter-base diode action.

GROUNDED (COMMON) COLLECTOR

The above compares the operation of the 3-terminal device with two different electrodes (terminals) regarded as a common between the input and output—the emitter and the base. There is a third possibility: the grounded collector (also called common collector, as it may not actually be connected to ground).

Here, with the same working beta of 124, a current gain of 125 will be achieved. Over a relatively wide range, this is almost unchanged, although going to extreme values can defeat the effective gain. The relationship between the output voltage and current is controlled by the output resistor connected between the emitter and ground (Fig. 12-7). If the resistor is 1000 ohms, then a 1-milliamp change in emitter current will produce a 1-volt change in emitter voltage.

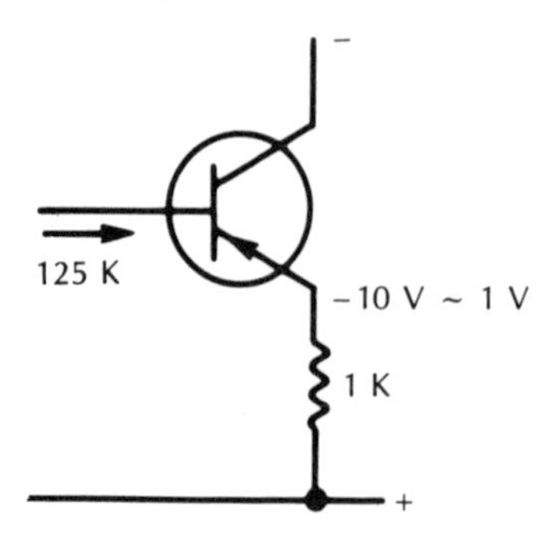

Fig. 12-7. Basis for operation of common-collector (emitter-follower) stage.

Because the base-emitter voltage changes negligibly for such a small change in current (the operating current may be 10 milliamps), the voltage change at the base is almost identical with the voltage change at the emitter, but the current change that goes with it is divided by 125, which will make it 8 microamps. And a 1-volt change accompanied by an 8-microamp change in current represents an ac resistance of 125,000 ohms. Thus, the actual resistance connected in the emitter circuit "looks like" the same value multiplied by the base-to-emitter current gain, which is beta plus one. Only if the chosen operating condition tends to either saturate or "strangle" the transistor, does the effective beta change so that this simple relationship is modified.

Notice that the grounded-base and grounded-emitter configurations are both discussed relative to a current input, applied to the emitter or base, respectively. This current input controls the output current and voltage. This means that the source circuit controls the current, although we have already mentioned the effect of the output load resistance on the impedance into which this current feeds, and which may thus affect the accompanying input voltage. The grounded- or common-collector configuration, often called an "emitter follower," uses a different reference. It is given the latter name because the emitter voltage "follows" the base input voltage, and impedances are reflected through the transistor based on this voltage-following property, coupled with current transformation.

Effects of Input Source Resistance

Because of these facts, the principles just stated are true only if certain facts are observed in regard to the input source resistance used. If the source resistance is not appreciably higher than the input (base or emitter) resistance, then the input is no longer a purely current function. This means that the source resistance now begins to affect the output resistance at the collector.

If the input is a voltage source of nearly zero resistance, the current drawn by the base-emitter diode will be determined by which element is "grounded." At the base, the current has one value, at the emitter another, but the voltage does not change, other things being equal. Either way, the current at the base and emitter have different values, which do not materially affect the voltage from which the currents are drawn. So at this condition (zero source resistance), the otherwise widely divergent collector resistance values, characteristic of current-controlled inputs to base or emitter, respectively, come together to an intermediate value (Fig. 12-8). But notice the different impedance range over which the change in collector resistance takes place in each configuration.

On the other hand, working with a common collector circuit, the voltage-following situation persists. The current taken by the base causes a drop determined by the resistance value of the source generator. On the emitter side, the reflected effective value is the actual value in the base circuit, divided by the working current gain. Thus, if the source resistance in the base circuit is 1 megohm, the effective source resistance at the emitter, when the current gain is 125, will be 1,000,000 divided by 125 or 8000 ohms.

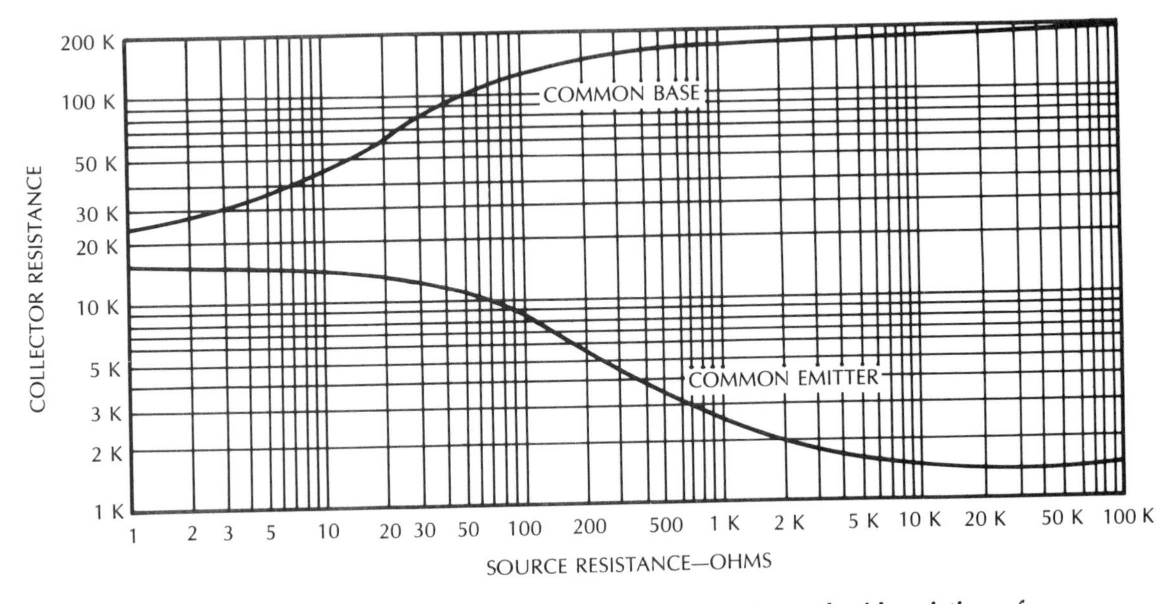

Fig. 12-8. Curves showing variation in collector resistance (internal) with variations of the source resistance presented to the input circuit, for both common-emitter and common-base circuits.

VARIETY OF CURRENT-AMPLIFYING TRANSISTORS

Transistors of the basic current type come in a tremendous variety of types. They are also made by a variety of processes, which are not considered in this book. Different impurities introduced into the pure elements cause alternative directions of semiconduction. This gives rise to the availability of positive and negative types, identified as pnp and npn.

In the pnp type, the direction of conduction in the emitter-base junction is from emitter to base, which means that the collector needs a negative supply voltage to function as an interacting device. If a positive supply is connected to the collector (through a resistance to limit current; otherwise, this polarity endangers the transistor), it behaves as a forward-biased diode, and is uncontrolled by the emitter or base current (according to configuration).

In the npn type, the statements in the previous paragraph are just reversed in polarity. The direction of conduction in the emitter-base junction is from base to emitter, and the collector needs a positive supply for normal operation (Fig. 12-9).

Both pnp and npn transistors are made from two materials: germanium and silicon. They can also be made with a variety of characteristics within each type, although in each the possibilities are limited. Germanium transistors, as germanium diodes, have a low forward-conduction voltage drop, which is useful for some applications. Silicon, on the other hand, produces a forward voltage drop in the order of 0.75 volt. This usually means that conduction does not commence at zero voltage, as it virtually does with germanium types, but at something around half a volt in the conducting direction.

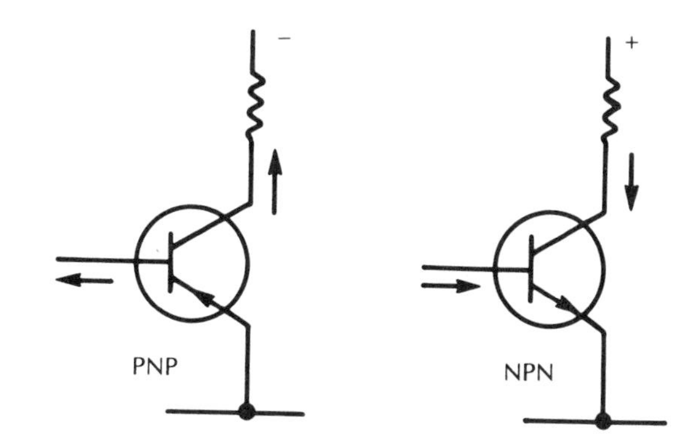

Fig. 12-9. Basic operating differences for normal interaction (amplification) using pnp and npn transistors.

The ratio of forward-to-reverse resistance is very much higher in silicon types. Otherwise expressed, the leakage current in the reverse biased direction is extremely low. It is usually measured in nanoamps (millimicroamps) and sometimes in picoamps (micromicroamps). Leakage current in germanium types is more commonly measured in microamps. Thus, the leakage in silicon types may be from a thousand to a million times smaller than in a corresponding germanium type. This factor makes for much "cleaner" circuit design with silicon types.

TRANSISTOR DATA

Transistor data can be presented in a variety of ways. The most concise is a tabulation of certain characteristics of the kinds already discussed, measured at certain specific values, which gives a rough idea of the general area of operation. These data may not be all that helpful about any specific instance, because few circuits operate at the precise voltage and current values used for measuring the transistor's parameters.

The more universal presentation for more complete data takes the form of collector voltage-current curves. These are usually given for grounded (or common) emitter operation and a sample is shown in Fig. 12-10. The slope of any one of the curves at a particular point represents the collector resistance at that operating point (Fig. 12-11).

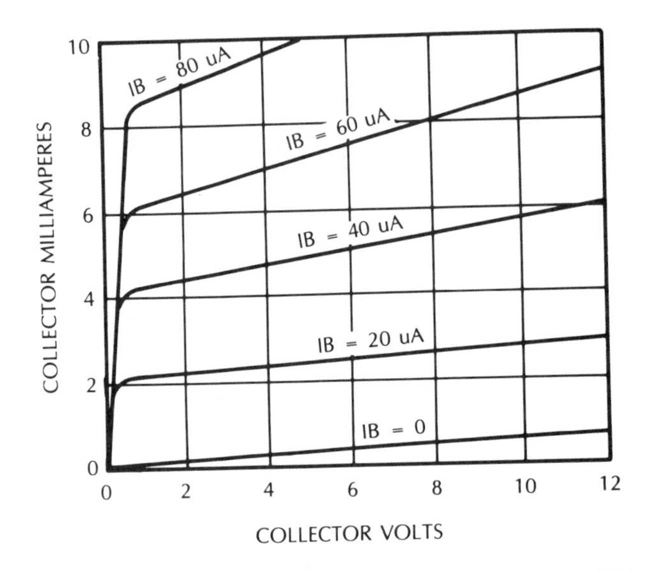

Fig. 12-10. Typical collector voltage-current curves for a current amplifying transistor in a common-emitter configuration.

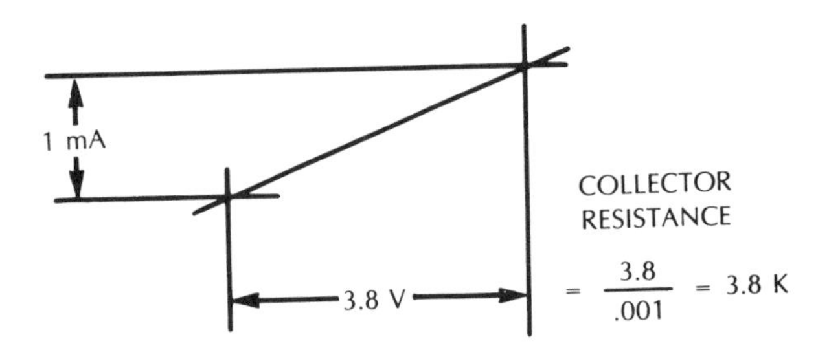

Fig. 12-11. Method of calculating the internal collector resistance at any point on the curves from the shape of the curve.

Notice the means of arriving at the resistance that a slope represents. If you take an interval along the curve (which should be such that the segment used is close to straight) that represents a convenient unit change of current, such as 1 milliamp, then notice the corresponding voltage change; this indicates the resistance value. If the current change is 1 milliamp, the voltage difference tells the resistance in thousands of ohms. As the curve slopes vary, both from curve to curve and along the length of any one curve, the collector resistance is thus seen to be inconsistent, even when the base current is controlled so the base voltage does not affect the result.

Input impedance is specified at a certain operating point, but curves are seldom given because this parameter varies in a quite complex manner. Some manufacturers give temperature coefficients for various parameters where such a variation may be critical to operation in specific uses. Or they may give curves that show the variation relative to a normalized value at some specific temperature, usually 25 degrees C. This kind of data is helpful in computing how changes of temperature will affect the behavior of a certain circuit.

In the foregoing discussion, we have ignored the matter of how the correct dc current is produced to ensure that the transistor is on a working part of its characteristic, and not either saturated or cut off (strangulated). This is discussed in the next chapter in some detail.

In this book, we do not follow the strict sequence in which devices were invented, but rather a logical sequence. On this basis the next one is the *Field Effect Transistor*, commonly called a FET. From that have developed other devices such as the MOSFET and CMOS. These devices have in common the fact that they are voltage, not current controlled.

Figure 12-12 shows the curves for an early type of FET. Note that the curves show how voltage and current vary between source and drain, as the voltage at the gate is held at various fixed values. For the current type of transistor (Fig. 12-10) the curves represent fixed values of current at the base, as the control electrode.

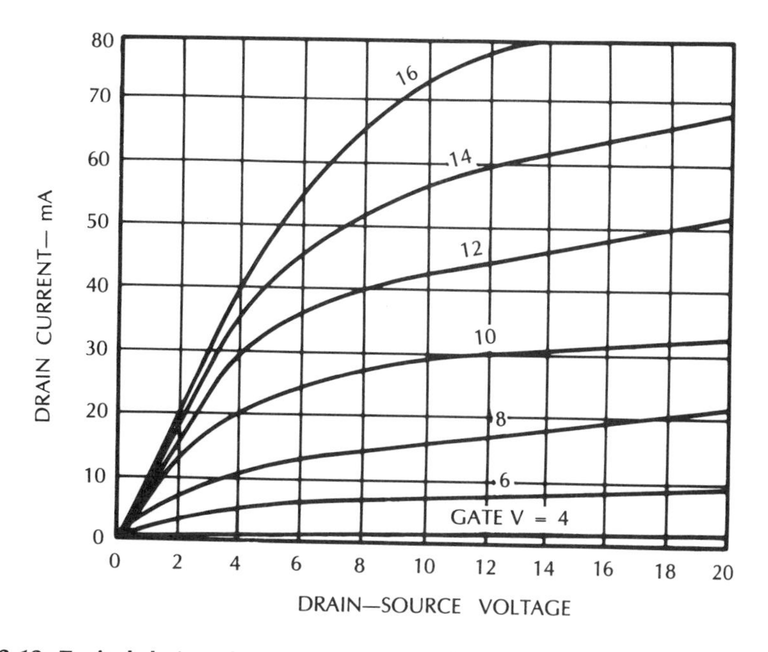

Fig. 12-12. Typical drain voltage-current curves for a field-effect transistor of early vintage.

The voltage applied to the gate of a FET is accompanied by such extremely small current that it is not measurable. It really amounts only to a charge, needed to change the voltage, and since the device is microscopic, so is the charge needed to change the voltage. This means that FETs are extremely sensitive to electrical charges and when FETs are available as separate entities (called discrete components) they are accompanied with a warning about keeping the gate and source connected together, to prevent stray charges which could easily destroy the device from collecting on the gate.

To understand the functional difference between current transistors and FETs, Fig. 12-13 illustrates a schematic arrangement of their structures. For the current transistor, the base is an extremely thin layer of silicon to which a controlled, very minute amount of impurity has been added, which may make that layer a "donor" or an "acceptor" base. A donor has surplus electrons, in its quiescent state, while an acceptor has a deficiency.

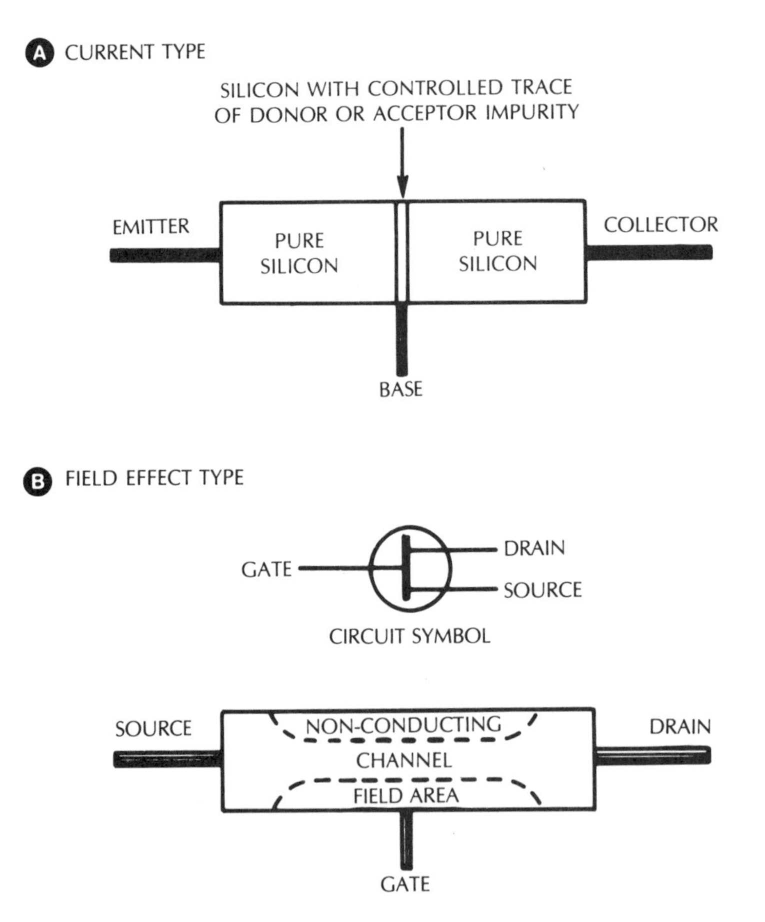

Fig. 12-13. The basic internal construction of current and field-effect type transistors.

The adjoining silicon is of high purity and thus normally non-conducting. But when a negative potential is applied to a donor base, electrons move from the base out into the adjoining silicon, rendering it conducting. One side is the emitter, so if the emitter is positive from the base (which is implied when we said a negative potential is applied to the base) current will flow between these two electrodes, as in the forward conduction mode of a diode, and if a negative potential much larger than that on the base is applied to the collector, a current almost equal to the base-emitter current will flow from collector to base, so that the resultant current at the base is then very small.

If the base is of the acceptor variety, it has "holes" where electrons would otherwise be. And if a positive potential is applied to the base (relative to the emitter) electrons are drawn out of the pure silicon adjoining, making it conductive, ready to receive replacement electrons from external circuitry. The effect is just the reverse of the donor type, so that the polarity of the external circuitry is reversed.

In either case, if a reverse polarity, that is positive for the donor, or negative for the acceptor, is applied to the base, this removes the surplus or deficiency in the base, so the whole device becomes non-conducting. That briefly outlines how a current type transistor works.

Now, in the case of an FET, the central part of the device has the small amount of impurity, making it conductive, while the gate connects to a band that surrounds the conductive path which is of high purity silicon, and thus non-conductive. But its surface can accept electrical charges, which have the effect of attracting or repelling electrons in the conducting channel, or the holes, as the case may be.

Thus a positive voltage applied to the gate will expand the conduction channel, while a negative one will constrict it, assuming the channel contains a donor impurity. A sufficiently negative charge will make the channel completely non-conductive. The important distinction from a current type transistor is that the gate is always in a non-conductive state, so that the controlling voltages are accompanied by virtually zero current. In this respect a FET is like the thermionic tube, which is introduced in Chapter 15, with two significant differences.

An FET is microscopically smaller than the smallest vacuum tube ever made, and thus, using modern microminiaturization, thousands of them can be built, along with their associated circuitry, into a tiny chip that can rest on your finger tip. And because of their tiny size, the charges needed to control them are many orders of magnitude smaller than those needed to control a vacuum tube. This is why such tiny chips, into which may be built a complete computer, clock mechanism, or whatever, can operate continuously for a year or more from a tiny battery.

So much for making things tiny and very sensitive. But semi-conductor technology has also gone the other way, to make it possible to control relatively large amounts of power, in a smaller space than was previously possible. This led to the SCR, *Silcon Controlled Rectifier*, which was the first of this type, from which many more of these have developed also.

An SCR must be triggered to start conducting. Actually, an SCR switches from one condition to the other when a certain control point is reached. An advantage of this operational mode is that dissipation is minimized because the device goes from nonconduction, with a voltage across it (dissipation zero) to conduction at saturation, where current passes liberally with virtually no voltage drop (dissipation quite low). The transition occurs, almost instantly, with a very short switch-on time.

The essential feature for a semiconductor device to be controllable by infinitely small increments (as is the current transistor) is that its alpha be less than unity (1) by however little. In the common-emitter connection, if the alpha is 0.9, then the beta is 9 and the current in the collector-emitter circuit is controlled by a current of one-ninth the current value in the base-emitter circuit. If the alpha is 0.99, then the factor is 1 over 99, and so on. But if the alpha exceeds 1, then a different situation prevails.

For example, if the alpha is 1.2, then the collector current will want to be 1.2 times the emitter current. This can happen only if the base current reverses as soon as the current starts. This means that a forward current, from base to emitter (whichever direction of flow that may be, according to whether the junction is pn or np) will initiate a collector-emitter current. Once this is initiated, the base current can reverse, but it will never regain control of the circuit because the collector-emitter current has taken over. The device has triggered from a nonconducting state to a conducting state. This happens at a specific trigger current at the base. After this has happened, the only way to stop conduction is to remove the current by turning off the voltage that is driving it. Then control can be re-established. The sequence, on a typical ac waveform, is shown in Fig. 12-14.

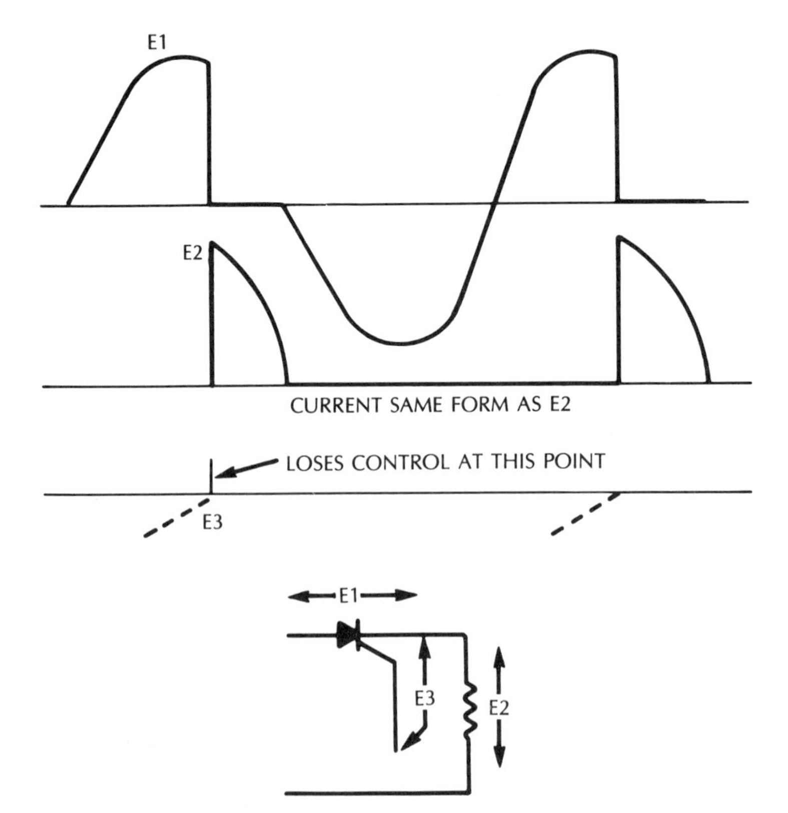

Fig. 12-14. How an SCR (silicon controlled rectifier) functions.

If the only difference between an SCR and a simple 3-electrode transistor was the fact that its alpha is greater than unity, then the reverse polarity, as required to extinguish the triggered conduction, would put the collector-base junction in the conducting mode, since the controlled conduction is in the nonconducting mode in the absence of the trigger. This would create problems because the device would never turn off. So incorporated into the same structure, during the building process that forms the device, is another diode that conducts only in the direction allowed (Fig. 12-15).

Fig. 12-15. Circuit and schematic symbol for an SCR.

Examination Questions

1. A succession of transistors are tested for dc alpha values and the readings obtained are as follows: 0.95, 0.92, 0.975, 0.99, 0.985, 0.992, 0.996, 0.995. Find the corresponding values of beta.
2. Following are a succession of dc beta values for a series of transistors tested: 88, 95, 98, 105, 112, 125, 135, 150. Find the corresponding values of dc alpha.
3. The following pairs of values represent the dc alpha, as measured at emitter currents of 1 milliamp and 2 milliamps, respectively. Find the value of ac alpha for each transistor over the range from 1 to 2 milliamps: 0.95, 0.98; 0.98, 0.95; 0.99, 0.98; 0.98, 0.99; 0.975, 0.985.
4. The following pairs of values represent the dc beta, as measured at emitter currents of 1 milliamp and 2 milliamps, respectively. Find the value of ac beta for each transistor over the range from 1 to 2 milliamps: 99, 101; 105, 95; 97, 103; 78, 88; 124, 135.
5. In a common-emitter circuit with a constant current (high-resistance base source) input, a transistor has a collector resistance of 12 K at a specific operating point. If the ac beta at that operating point is 95, what will be the collector resistance when connected in the common-base configuration using the same operating point?
6. A transistor's collector admittance is quoted as 80 micromhos at a specified operating point in a common-emitter circuit. What will be the admittance if the transistor is used in a common-base circuit if the ac beta at the operating point chosen is 65?
7. A transistor operates with an ac beta of 120, has a collector admittance of 75 micromhos, and uses a collector resistor of 10 K. Find the working current gain from the base into the 10 K load. What will be the output signal voltage for a 1-microamp signal input at the base?
8. A transistor's characteristics in a common-base connection are listed as: ac alpha, 0.985; collector admittance, 1.25 micromhos; base input impedance, 12.5 ohms. Find the working gain, if it is used in a common-emitter circuit with the same operating conditions (collector current and voltage) when a collector resistance of 15 K is used (interpret gain to mean current output-current input), and the input impedance.
9. A common-collector stage (emitter-follower) uses an emitter resistor of 1.8 K. The ac beta at the operating point used is 115. Find the input impedance at the base, due to the reflected emitter resistance.
10. In the stage described in Question 9, the signal is fed through a 270 K resistor to the base. If the signal input is 1 volt, assuming the reflected resistance at the base is the only load at the other end of the 270 K resistor, find the signal voltage appearing across the 1.8 K resistor in the emitter circuit.

13
Linear Amplification

In a grounded- or common-base transistor circuit, the collector circuit reproduces the current input presented to the emitter circuit at a much increased voltage swing. In the grounded- or common-emitter circuit, the collector circuit reproduces variations presented as an input to the base circuit, with a considerably increased current and possibly also a voltage swing. In the common-collector or emitter-follower circuit, the emitter circuit produces substantially the same voltage variations presented at the input to the base circuit, but with much greater available current variations.

Linear amplification requires more than that the variations in voltage and-or current at the input shall produce larger voltage and-or current variations at the output terminals. To be linear, the larger variations at the output must be precisely proportionate to the input. In other words, the waves must be identical in form, just different in size.

Suppose that a transistor, working with a 1000-ohm collector resistor from a 12-volt supply, sustains a steady current flow of 8 milliamps. The 8-milliamp current in the 1000-ohm resistor will produce an 8-volt drop, leaving 4 volts across the transistor (Fig. 13-1). Now suppose that the type of transistor used has an average ac beta of 100, which means that a 10-microamp input current change will produce a 1-milliamp change in the current output.

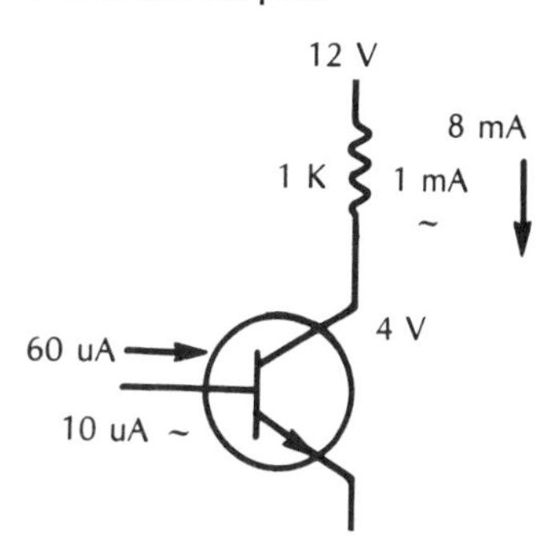

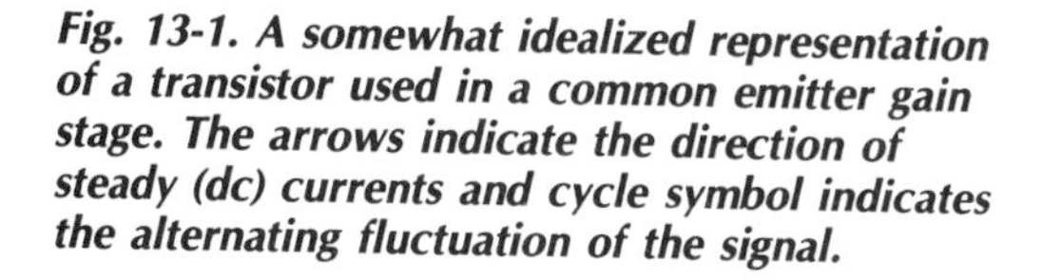

Fig. 13-1. A somewhat idealized representation of a transistor used in a common emitter gain stage. The arrows indicate the direction of steady (dc) currents and cycle symbol indicates the alternating fluctuation of the signal.

In fact, even in the same transistor, beta is not constant. Suppose that an input current of 60 microamps, opposing the steady base current required to produce the 8-milliamp operating condition, just reduces the collector current to zero, while an input current of 60 microamps in the opposite direction (making the base current a total of 120 microamps) just raises the collector current to its saturation value of 12 milliamps (as controlled by the 1000-ohm resistor across the 12-volt supply).

The output will be something that looks a little like a sine wave, but quite distorted (Fig. 13-2).

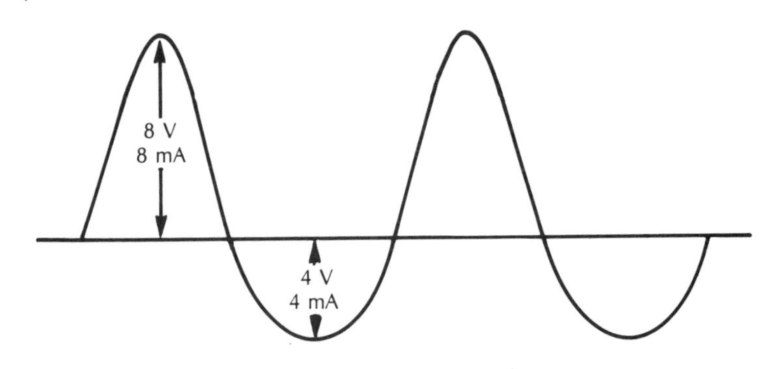

Fig. 13-2. An exaggerated example of distorted waveform (without clipping).

Actually, no modern transistor would produce an output wave quite this distorted (unless it was biased completely wrong). At the same time, unless care is exercised in the design of the circuit, the output wave will not be perfect. That somewhat exaggerated distortion is due to the nonlinearity of the transfer characteristic—output current against input current. This can be drawn as a transfer curve (Fig. 13-3).

ACHIEVING LINEARITY

If both the input and the output are purely current, such an amount of distortion from linearity, whatever it is, will be characteristic to the transistor in question, and nothing can be done to reduce the nonlinearity of this characteristic. However, other things may be possible: both input and output have an accompanying voltage as well as current. Let us consider each of these properties separately.

The input can be viewed as purely a current by feeding a sinusoidal voltage source through a very high resistance, so that the changing input resistance of the base, over the entire waveform, does not materially change the total resistance through which this input voltage drives current. Suppose that the 60 microamps needed to offset the steady bias current of 60 microamps requires a voltage change of 0.24 volt, while the extra voltage to raise the base current to 120 microamps requires a change of only 0.06 volt in the opposite direction, making the total base-emitter voltage 0.3 at this point.

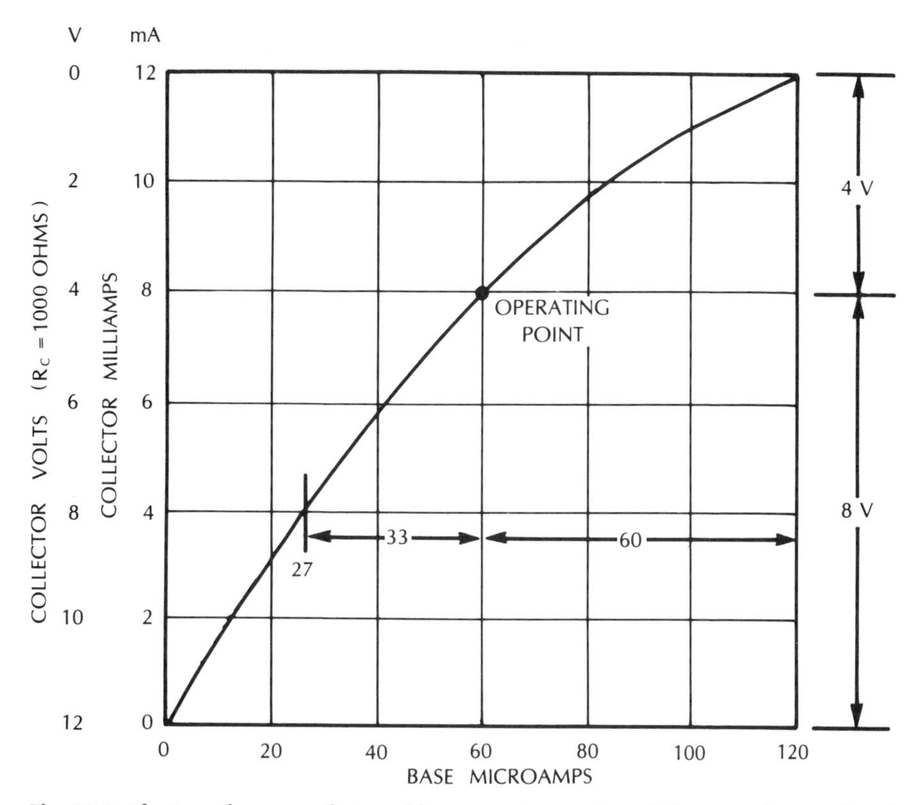

Fig. 13-3. The transfer curve that could represent operation of the state that produced the waveform in Fig. 13-2.

The input resistance toward cutoff is 0.24 divided by 0.06 or 4000 ohms. Up toward saturation, it is 60 divided by 0.06 or 1000 ohms. At the operating current, the dc resistance of the base, which may not be the same as the small-signal ac resistance (change of input voltage divided by corresponding change of input current) is 240 divided by 0.06 or 4 K.

Actually, the input resistance at cutoff will be much higher than 4 K, because this is the average over the 60-microamp change from the operating point down to zero current. At cutoff, the input resistance approaches infinity, because beyond cutoff it is merely leakage current. Based on the slope of the curve in Fig. 13-4, the resistance at the zero point is about 10 K. And the input resistance at saturation may be lower than 1000 ohms. Also, based on the slope of the curve in Fig. 13-4, it is about 833 ohms. And the slope at the operating point is about 1.5 K.

INPUT: CURRENT, VOLTAGE, OR COMBINATION

The point here is that to linearize a current input, a series resistor of 100 K or perhaps 1 megohm should be used. If the peak signal is to be the 60 microamps already mentioned, then with a 100 K resistance, this needs a 6-volt peak signal. With a 1-megohm input resistance, the peak signal would need to be 60 volts. Either is a little unlikely, when the supply voltage available is only 12 volts.

If no series input resistance is used, one excursion of input signal needs to be 0.24 volt, while the other is only 0.06 volts, each providing an input current excursion of 60 microamps in its respective direction. If the input voltage is sinusoidal, at, say, 0.06 volt peak, saturation will be reached in one direction, but in the other, 0.18 volt corresponds with 30 microamps base current (Fig. 13-4) which yields a collector current (Fig. 13-3) of about 4.4 milliamps. Thus, with a peak current input of plus or minus 60 microamps, the output swings 4 volts one way and 8 volts the other. With a voltage input of 0.06 volt peak (plus or minus 60 millivolts) the output still swings 4 volts that way, but only 3.6 volts the other. Admittedly, this is an exaggeratedly poor transistor, not likely to be found in practice.

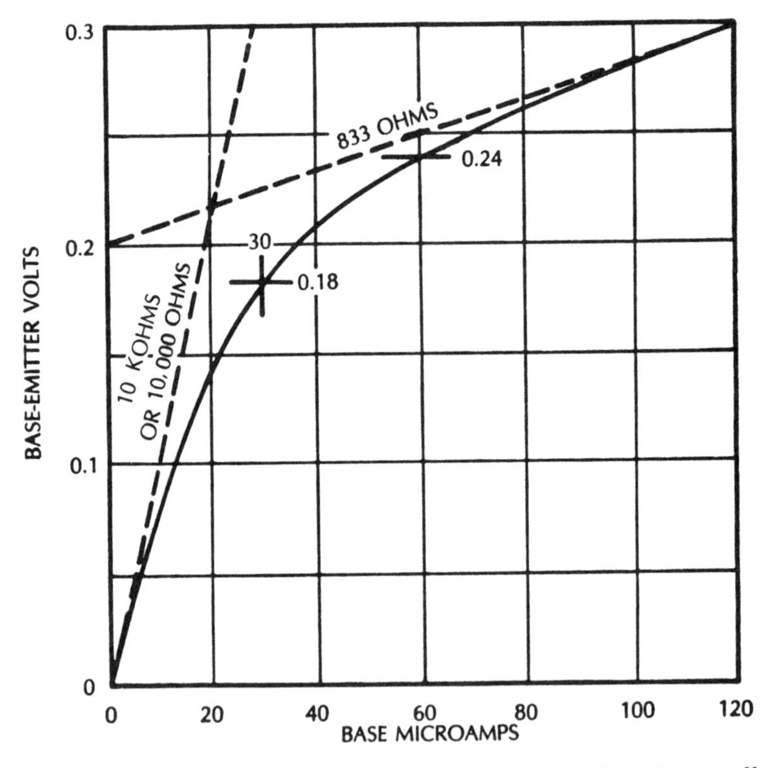

Fig. 13-4. Typical base input resistance curve showing how non-linearity can affect the transfer curve.

The trick, apparently, would be to put just enough resistance in series with the base to make the input voltage swing equally each way and at the same time produce an equal output voltage swing each way. To get an equal output swing each way would require a 60-microamp swing in one direction and about a 33-microamp swing in the other direction. The accompanying voltage, at the base, will be 0.06 volts in the first direction and about 0.07 volt in the other direction (Fig. 13-4).

So the 60-microamp swing in one direction must produce, in the series resistance, 0.01 volt more than the 33-microamp swing does in the other direction.

If a difference of 27 microamps produces a difference of 0.01 volt, this series resistance needs to be 10 divided by 0.027 or about 370 ohms. With this value, the input signal swing toward cutoff will need to be 0.07 volt at the base, plus 33 microamps in 370 ohms, which produces 0.012 volt in the series resistor, making a 0.082 volt peak. In the other direction, the change at the base is 0.06 volt, while the 60 microamps in the 370 ohm resistor will produce .022 volt, again making a total of .082 volt peak.

The two forms of nonlinearity tend to neutralize one another when correctly applied. The foregoing is a simplistic treatment. In fact, making sure that both halves have equal peak amplitudes does not ensure that they both have the same or even the correct shape. But it is a first step in the right direction.

AC COLLECTOR LOADING

The above assumes that the resistor in the collector circuit produces both a voltage and a current output swing. In some circuits, the resistor used to provide the dc operating point (e.g., 4 volts at 8 milliamps in the example just discussed) may not be the ac load for the collector when signal is handled by the same transistor.

This can best be seen by referring to the collector voltage-current curves (Fig. 13-5). These are fairly representative of such curves for a lower power

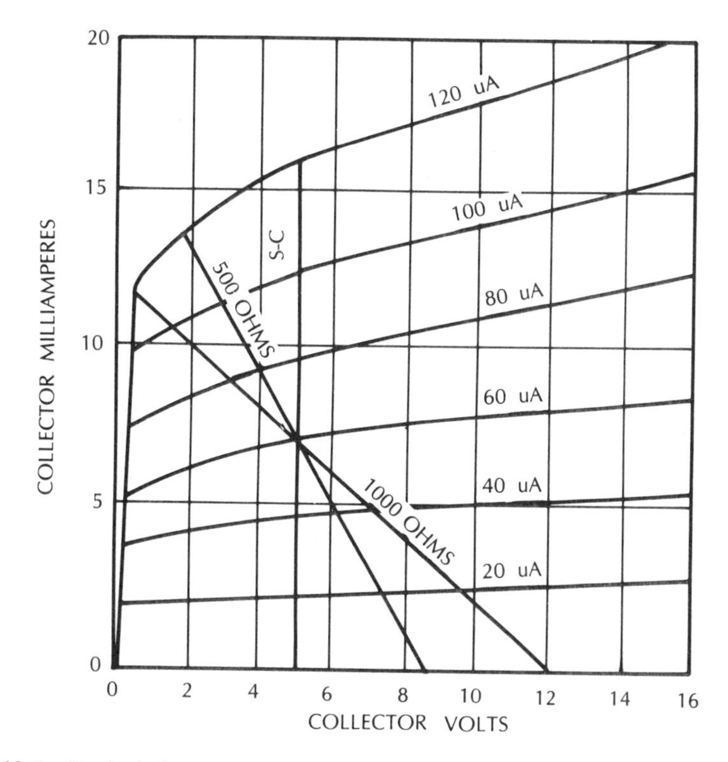

Fig. 13-5. Typical low-power transistor collector voltage-current curves, with superimposed load lines to illustrate the dependency of linearity on the load value.

transistor. Using a dc resistor of 1 K, with a base input bias (steady dc component of) current of 60 microamps, yields an operating point of 5 volts at 7 milliamps on a 12-volt supply. This is found by the "load-line" construction shown. Starting at 12 volts, zero current, and marking the point representing 12 milliamps at zero voltage, which is the current a 1 K resistor will take across the full 12 volts, these points are joined. Every dc combination of voltage and current on the transistor collector must lie on this line.

Assuming the base input is controlled in some way to be plus or minus 60 microamps; +60 microamps (making 120 total) swings the collector to saturation; zero volts at 12 milliamps (or close to that point). In the other direction, zero base current swings the collector to cutoff: 12 volts at zero current.

This swing possesses the same nonlinearity just described, but not so severe. Now assume that the collector is coupled by an adequately large capacitor to a really low resistance in the next circuit, such as the base input to a power transistor. This means the operating point of 5 volts will hold, with very little deviation, as the current swings up and down due to signal amplification. The almost vertical load line drawn through the 5-volt 7-milliamp point shows how the transistor behaves with signal passing through.

When the base current now swings up to 120 microamps, the voltage stays close to 5, but the current swings up to about 16 milliamps, instead of 12. Yet, when the base current swings down to zero, because the signal offsets the bias current, the collector swings down to zero current, still at 5 volts (or very little more) on the collector.

The difference can be seen by comparing the collector current change. Without the almost short-circuit load coupled, the swing goes from 7 milliamps to 12 milliamps and zero. The zero direction induces the bigger change—by a 7:5 ratio—than the saturation direction. With the load coupled, the swing goes from the same 7 milliamps to 16 milliamps and zero. Now the upward swing has a bigger current change—by 9:7—than the zero direction.

Now suppose that, instead of coupling to a very low resistance load, the capacitor couples to a load of another 1000 ohms, so the combined ac load for the collector is 500 ohms—the two 1000-ohms in parallel (Fig. 13-6).

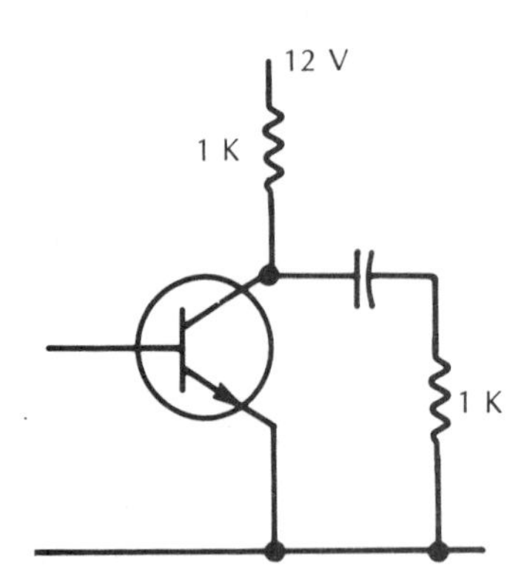

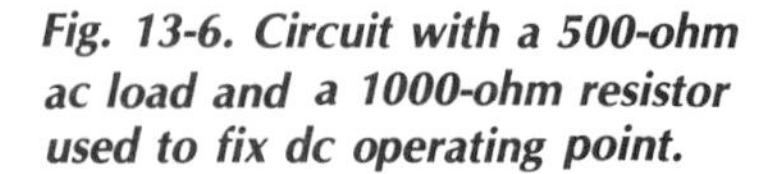

Fig. 13-6. Circuit with a 500-ohm ac load and a 1000-ohm resistor used to fix dc operating point.

This results in an intermediate condition, where the signal current swings the collector voltage and current up to 14 milliamps at 1.5 volts and down to zero current at 8.5 volts (also shown on Fig. 13-5). This again balances the two halves of the waveform, at least for size, and probably makes them quite close to the same shape.

If we put together all the variables of input and output resistance, it is obvious that they are interdependent. Any one parameter, input resistance or output load (ac) can be changed to make the operation closer to linear. But changing a second parameter may again increase the deviation from linear. However, careful selection of all parameters with this interdependence in mind should result in the best possible linearity.

BIAS

So far we have made what seemed like a fairly simple assumption: that the transistor is biased to a suitable dc operating point. This is not quite so easy as it sounds. The simplest arrangement is shown at Fig. 13-7. The bias resistor is chosen so that the correct operating point, in regard to collector voltage and current, is achieved. That is more easily said than done. Suppose a transistor type has an average dc beta of 120. For a desired operating point of 5 volts at 7 milliamps on a 12-volt supply, with a 1 K collector resistor, the bias current needs to be 7 divided by 120 or 58.5 microamps. Using the 12-volt supply as the source, the bias resistor needs to be 12 divided by 0.0000585 or 205 K.

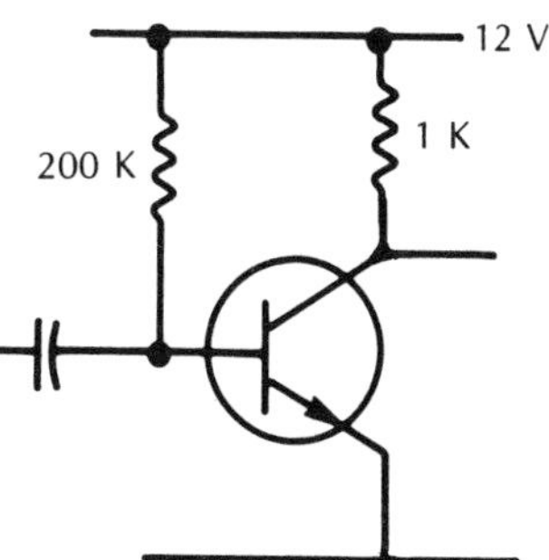

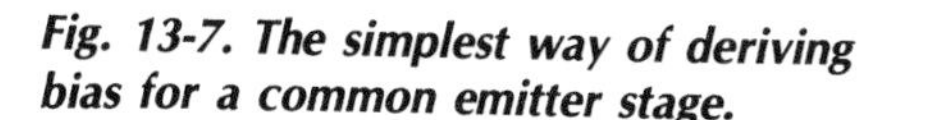

Fig. 13-7. The simplest way of deriving bias for a common emitter stage.

Here is the problem: A transistor with an average beta of 120 may have a production variation between individual samples of 60 to 180. Possibly, at higher cost, the same transistor can be supplied within a variation restricted to from 90 to 150. but even at this, they all have a temperature coefficient (or more than one). For this reason, a transistor that has a beta of precisely 120 at some temperature may vary say from 90 to 150 at other temperatures.

What does this do to the operating point? For simplicity, assume the bias current, as fixed by the resistor, which is 200 K, is set to 60 microamps. With a beta of 120, the collector current will be 7.2 milliamps, dropping 7.2 volts across the 1 K resistor, which leaves a collector voltage of 4.8 volts. With a beta of 90, the collector current will be 5.4 milliamps, dropping 5.4 volts across the 1 K resistor, which leaves a collector voltage of 6.6 volts. With a beta of 150, the collector current will be 9 milliamps, leaving a collector voltage of 3 volts. With a beta of 60, the collector current will be 3.6 milliamps, and voltage 8.4 volts. With a beta of 180, the collector current will be 10.8 milliamps and the voltage 1.2.

In most circuits these last variations are too much. The first ones may be too much in some circuits. They seriously limit the signal-handling capacity. How can this situation be improved? One trick is to employ dc feedback from the collector. If a supply of 12 volts requires a bias resistor of 200 K, then taking bias from the collector, which should be at 5 volts, will require a resistor of 200 K × 5 divided by 12 or 83 K. 82 K is a stock value, so we will assume this value is used (Fig. 13-8).

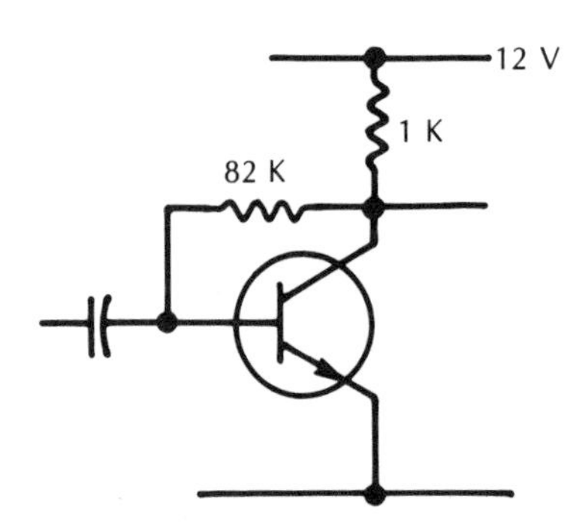

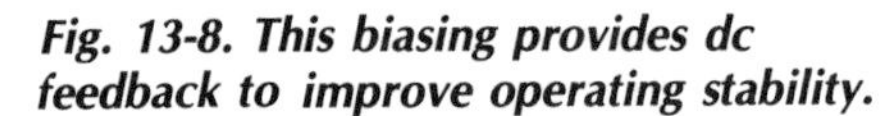
Fig. 13-8. This biasing provides dc feedback to improve operating stability.

Now we have the problem of calculating the operating point, with everything else given. Here's the easiest way to do this: Assume that as a result of the dc beta the transistor has a dc resistance from collector to emitter equivalent to the bias resistor value divided by the dc beta. With the nominal gain of 120, the transistor thus appears as 82 K divided by 120 or 680 ohms. The 1 K collector resistor with the transistor equivalent (on dc only) to 680 ohms divides the 12-volt supply so the collector gets 12 × 680 divided by 1680 or 4.85 volts.

If the beta goes to 90, the equivalent dc resistance of 82 K divided by 90 is 910 ohms. The 1 K resistor with 910 ohms produces a collector voltage of 12 × 910 divided by 1910 or 5.72 volts. If the beta goes to 150, the equivalent dc resistance of 82 K divided by 150 is 547 ohms, producing a collector voltage of 12 × 547 divided by 1547 or 4.25 volts. If beta goes to 60, the equivalent dc resistance of 82 K divided by 60 is 1370 ohms, yielding a collector voltage of 12 × 1370 divided by 2370 or 6.95 volts. If beta goes to 180, the equivalent dc resistance of 82 K divided by 180 is 455 ohms, yielding a collector voltage of 12 × 455 divided by 1455 or 3.75 volts.

With the extremes of beta variation considered, the variation in collector voltage is reduced, by changing from the circuit in Fig. 13-7 to that in Fig. 13-8, from a range 1.2 volts to 8.4 volts down to a range from 3.75 volts to 6.95 volts, which is a considerable improvement.

But what does this do to amplification? If the 1 K resistor is the collector load (it is not coupled to some lower value as in Fig. 13-6) then the signal produces the voltage swings described earlier. And an input swing that would produce a change of 8.4 - 1.2 or 7.2 volts peak to peak will be reduced to a swing of 6.95 - 3.75 or 3.2 volts peak to peak.

That little calculation assumes that the ac beta and dc beta are the same, in which case the effect of the beta on the signal will be the same as the change of beta on the collector current. In practice, the changes will be of the same general order, so the calculation gives a good idea about the change, although it may not be precise. This reduction in gain can probably be overcome, in some circuits at least, by increasing the input signal to more than twice as much, so the same output is achieved. But this increase will require a similar increase in the input voltage swing, which it may be difficult to accommodate.

REMOVING THE AC FEEDBACK

However, the above mentioned reduction in ac amplification or gain occurs only where the collector load impedance is the same as its dc resistance. Where the signal currents are coupled by a capacitor to a much lower impedance or resistance (Fig. 13-9), the signal voltage does not develop, as stated earlier. Only the output current varies; the voltage stays nearly constant at its operating value. This means that no signal current is fed back through the bias resistor. Thus, the stability of the operating point on dc is improved by more than 2:1 by the change from the circuit in Fig. 13-7 to that in Fig. 13-9, but the amplification remains the same in the latter circuit. Notice how important it is to take into account all the parts of the circuit. Every circuit has its own advantages.

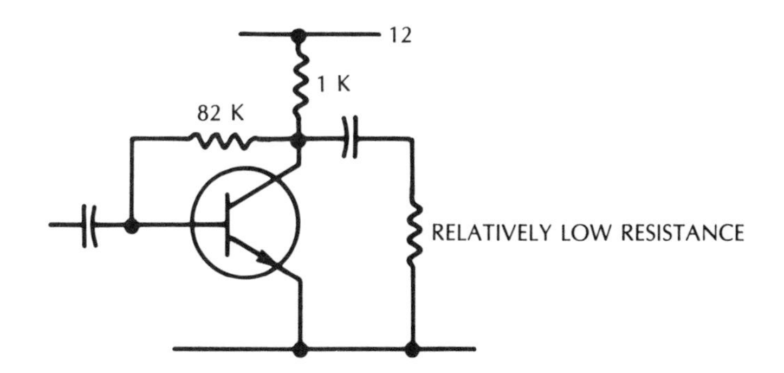

Fig. 13-9. Low impedance ac load reduces signal in the dc feedback path.

One more thing should be mentioned, before leaving the group of circuits in Figs. 13-7, 13-8 and 13-9. This is the fact that no base-to-ground resistor is shown. A circuit without such a resistor will operate successfully—provided the signal never swings the transistor into its cutoff region. When that happens, the base-emitter junction acts as a rectifier, biasing the transistor further into cutoff.

All the while the transistor is conducting for the entire signal cycle, the average base current stays reasonably constant, and the operating condition does not shift. But as soon as the base-emitter junction ceases to conduct for part of the signal cycle, it is acting partially as a rectifier and thus shifts the operating point by developing a different voltage on the base side of the coupling capacitor, shown in each circuit as a means for coupling the input signal.

When such rectification begins to happen, the transistor is biased nearer to cutoff, which means, unless the signal amplitude is reduced or removed by this time, that more of the cycle puts the transistor into its cutoff region, causing a bigger "off" bias. A few cycles of signal will run the transistor almost completely into cutoff.

Thus, this circuit should never be operated without a base-to-ground resistor to help restore the correct operating condition as quickly as possible after such an error occurs, as well as to minimize the effect in the first place. Without such a resistor, which can be large enough so it has a negligible effect in normal operation (e.g., several times the working base input resistance), the capacitor acting with the high-value bias resistor (82 K or 200 K in the examples discussed) will keep the transistor cut off for some time before normal operation is resumed, even after the signal causing the error is reduced to well within a normally safe operating range.

A VOLTAGE-GAIN STAGE

Each of the above circuits utilizes the transistor essentially in its basic mode as a current amplifier. In another circuit the transistor still operates as a current amplifier, but swamps its nonlinear effects and variations by the associated circuit resistance values, so the circuit, rather than merely the transistor, behaves as an amplifier of voltage.

Figure 13-10 shows the basic elements for a stage of this kind of amplification. The collector feeds into a predetermined load, and the gain of the stage is determined, basically, by the relationship between the emitter resistor and the collector load, which must be its ac value.

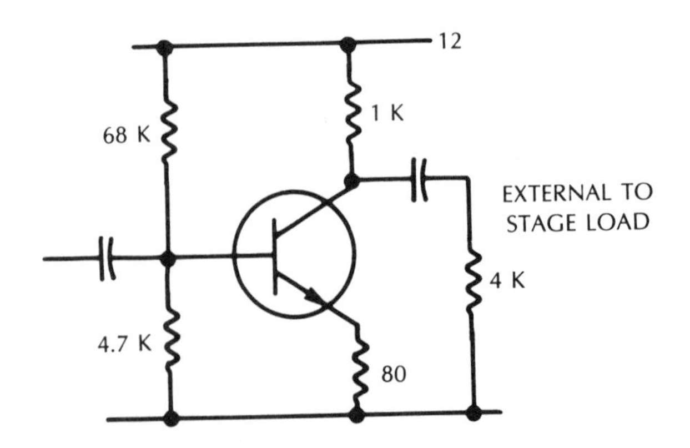

Fig. 13-10. An alternative amplifying circuit for the stage in Fig. 13-9.

If the collector resistor is 1 K and it is coupled to a load of 4 K, the ac load is 1 K and 4 K in parallel, or 1000 × 4000 divided by (1000 + 4000) or 800 ohms. If the emitter resistor is now made 80 ohms, the same signal current (within close limits) passes through both the ac collector load impedance and the emitter resistor, so the voltage gain from the emitter to the collector is 10 times.

Now to set an operating point. Suppose the beta of this transistor also rates between 60 and 180, with an average of 120. If the supply is 12 volts, making the collector voltage about 6.6 will give the maximum available signal swing, because a tenth of the available swing appears between emitter and ground.

Without the part resistance included at the emitter, making the collector voltage half of the supply voltage would allow the output to swing over almost the entire supply voltage range. But the collector cannot swing below the emitter. The collector's down swing (we use "up" and "down," rather than "positive" and "negative" here, because polarity may go either way, according to whether the transistor is pnp or npn) coincides with the emitter's upswing, so the collector can have only the upper ten-elements of the supply voltage to swing in. If the collector voltage is 6.6, the drop across the 1 K resistor, due to operating point collector current, will be 5.4 volts, with the collector current at 5.4 milliamps. The emitter voltage is given by the same current (actually, the emitter current is a little more than the collector current by from one-sixtieth to one one-hundred eightieth) through 80 ohms, which produces 0.43 volt at the emitter.

The base voltage will be a little higher than this, in the case of a germanium transistor by from 0.1 to 0.25 volt (varying with type), or in the case of a silicon transistor by about 0.75 volt. Assume this is a germanium transistor with a 0.15-volt drop from base to emitter. This makes the base voltage 0.58 volt. The base current will vary from 5.4 milliamps divided by 60 to the same divided by 180, with an average of 5.4 divided by 120. These values work out to 90 microamps, 30 microamps and 45 microamps, respectively.

Bias is provided by a voltage divider, allowing for the base current. If the bottom resistor is 4.7 K, with 0.58 volt across it, the current in it will be 0.58 divided by 4.7 K or 123 microamps. Taking the average beta, the base current is 45 microamps, so the top resistor must pass 123 + 45 or 168 microamps and drop 12 − 0.58 or 11.42 volts, requiring a resistor of 11.42 divided by 0.000168 or 68 K (almost exactly).

Two more things need calculating: the variation with beta tolerance deviation and the input impedance presented by the base to the signal voltages. The easiest way to calculate variations with the beta tolerance is to allow the 0.15-volt drop and calculate the emitter voltage provided by the current taken by a base input resistance consisting of 80 ohms multiplied by the respective beta values.

By Thevenin's principle, the voltage provided by 68 K and 4.7 K across a 12-volt supply, with no base current taken, is 12 × 4.7 divided by 72.7 or 0.775 volt. Subtract the 0.15-volt base-emitter drop, which is virtually constant, and that leaves 0.625 volt. The source resistance at the 68 K and 4.7 K junction, also determined by Thevenin's principle is 4.7 K in parallel with 68 K, which is 4.4 K. For a beta of 60, 80 ohms multiplied by 60 makes 4.8 K and for a beta of 180, 80 ohms multiplied by 180 makes 14.4 K.

Thus, the emitter voltage can be found by regarding it as the result of a voltage divider across 0.625 volt, with the top resistor at 4.4 K and the bottom one variable between 4.8 K and 14.4 K. This yields a variation from 0.325 to 0.48 volt. Across 80 ohms, these values put the collector current variation between 4.06 and 6 milliamps. The drop across the collector resistor will be between 4.06 and 6 volts, resulting in a collector voltage variation between 7.94 and 6 volts. This is better control than the circuit in Fig. 13-9 offers.

Now, for the input impedance. Without a better knowledge about base input impedance, as an entity separate from the network feeding it, it will look like the emitter resistor multiplied by the beta, which varies from 4.8 K to 14.4 K. The voltage divider providing bias produces a combination value, already calculated at 4.4 K. So the overall combination is 4.4 K in parallel with a resistance ranging from 4.8 K to 14.4 K, which calculates to from 2.3 K to 3.4 K.

If this is to be a load for an earlier stage similar to this one, then the value is too low; we set that as 4 K. The whole thing could be recalculated for a resistor value higher than 4.7 K for the bottom one of the voltage divider.

The result of using an input impedance lower than that specified will be a reduction in the gain of the previous stage, because its collector load impedance would be 1 K paralleled by from 2.3 K to 3.4 K, instead of by 4 K, resulting in an ac load value of from 700 to 770 ohms instead of 800 ohms. Notice that even this is not a serious departure, because a much higher value than 1 K was chosen as the design value. An alternative would be to accept 700 ohms as an acceptable ac load, drop the emitter resistor to 70 ohms, and recalculate that way.

Another aspect of linear amplification is the assurance of an adequate swing. As stated earlier, when the collector resistor is the only load, the best operating point to provide maximum swing is at about half the supply voltage. But when the collector is ac-coupled to another load, that is no longer true.

Suppose the collector resistor is 1 K, coupled by a capacitor to another resistance of 330 ohms (Fig. 13-11). The combined parallel value is 330 × 1000 divided by 1330 or 250 ohms. If the supply voltage is 12, biasing the operating point to 6 volts will yield a collector current of 6 milliamps. With an ac load of 250 ohms, an output of 6 volts peak would correspond with a peak current of 24 milliamps. But the downward peak current only can neutralize the transistor's collector current as far as zero.

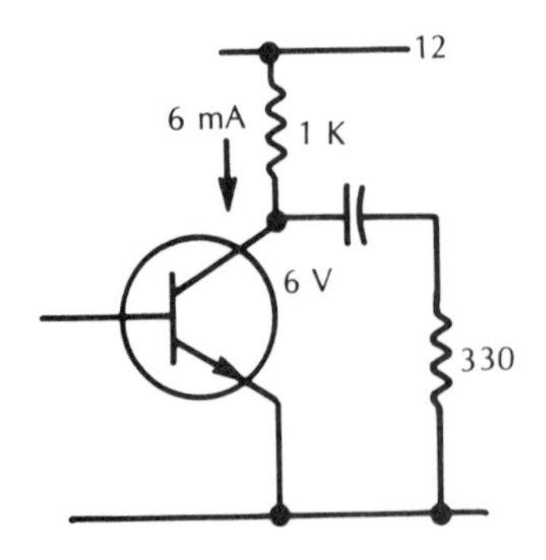

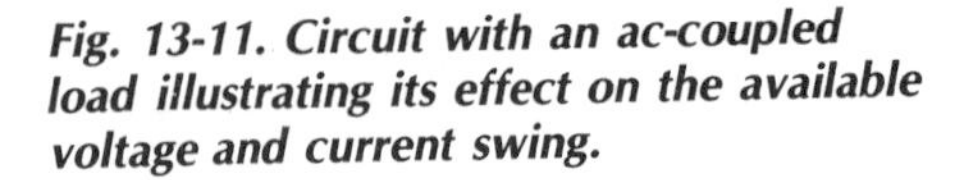

Fig. 13-11. Circuit with an ac-coupled load illustrating its effect on the available voltage and current swing.

So with this operating point, the transistor can deliver only 6 milliamps peak in the downward direction, although the current can increase until the transistor reaches saturation, when the voltage peak in that direction is 6 volts at 24 milliamps.

Thus, the limits in opposite directions are quite different when the ac load is coupled. This difference can be reduced by changing the operating point. Because the ac load is one fourth the value of the dc resistance, the available current swing one way is 4 times the available current swing the other way. Without the extra ac load, the swing available both ways was equal, which led to the half-supply-voltage choice of operating point.

So, the ideal operating point would be one-fifth of the supply voltage (Fig. 13-12). Assuming the collector voltage is 2.4 volts, the current will be 9.6 milliamps. The ac load of 250 ohms will accept 2.4 volts at 9.6 milliamps peak in each direction. The 2.4 volts downward swing will take the transistor to saturation at zero volts (actually this is ideal; a little margin should be allowed for the saturation voltage drop) at 19.2 milliamps (twice 9.6) and will reach cut-off with the collector voltage at 4.8, at which moment, the capacitor will be delivering 7.2 milliamps into the 330-ohm load, while the current through the 1 K resistor is reduced from 9.6 milliamps by 2.4 milliamps to 7.2 milliamps, which checks by providing the 7.2-volt drop when the collector voltage is 4.8.

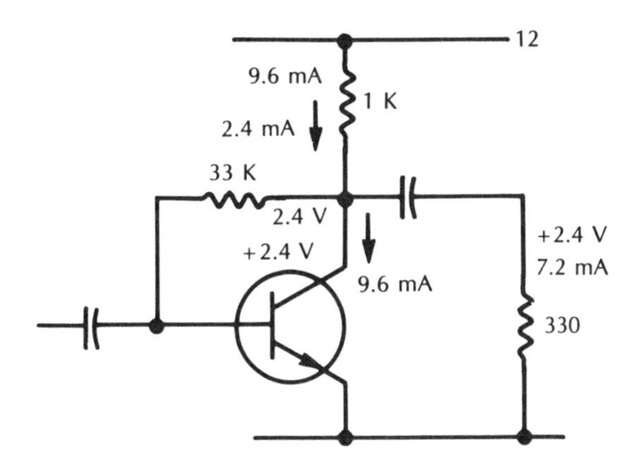

Fig. 13-12. An optimized circuit for the coupling arrangement in Fig. 13-11.

Again assuming that the beta is 120 average, with limits of 60 and 180, some stabilization can be achieved by running the bias resistor from the collector. The collector voltage is ideally 2.4 and the current is 9.6 milliamps. With a beta of 120, the base current needs to be 9.6 divided by 120 or 80 microamps. 2.4 volts at 80 microamps requires a resistor of 2.4 divided by 0.00008 or 30 K. A 33 K resistor will allow a little margin for the saturation voltage.

This 33 K parallels an ac load of 250 ohms at the collector, and thus accepts 250 divided by 33,000 or one one-hundred-thirty-seconds of the output signal current. With a gain of 120, the feedback signal current is slightly less than input current, so the gain will be reduced by slightly less than a 2:1 factor (which is 6 dB). If this loss of gain is undesirable, the resistor can be divided into two (e.g., 15 K and 18 K) and decoupled, so the ac signal is not fed back (Fig. 13-13).

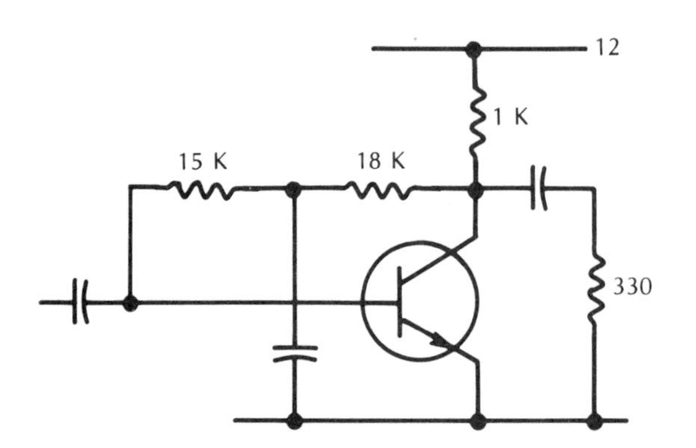

Fig. 13-13. The decoupling capacitor removes the signal from the dc feedback in Fig. 13-12.

Assume the bias resistor is 33 K. With a beta of 60, the transistor looks like a dc resistance of 33 K divided by 60 or 550 ohms. With 1000-ohm collector resistance, the collector voltage is 12 × 550 divided by 1550 or 4.25 volts. With a beta of 180, the transistor looks like a dc resistance of 33 K divided by 180 or 183 ohms. With a 1000-ohm collector resistor, the collector voltage is 12 × 183 divided by 1183 or 1.85 volts.

Either of the above is a serious deviation from 2.4 volts, which reduces the available current swing somewhat. But the alternative is to take the bias from the 12-volt supply, which would require 150 K to give 80 microamps. With this arrangement (Fig. 13-14) the collector current could vary between 80 microamps multiplied by 60 or 180, which makes 4.8 or 14.4 milliamps.

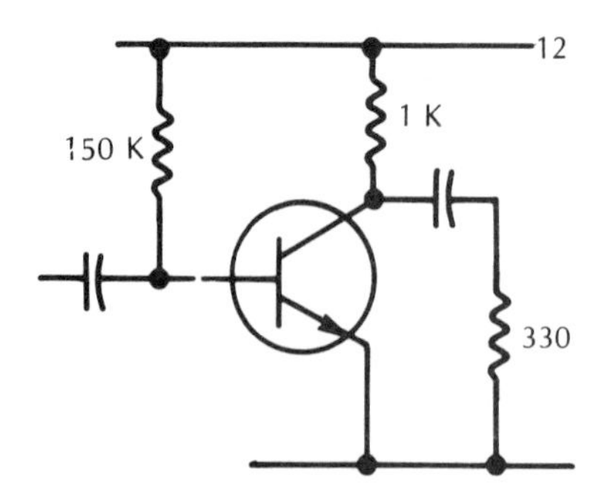

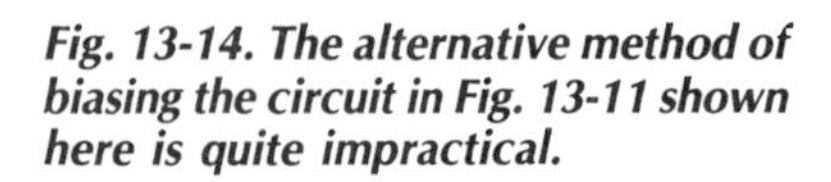

Fig. 13-14. The alternative method of biasing the circuit in Fig. 13-11 shown here is quite impractical.

At 4.8 milliamps, the collector voltage would be 12 − 4.8 or 7.2 volts, and 12 milliamps is saturation with a 1 K collector resistor, so 14.4 milliamps would never be reached; a transistor approaching a beta of 180 would run into permanent saturation. The circuit in Fig. 13-13 is unquestionably better than this!

A similar shift must be made when a common-collector (emitter-follower) connection is used. This is a fact that can get overlooked, because the ac resistance presented by the transistor may be much lower than the emitter resistor used. Suppose, for example, that the circuit resistance from which the signal is fed into the base is 7.2 K, that the beta is 120 and the emitter resistor is 1 K. By calculation, the source resistance presented by the emitter is 7.2 K divided by 120, which is 60 ohms. Now suppose this is connected through a large value capacitor (Fig. 13-15) to a matching load of 60 ohms.

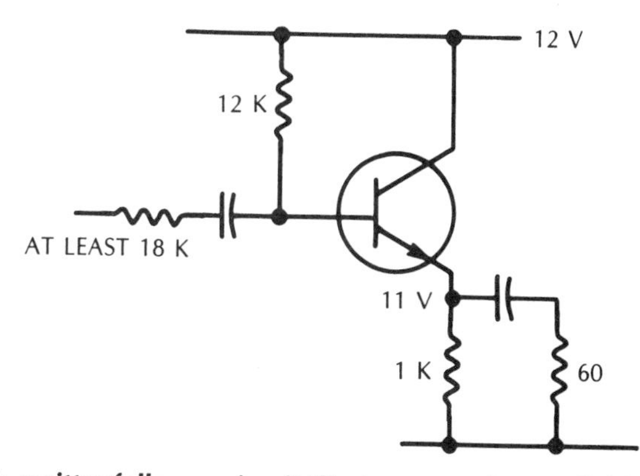

Fig. 13-15. This emitter-follower circuit illustrates another available swing problem.

The emitter load is 60 ohms in parallel with 1 K, which calculates to 56.5 ohms. The ratio of the dc to the ac load is 1000 divided by 56.5, or 17.7:1. So, theoretically, the drop across the transistor should be 1 over 18.7 of the supply. It is not feasible to use less than a volt, so we may allocate 1 volt across the transistor, 11 volts across the 1 K resistor, allowing a collector current of 11 milliamps.

The base current needs to be, assuming a beta of 120, 11 divided by 120 or 92 microamps. To get a 1-volt drop at this current requires a resistor of 11 K. A 12 K resistor will allow a margin. To make the source 7.2 K will require the ac source to be 7.2 K × 12 divided by (12 − 7.2) or 18 K.

Into an ac load of 56.5 ohms, a current of 11 milliamps will produce a voltage of 0.62 peak, or 0.44 volt rms. This is the best possible, using a 1 K emitter resistor with 12-volt supply, working into the 60-ohm load.

Possibly the 1 K resistor choice was based on transistor dissipation, and a supposition of 6 volts at 6 milliamps on the collector, which makes 36 milliwatts. Say the transistor is rated at 40 milliwatts, but 11 milliamps at 1 volt is only 11 milliwatts. Lowering the emitter resistor value will increase the proportion of voltage required across the transistor, as well as increase the dc operating current. Assume the emitter resistor is changed to 500 ohms. Now, the ac load is 500 ohms in parallel with 60 ohms, which calculates to 53.5 ohms. At least one tenth of the supply voltage should be present across the transistor.

With 2 volts (to allow a little margin) across the transistor, the 500-ohm resistor takes 20 milliamps. At a beta of 120, this requires a base current of 20 divided by 120 or 167 microamps. To drop 2 volts, the resistor needs to be 12 K again, so the same base circuit serves. Into an ac load of 53.5 ohms, a 20-milliamp peak current will deliver 1.07 volts peak, which is an rms of 0.76 volt.

If a similar transistor with a larger rating could be used, the emitter resistor could be reduced to, say, 120 ohms. Then the ac load is 120 ohms in parallel with 60 ohms, which makes 40 ohms. For these values, one fourth of the supply voltage (or 3 volts) should be applied across the transistor, leaving 9 volts across the 120-ohm resistor, which amounts to a collector current of 75 milliamps.

A peak current of 75 milliamps into an ac load of 40 ohms will result in a peak voltage of 3 volts or an rms of 2.12 volts. The dissipation in the transistor will be 3 volts at 75 milliamps, which is 225 milliwatts, and the 120-ohm resistor passes 75 milliamps at 9 volts, which represents a dissipation of 675 milliwatts. The ac signal output into 60 ohms peaks at 3 volts at 75 milliamps or 225 milliwatts, which is an average (rms current times rms volts) of 112.5 milliwatts.

SPLIT LOAD STAGES

Calculations of this sort are complicated a little when equal loads are coupled to both the collector and emitter of the same transistor (Fig. 13-16). Now, both dc and ac values in both collector and emitter must match if equal power is to be delivered at each point. Possibly the easiest way to think of this circuit is to imagine the supply divided into two equal parts.

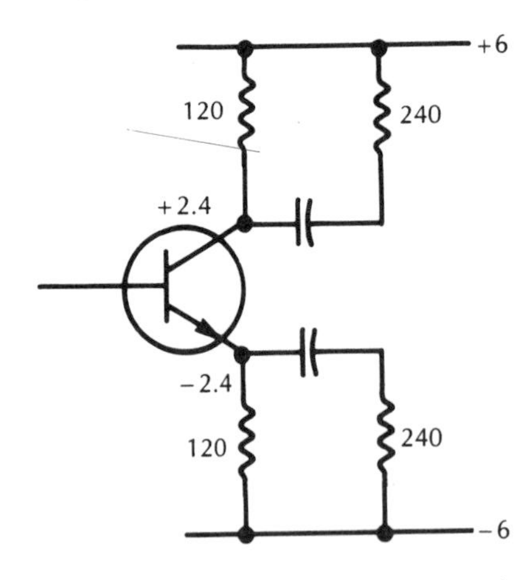

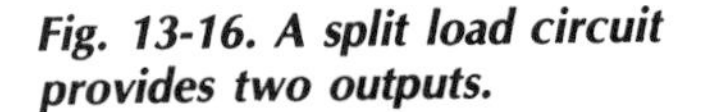

Fig. 13-16. A split load circuit provides two outputs.

If the supply is 12 volts, then 6 volts is available for the collector output and 6 volts for the emitter output, with the transistor receiving part of its working voltage from each half of the supply. For dc coupling purposes, if each resistor is 120 ohms and the load coupled is 240 ohms, the ac load at each point is 80 ohms. Since the ratio of the ac to the dc load is 2:3 (80:120 actual), two-fifths of the supply voltage should be applied across the transistor. This means that 4.8 volts should exist across the transistor (2.4 volts from each 6-volt half), with 3.6 volts across each of the 120-ohm resistors. So the collector current will be 3.6 divided by 120 or 30 milliamps.

This means the peak ac into each load of 80 ohms will be 30 milliamps, yielding a peak signal voltage of 2.4, representing an rms of 1.7 volts. The voltage-divider bias arrangement serves well here to give a base voltage of 3.6 volts (above the supply negative) at an average current of 250 microamps. A 3 K value for the lower resistor will take 1.2 milliamps, so the top resistor needs to pass 1.45 milliamps at 8.4 volts,and that takes a 5.8 K resistor (Fig. 13-17).

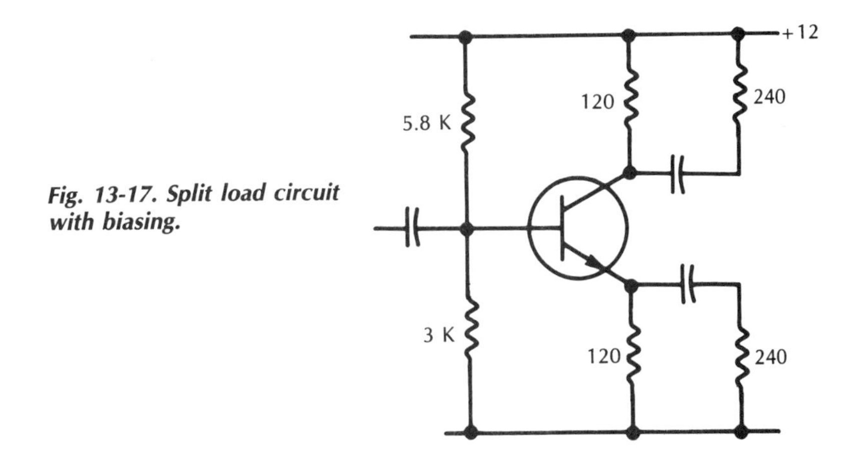

Fig. 13-17. Split load circuit with biasing.

With split-load stages, the collector and emitter loads must be equal *throughout the amplification cycle*. For example, one possible use of a split load might be to drive the bases of push-pull power stages (Fig. 13-18) where each may have an equivalent input resistance of 240 ohms. Since both power stages are identical, it would seem that the requirement for identical loads has been met. But it may not be. If the power stages are common-emitter connected, with their loads collector-coupled, the base input resistance varies throughout the signal cycle, as described earlier.

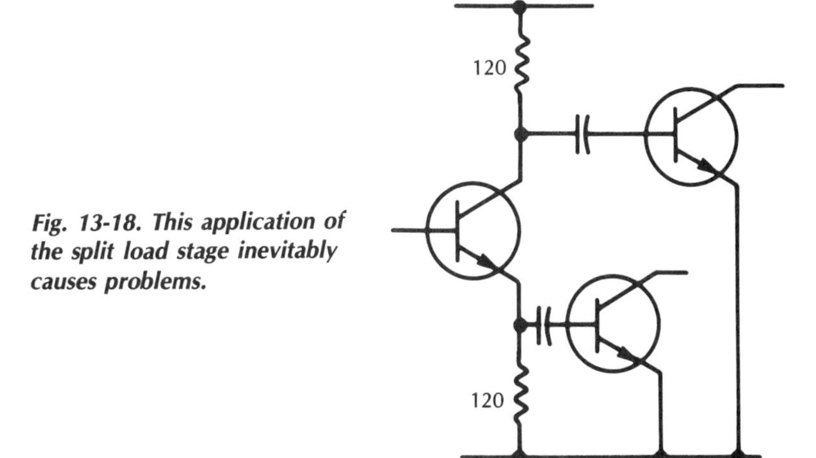

Fig. 13-18. This application of the split load stage inevitably causes problems.

At the operating point of all transistors (with no signal being amplified) the loads are identical (or should be). But when the circuit handles signal, one output transistor is driven to a higher current while the other is driven to a lower current. The one momentarily driven to higher current will have a lower base input resistance and the one momentarily driven to a lower current will have a higher input resistance. This means that the ac-coupled load of the split-load stage becomes unequal at all points except the operating point. What is the effect of such unbalance?

Obviously, the output voltage will divide between the collector and emitter loads in proportion to the instantaneous resistance values, since the same signal current flows through both emitter and collector loads at every instant. Since the load is split, this would suggest that the shift of voltage division between the loads would be symmetrical over the cycle, but this is not so.

If at one peak, for example, one load dropped to one-third of the resistance of the other load, then at the other peak the change would be precisely vice versa. That much is true, but the effect on waveforms is not symmetrical. This is because the input is virtually tied to the emitter side, since the voltage difference between the base and the emitter is always small. So the emitter output side tends to work like an emitter follower. The result is that the waveform delivered to the emitter load follows the input waveform quite closely.

When the emitter load becomes one-third of the collector load, the collector signal voltage becomes three times the emitter signal voltage. But when the collector load becomes one third of the emitter load, and collector voltage becomes one third of the emitter signal voltage. The emitter load voltage tends to faithfully follow the input, while the collector signal voltage fluctuates from one-third to three times what it ought to be at different points on the waveform.

The way to overcome this is to revise the circuit design so that the loads do not change during the waveform. A relatively small emitter resistor, in each of the collector-coupled power stages, will achieve this (Fig. 13-19), because its effective value in the base circuit will be multiplied by the power stage current gain.

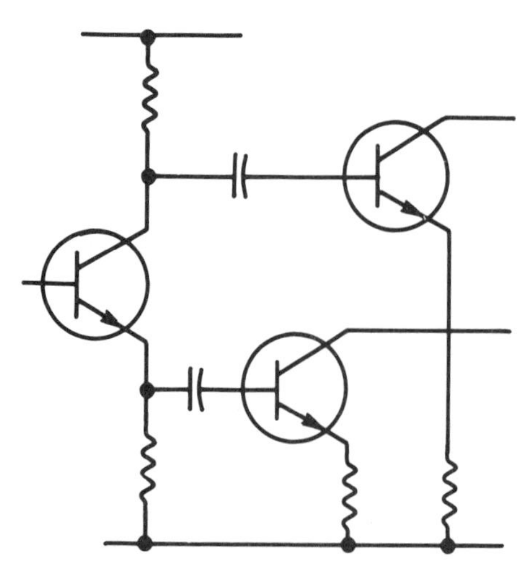

Fig. 13-19. This circuit shows one way to improve on the arrangement shown in Fig. 13-18.

This is also a good way to divide current loading between two power transistors connected in parallel to double their available current (Fig. 13-20). Without some provision for equalizing the current handled by each, one of them tends to 'grab' the whole load. This happens because of the temperature coefficients of these devices. When they get warmer, the gain usually increases. If two of a pair operate in parallel at different gains, the one with the greater gain takes the lion's share of the current load. This, in turn, heats that one a little more than the one with less load, until the first gets really hot and grabs the whole load—until it destroys itself.

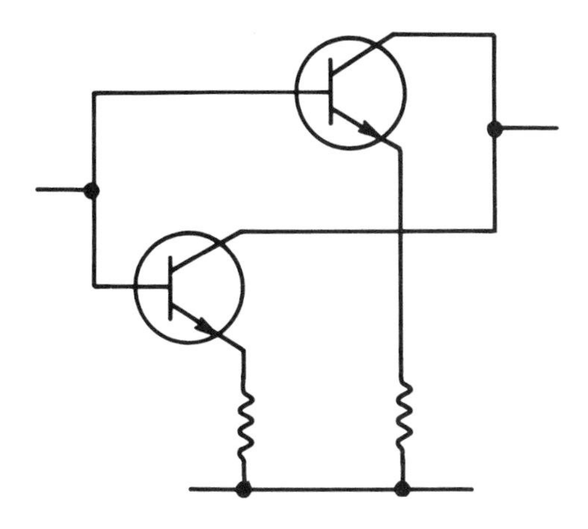

Fig. 13-20. The emitter resistors equalize the load shared with transistors connected in parallel.

DARLINGTON CONNECTION

One more little trick enables two transistors to do more than one: this is known as the Darlington circuit. By strapping two transistors together, the current gain available is the product of their individual betas. When trying to design a circuit, such as the one in Fig. 13-10, to have a high gain, a certain amount must be thrown away to achieve stability. This limits the gain that can be achieved with one transistor at satisfactory stability. Putting two together in a Darlington circuit enables higher gain to be achieved after stabilization (Fig. 13-21).

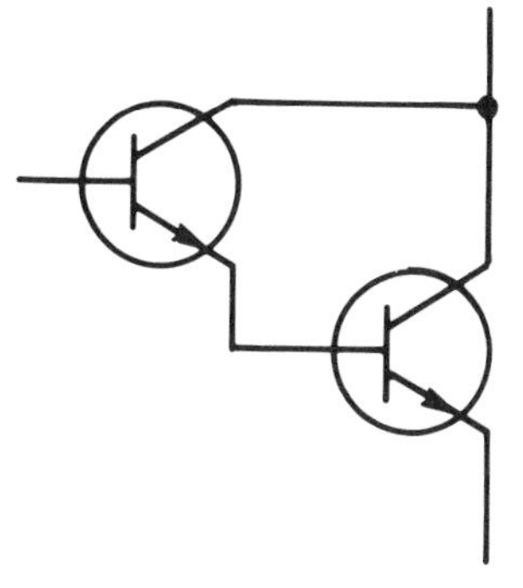

Fig. 13-21. The Darlington connected circuit puts two transistors together to multiply the gain.

OP AMPS

Operational amplifiers are basically the first form of linear integrated circuit. Figure 13-22 shows how they are indicated schematically. It must have a positive and negative supply, of appropriate voltage, like every other electronic device. Then its minimum other connections are a positive and negative input, and an output.

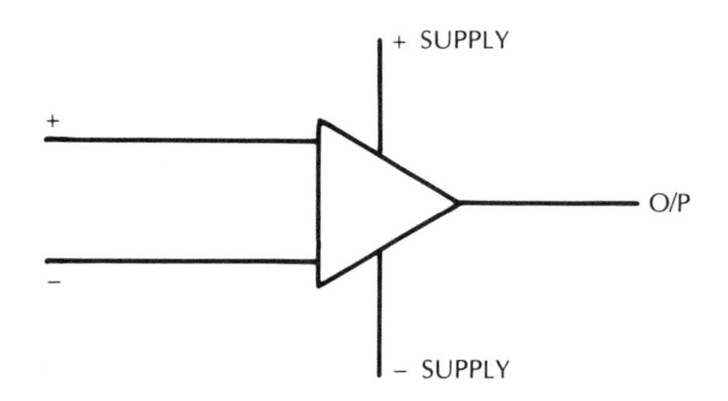

Fig. 13-22. Schematic representation of an op amp (operational amplifier).

Op amps come in tremendous variety, so in this book we can only tell the general details of how they work. When supplied with the correct supply voltages, a positive input makes the output change in the same direction, and a negative input makes the output change in the opposite direction.

Op amps have such enormous gain that for all intents and purposes, it is infinite. This means that an infinitely small difference in voltage between the plus and minus inputs will swing the output all the way from positive to negative, or vice versa. In practice, the gain is so great that just connecting one up as in Fig. 13-22 will probably be unstable and oscillate. In other words, an op amp needs an external circuit to make it behave.

One such circuit is shown at Fig. 13-23. Here an input is supplied from a source having resistive impedance (perhaps an actual resistance, but not necessarily) r and feedback is connected from the output to the same (negative) input through resistance R. Bias is provided by a potentiometer to the other input, so it can be adjusted.

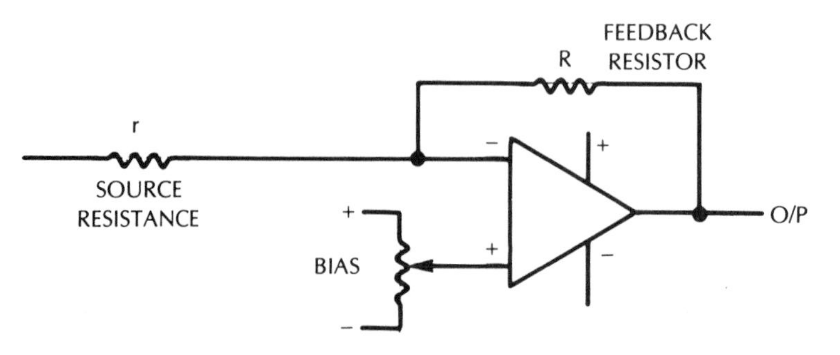

Fig. 13-23. A simple way of connecting an op amp to achieve controlled gain and biasing.

Without the feedback, an incredibly small change in input voltage in the vicinity of the bias voltage will make the output go all the way from positive to negative, or vice versa. That is what the infinite gain means.

Now consider what happens with the feedback in place. If the voltage at the input end of r is the same as the bias voltage, the output voltage must be the same as the bias voltage too, for everything to balance. Now, suppose the input voltage moves positive by one unit (volt, millivolt, or whatever). For balance still to be preserved, the output must move negative by an amount determined by R/r.

If r and R are equal, then the op amp is just a phase inverter. If R is 100 times r, then it has a gain of 100.

So what's inside the op amp? Actually, modern op amps have a lot of elements inside them, but Fig. 13-24 shows a very much simplified, primitive type. Transistor Q1 has a bias resistor R1 which causes it to have a constant collector current, regardless of collector voltage, determined by beta times the current in R1.

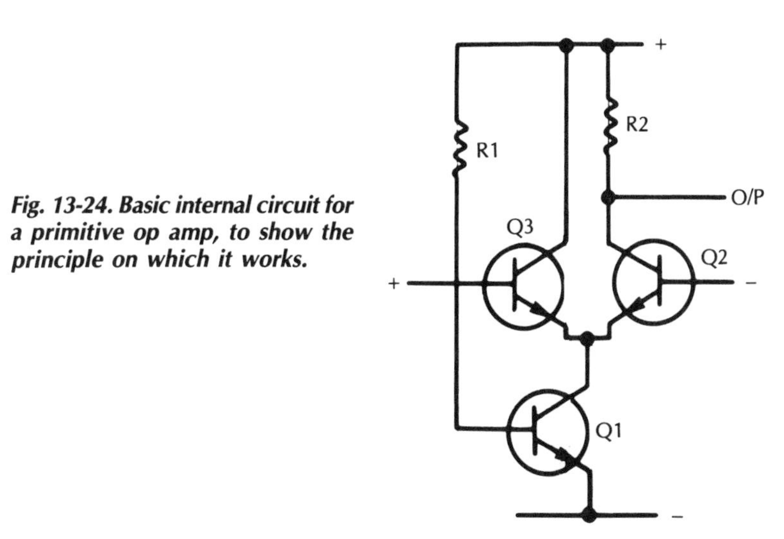

Fig. 13-24. Basic internal circuit for a primitive op amp, to show the principle on which it works.

This means that the total current in Q2 and Q3 is fixed by Q1. If one of them takes more, the other one must take less, by the same amount. The current each of them takes is determined by their respective base inputs. Being current type transistors, the base-to-emitter voltage doesn't change much; it's essentially constant. So slight differences between the positive and negative inputs will switch all the constant current from Q2 to Q3 or vice versa.

Very small changes in relative input will cause a change in the balance of currents, resulting in change in output, developed across R2. In this simple circuit, obviously there is some input current taken by Q2 and Q3. Making both of them Darlingtons would reduce this considerably.

There are many more additions that are made to actual op amp "innards" that you won't need to bother with. The important thing is to know what they do. The use of FETs further improves the input impedance—making it very high indeed, and being all built on the same microchip eliminates variations with temperature, by making all the temperature coefficient changes move together.

Examination Questions

1. The operating condition discussed with respect to Figs. 13-1 to 13-4 is changed from 4 volts at 8 milliamps to 5 volts at 7 milliamps, using a 50-microamp bias current instead of 60 microamps. Find the value of resistor needed in series with the base input that will cause equal swing either way at the input to produce the maximum equal swing at the output, from this operating point, and the peak input voltage needed.
2. Extend the correction started in Question 1 so that the operating point can use the full 12-volt peak-to-peak swing, symmetrically. What bias current is needed, and what series resistor in the base input? Also, what peak input voltage is required to secure a 6-volt peak output?
3. Assuming the base input current-voltage curves in Fig. 13-4 also apply to the transistor characteristics shown in Fig. 13-5 when it is operated with a 1 K collector resistor with no ac-coupled load, find the base current needed to produce a symmetrical 6-volt peak output and the series resistance needed at the base so a symmetrical input produces a symmetrical output. Also the peak input voltage necessary.
4. Find the bias resistor, used to connect from the collector to the base, to provide the operating conditions described in Question 1, 2 and 3.
5. Assume the voltage divider in the circuit in Fig. 13-10 is changed from 68 K and 4.7 K to 100 K and 10 K. Find the new extremes of input resistance for betas of 60 and 180 and the corresponding collector voltages.
6. Recalculate the input resistance values and the working collector voltages for the same range of beta values when the emitter resistor is changed from 80 ohms to 70 ohms and the base bias divider is made up of 82 K and 5.6 K resistors.
7. In the circuit in Fig. 13-11, if the coupled load is 500 ohms instead of 330 ohms, calculate the collector voltage needed to enable the stage to generate the maximum swing and, assuming a beta of 120, the collector-to-base bias resistor that will provide that operating condition. Also calculate the variation in collector voltage as beta varies from 80 to 160.
8. An emitter follower is to be used on a 20-volt supply to deliver the maximum signal into a 500-ohm load, using an emitter resistor of 1000 ohms. Calculate the bias resistor to be used, assuming a beta of 90, and the dissipation in the transistor and the 1000-ohm resistor.
9. A coupling capacitor produces a 1 dB loss (about 12 percent, voltage or current) at a frequency where its reactance is half the sum of the resistances or impedances between which it couples (i.e., half of 2 K in Fig. 13-6). Calculate the coupling capacitors that must be used to produce not more than a 1 dB loss at 20 Hz in the output circuits of Figs. 13-6, 13-9, 13-10, 13-11, 13-15 and 13-16.
10. Assuming that the beta is 120 in the circuit in Fig. 13-15, find the input coupling capacitance needed, so the loss will not exceed 1 dB at 20 Hz.

14
Solid-State Switching Functions

In this chapter we will cover two basic aspects that come under this heading. The first is how switching devices, particularly those that handle power, differ from linear devices. This is concerned with how the device handles its "on" and "off" transitions. The second is the basis for a much bigger part of modern electronics, but is much simpler: the way in which combinations of two-state devices can be built up into the "logic" that forms the basis of computer systems.

HOW SWITCHING TYPES DIFFER

Some solid-state devices, such as SCRs, are made specifically for two-state operation. They are incapable of being held between states. Others are basically no different from the transistors discussed in the previous chapter, but the circuits are different. Some transistors are listed as switching types, mainly because they were designed with that in mind. They may be used in linear circuits, of course, but only under appropriate circumstances.

What this last statement means is that, by having a low saturation voltage and possibly a high reverse (non-conducting) voltage, they can handle both high currents and high voltages. If, for example, a transistor can pass 20 amps and withstand 100 volts when it is nonconducting, the mid position between these extremes of current and voltage is 10 amps at 50 volts, which represents a dissipation of 500 watts.

However, at the 20-amp saturation current the voltage drop may be 0.7 volts, while the leakage current at 100 volts is perhaps 10 microamps. Thus, when it is fully conducting, it dissipates a maximum of 20×0.7 or 14 watts; when it is non-conducting, it dissipates a maximum of 100×0.00001 or 1 milliwatt. Forget the 1 milliwatt. The important figure is the 14 watts at saturation. Possibly its maximum safe dissipation is 25 watts when suitably mounted to get rid of the heat.

Obviously, if such a transistor were used in linear mode, assuming a 25-watt maximum dissipation rating and a 100-volt supply with a 50-volt operating point, the maximum current would be 25 divided by 50 or 0.5 amp (500 milliamps). So it might swing from 1 amp at zero voltage to zero current at 100 volts. But this swing would be using only about one twentieth of its handling capability, based on a saturation current of 20 amps.

If the transistor handles 20 amps at saturation, the saturation voltage must never exceed 1.25 volt, because then its dissipation would exceed 25 watts. Also, it must switch almost instantly from 20 amps at a 0.7 volt drop to nonconducting at any voltage up to 100 volts. The maximum dissipation, for a very brief peak during the switching action, still goes through at least 500 watts. This time interval must be kept short—almost instantaneous—so the thermal capacity of the junction can absorb it without overheating.

Another reason for designating a transistor as a switching type is its capability to effect a quick transition from saturation to cutoff and vice versa, which involves quite different characteristics from those needed for linear amplification.

HIGH-FREQUENCY CUTOFF

Linear transistors have what is known as a cutoff frequency. Because of the finite but very short time taken for the output current change to follow the input current change, there comes a frequency where it does not follow completely, because the cyclic period does not allow enough time. In linear transistors, this property is not materially dependent on signal amplitude. Whether the signal fluctuation is complete, full amplitude, or merely a small fraction of the available operating voltage and current, the cutoff frequency remains about the same.

The cutoff frequency in linear transistors is a function of the circuit mode because of the relationship between alpha and beta. Used in the common- or grounded-base mode, alpha is the relevant quantity, and the designated cutoff frequency is the point where the alpha drops to half its operating value with a rise in frequency. Used in a common- or grounded-emitter mode, beta is the relevant quantity, and cutoff is where beta drops to half its operating value.

Suppose alpha is 0.992, which makes beta 0.992 divided by 0.008 or 124. Cutoff in the grounded-base mode occurs when the alpha drops to 0.496 (probably 0.5 would be taken as close enough for practical purposes). But cutoff in the grounded-emitter mode occurs when the beta drops to 62, at which point the alpha is still 62 divided by 63 or 0.984. An easy formula for deriving alpha from beta is:

$$\alpha = \frac{\beta}{1 + \beta}$$

There is no fixed relationship between the frequency and alpha or beta deterioration. At worst, the increase in the reciprocal of beta is proportional to frequency. In this case, the working value of one over beta is approximately 0.008. If reciprocal beta rises from 0.008 to 0.016 at 20,000 Hz, this is an increase of 0.008 from its working value. To get to 1, which is the value of beta required to make alpha 0.5, the frequency must be increased 125 times, from 20,000 Hz to 2.5 MHz. So it is a safe rule that the alpha cutoff is beta times the beta cutoff. The ratio may be more than that. This means that if the alpha cutoff is quoted, as it is in some transistor specifications, the beta cutoff will not be more than this frequency divided by beta; in fact, it may be lower than that.

SWITCHING CHARACTERISTICS

The characteristics relating to the speed of switching deal with more than the simple cutoff frequency. Since they relate to full switching from off (cutoff) to on (saturation) and vice versa, a switching function is divided into several parts (Fig. 14-1). In the absence of further data, two characteristics are important: switch-on time and switch-off time. These times further break down into two

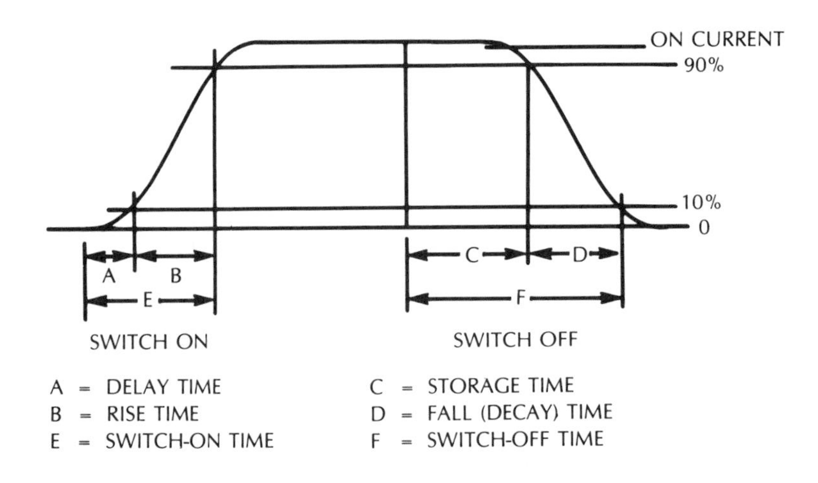

Fig. 14-1. Curve showing the basic switch-on and switch-off intervals.

components each. Switch-on time includes two intervals—delay time and rise time; switch-off time also involves two segments—storage time and fall time (sometimes called decay time).

For measurement purposes, the delay time is defined as the time between the moment that the base turn-on current is applied and the collector current rises to 10 percent of its saturation value. Rise time is the interval required for the current to rise from 10 to 90 percent of its saturation value.

Turning the transistor off involves two similar times: The storage time begins with base current cutoff and lasts until the collector current falls to 90 percent of its saturation value. The fall (or decay) time is the time taken for the current to drop from 90 to 10 percent of that value.

Generally speaking, the higher the current the transistor is built to handle, the longer the delays. With a low-current transistor that switches only milliamps, all times may be shorter than a microsecond. A transistor built to handle amps may take several microseconds switching time, which can create problems.

Much work has gone into trying to shorten switching time. For example, both delay time and rise time can be shortened by increasing the base current used to perform the switch-on. However, during the rise time, the base current should not appreciably exceed the saturation value needed to hold the current on. It may be possible to materially shorten the delay time by pulsing the current to a somewhat higher value than required to maintain the on current.

In discussing this, it should be realized that, while the saturation condition may be defined in terms of a base current that is more than a specified fraction of the saturation collector current, increasing the base current beyond this value will usually lower the collector-emitter saturation drop a little. An expanded voltage scale, near zero, for the collector voltage-current curves illustrates this fact (Fig. 14-2).

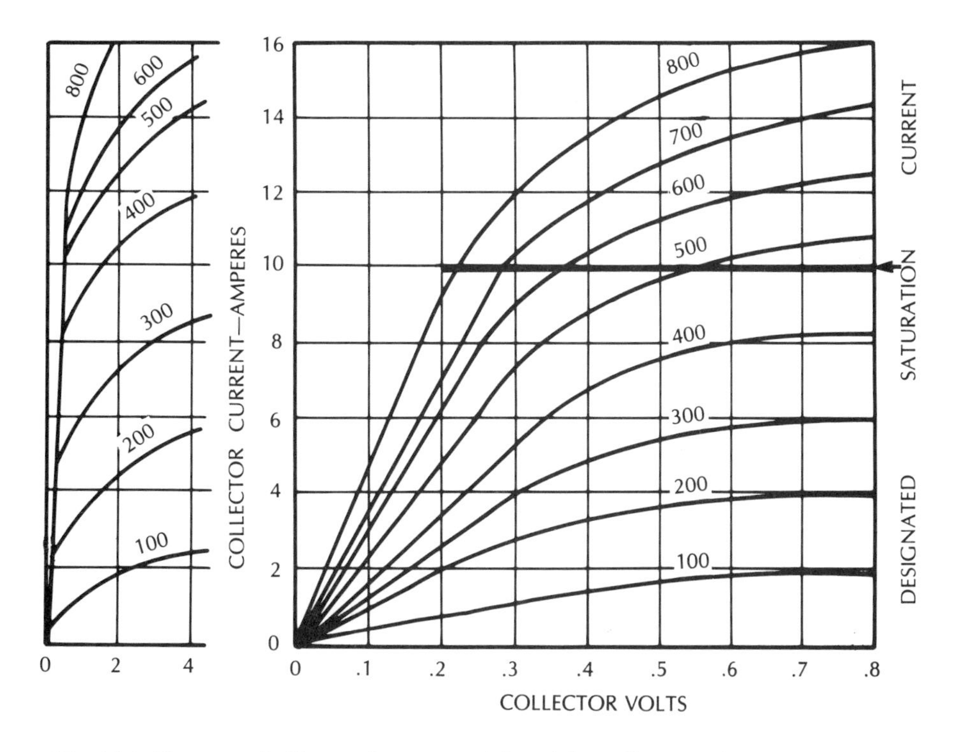

Fig. 14-2. The expanded low-voltage part (right) of the collector voltage-current curves aids in visualizing the range of saturation conditions.

For example, a transistor may have a working beta of 30 (which may not be specified as such for a switching type, because beta is not used as such). If the on current is to be 10 amps, a base current of 10 divided by 30, 330 milliamps, may hold the current on, with considerably more than a 1-volt drop across the transistor. Increasing this to 600 milliamps may reduce the drop to less than 0.4 volt. Increasing it to 800 milliamps might reduce the drop to a little over 0.2 volt.

Obviously, there comes a point where a further increase in the base switch-on control current produces diminishing returns in the collector circuit. Even if this did not happen (presumably, some reduction in the drop will continue with these diminishing returns) the base current reaches a safety limit beyond which the base-emitter junction is in danger.

The importance of keeping the rise time short may be two-fold: to get the current on as quickly as possible because the circuit requires it; and to protect the transistor from dangerous dissipation, because the longer it takes passing through that 500-watt peak, the more heat the current develops.

Whether the delay time matters or not may depend on the application. If the only reason for wanting a fast switch is to prevent the transistor from blowing during the rise time, then the fact that the rise itself is delayed may not matter. In other applications, it may be important for the switch-on to happen with a minimum delay after the control signal initiates the action. Then, it is of value to consider ways of expediting the rise.

Storage time occurs because saturation builds up a charge in and around the junction, which has to be removed before the current starts to fall. Since this is due to an actual charge inside the transistor, the storage time can be shortened by reducing the maintain-on current (which also increases the saturation drop a little, and probably lengthens one or both of the switch-on times). By reversing the base current immediately, switch-off is initiated. Of course, when switch-off has been completed, the base emitter junction is non-conducting. So it may have a reverse voltage connected to ensure the transistor is well and truly switched off—nonconducting.

All these time constants are specified in the complete data for switching-type transistors. But the way they are specified varies. For a fair comparison of type, it might be best to specify each of them with the simplest possible unexpedited switching drive—just on and off (Fig. 14-3).

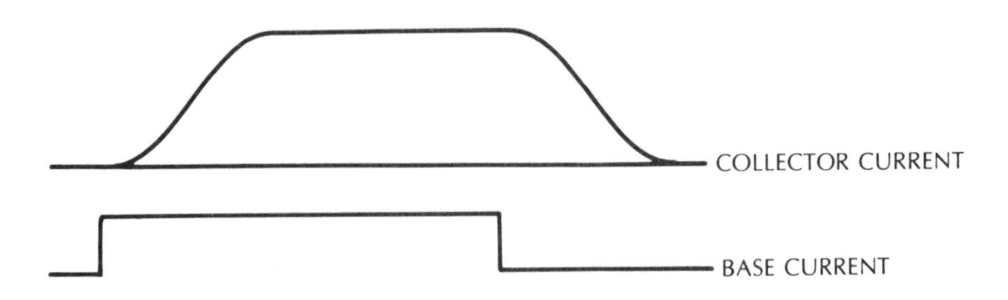

Fig. 14-3. Switching in some transistor types is specified by curves of this type.

Then, of course, methods appropriate to each type of transistor could be used to expedite the desired transitions, and such conditions differ with individual types. However, because of the differences between individual types, different modifier circuits are recommended for the expediting process.

TEST CIRCUITS

Accordingly, the most usual practice is to picture the complete switching circuit, or a test version of it (Fig. 14-4), and then show or list the times that are obtained typically, or as maximum values) with this transistor type for specified values of on current. Reducing the on current almost invariably shortens all the times involved, provided the base control current is also adjusted accordingly.

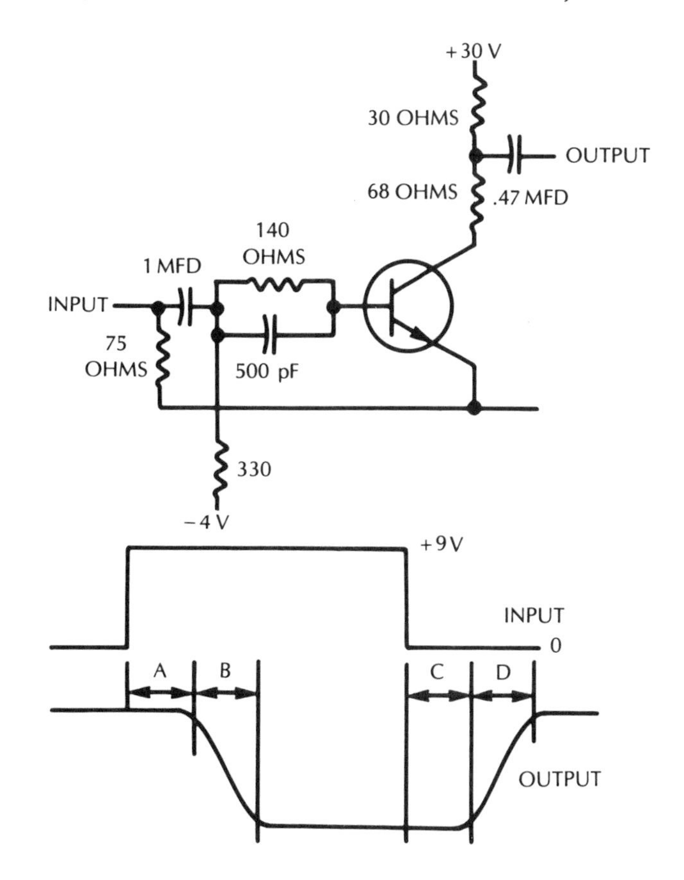

Fig. 14-4. Other types give test circuit, like this, and specifications to be met.

In Fig. 14-4, when the pulse is absent, the reverse voltage in the base circuit puts the base 4 volts negative. The +9-volt pulse overcomes this negative bias by 5 volts, which is fed through the 140-ohm resistor to the base. With the input resistance and voltage drop, the base current is about 30 milliamps. The result would be the steady on base current (if the 1-mfd capacitor did not prevent a steady on current). However, the 500-pf capacitor causes the base current to exceed 30 milliamps by several times at the instant of switching to help overcome the negative charge at the base and the delay time that it causes.

At switch-off, the pulse is short enough so that little change occurs to the charge on the 1-mfd coupling capacitor; therefore, the 4-volt negative reverse voltage is again applied to the base. But at the initial instant of the input signal's negative swing, the 4-volts-odd across the 500-pf capacitor, due to the drop across the 140-ohm resistor while the base current is flowing, is added to the 4 volts negative derived from the external supply. This further shortens the storage time.

When the transistor is operating below its rated maximum on current, there may be some choice of control current value. The same on current could be used as is required for maximum collector current, or a proportionately reduced value could be used, or perhaps some value in between.

Increasing the control on current will expedite the switch-on, at a particular main current (collector saturation value), but it usually will extend the storage time also, which is more a function of control current than of main-circuit current. Reducing the control current, thus reducing the degree of saturation present in the on condition, will expedite the switch-off time, particularly the storage interval.

PRACTICAL CIRCUITS

The "kicks" needed to expedite switch-on and switch-off are frequently attained by adding "shaping" to the control current changes with a pulse-shaping network in the base circuit (as in Fig. 14-4) or by a circuit that employs trigger action, with feedback from the collector circuit of the transistor itself. Two stages are usually involved to get positive feedback of a form that is achieved only in transitional modes by using an appropriate choice of capacitor values. But the action is different each way and can be different, also, depending on the drive circuit chosen.

For example, the previous stage may be designed to be on when the controlled stage is off, and vice versa (Fig. 14-5).

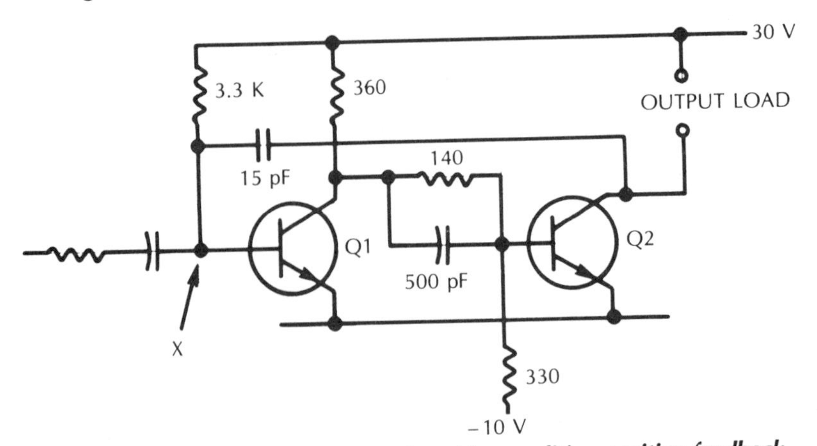

Fig. 14-5. Two stage switching circuit, with expediting positive feedback.

In another situation, the controlled transistor, Q2, is normally "on" and its control stage, Q1, serves to switch it off. With the control stage normally on, the controlled stage is normally off. Two stages cascaded, without any expediting switching, are likely to be quite slow because the respective on and off switching times are added together in the overall action.

In the off condition, the 3.3 K resistor in the Q1 base allows about 9 milliamps of base current, which keeps Q1 at saturation. In this case, to find what saturation means, consider first the current through the 140-ohm and 330-ohm resistors in series between the Q1 collector (which is effectively grounded when Q1 is saturated) and the −10 volt supply. This is about 21 milliamps, providing the Q2 base with −3 volts to hold it at cutoff. In this condition, the 30 volts across the 360-ohm resistor in the collector of Q1 will cause it to pass about 83 milliamps, so Q1 will take the difference 83 −21 or 62 milliamps at saturation.

In the on condition, a pulse at the base of Q1 takes it to cutoff, expedited in the way we discuss in a moment. When Q1 is cut off, Q2 conducts, so the base of Q2 is just above ground (by a fractional voltage). This means the 330-ohm resistor connected to the −10 volt supply will pass 30 milliamps.

The 360-ohm and 140-ohm resistors in series from the positive supply (30 volts) to the Q2 base (since Q1 is now cut off) will pass about 60 milliamps, which will provide the Q2 base with 30 milliamps, in addition to the 30 milliamps going to the −10 volt supply point through the 330-ohm resistor. That sets the on and off conditions. Now, the actual triggering of Q2 is achieved by the 500-pf capacitor which bypasses the resistor between the Q1 collector and base. When Q2 is off, it is because Q1 is saturated, allowing the Q2 base to be at −3 volts.

When it is switching on, the control stage, Q1, suddenly becomes nonconducting, which allows the voltage at the collector to rise. But because this stage has been saturated, even though the transistor may not be a switching type and it is not of a high-current type, its cutoff is not sudden. If the cutoff can be made more sudden, then the capacitor will increase the initial turn-on current for Q2 to shorten its rise time.

To facilitate the Q1 cutoff, some feedback from the collector of Q2 can assist by applying a negative pulse to the Q1 base (assuming the polarity of the transistors is npn, as shown). This does little or nothing to shorten the delay time, which in this case is due to the storage time of Q1, which runs normally saturated. To shorten the delay time a pulse-shaping circuit, such as a capacitor, is connected across the series resistor through which the negative pulse initiating the action is fed. When switch-off is needed, control stage Q1 is pulsed back into saturation. This momentarily pulses the controlled stage base negatively and the capacitors assist in both actions again in the reverse direction.

DUTY CYCLE PROBLEMS

Further problems that can arise in circuits of this type may be due to the duty cycle times; that is, how much of the time each transistor spends switched on and off (Fig. 14-6). Effective expediting of the next switching function is usually based on the assumption that adequate time has elapsed since the last switch, in either the same or opposite direction, for the circuit to have approximately reached a steady-state, either on or off.

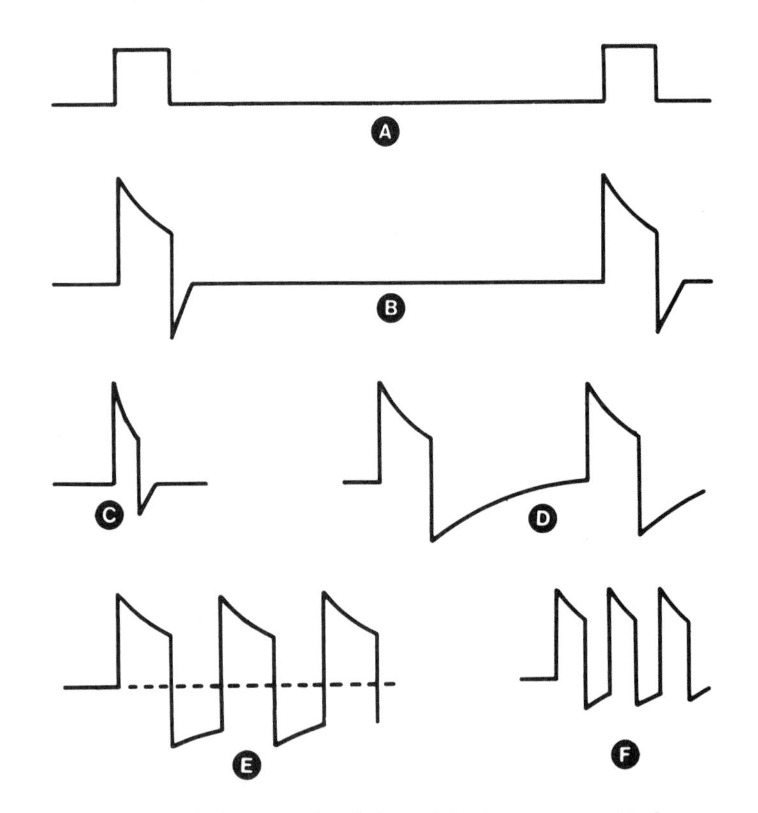

Fig. 14-6. The effects of changing the timing of the input wave: (A) the wave without modification by an expediting network; (B) as modified normally; (C) on time too short to allow recovery by the time the "off" pulse arrives; (D) longer recovery time following the off pulse, due to circuit time constants; (E) effect of the off time being too short, when the recovery time constant interval is too long; (F) effect of both times being too short for the respective recovery times.

That term, steady state, may be relative, because, as in Fig. 14-4, a steady-state on is impossible, except for the short-term pulsing sense. If the reverse action comes before the circuit has had time to settle down after the expediting pulse (which is extremely short-term—microseconds or fractions of microseconds), the next pulse may be inadequately expedited for its purpose.

Alternatively, if the circuit values (particularly the capacitors) are beefed up to allow for such shortened intervals, something else can happen if the intervals are too long. Then, the beef up may be too much, overdoing it and maybe causing instability. For example, if too much feedback is used to make sure there is enough when the action is extra quick-firing, when time allows steady on and off states to be reached the circuit may overshoot, resulting in a spurious trigger in the opposite direction before the input intended to make that next trigger actually arrives. This can lead to multiple switch-on-switch-off sequences, at a very high rate, until the opposite steady state settles down (Fig. 14-7).

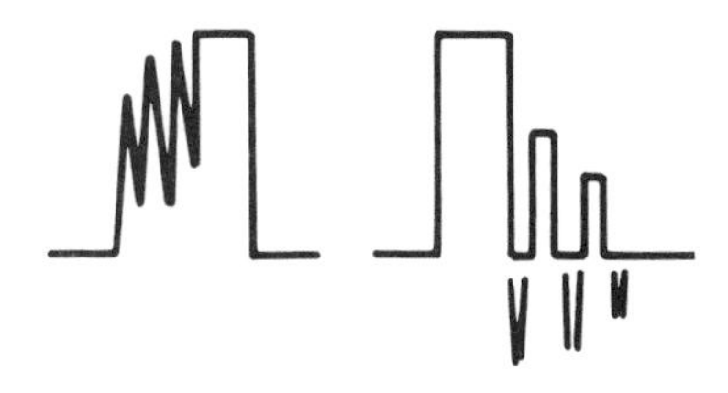

Fig. 14-7. These waveforms show two possible effects of over-compensation for quicker than normal pulsing.

Where this can happen, a more complicated design (beyond the scope of this book) aims to reset the circuit, so it is ready for full pulse action sooner than the normal recovery time after switching the simple circuit.

SCR CIRCUITS

In some respects, an SCR is easier to operate in this fashion than an ordinary transistor, because its switching action is internal to itself: you cannot stop it switching, once initiated. A disadvantage is that it cannot be switched off by the same control that switched it on, as is the case with an ordinary transistor. Switch-off must be accomplished in the collector circuit.

This feature makes the SCR an admirable device for continually controlling current flow during the timing of an ac waveform, so that it strikes at the same point on every cycle and is extinguished by termination of that direction of the ac pulsation (Fig. 14-13). A circuit that will control current during one half of an ac waveform is shown in Fig. 14-8.

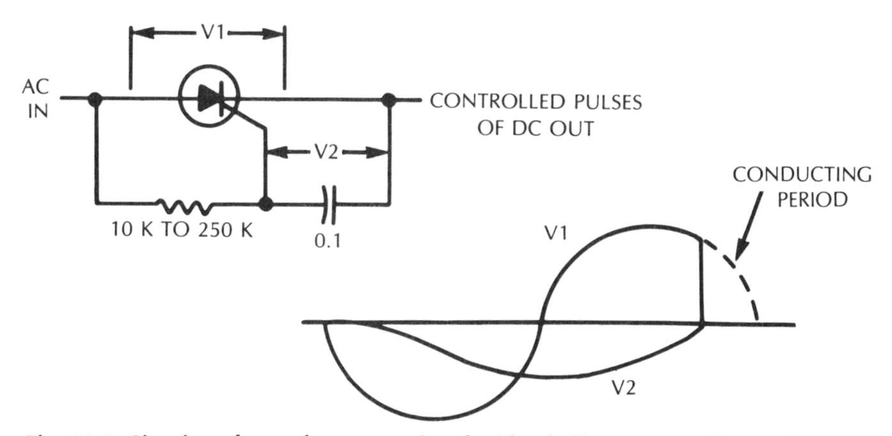

Fig. 14-8. Circuit and waveforms associated with a half-wave SCR, designed to switch on at a point past the mid-point of the "on" half wave.

The capacitor is reverse charged during the nonconducting half cycle, and as the normally conducting half-cycle commences, the capacitor discharges through the same resistor until the voltage reaches zero (or a little positive) to trigger the gate.

The time at which the SCR strikes during the half cycle can be controlled by varying the resistance value. Because of the relatively small control current, a resistor with a low power rating can be used, since the current is used only to trigger, after which the SCR conducts until the termination of the half cycle extinguishes it. This action varies the effective current delivered to the load by changing the proportion of the cycle during which it is delivered. Thus, instead of delivering half current or half voltage, this method of control delivers full voltage or current for only half the time per half cycle.

It is capable of a little more variation than this, though. If striking is postponed until after the maximum point in the half cycle, it will occur on the downward part of the curve. Accordingly, there is a reduction in the average of the time it is struck, as well as a reduction in the duration of the fraction of the cycle. Figure 14-9 shows a succession of such averages over the half wave.

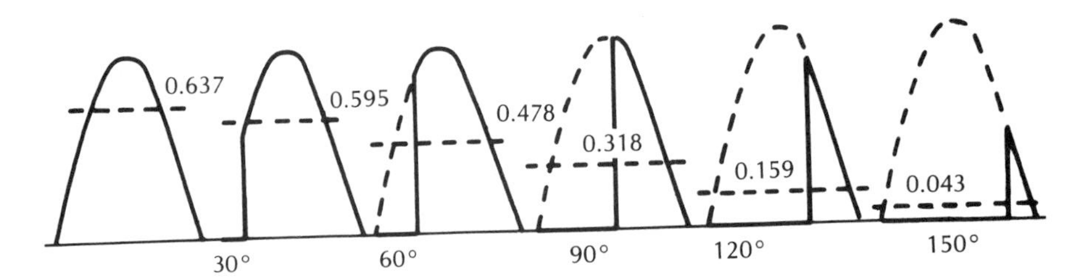

Fig. 14-9. The effect of switching at different points on the half waves. The dashed horizontal lines are the averaged outputs, over the full half wave in each case.

When the conduction period is short, the precise control of the time is dependent on the double-action use of the time-constant between the resistor and the capacitor, which extends over most of the cycle. However, the important part is the little bit that remains. It spends almost half the cycle charging, then almost half the cycle discharging; the remainder is the conduction time.

The fact that the switch-off point is somewhat critical can make the conduction period unstable when the device is "throttled down" this far. These devices cease to conduct, not at precisely zero voltage and current, but at a small forward current that is insufficient to maintain conduction. And the precise value of the cease-conduct point may vary and react with a varying conduction time to cause the instability referred to in the previous paragraph.

For example, if the device is designed to conduct 10 amps, and has a striking current at the gate on the order of 200 milliamps, the point at which current extinguishes may be somewhere between 100 and 300 milliamps (which is from 1 to 3 percent of maximum current).

When the current is low, because only the lowest part of the forward voltage curve as its "tail" is used, the current stop point can become indeterminate, so that an interaction may occur between the charge-discharge period of the RC combination and the extinguishing point of conduction, resulting in a current "bibber" (e.g., when used for a light dimmer control).

One remedy for this is to employ a clamp diode to more accurately control the point where the charge-discharge cycle commences, regardless of where conduction ceases, thus isolating the interacting effects. Further, an avalanche diode, which has the characteristic similar to a zener, allows a small forward voltage to build up before the striking current is reached. Figure 14-10 shows these additions to the half-wave circuit and Fig. 14-11 shows typical characteristics for an avalanche diode.

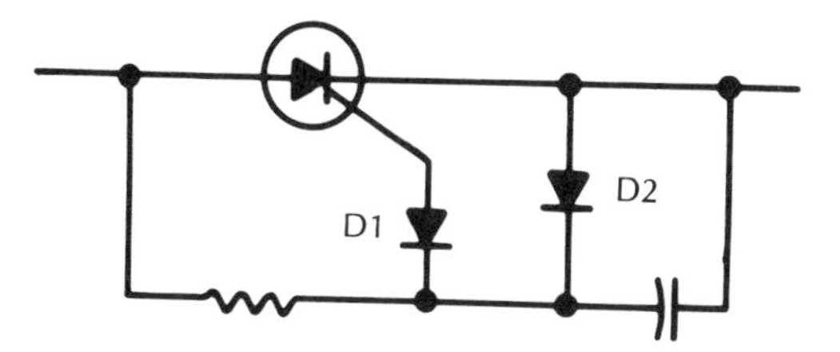

Fig. 14-10. Circuit using an avalanche diode (D1) and a clamping diode (D2) to improve the performance of the circuit in Fig. 14-9, particularly when the current "on" periods are small.

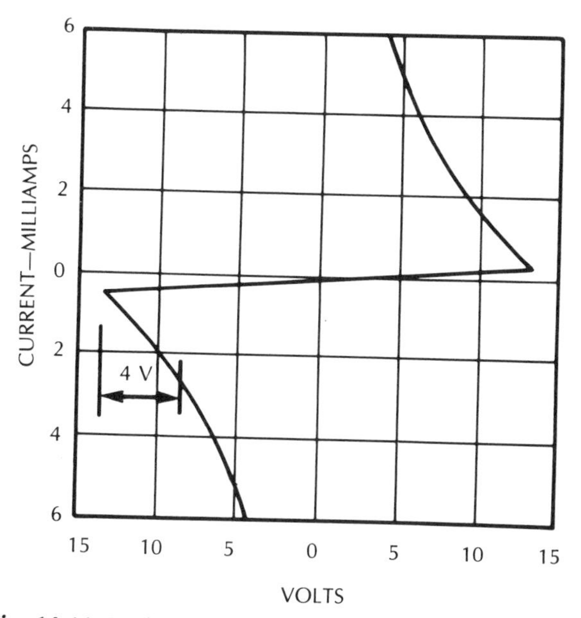

Fig. 14-11. Typical characteristics of an avalanche diode.

Putting two such circuits together, with a common control resistor, enables both half waves of the alternating cycle to be similarly controlled (Fig. 14-12). It is obviously inefficient to use only one half of an alternating current. Utilizing both halves increases the efficiency of the system in which such a device is installed.

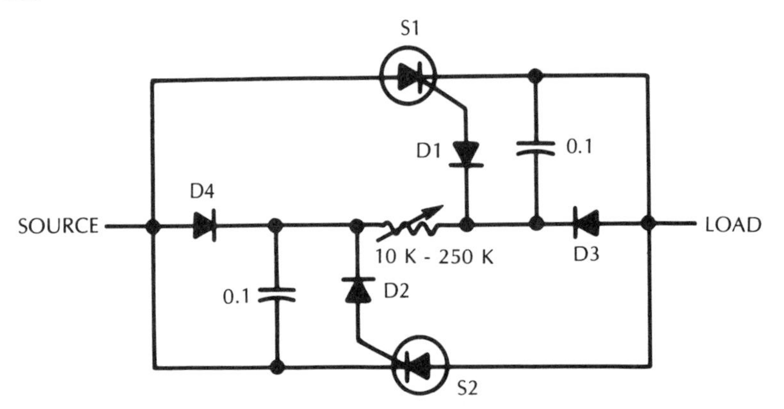

Fig. 14-12. This full-wave SCR circuit is an expanded version of the half-wave schematic in Fig. 14-10.

A more recent device is the bidirectional thyristor, which has properties identical to the SCR, but it works in both directions, simplifies the circuitry somewhat, and requires fewer attendant components (Fig. 14-13).

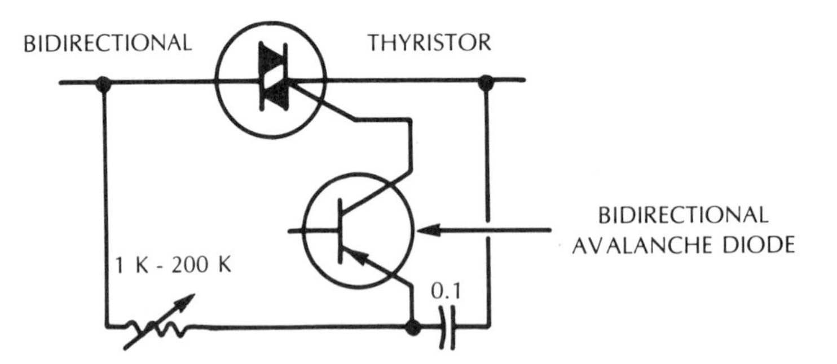

Fig. 14-13. A bidirectional thyristor with a bidirectional avalanche diode reduces the number of components and simplifies the circuit.

TWO-STATE LOGIC SYSTEMS

Now we come to the basics for the logic systems that have become the basis for modern computers and microprocessors. Later chapters will develop methods of design and system building. Here we want to get an understanding of the basic elements and how they go together. The most basic element is a bistable unit, also called a "flip-flop," of which the simplest form, using the by-now well understood current controlled transistor, is shown at Fig. 14-14.

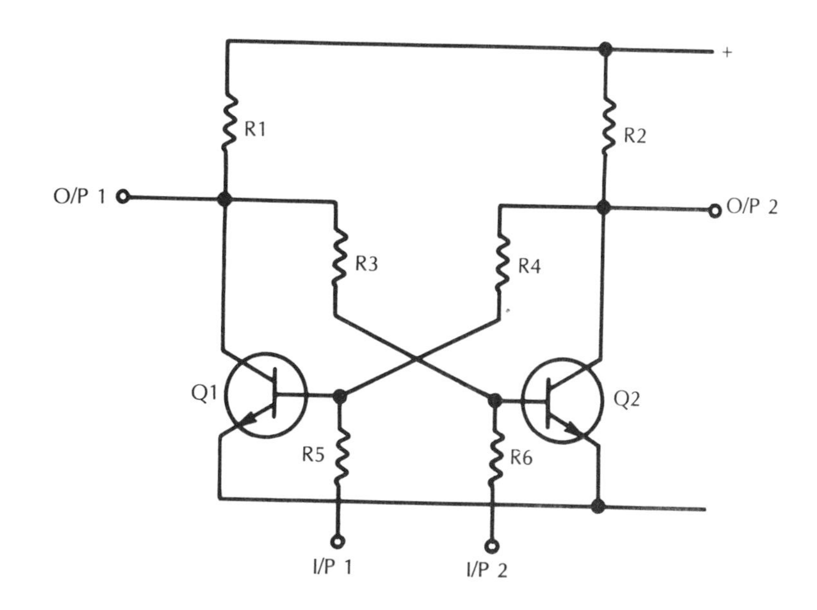

Fig. 14-14. A simple bi-stable, or flip-flop circuit, from which all later variations were more or less derived.

To identify elements we have put resistor numbers, which will have values dependent on parameters, voltages and use. Consider a condition in which the left transistor is conducting. This means the junction of R1 and R3 will be close to supply negative, so the right transistor is cut off. R2, whose value is much lower than R4, will drop very little voltage from supply plus, so the current in R4 will maintain the left transistor in a conducting state.

That condition is stable, provided that no external voltages are applied. Output 1 is "low", meaning near to supply negative, while output 2 is "high", meaning near to supply positive. In this kind of system, "signals" are either "high" or "low." With both inputs "low" the condition is stable.

A "high" signal on input 1 will make no difference. The left transistor is already conducting and a high signal will not make it conduct any more, because it is already saturated. But a "high" signal on input 2 will cause the right transistor to conduct, making output 2, which was "high," go "low," and because R4 cuts off the left transistor, now output 1 will go "high."

Removal of the "high" signal from input 2 will not change anything, because now the circuit will be stable in this set of conditions: left transistor cut off, right transistor saturated. By proper choice of values, only these two stable conditions are possible for the bistable circuit, which can be changed from one to the other by appropriate inputs.

One way to make alternate inputs flip the circuit either way would be to just tie them together. However, then the circuit wouldn't "know" whether to stay "as is" in response to the "high" supplied to the base that's already on, or whether to flip to the other. Figure 14-15 shows a circuit that overcomes this dilemma. The additional components are lettered, the others having the same numbers as those in Fig. 14-14.

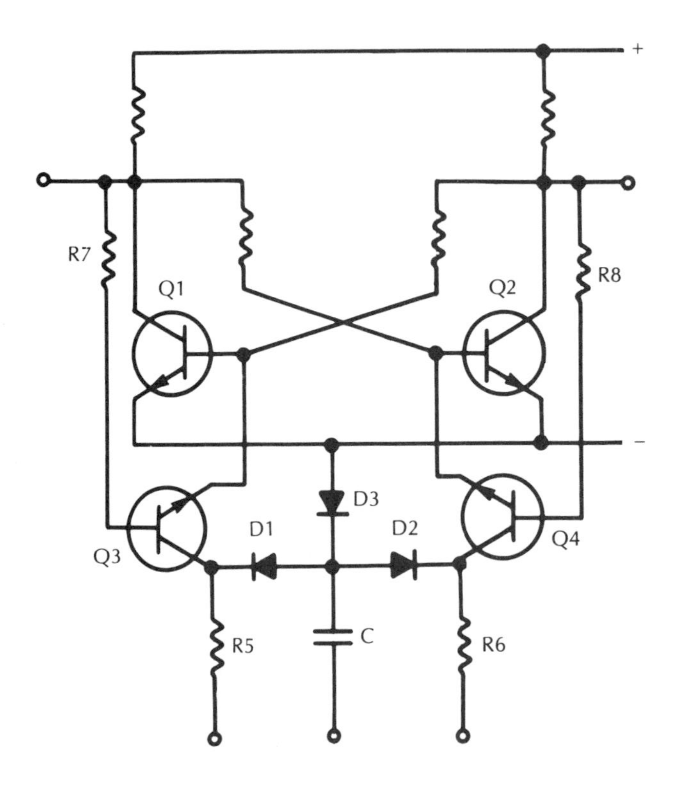

Fig. 14-15. A bistable circuit with inputs modified to accept different kinds of input.

Suppose Q1 is conducting, making O/P 1 low and O/P 2 high. This will provide base current to Q4, so it is ready to receive a "high" through R6, to bring Q2 into conducting, flipping the circuit. As Q3 is nonconducting when this flip is initiated, Q1 will not be held "on" by it. Similarly, a "high" through R5 can now flip the circuit back to its original condition.

The center input in Fig. 14-15 is applied through a small capacitor C (usually only picafarads in size) that is just big enough to deliver a charge through whichever of Q3 or Q4 is on, through diodes D1 or D2, to bring about a flip. If the center terminal remains high for longer than needed for that to happen, C becomes charged with the input side high (+) and the other side minus. Now, when it returns to its low state, D3 will discharge it, making it ready to receive the next high.

One thing that circuits for computers are always designed for is speed. In the circuits of Fig. 14-14 and 14-15, the down action, from 1 to 0, or high to low, is fast, because the transistor causing it goes from non-conducting to conducting. The up action, from 0 to 1, is slower, because any circuit capacitance has to be charged up through the resistor R1 or R2 (according to which one you take the output from).

A start toward changing this situation, so both actions are equally fast, is shown at Fig. 14-16. Here, whichever way the output moves, either Q2 or Q4 becomes conducting, the other one ceasing to conduct at the same time. If Q4 is conducting, the output is low, making Q1 conduct, which maintains Q4 conducting. Each keeps the other conducting. At the same time, Q2 and Q3 are both non-conducting. We haven't devised an input for this circuit yet, but if Q2 is conducting, then the output is high, keeping Q2 and Q3 conducting and Q1 and Q4 in the off state.

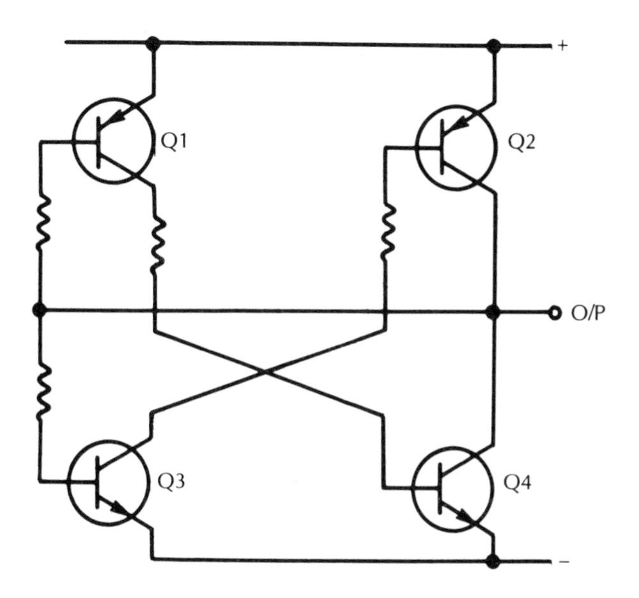

Fig. 14-16. A bistable circuit, in which positive "pull down" is provided for both high and low states of the output.

In Fig. 14-17, we show a combination of these circuits, with some extra features. The output transistors are Q1 and Q3, but Q2 functions more like Q1 of Fig. 14-14. The real flip-flop action is between Q1 and Q2, but when Q2 comes on, it also pulls Q3 on with it. Q4, Q5 and the associated diodes serve the same purpose as those in Fig. 14-15. But Q6 and Q7 serve an extra purpose that may be necessary when this circuit is employed.

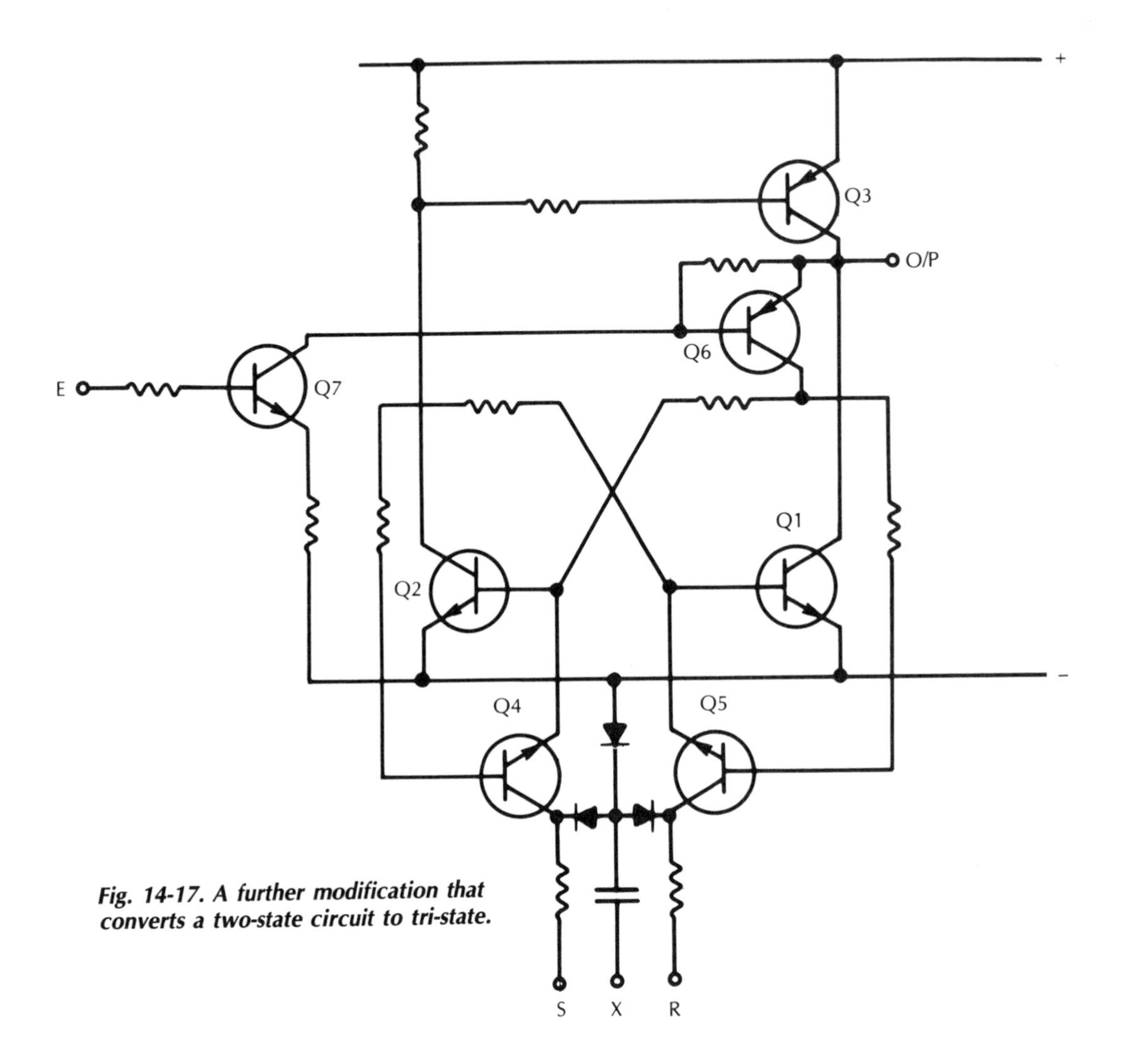

Fig. 14-17. A further modification that converts a two-state circuit to tri-state.

In computer circuitry, inputs and outputs are connected and interconnected with circuits called "buses." This means that a number of inputs and/or outputs may all be connected to the same "bus" at the same time. So when the output uses transistors that force the bus to either 1 or 0, with only the saturation voltage of the transistor separating them, it is possible that in one location a Q3 is tying it to positive, while in another a Q1 is tying it to negative. The whole supply is across two transistors, both in saturation. Something has got to give. Usually one or both transistors will blow.

It is impractical to disconnect all circuits not in use, so they are left connected, but the flip-flop is disabled, so neither out transistor is conducting until that flip-flop is wanted in use. This is done with transistors Q6 and Q7. When input E (for "enable") is low, Q7 is off, turning Q6 off with it. The latter can "float" at whatever level the output may go to, due to other circuits, while Q1, Q2 and Q3—in fact all the transistors in this whole unit, remain "off."

To bring them on, E goes high, which makes Q6 conducting too and activates the flip flop, when the inputs S, X and R behave just as they did in Fig. 14-15.

Coupled with flip-flops a whole bunch of other logic circuits are needed, to which we have already referred. These are ANDs, ORs, NANDs, and NORs. We'll show how ANDs and ORs can be constructed, and some of their properties. NANDs and NORs are quite similar. Figure 14-18 shows a simplified form of RTL logic—Resistance Transistor Logic.

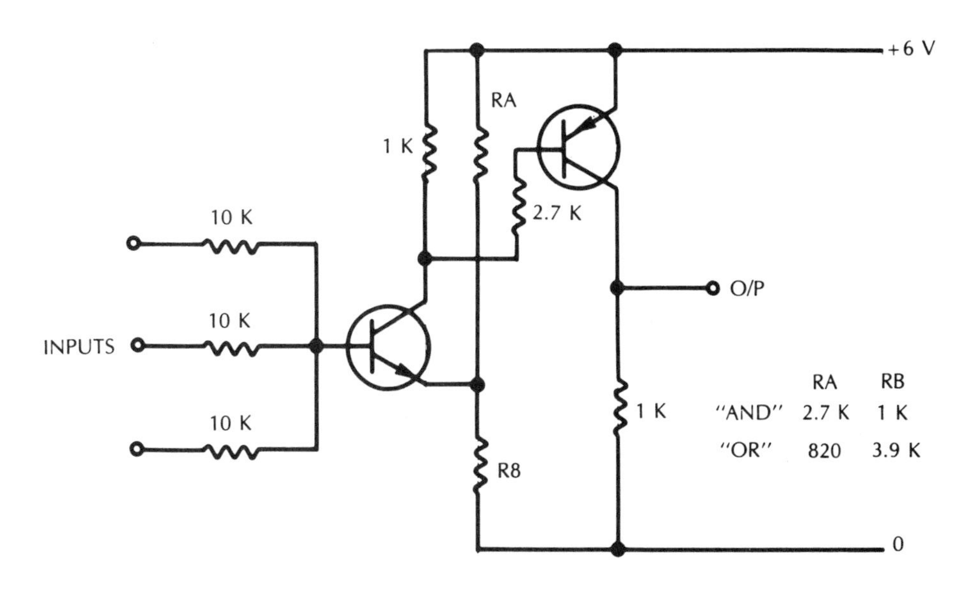

Fig. 14-18. This schematic is a simple DTL logic gate, in which only value changes make it an AND or an OR gate.

In this circuit, whether the output responds as an AND or an OR gate depends on the values of RA and RB. With the values shown for AND, the bias of the first transistor is such that with the 3-10 K resistors terminated low, both transistors are "off," resulting in a 0 output. Only when all three 10 K inputs go high, do the transistors turn on. With the values shown for OR, any one 10 K input going high will turn the transistors on, and all three going high can do no more than turn them on. From here on we'll call low 0 and high 1.

Rather obviously, RTL logic has to be carefully calculated, according to the number of circuits coupled by AND or OR inputs, and the action will sometimes be critical, and never as fast as the other types we are about to describe. Next comes the DTL type, "Diode Transistor Logic" exemplified in Fig. 14-19. In the AND configuration, with all three diode inputs at 0, the base of the input transistor is kept low, and hence nonconducting.

In fact, if any one of the diode inputs remains at 0, the input transistor base is kept negative, which keeps it cut-off. Only when all three go high, do all diodes cease to conduct, and the resistance bias to the base of the input transistor takes over, turning it on, so the output goes high.

In the OR version, with the diode inputs low, the diodes are nonconducting, unlike the AND circuit, where they are conducting. So the input transistor is off, and the output low. But any diode going to 1 biases the input transistor on, and all of them going to 1 doesn't increase the bias any. Note the difference from the RTL, where the transistor bias depends on how many 10 K resistors are paralleled to the high voltage.

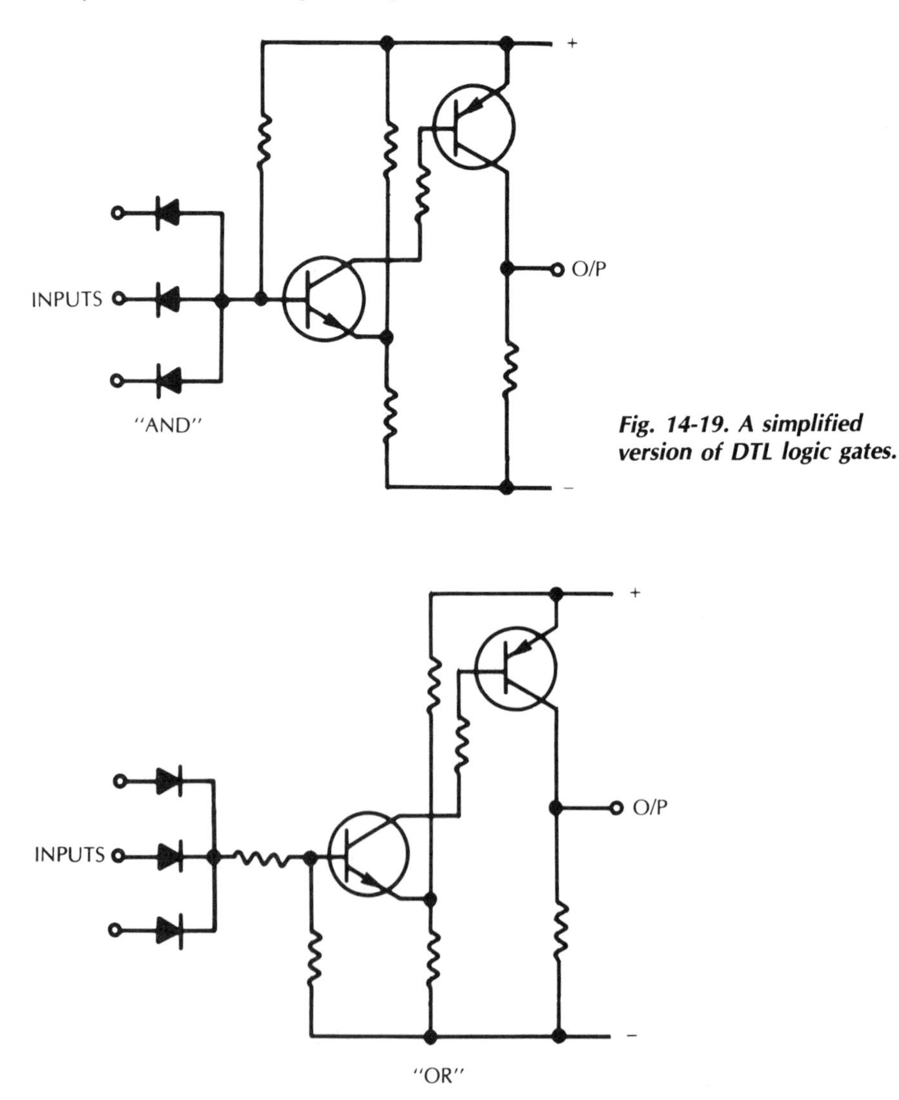

Fig. 14-19. A simplified version of DTL logic gates.

Finally, the fastest type of logic is TTL, Transistor Transistor Logic, for which AND and OR gates are shown at Fig. 14-20. As with the DTL, in the AND gate, inputs at 0 are conducting. As with the DTL, even one such input ties the common input transistor's base to negative, cutting it off too. But when all inputs go to 1, all three ANDing transistors are cut off, and the main input transistor comes on, giving the 1 AND signal at the output.

For the OR circuit, with all inputs at 0, all input transistors are cut off. Only one input going to 1 will cause the input transistor to go on and, as with the DTL, all of them going to 1 doesn't increase the input transistor's base current, it being limited by the one resistor for all three inputs.

Fig. 14-20. Simplified TTL gates.

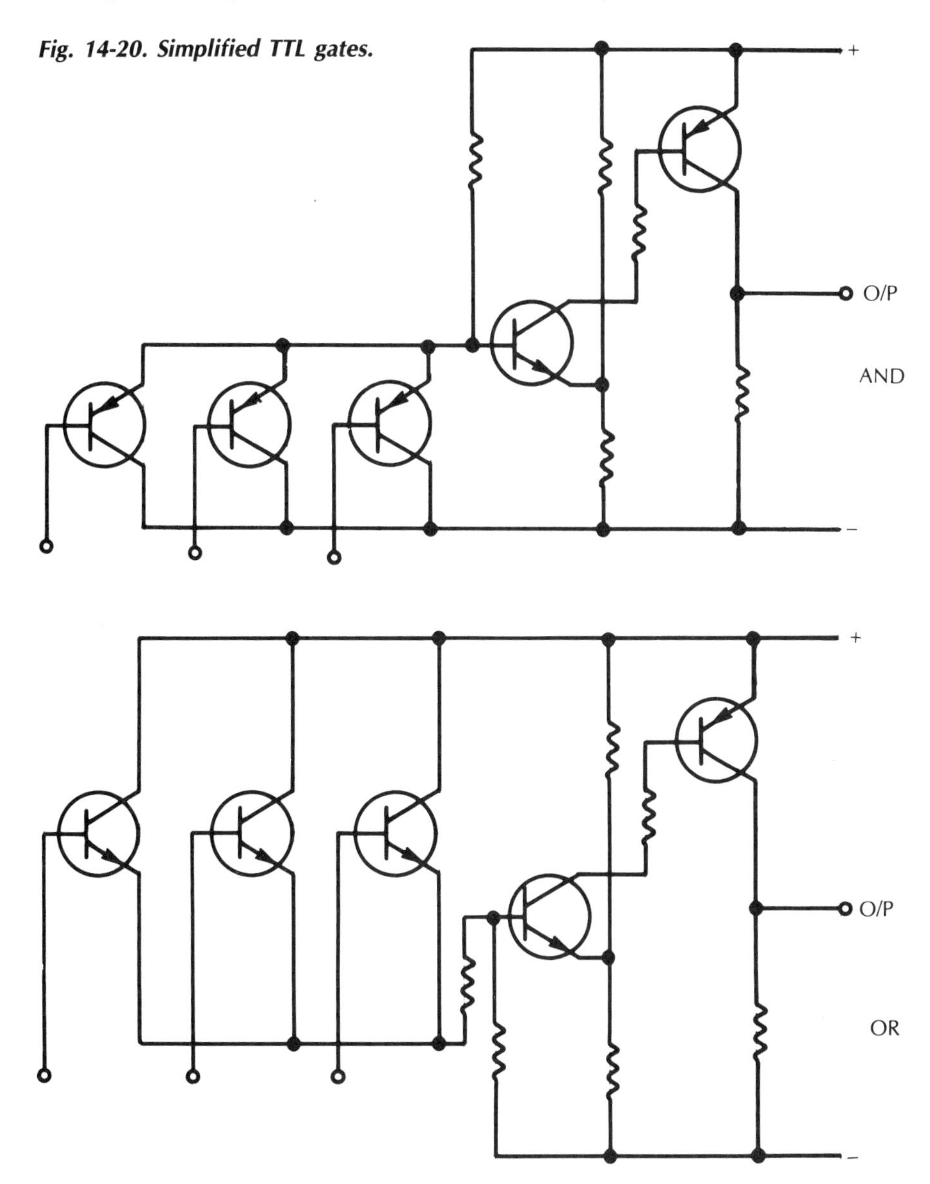

Before leaving this, we should stress that these circuits are simplified considerably, to make the principle of operation easy to follow. Actual circuits increase speed in the same way we showed for flip-flops. The circuits we have shown would have a fast rise, from zero to 1 at the output, and a slower fall time, from 1 to 0.

Also, we have not touched on other factors that deal with both speed and power consumption, very important factors in the design of such systems, when they use thousands of these devices. This is another place where the more complicated design of flip-flops in Fig. 14-17 helps. The biggest power drain is in the output circuitry. The internal controls can be kept down to microamps, but switching speeds in outputs necessitate low resistances, the kind that are presented when transistors saturate.

With the circuits of Figs. 14-18 to 14-20, the current at 0 output may be low, but at 1 it is at least what the collector resistance of the output transistor takes. If the inputs that this output feeds take more current, then that is added. The circuit of Fig. 14-17 eliminates the internal drain in the output resistor by substituting a cut-off transistor. In both output positions, one transistor is short circuited, and the other is open circuited, taking no current.

Current taken at inputs, which has to be supplied by outputs from somewhere, is larger with RTL than with the others. It is least with TTL, meaning that as well as being the fastest, it also consumes less current. In flip-flops and other circuits, there are other factors that more advanced courses will go into. For example, as we showed in the earlier part of this chapter, switching time (for small transistors as well as big ones) is faster where the margin of saturation (in the "on" condition) is smaller. Also, speed may be increased by reducing cut-off bias, such as by keeping them always conducting a little, but so little that the circuit behaves in similar way to cut-off. This is what is referred to as "current mode" or CMOS operation.

Op amps, introduced in the chapter on linear amplification, also find much use in digital. Having practically infinite gain, they operate quite well going from "floor to ceiling" and vice versa. The high gain merely expedites that transition. In digital circuitry, op amps are often used as comparators, as we shall see later in this course.

To conclude this chapter, we turn to an interesting application of this technology, that illustrates how technology has advanced. In the early days of electronic organs, at least 12 individual oscillators, one for each named note in the octave scale, had to be tuned. In more complex organs, much more complicated tuning was necessary.

But digital programming has made possible organs in which the whole organ can be tuned in one simple adjustment. Figure 14-21 shows a sample taken from such a circuit. This represents a counter designed to deliver a regular output pulse, for precisely every 1433 cycles of a 3 MegaHertz (or thereabouts) input. Assuming the "clock" frequency is 3 MegaHertz precisely, a count of 1433 will deliver output pulses at a frequency of 2093 Hertz.

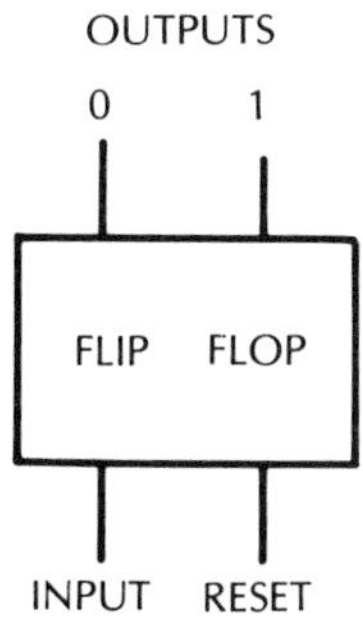

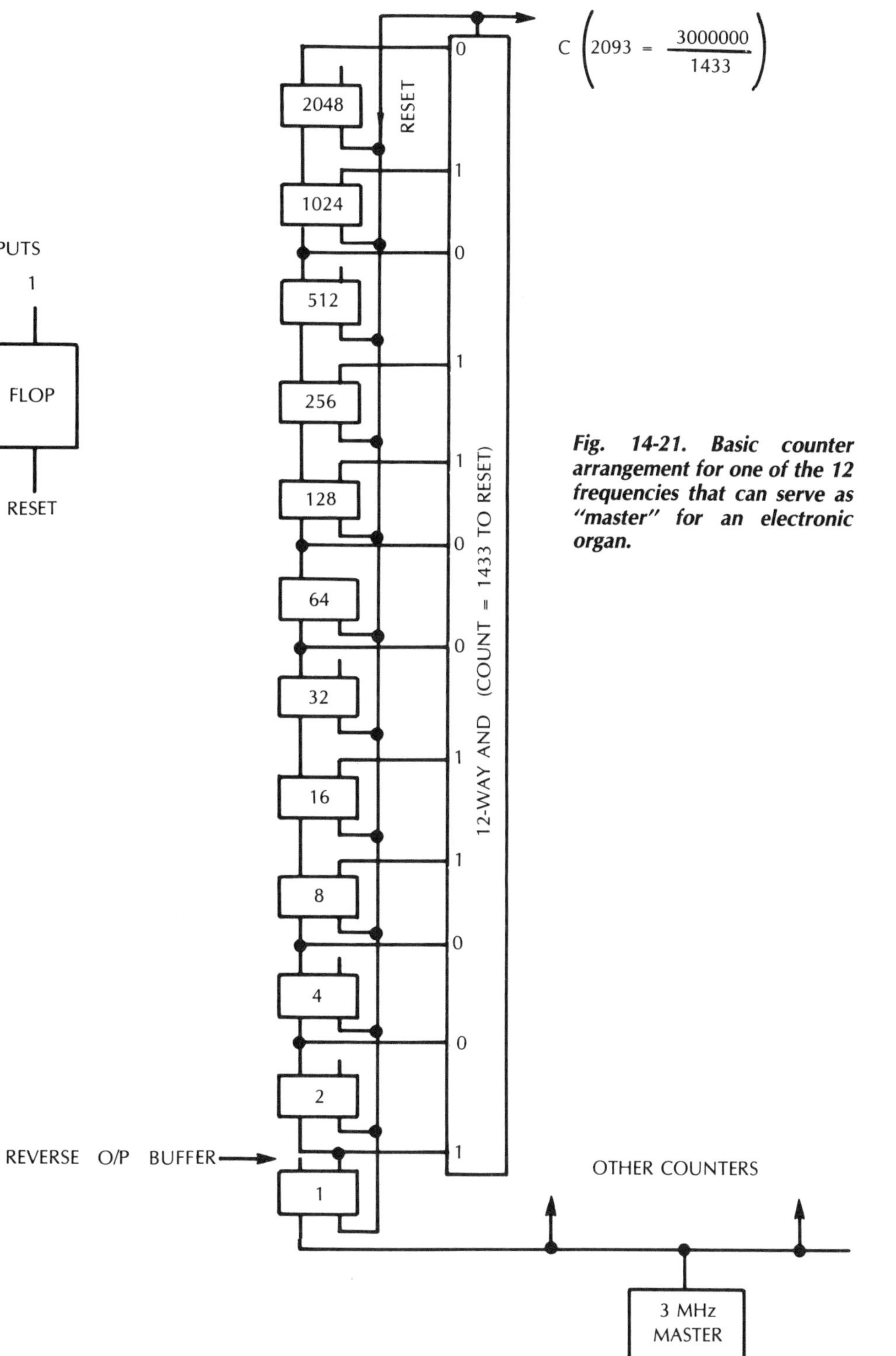

Fig. 14-21. Basic counter arrangement for one of the 12 frequencies that can serve as "master" for an electronic organ.

NOTE	FREQUENCY	COUNT	BINARY
C	2093	1433	010110011001
B	1975.5	1519	010111100101
#	1864.7	1609	011001001001
A	1760	1705	011010101001
#	1661.2	1806	011100001110
G	1568	1913	011101111001
#	1480	2027	011111101011
F	1396.9	2148	100001100100
E	1318.5	2275	100011100011
#	1244.5	2411	100101101011
D	1174.7	2554	100111111010
#	1108.7	2706	101010010010
C	1046.5		

Table 14-1. Frequencies for Organ Notes.

Each counter is designed to recycle according to the schedule of counts shown in Table 14-1. The frequencies delivered as outputs are closer to correct tuning than most organ tuners could achieve. In fact the worst possible error is less than 1/30 of 1%. Add to this the fact that the whole organ can be tuned up or down a semitone or more if necessary, just by changing the 3 MegaHertz input frequency and you see what a tremendous advance this signifies.

Examination Questions

1. A transistor has maximum ratings of 75 volts at 10 amps and is capable of dissipating 25 watts. If this transistor is operated in the linear mode, what is the maximum operating current if the voltage swing is maximum, and what is the maximum voltage if the current swing is maximum?
2. If the transistor in Question 1 can switch 75 volts at 10 amps with its maximum expedited switching time of 10 microseconds (for each, on and off) what would be the maximum voltage and current ratings if removing the expediting components results in switching times of 25 microseconds?
3. The alpha cutoff frequency of a transistor is rated at 2 MHz and its normal operating beta is 150. What is the highest value of beta cutoff you would expect for this transistor?
4. A transistor's cut-off frequency in the common-emitter mode is rated as 50 kHz. If the transistor's beta is 50, what is the lowest cutoff frequency you would expect with the same transistor operated in the common-base mode?
5. Assuming the base input resistance of the transistor in Fig. 14-4 is 14 ohms at saturation, what is the peak current when the +9-volt leading edge is first applied?
6. In Question 5, how long does the pulse of base current take to drop to about 0.37 of the initial value? (See Fig. 9-10 for a hint about this.)
7. Estimate the length of a pulse that the circuit in Fig. 14-4 could hold before it would cut off without waiting for the input signal to cut off.
8. In Fig. 14-5, what is the peak base current delivered to Q2 at the instant of switch-on?
9. After a transistor has been pulsed off, a good estimate of the time for the pulsing circuit to recover, for full expediting of the next on pulse, is twice the time constant. What will this be for Q2 in Fig. 14-5?
10. After Q1 has pulsed Q2 on, how long will it be before the 15-pf capacitor will be ready to give substantially full boost to the off pulse?
11. In an AND circuit, operation is built round a central transistor or other semiconductor that is normally conducting to saturation, so only when a 1 signal is present on all inputs, does it go to cutoff. Tabulate the appropriate usage for the other kinds of gates:

TYPE OF GATE	NORMALLY	CHANGES WHEN
AND	Saturated	All Inputs "1"
OR		
NAND		
NOR		

12. NAND and NOR gates can be synthesized, if not available as separate types, by combining AND and OR gates with a buffer (of which Fig. 14-18 shows a primitive type). Show the appropriate pairs for each synthesis.

15 Thermionic Tubes

Before the advent of semiconductors in the variety now available, a much earlier discovery, that a heated electrode in an evacuated space (vacuum) emits free electrons, led to the development of the vacuum "valve" or diode (Fig. 15-1). As with a semiconductor diode, current will flow one way but not the other.

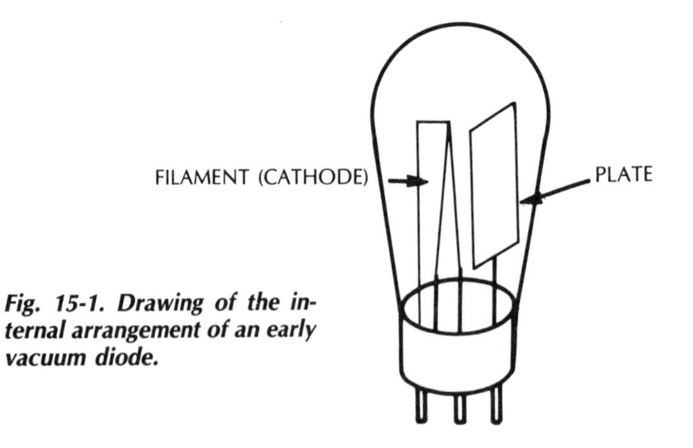

Fig. 15-1. Drawing of the internal arrangement of an early vacuum diode.

In this respect, as well as others, the vacuum diode is similar to its semiconductor cousin, except that it does not possess a zener voltage characteristic. Instead, it has a vacuum flash-over potential rating which if exceeded can result in damage to the diode.

Electrons emitted by the heated negative electrode, called the filament or cathode, are attracted by a positive voltage applied to the other electrode, called the plate or anode; thus, current flows in this direction. However, if the cold electrode is made negative and the hot one positive, no current flows.

A further discovery revealed that an electrode in the form of a grid, located between the negative and positive electrodes, with spaces between for electrons to pass through, can control the electron flow between cathode or filament and plate. Thus, the amplifying valve (British term) or tube was born. A negative voltage applied to the grid established a field that repels the electrons, just as the positive voltage on the plate attracts them (Fig. 15-2). The electric field contours from plate and grid, depending on which exerts the stronger influence at the filament or cathode, control the flow between cathode and plate for a given voltage applied to each electrode. Since it had three elements the first amplifying tube was called a *triode*.

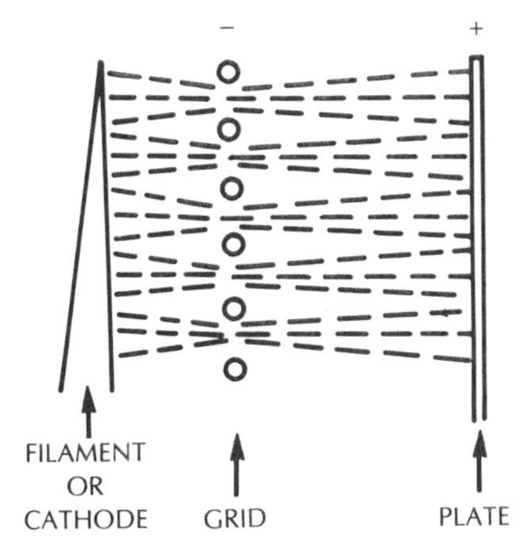

Fig. 15-2. Drawing of the internal cross-section of a triode tube.

The earliest tubes, diode or triode, used a filament as the heated electrode, which meant that a current had to be passed through it. The result was that different voltages existed along its length, just as in any other conductor possessing resistance. This led to an uneven distribution of electrons attracted to the plate (Fig. 15-3), with more coming from the negative end of the filament than from the positive end. It also meant that some form of supply had to be tied to the circuit to provide the higher voltages needed for amplification or other functions.

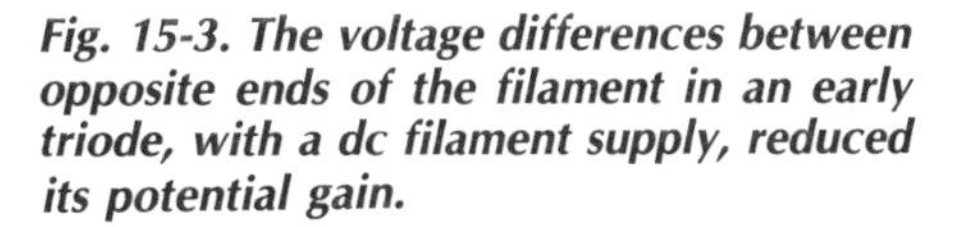

Fig. 15-3. The voltage differences between opposite ends of the filament in an early triode, with a dc filament supply, reduced its potential gain.

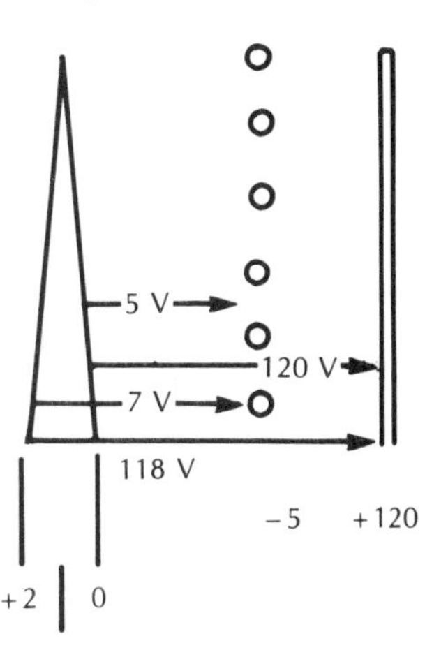

INDIRECTLY HEATED TYPES

The above feature proved inconvenient for various reasons, some of which appear later, so an indirectly heated cathode was developed. It has a separate heater and cathode. The heater provides the needed heat, while the cathode has only one connection instead of the two needed for a filament. The heater is insulated from the cathode, so that voltage differences can exist between cathode and heater. With this arrangement, the entire cathode is at a uniform temperature and voltage, which enables more precise characteristics to be obtained.

Before getting to other reasons for preferring the indirectly heated cathode, there is one advantage with the directly heated type: greater emission for the power used to heat the emitting device or less heating power to obtain the same emission. For example, a typical tube of the directly-heated era (you might have difficulty finding one today) used a 2-volt 300-milliamp filament which requires 0.6 watt to heat it sufficiently to allow a plate current of, say, 20 milliamps at a plate voltage of 120 for a plate dissipation approaching 2.5 watts (actually 2.4). An equivalent indirectly heated tube needs a heater supply of 4 volts and 1 amp, a heater power of 4 watts.

Both filaments and heaters can be designed for any voltage and current combination that generates enough heat to yield adequate electron emission. However, directly heated tubes usually used 2 volts at various currents, mainly because this was a convenient voltage for operating from a lead-acid rechargeable cell, which was the predominant power source for that purpose in those days.

Later, directly heated tubes with 4-volt (and sometimes other voltages) were used for higher power applications, because of the greater efficiency realized. These were usually ac energized. If ac merely replaced the dc circuit in Fig. 15-4 problems would be encountered.

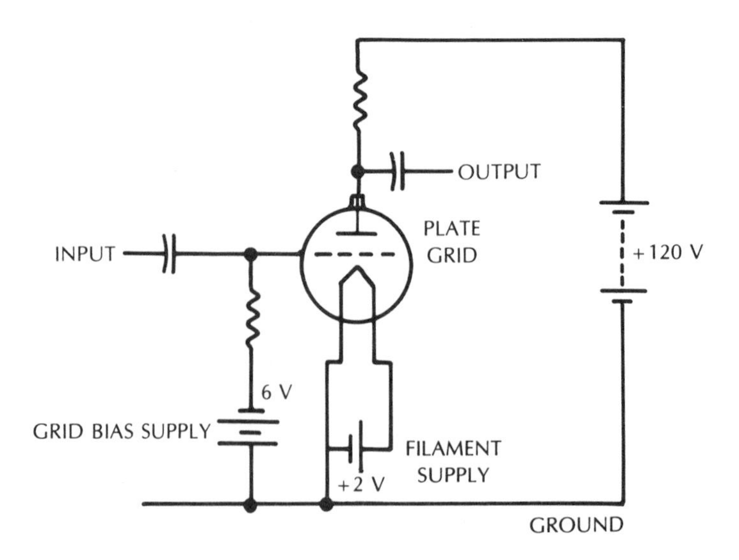

Fig. 15-4. Various supplies needed to operate an early triode tube as an amplifier stage.

One side of the filament supply is tied to one side of the high-voltage (plate) supply. If ac were applied in place of dc for the filament, ac would be injected into the plate circuit because the average cathode voltage would include an ac component.

Actually, there were two reasons for this ac injection: (1) the average voltage on the filament included an ac fluctuation; (2) the old filament usually heated almost instantly, so the heat fluctuation during the ac cycle would cause fluctuating emission.

To overcome these deficiencies, two changes were needed for ac filament operation—one internal and one external. The internal change consisted of providing the filament with sufficient thermal capacity to ensure that its temperature did not vary appreciably during the ac cycle. The external change called for connecting the filament in such a way the voltage varied in equal portions, positively and negatively, in regard to the high-voltage negative.

If plate voltage-current curves are plotted for a triode tube, they look typically like Fig. 15-5.

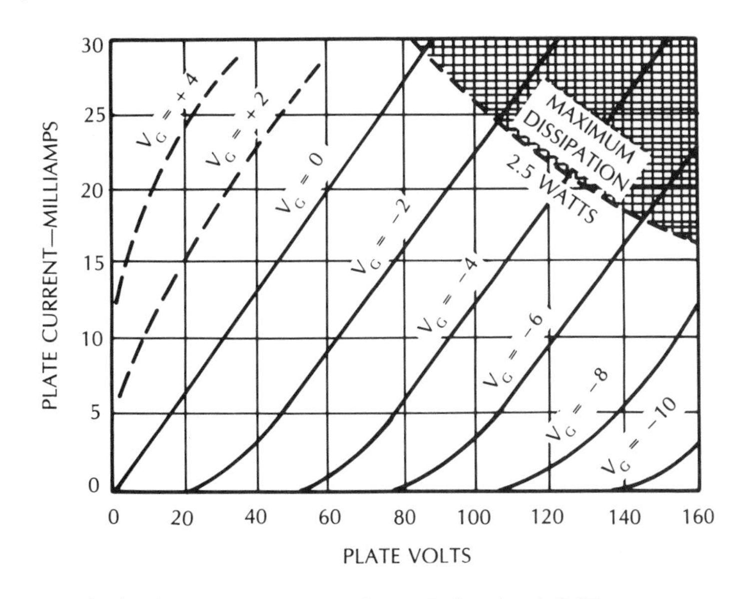

Fig. 15-5. Typical voltage-current curves for a triode tube. Solid lines represent negative grid voltage values, dashed lines indicate the positive grid region.

Curves are usually plotted only for grid voltage values (referred to cathode voltage, as zero) that run from zero negative. More than a certain negative voltage on the grid, for a specific plate voltage, cuts the plate current off.

LIMITS OF OPERATION

Positive grid voltages are not normally used because when the grid is positive in respect with the cathode it behaves as another plate and, being much closer to the cathode, conducts current quite heavily, considering its restricted surface area and consequent inability to dissipate the corresponding heat generated. Unless it is specially constructed for this operation, positive voltage on the grid is dangerous to the tube.

With negative grid voltages, an extremely small current flows. And most of the current in the grid circuit is due, not to conduction, but to a small capacitance between the grid and other electrodes. Thus, when the tube is operated with negative grid voltages, the grid input is essentially only a voltage; no current flows with it. Extending the grid voltage range into the positive region spoils this by causing a relatively heavy current to start flowing rather suddenly in the region of zero grid voltage. If a triode tube is specially constructed to allow positive grid voltages for part of its operating cycle only, the characteristic curves bend the other way, so a greater portion of the working area is used.

In thermionic tubes, two factors usually limit operating values: maximum plate voltage and maximum dissipation. Maximum voltage is a vertical line on the plate voltage-current curves. Maximum dissipation is an exponential curve, joining all the points where multiplying voltage by current results in the specified maximum wattage dissipation.

The operating point should be below the maximum dissipation curve. The signal swing may take the actual dissipation over the dissipation line for parts of the signal cycle, so long as the average stays below the line. But the signal should never swing above the maximum voltage line (Fig.15-6).

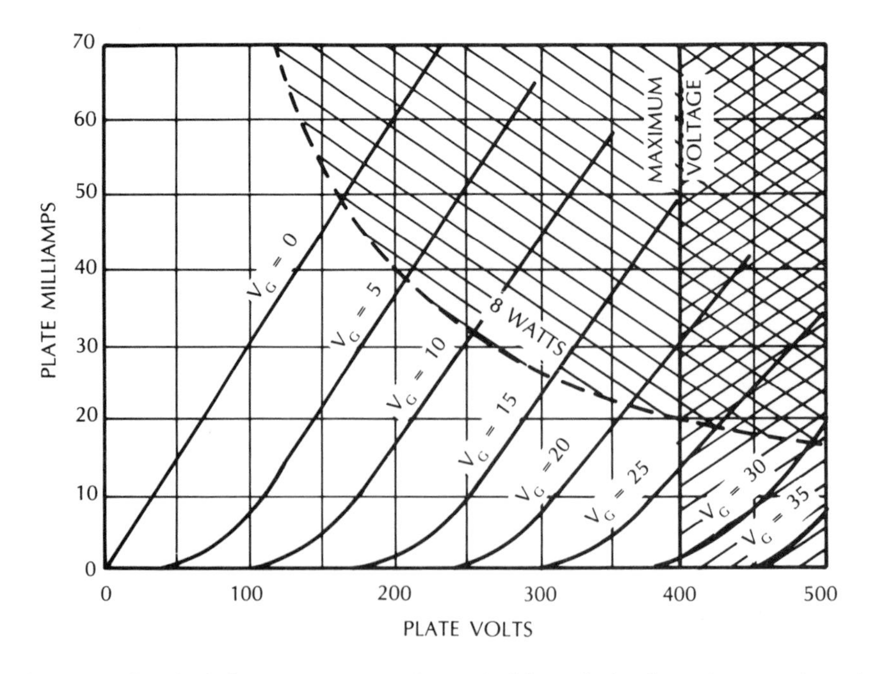

Fig. 15-6. The shaded area represents the typical boundaries for safe operation of a triode tube.

FROM TRIODE TO PENTODE

In studying the properties of the triode tube, the fact that the signal swing does not go above zero grid voltage limits the range that the tube can handle. Also, the amplification is seriously limited by the resistance value connected into the plate circuit. If a low-value resistance is used, a relatively high current swing can be achieved. But if a high-value resistance is used, even though a higher voltage swing can be achieved, the current swing is then seriously limited (Fig. 15-7).

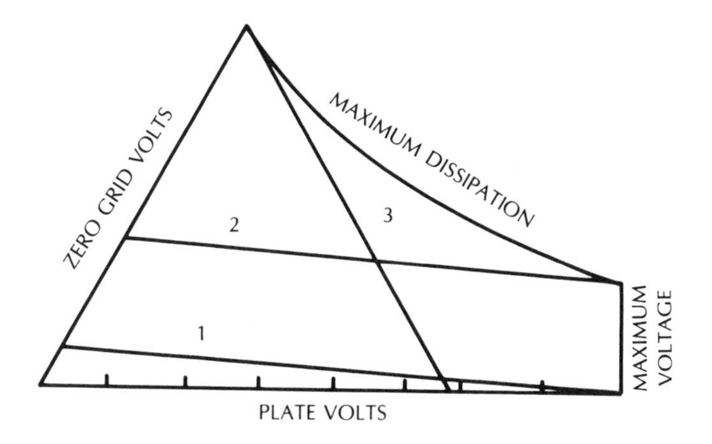

Fig. 15-7. Operating limitations of a triode tube: 1. high value load working at low plate current; 2. similar high value at higher plate current; 3. lower value load with a greater current swing, but much lower voltage swing.

To overcome this combination of deficiencies of the triode tube, the next step was to introduce the *tetrode*, or 4-electrode tube. By using another grid, maintained at a constant (positive) voltage, between the original (first) grid and the plate, the field at the grid and cathode is practically the same as when plate voltage does not change, resulting in a higher available current swing. Then, putting a lower resistance value in the plate circuit will maintain most of the voltage swing, while gaining this higher available current swing.

That is *almost* the way a tetrode works. A limitation occurs when plate voltage falls below the voltage applied to the second grid (usually called a "screen" or a "screen grid") because secondary emission takes place, current drops almost to zero and screen current correspondingly rises (Fig. 15-8). The consequence is a discontinuity in the curvature, rendering the tube unusable for linear amplification under those circumstances.

Such a discontinuity can be useful in some circuits, which we do not discuss in this book. One method used to remove these "kinks" was to put in yet another (third) grid, making it a 5-electrode tube (called a *pentode*), in which case this third grid (called the *suppressor*) is usually connected to the cathode, thus screening the screen from the plate to stop secondary emission (which is a bouncing of electrons from the plate back to the screen) (Fig. 15-9). Another tube developed to eliminate secondary emission was the beam tetrode, in which another set of electrodes, arranged differently from a grid, direct the electron stream in a way that also prevents secondary emission. The latter method is a little more efficient.

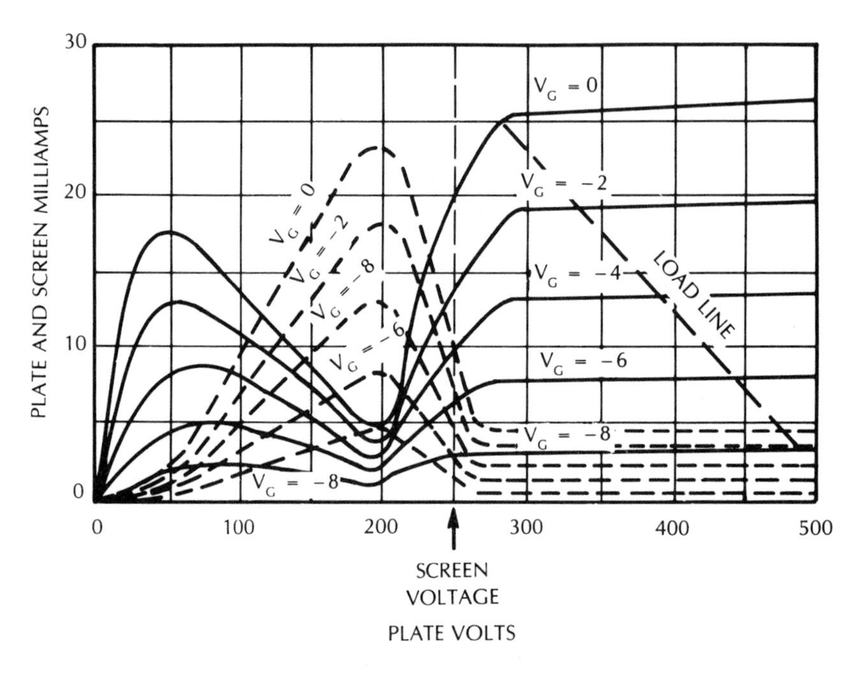

Fig. 15-8. Typical curves for a simple tetrode or screen grid tube.

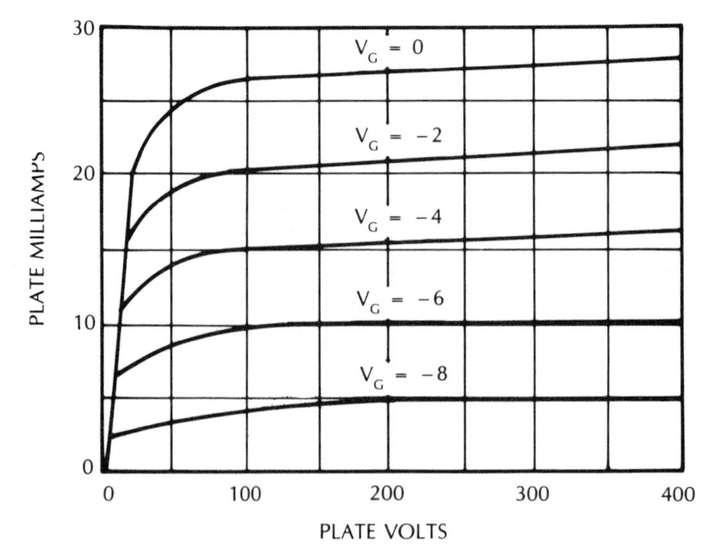

Fig. 15-9. Typical curves for a pentode tube.

PARAMETERS

In all thermionic tubes used as simple amplifiers (they are sometimes used for other purposes, as are transistors, but those uses will not be treated in this chapter) there are three parameters of importance:

- **Mutual conductance or transconductance** is the change in plate current occurring for a unit change of grid voltage when the plate voltage is held steady. The symbol is g_m.
- **Plate resistance** is the effective internal resistance of the plate, determined by comparing the voltage and current change that go together when the grid voltage is held constant. The symbol is r_a.
- **Amplification factor** is the voltage change produced in the plate circuit for a unit change of grid voltage when the plate current is held constant. The symbol is μ.

These three parameters are related by the simple formula:

$$\mu = g_m r_a \qquad (28)$$

This relationship is illustrated in Fig. 15-10, related to some idealized plate voltage-current curves (idealized, because they are straight lines). From that formula, you might assume that, given any two, the third can be calculated, which is academically true. But in practice, not all three quantities hold reasonably constant over the working range used. In a triode tube, the amplification factor holds more constant than either of the other two parameters. In a pentode tube, mutual conductance stays closer to a fixed figure.

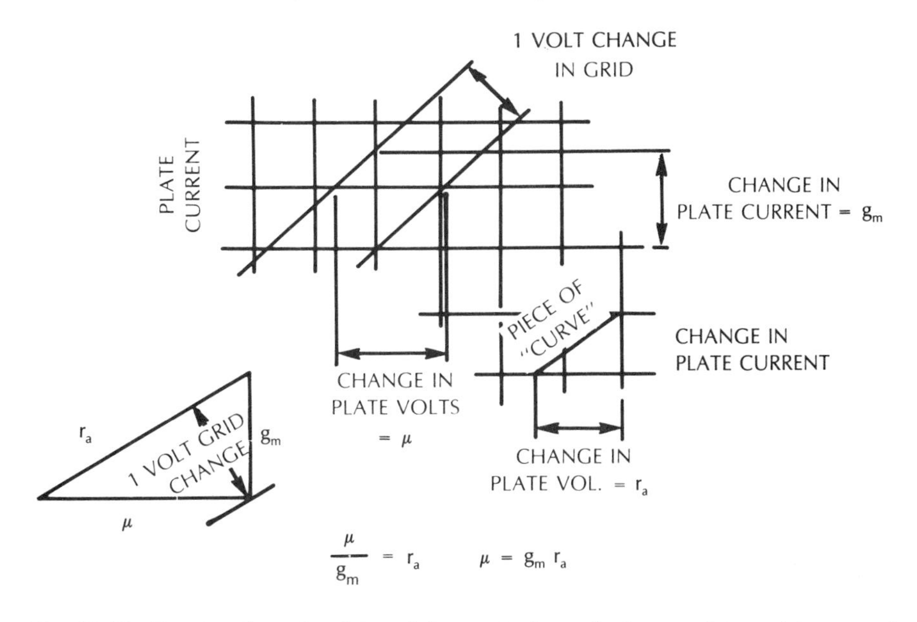

Fig. 15-10. Constructions for determining mutual conductance, plate resistance and amplification factor from plate voltage-current curves.

Typical values for a triode tube are: amplification factor, 40; plate resistance, 6 K. The amplification factor may vary usefully from, say, 35 to 45, over a working range. This will show the highest value near zero grid voltage, lowest when the negative grid voltage biases the plate current almost to cutoff. In the same triode, the plate resistance might vary from 9 K (and higher, as cutoff is reached and linear amplification ceases) to 4 K. The plate resistance drops as the plate current increases, almost independently of plate and grid voltages which, while related to each other, affect only the plate resistance as the combination of values affects plate current.

At the operating point, the mutual conductance is 40 divided by 6 K or 6.6 milliamps per volt (also called 6600 micromhos). At one extreme, near cutoff, the value changes to 35 divided by 9 K or 3900 micromhos, while at the other extreme (high current), near zero grid voltage, it is 45 divided by 4 K or 11,250 micromhos.

The plate resistor for such a tube might commonly be 30 K. This means that a stage of amplification will have a gain of 40 × 30 divided by 36 or 33.3 at the operating point. At one end of the swing, it changes to 35 × 30 divided by 39 or 26.9, while at the other end it is 45 × 30 divided by 34 or 39.1. Notice that the reduction at the bottom end is greater than the increase at the top end.

Typical values for a pentode tube of similar proportions are: mutual conductance, 5 milliamps per volt (5000 micromhos), with an operating point plate resistance of 750 K. With these values the amplification factor (by formula 28) is 5 × 750 or 3750. This is not very meaningful, because it is virtually impossible to realize anywhere near that much gain in a practical circuit.

The typical triode tube has an amplification factor of 40 and can be used to produce a reliable gain of over 30 fairly easily. If the pentrode used the same plate resistor, its gain is obtained by multiplying g_m by the plate resistor and then allowing for the loss due to the voltage swing (due to the fact that output is not purely a current swing). Multiplying 5 × 30 gives gain as 150, reduced by the factor of 750 divided by 780 to 144. This is a practical figure.

The fluctuation of parameters over a pentode's range of operation is greater than in triodes, and the mutual conductance is the parameter that shows the least variation. Plate resistance may vary, from 100 K to 10 megohms or so, with a corresponding g_m variation from 5 down to 2, maybe.

Obviously, computing the amplification factor does not mean much. Near zero grid, at a g_m of 5 with an r_a of 100 K, it is 500; nearer cutoff, at a g_m of 2 with r_a of 10 meg, it is 20,000. Anomalously, the area where amplification calculates the lowest—500—is where practical gain is usually the highest.

DISSIPATION PROBLEMS

With semiconductors, since techniques have been developed for making larger junctions, getting the heat away from the junction is relatively simple: merely provide a good heat conducting path, largely through metal, to an adequate form of heat radiator. In a large device, that may dissipate 100 watts, a fairly large assembly of radiating fins is needed. With a vacuum tube, dissipation is complicated by the fact that a vacuum does not conduct heat at all. For the first step of its journey, heat must escape by radiation.

With few exceptions, it is not practical to operate plates or other electrodes at a temperature that is high enough to radiate freely, a point where such elements glow brilliantly. With most plates, operating at such an elevated temperature causes the metal of which the plate is made to release gases, which deteriorates the high vacuum necessary for proper operation.

Therefore, it is desirable to find ways of handling more signal power for a corresponding dissipation in the tube(s). There is a way of operating tubes which keeps dissipation very low until the signal power is handled, and even then the signal power is higher than the tube dissipation. Essentially, the tube is biased to a point right at the edge of conduction. The tube conducts maybe one-tenth of the average current that flows at full output for that tube.

Another problem in the larger power-handling type tubes of the multi-electrode type is the dissipation at electrodes other than the plate. For example, a good pentode tube will have, in Class A operation, a screen current from one-tenth to one-fifth of the plate current. This means, assuming both have the same working voltage, that the dissipation splits in the same ratio. In Class A operation, the screen may have problems dissipating from one-tenth to one-fifth of the plate dissipation, because it consists of a rather thin wire, with a poor means of getting the heat away from the zone where it is generated. The plate can fairly easily get rid of from 5 to 10 times the heat that the screen can.

However, when pentodes are operated in Class B, the plate dissipation is drastically reduced in comparison with the power output, where the screen is not, because its voltage stays constant when the plate voltage dips due to power delivered to the load at peaks. It is feasible, from the viewpoint of plate dissipation, to raise the operating condition so more power is available, without exceeding the permissible plate dissipation. But to get this plate dissipation requires a higher screen voltage, which creates screen dissipation problems.

This general compromise dilemma made the introduction of the beam tetrode tube a great step forward. This design orients a third electrode differently from the intervening two (Fig. 15-11).

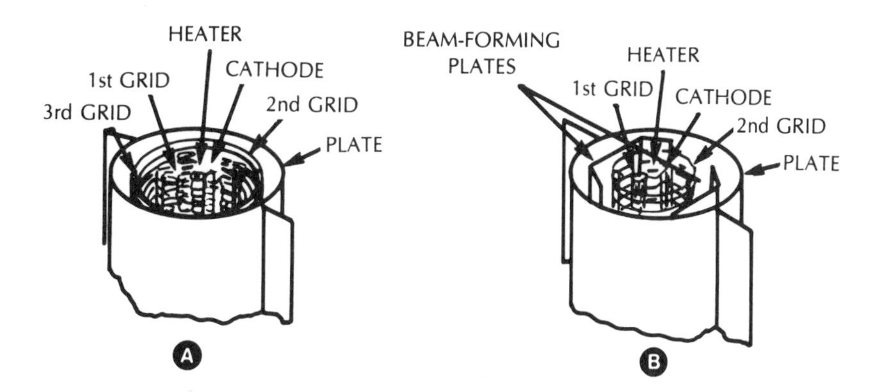

Fig. 15-11. The difference in construction between a pentode (A) and a beam tetrode (B).

The result has the effect of beaming the electrons in a way different from that produced by the third grid. A further step is the use of aligned grids: putting the second grid so each wire is precisely in the shadow of a wire on the first grid. This reduces the proportion of current collected from the electron stream by the second or screen grid, and makes the screen current the smallest possible fraction of the plate current.

Beaming the electron flow by the third electrode makes the current distribution on the second grid nonuniform, so the heat flow to get rid of dissipation can be better controlled. A final advantage to the beam tetrode is a form of "kink" that appears in the plate current-voltage curves (Fig. 15-12).

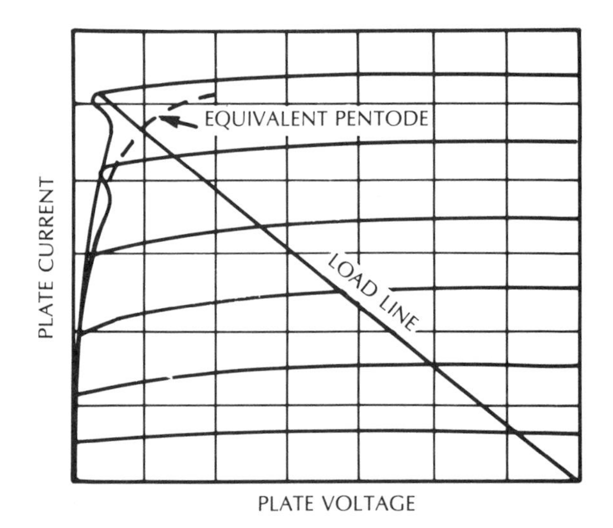

Fig. 15-12. Typical beam tetrode curves show, by the load line drawn, how a greater swing can be achieved that with a corresponding pentode.

The plate voltage can swing lower, nearer to zero, than with corresponding pentode designs.

BIASING

Part of the circuit must provide a way of setting the tube to its correct operating point. With the recommended voltage applied to the plate (and screen), the grid is maintained at an appropriate dc voltage which is invariably negative from the cathode. In the early battery-operated radio sets and amplifiers, this was provided by a separate battery (Fig. 15-4). In modern equipment, the bias voltage for a tube is developed in a variety of ways from the same supply that serves the plate circuit. The usual way is to include a resistor between cathode and ground or supply negative (Fig. 15-13), while the grid is returned through a different resistor, which carries no current, to ground. Because of the plate current flowing through the cathode resistor, the cathode becomes positive in respect to ground. Thus, the grid is negative to the cathode by the same voltage that the cathode is positive to ground.

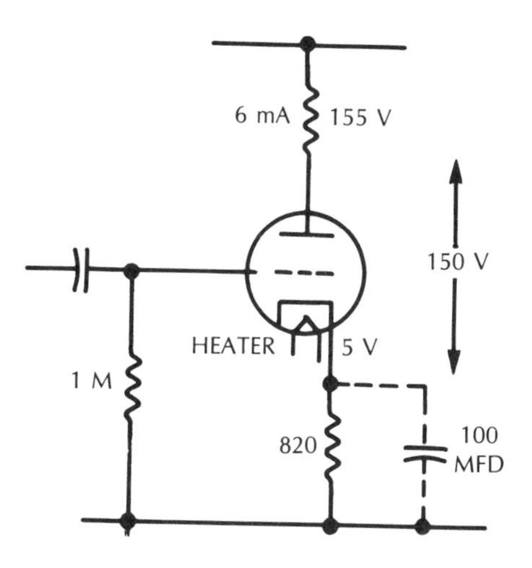

Fig. 15-13. Method of deriving grid bias with a cathode resistor.

For example, suppose the desired operating point is 150 volts with a 6-milliamp plate current, and that this corresponds to a point on the curve representing a grid voltage of −5. To get this bias, the cathode must be made 5 volts positive to ground with a plate current of 6 milliamps. This requires a resistor of 5 divided by 0.006 or 833 ohms (an 820-ohm resistor will serve). To maintain the calculated operating point, the plate voltage should actually measure 155 volts positive from ground, because the cathode is 5 volts positive, leaving 150 volts across the tube.

Such an arrangement is to some degree self-compensating. If the supply runs high, the plate voltage could be 175 instead of 150. If the grid bias were somehow fixed so the grid was 5 volts negative of cathode, the plate current might rise to maybe 10 milliamps. But with 10 milliamps flowing in the 820-ohm resistor, the cathode voltage would be not +5, but +8.2 volts. If the tube had a grid voltage of −8.2, even with 175 volts on the plate, the current would drop below 6 milliamps. However, the voltage and current in the 820-ohm resistor are related by that resistance value. So a bias voltage is obtained that automatically sets the current at a point not very much higher than when the plate voltage was 150.

This is dc stabilization or feedback. It can also result in signal feedback. For example, suppose the operating point at −5 volts is in the middle of a swing that goes from 250 volts plate to cathode, with zero current (or close to it) in one direction and to 50 volts on the plate at 12 milliamps in the other direction. If the 5-volt bias does not change, as can happen if a large electrolytic capacitor is connected across the bias resistor so the variation in plate current is absorbed without allowing the voltage to vary, the requisite input swing of plus or minus 5 volts will produce this output.

But if such a capacitor is not used, the change in plate current will also produce a change in voltage across the bias resistor. When the grid-to-cathode voltage is zero, which means it is +5 from its bias value, the plate current doubles from 6 milliamps to 12, producing 10 volts across the bias resistor instead of 5. This means the signal input will have to swing 10 volts positive, 5 to offset the original bias, and 5 to offset the extra voltage across the bias resistor due to the increase in plate current.

In the other direction of swing, assuming the tube to be linear (perfectly, which no real tube is), it will take 10 volts negative to cut off plate current. With zero plate current, the bias resistor will have a zero voltage left across it. Consequently, the signal voltage must now go to 10 volts negative to ground to achieve this cutoff.

When the bias resistor is bypassed with a large capacitor, the swing needs to be plus or minus 5 volts. Without that, the swing needs to be increased to plus or minus 10 volts. This represents a 6 dB loss in gain. This is typical. Practical circuits may produce more or less than this particular gain reduction when the bias resistor is not bypassed.

There is an alternate way of providing bias for tube operation. This is grid current biasing. It depends on the fact that as grid voltage approaches zero, it begins to conduct. An ideal operating condition in such circuits is as close to zero grid voltage as possible, without running into the positive region. So a grid return resistor on the order of 10 megohms is used, in conjunction with a coupling capacitor from the previous stage or the input circuit (Fig. 15-14). Whenever the grid goes to positive, the result is that the grid current so produced charges the output side of the coupling capacitor, so the bias is just negative enough to offset this positive excursion.

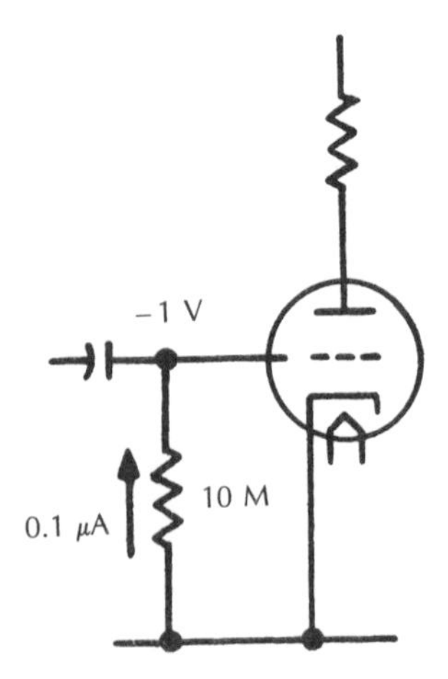

Fig. 15-14. Schematic of a stage using grid-current biasing.

GAS-FILLED TUBES

In a vacuum, electrons move freely under the influence of an electric field, but the space itself is nonconducting. Introducing a small quantity of gas into the space (at a pressure that is usually a fraction of atmospheric pressure) does not materially change this situation while the electric field is relatively small. But when more than a specific potential is reached for the particular gas and pressure involved, the gas commences to ionize. This is a process whereby positive and negative elements of the gas molecules, normally bonded by intermolecular attraction, separate and move in opposite directions, exchanging places with corresponding elements of adjacent molecules as long as the requisite release potential exists across the space.

This action is similar to the process of electrolysis, by which liquids can be separated into component elements. Because of this property, when a gas ionizes, it quickly changes its state from nonconducting, not materially unlike the condition in a vacuum, to conducting, whereby electrons flow freely through the space occupied by the ionized gas, with only a voltage difference across the space that is characteristic of the gas and its pressure.

This effect is analogous to the zener breakdown effect in a semiconductor diode. Once such a breakdown has occurred, current is controlled only by external circuit values, and the current flow can be extinguished only by dropping the applied voltage below the ionization potential. Actually, there are two potentials or voltages quite close together in value: one at which ionization starts and the other at which it ceases. Ionization is initiated by electrons striking the gas molecules at more than a critical velocity. The velocity of electrons is measured in volts that accelerate the electrons. So ionization in a given gas commences when that critical velocity and its corresponding voltage is exceeded.

For this form of ionization to occur, there must be a heated cathode to provide free electrons that can be accelerated by the voltage field. This should not be confused with another type of gas-filled tube that does not use a heated cathode. In the latter instance, ionization is initiated when the electrical potential gradient in the gas exceeds a critical value.

Ionization starts at one of the electrodes (Fig. 15-15). An electric field builds up in any space between electrodes at different potentials (voltages), but the gradient is greatest where the lines of force converge toward an electrode.

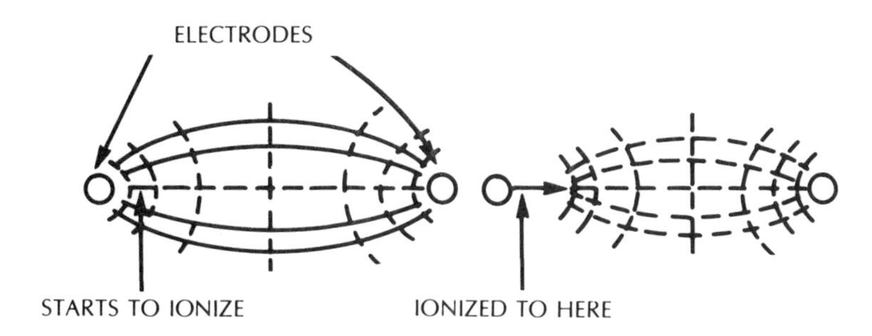

Fig. 15-15. The progress of ionization, once started, results in increased potential gradient and accelerated breakdown of the remaining space.

As soon as ionization starts in the vicinity of one electrode (the one that is momentarily the cathode or negative), the gas in the immediate area begins to conduct and the gradient increases on the next adjoining gas, until the whole tube strikes, virtually instantaneously.

The important difference between the hot and cold cathode tube types is that in the latter the strike voltage depends on the geometry of the tube electrodes and the space between, and thus on the voltage gradient that develops in their immediate vicinity before ionization commences.

With either type of gas-filled tube, once the striking potential has been exceeded, the current is limited only by external components. Ordinary fluorescent lights are in this category. If they were connected directly to the line voltage, the tube would strike when the striking voltage, say 100 volts, was reached, well before the line voltage reached its peak of 1.414 × 115 or 162.5 volts. Consequently, the current would be excessive enough to destroy the tube. As a protection against this, a current-limiting choke (inductor) connected in series with the tube (Fig. 15-16).

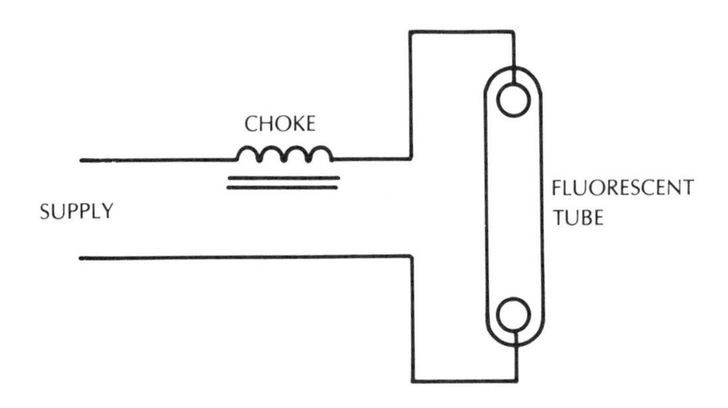

Fig. 15-16. Use a choke (or inductor) to control gas discharge tube current.

The cold electrode type tube strikes both ways, and is very suitable for ac operation. If a resistance were used to control current, instead of an inductance, the voltage across the resistance would be a relatively simple modification of the sine-wave input (Fig. 15-17). But with a choke, while the voltage across the tube must still be flat-topped (not necessarily square, since the sides may be sloping), the voltage-current relationship during the conducting portions of the period or cycle are controlled by the inductance—the voltage across it must always be proportional to the *rate of change of* current (Fig. 15-18).

Notice that the inductance maintains current flow for a longer part of the cycle than resistance does; it also avoids dissipation in addition to that in the tube. The sudden change of voltage, when the current reaches zero and before the tube strikes in the opposite direction, causes serious "static" radiation, which is the cause of the annoyance that fluorescent lights often inject into radio and other electronic appliances.

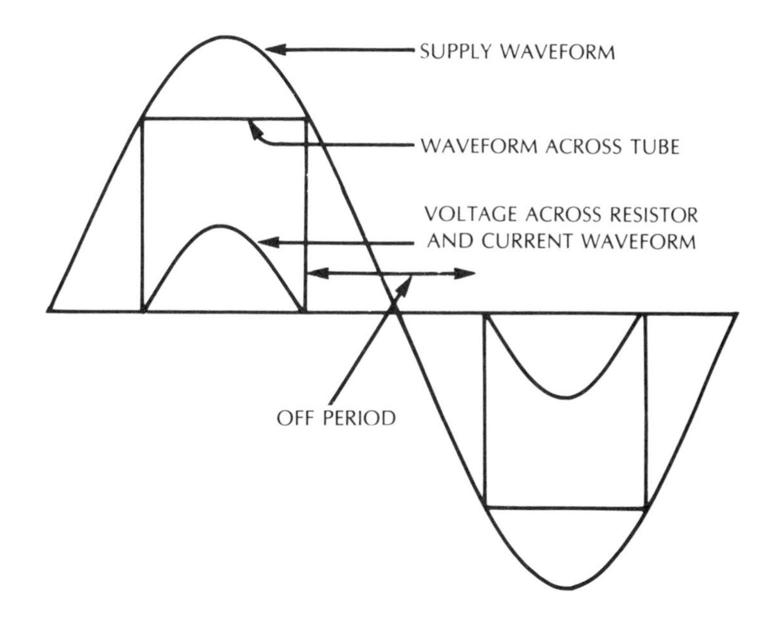

Fig. 15-17. Waveforms associated with a gas discharge tube operating on ac with a resistance "ballast" instead of a choke.

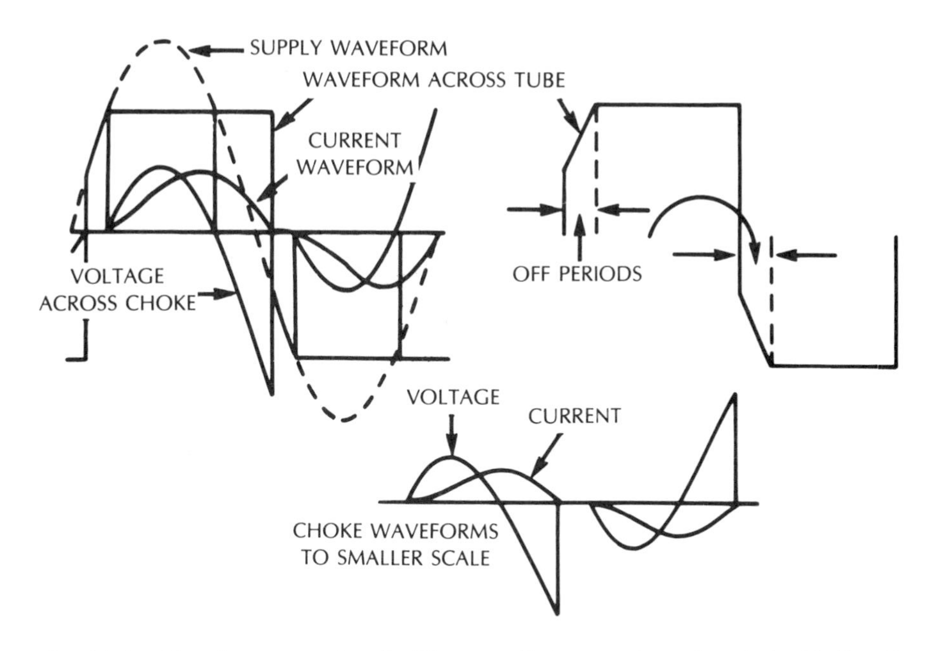

Fig. 15-18. Waveforms associated with a gas discharge tube using choke "ballast."

The thermionic type gas-filled tube, often called the *thyratron*, was the forerunner of the SCR. (Thyratrons are still used in heavy-current industrial applications.) Due to the nature of such devices, the thyratron is characterized by a *control ratio*. It is held in a nonconducting state by an appropriate grid voltage determined by the plate voltage current cutoff characteristics. The ratio of plate voltage to grid voltage needed to bias the plate current to cutoff is close to constant over a fairly wide range (Fig. 15-19) above quite low voltages where the ionization potential sets the limit.

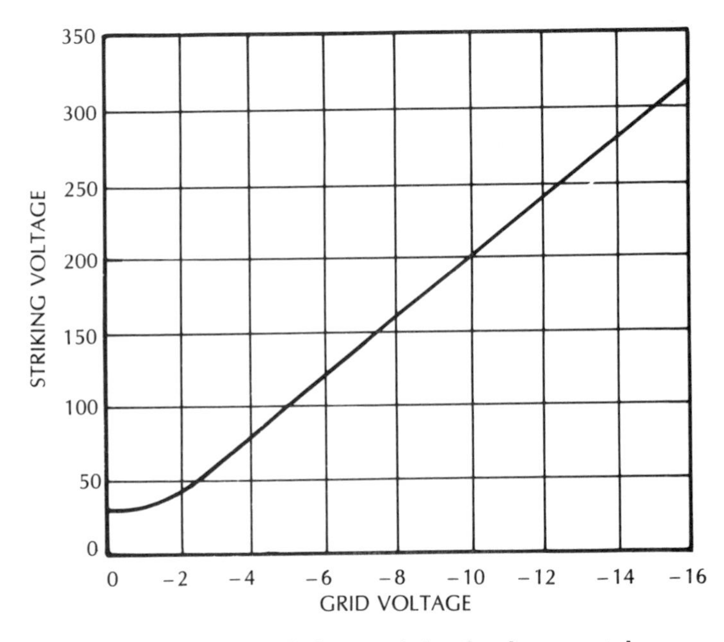

Fig. 15-19. Essential characteristic of a thyratron tube.

This can be regarded as a plot of the bottoms of the plate voltage-current curves for the same tube operated under vacuum conditions, except that when the gas is introduced, the tube conducts fully as soon as plate current starts at all (beyond a minimum ionization current of a few microamps). At this point, the gas ionizes. Once the tube is conducting, the only way to stop the current flow is to allow it to deionize, which involves dropping the plate voltage below the ionization potential, corresponding with a current below a minimum needed to maintain ionization. While the tube is conducting, the current flow must be controlled by the external circuit, because once the tube is struck the current is unlimited by the tube itself.

At one time, mercury vapor rectifiers were one of the most efficient types. With a heated cathode, the drop from cathode to plate is only 15 to 20 volts. Any high-voltage vacuum diode (the alternative in those days) introduces a much greater voltage drop than this to enable appreciable current (say more than 100 milliamps) to flow. So the first high-voltage, high-current rectifiers were mercury vapor diodes.

The danger point with vacuum rectifiers is greatest at the maximum current rating. If exceeded, the high current bombarding the plate (with only 15 volts drop at, say, 5 amps, it represents 75 watts) heats the plate until it reaches an adequate temperature to emit electrons, resulting in a reverse current flow. The reverse current blows a fuse or destroys the rectifier, whatever happens first (and sometimes second, as well).

A more common type of tube, still only slowly being replaced by solid-state devices, is the neon tube and similar devices containing other type gases. These have a virtually constant voltage drop, in either direction, once the current has been struck. So they behave in a manner very similar to two zener diodes connected together back-to-back. The main difference is that the gas-filled tube is suited for higher dissipations.

Zener diodes work at voltages on the order of 3 volts or so, extending up to almost any voltage. Some may even work at less than 3 volts. Gas-filled tubes have striking voltages on the order of 50 to 100 volts. So the choice depends largely on the operating condition required. As a practical consideration, many circuits can be designed so that any convenient voltage may be used. Thus, the deciding factor may become a matter of either size or cost, rather than the precise voltage used.

BEAM-TYPE TUBES

A variety of other effects can be utilized within tubes. The principles discussed in Chapter 7, which show the interaction between electric fields and magnetic fields, and the motions and forces produced, can be applied to moving electrons in a free space (vacuum). Beam tubes, such as the beam tetrode already discussed in the previous chapter, work on this principle. So do the cathode ray tubes, used for television and many other purposes, as well as special-purpose tubes, such as the klystron, magnetron, electron-multiplier, etc.

The cathode ray tube, which was one of the first tubes to apply beam principles, based on the phenomenon observed by Sir William Crookes, demonstrates the beam principle quite well. Electrons emitted by a cathode are met by an electrostatic field that concentrates them into a beam. The electrodes that do this are called an "electron gun" (Fig. 15-20). The gun also focuses the beam to a spot at the surface of the tube where a fluorescent screen makes their arrival visible.

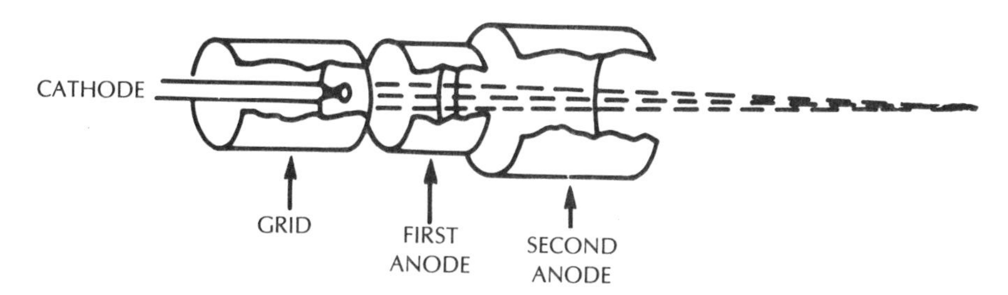

Fig. 15-20. Cross-section of a CRT (Cathode Ray Tube) electron "gun" with electrostatic focusing.

The focusing capability exists because the electric field has curved contours, like an optical lens, that are symmetrical about the axis. An alternative method of focusing uses a magnetic field, but it bends individual rays back inward in a twisting fashion, because the force on the moving electron is always mutually at right angles to both its own direction of movement and the direction of the magnetic field it intersects. (Fig. 15-21).

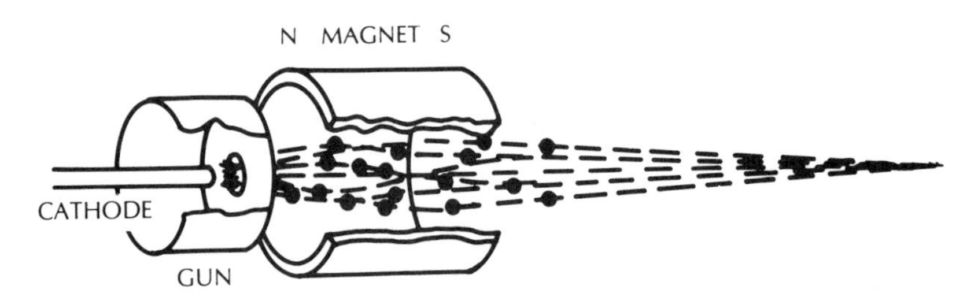

Fig. 15-21. Cross-section of a CRT gun with magnetic focusing.

Once the beam has been focused (by whichever method), it can be deflected by a more linear field, again either electric or magnetic. The deflection field is introduced by two plates, one of which is made positive and the other negative in respect to the voltage at the final accelerating electrode (usually called the anode and, where more than one is used, the second, or third anode), which brings the electron stream to its final velocity (Fig. 15-22). Successive sets of plates at right angles are found in the cathode ray tube most used for measurement purposes. The beam is deflected on "X" and "Y" axes (horizontal and vertical, respectively).

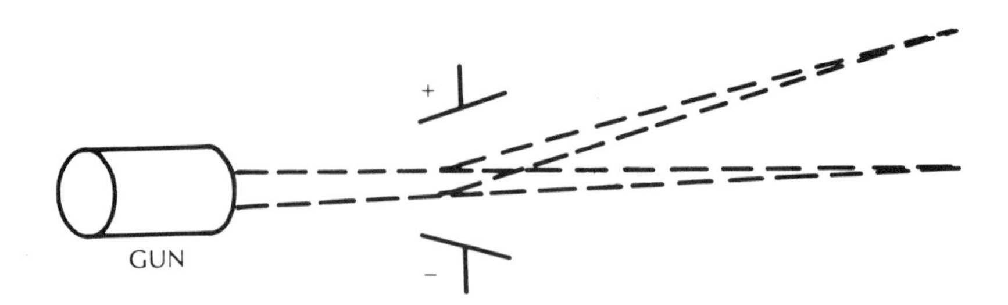

Fig. 15-22. How electrostatic deflection works.

Magnetic fields introduced into the beam by coils outside the tube (Fig. 15-23) can also deflect the spot, but at right angles to the axis of the field the coils produce. Pairs of coils thus disposed are used in most television receivers to control the movement of the spot. A grid modulates the beam to get the variation in intensity that produces an intelligible picture.

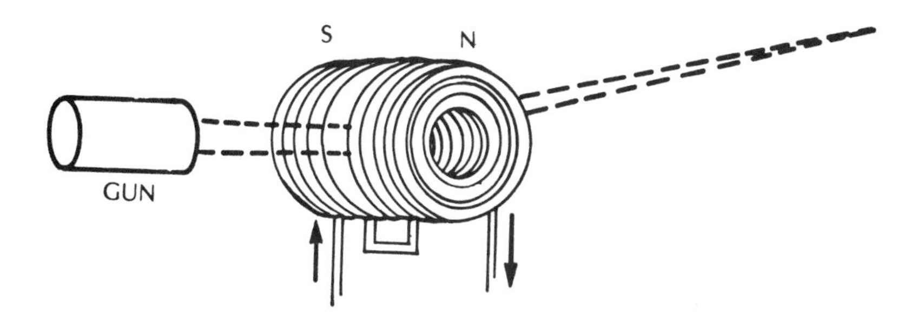

Fig. 15-23. How magnetic deflection works.

An important feature in all these tubes is its sensitivity: how much or little voltage, current, or energy is required to move the spot at the rate or over the distances required. In the electrically deflected tubes, the divergence of the spot from its undeflected course is proportional to the deflection voltage applied. It is also inversely proportional to the final anode voltage, which controls the velocity of the electrons in the beam. Thus, if the final anode voltage is doubled, the required deflection voltage must also be doubled. The same is true of the magnetic deflection method. The field strength must be proportional to the final accelerating voltage.

All of the tubes described so far use electric and magnetic forces in an essentially "static" way. True, the electrons under control are moving, and their rate of movement (as controlled by the final accelerating voltage) interacts with the deflection produced. But these relationships themselves are fixed. They do not involve the propagation velocity of electromagnetic waves, merely relationships on which the electromagnetic waves are eventually based by mutual interaction.

The more sophisticated electronic devices use successive changes of voltage or magnetic field to build up more than is possible with such a static system. For example, in a static system, the final velocity of the moving electrons is strictly a function of the final anode supply voltage. To make electrons travel faster requires a higher final anode voltage.

The more sophisticated devices manage to escape this limitation so that electron velocities equivalent to many millions of electron volts (an *electron-volt* being the velocity achieved by a 1-volt electric field being applied to an electron) can be achieved without necessitating the use of such high voltages.

A variety of such devices divide into two categories: devices with the primary purpose of accelerating electrons so they achieve an energy level much more difficult to attain with a static device (actual high-voltage equivalent to the required velocity); and devices that apply the movement of electrons in such a way to utilize the time of movement rather than the rate to offset problems that arise with static devices in which electrons do not move fast enough to maintain the device's normal characteristics when dealing with extremely high frequencies.

In the latter case, ordinary tubes cease to amplify because the duration of one cycle or Hertz of signal energy is less than the time needed for an electron to travel from cathode to plate. Tubes that allow for the shorter transit time associated with extremely high frequencies overcome these deficiencies by using the time taken for electrons to move as part of the design, rather than having to fight the fact that they do take time to move.

ACCELERATOR DEVICES

The simplest accelerator device is the linear accelerator (Fig. 15-24). Electrons are emitted by a heated cathode and started on their journey by a positive potential in a gun like that of a cathode ray tube. Additional acceleration is achieved by alternating voltages, obtained from a high-frequency high-voltage supply.

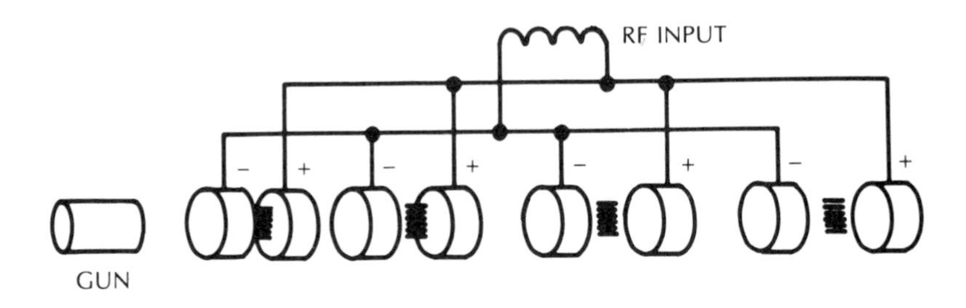

Fig. 15-24. Essential arrangement of a linear accelerator tube.

Electrons will always accelerate "up" a field, from negative to positive. So bunches of electrons move in such a way that they are always moving from a ring electrode that is momentarily negative toward one that is momentarily positive. When they pass through each ring, its potential reverses, as part of the high-frequency cycle, so that they emerge from the opposite side, once more leaving a negative electrode, headed toward a positive one.

Because of the increasing acceleration, the spacing of successive electrode rings is increased, since the frequency of the accelerating voltage remains constant. The gun emits electrons at a steady rate, but they are accelerated in bunches or groups. As each group moves forward, some electrons are retarded instead of accelerated by the field until they join the next bunch. Consequently, the electron stream forms into discrete bunches that are accelerated down the tube.

If electrons move in a magnetic field, the path curves at right angles to both the momentary direction of movement and the field. Thus, if a cathode emits electrons between two poles of a magnet, with a cylindrical plate placed between the magnet poles, the electrons will move in a circular or spiral path toward the positive plate (Fig. 15-25).

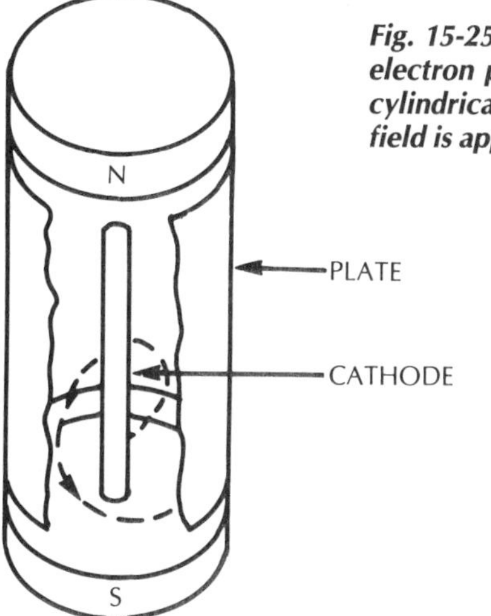

Fig. 15-25. The dashed arrow indicates the electron path from a central cathode to a cylindrical plate when a strong magnetic field is applied along the axis of the device.

But if the plate is replaced by two segments that are D-shaped, connected to a high-frequency supply as well as a positive dc voltage, the electrons will again form into bunches and rotate in an expanding circle as they are accelerated by alternations of the high frequency supply (Fig. 15-26). When the electrons reach the escape velocity, they pass through a window provided. The above describes the operating principle of the cyclotron.

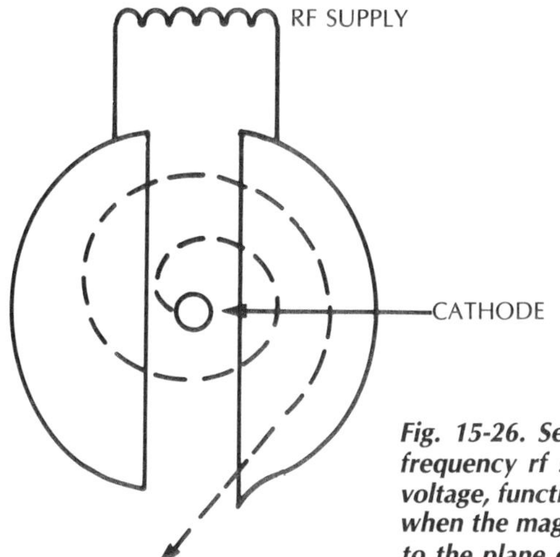

Fig. 15-26. Semicircular plates with high-frequency rf superimposed on the supply voltage, function as an electron accelerator when the magnetic field is applied vertical to the plane of the diagram.

It is important to realize that accelerating electrons requires energy, just as the energy is dissipated when those electrons strike something such as a plate. So the high-frequency supplies used with linear accelerators or cyclotrons must provide the necessary energy to reach the electron velocities attained. The advantage in using an alternating supply is that it is much more convenient to produce extremely high electron velocities this way than by providing equivalent energy at the extremely high dc voltages that such electron velocities represent.

Another electron-accelerating device, invented during World War 2, is the klystron. Each version consisted of two doughnut-shaped resonators with a space between, an electron gun at one end and a collector at the other (Fig. 15-27).

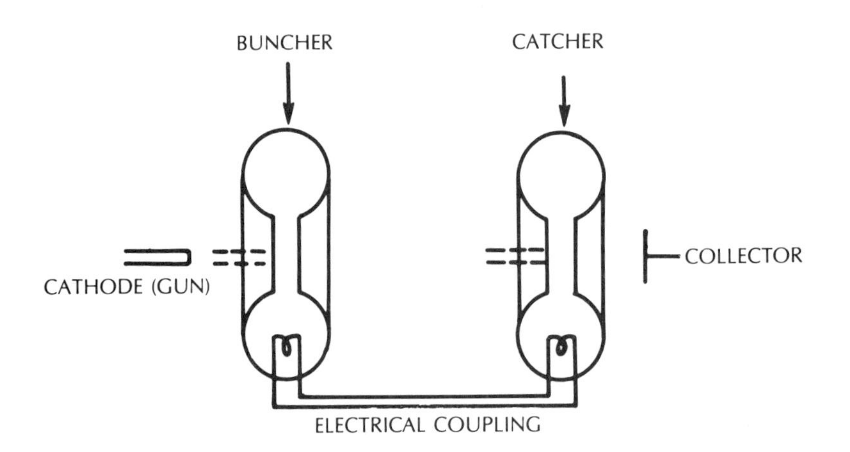

Fig. 15-27. Cross-section of the element arrangement inside the earliest, "simple" klystron.

The inside edges of doughnuts, where the electron stream passed through, were split. The moving electrons acted very much like a current of air going past the lips of an acoustic resonator, such as an organ pipe, with the result that the first opening bunched the electrons, while the second one retrieved the energy developed by the now well-separated electrons. A feedback circuit was used, working at the resonant frequency of both doughnut cavities, to supply the energy needed to produce bunching, so the thing could work as a self-maintained oscillator.

Since electrical coupling is inefficient at these extremely high frequencies, the device is much more efficient if the same cavity can serve both functions and thus eliminate the coupling. This was finally achieved by using a reflex electrode instead of a collector to send the electrons back into the same cavity gap (Fig. 15-28).

Just as the accelerator can be wrapped into circular form by putting it between the poles of a magnet, so the action of a klystron tube can be turned into a circular path, with cylinderical cavities around a central chamber to produce the magnetron (Fig. 15-29).

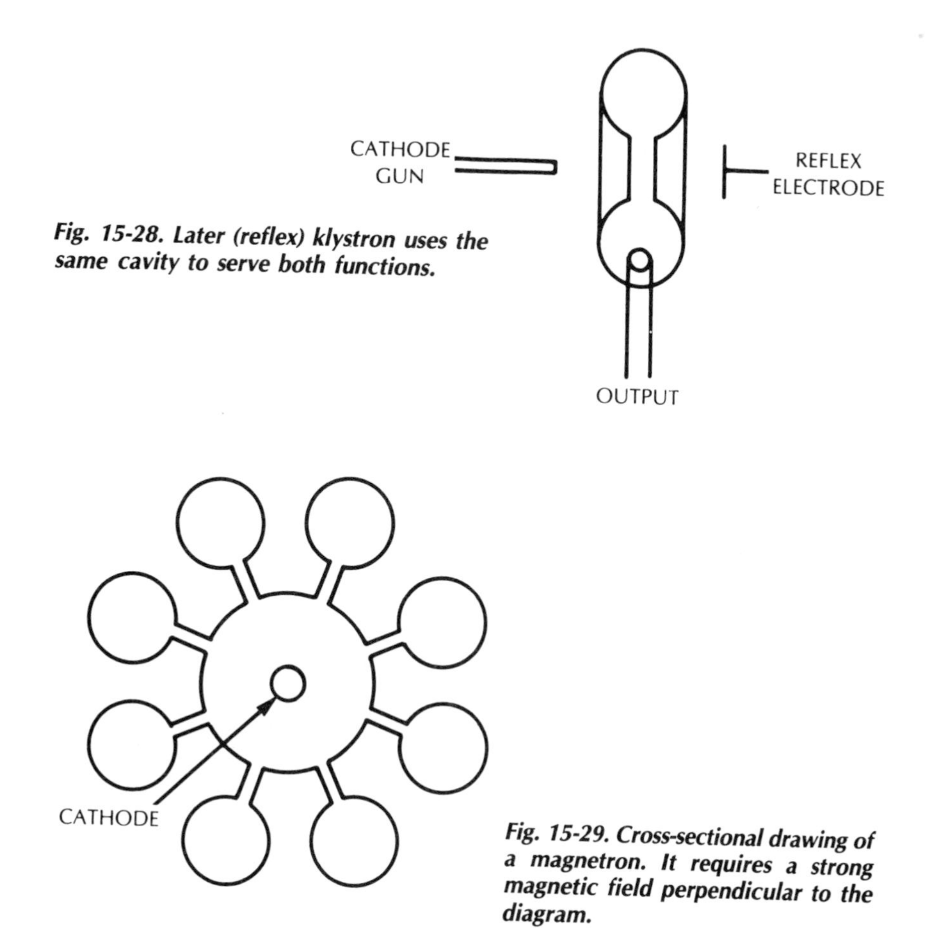

Fig. 15-28. Later (reflex) klystron uses the same cavity to serve both functions.

Fig. 15-29. Cross-sectional drawing of a magnetron. It requires a strong magnetic field perpendicular to the diagram.

An important point to grasp is that all these devices possess unique relationships: they work at a specific resonant frequency for which they are designed. Frequency can be correlated with the initial accelerating voltage used, but the resonant cavities are uniquely controlled by the propagation velocities discussed in Chapter 8.

Examination Questions

1. A triode tube has a mutual conductance of 1.25 milliamps per volt with a plate resistance of 80,000 ohms. When used with a plate resistor of 220 K and coupled to the grid of a following stage, where the grid-to-ground resistor is 1 megohm, find the stage gain.
2. In the previous question, a stage gain of 80 is required. The grid resistor of the following stage is at its maximum value, so the only value that may be changed is the plate resistor. Find the value of plate resistor required, assuming that the parameters quoted do not change with the lower operating point this will cause.

3. A pentode has a mutual conductance of 3000 micromhos and a plate resistance of 600 K. Find its gain when a plate resistor of 100 K is used.
4. A tube uses an operating point of 125 volts at 10 milliamps, with a plate load resistance of 12 K. The required bias voltage is −8 volts. Calculate the cathode bias resistor needed and the high-voltage supply required.
5. In Question 4, an input voltage of plus or minus 3.5 volts peak, grid to cathode, produces an output swing of plus or minus 5 milliamps, plus or minus 60 volts. Calculate the grid-to-ground input required without a cathode bias resistor bypass capacitor.
6. The ionization level of gas-discharge tube is 105 volts. It is supplied by a square-wave voltage with a peak value of 140 volts. The frequency of the square wave is 250 Hz. Figure 15-30 shows the relevant waveforms. Find the delay between the input square wave and that appearing across the tube.

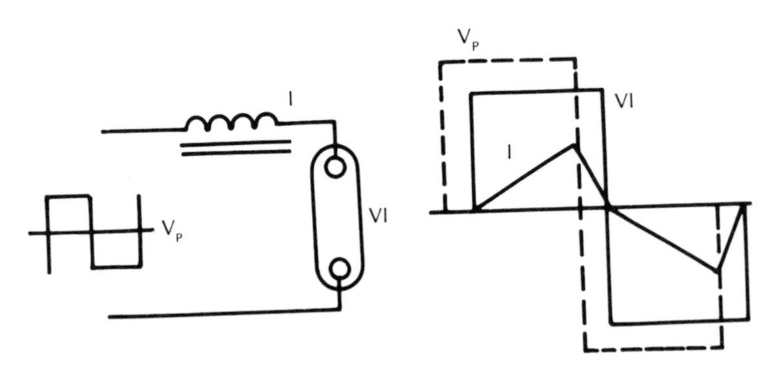

Fig. 15-30. Circuit and waveforms related to Questions 6 through 11.

7. If the inductance value is 50 millihenries, find the peak and average current values.
8. The square wave voltage of Question 6 increases to 180 volts peak. Find the new delay time and the revised peak and average currents.
9. If the inductance value is changed to 100-millihenries find the peak and average currents for supply voltages of 140 and 180 peak.
10. Find the tube dissipation in Questions 7, 8 and 9.
11. Find the value of inductance that must be used to ensure that dissipation will not exceed 50 watts when the supply voltage is a 180-volt peak square wave.
12. A triangular waveform supply is used for the same 105-volt ionization tube (Fig. 15-31). Find the total duration of strike on each half cycle when the frequency is 250 Hz and the peak voltage of the triangular waveform is 155 volts.

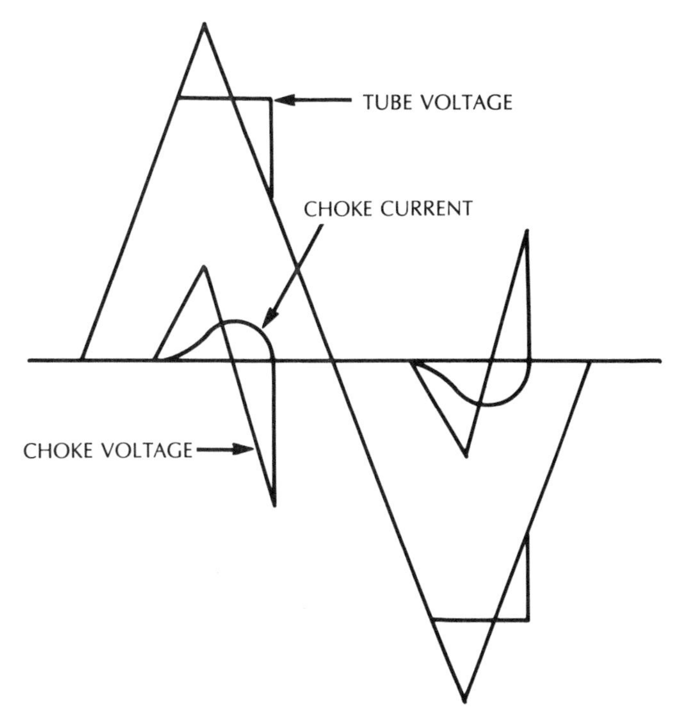

Fig. 15-31. Waveforms associated with Questions 12 and 13.

13. Find the peak voltage value that will just eliminate the off time, so the gas-discharge tube strikes in the opposite direction the instant it extinguishes in the first direction.
14. A certain input signal voltage applied to a cathode ray tube produces a 4-inch peak-to-peak deflection. The final plate voltage is 5000 volts. For a specific purpose, the deflection should be increased to 5 inches, but it is not feasible to increase the deflection voltage. At what plate voltage would this deflection be possible?
15. To obtain increased brilliance of the trace, the final plate voltage of an oscilloscope is stepped up from 6000 to 8000 volts. At the lower voltage, the deflection needed to utilize the full useful area of the screen was plus or minus 150 volts. What deflection voltage will be needed at the higher plate voltage?

16

Other Devices

In this chapter, we gather together some of the various "odd" pieces that have accumulated in electronics. Over the years, since electrical and magnetic phenomena were first discovered, many effects have been found, some of which are still "curiosities." Many of them eventually find use. So whether the devices are actively used yet, or remain as an interesting demonstration in physics (as the semiconductor properties that form the basis for the whole solid state "family" once were) we introduce you to some of them here.

PHOTOSENSITIVE DEVICES

The effects of electron action, inter-related with other phenomena, are responsible for the operation of other devices. The easiest way to understand this is to recognize that electrons in movement constitute one form of energy and that there is usually, somewhere, a way of converting energy in any given form to energy in any other given form.

Light and heat are alternate forms of energy. Conversion of light or heat into electronic energy occurs with photoelectric devices. The reverse conversion takes place on fluorescent screens, used on the inside of fluorescent lighting tubes and on the inside face of a cathode ray tube. Some devices convert energy directly, others merely modulate energy of some other form, so an effective conversion is achieved. Photovoltaic cells, used on photoelectric exposure meters, generate an output voltage or current directly dependent of the amount of light striking the light-sensitive surface (Fig. 16-1). This device needs no external source of energy, other than the light that is the source of the energy converted.

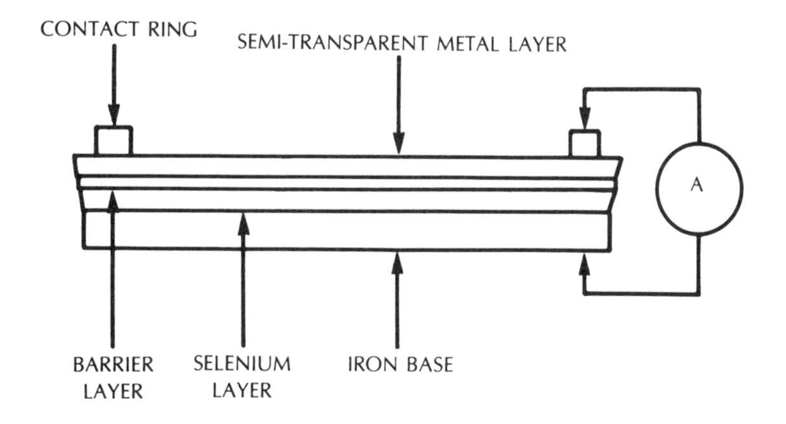

Fig. 16-1. Cross-section of a photo-voltaic cell of the type used in exposure meters.

Figure 16-2 shows typical characteristics of such a cell. If the circuit to which the cell is connected has a very low resistance, the current in microamps is directly proportional to the illumination from which it is derived. But a meter to measure microamps has appreciable resistance, causing the cell to generate voltage as well as current at its terminals. What the curves of Fig. 16-2 indicate is that under these circumstances the cell reaches a maximum voltage, beyond which further illumination will not produce more output. With a 1000-ohm external resistor, this appears to be represented by 160 microamps, which produces 0.16 volt.

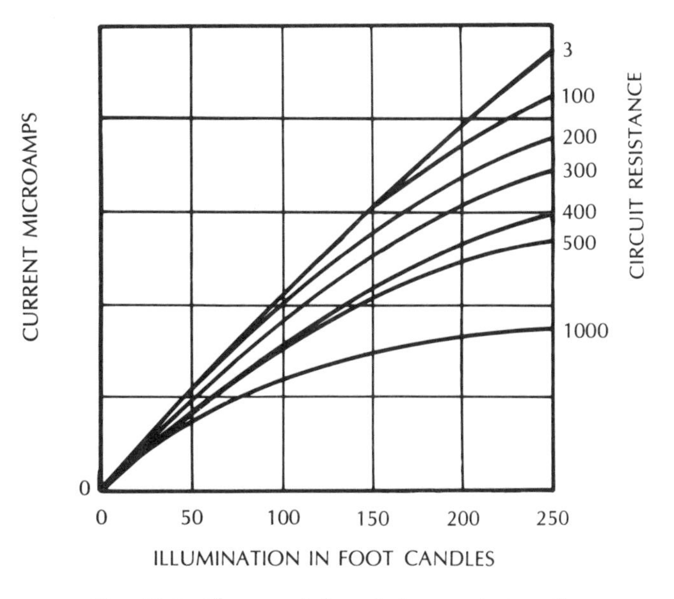

Fig. 16-2. Characteristics of photo-voltaic cell.

This is the maximum voltage the cell will produce. Thus, while the cell is called "photovoltaic," meaning it generates electrical output for light input, it is better understood as generating a current related to the light; the voltage is developed by the resistance load acting as a current limiter.

Photoconductive devices work differently. When light strikes the sensitive surface, its resistance changes (Fig. 16-3). Usually, the device is virtually an insulator in the dark, but the incidence of light on the sensitive surface causes it to conduct to a degree dependent on the illumination provided. Output is then dependent on the presence of a voltage to drive current through the conductance of the device, and the current passed is a measure of the illumination.

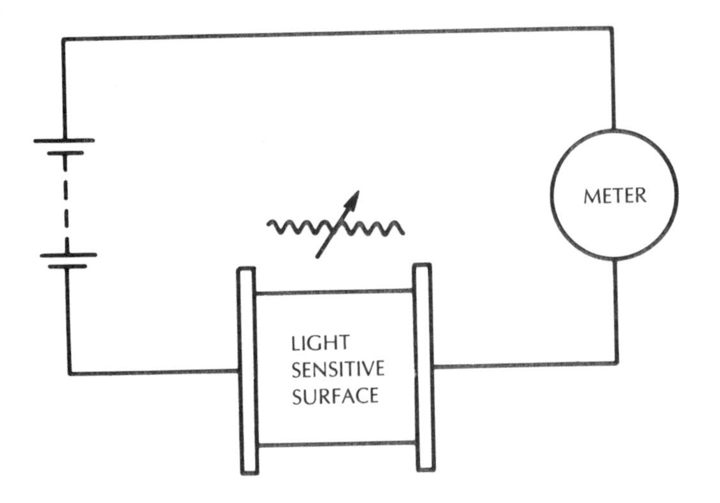

Fig. 16-3. Basic circuit for operating a photoconductive cell.

A far more sophisticated photoelectric device is the light-sensitive tube used in a television camera. In simplest terms, it is like a picture-tube working in reverse, but in reality it is much more complicated than that, because fluorescence is a non-reversible conversion. The camera tube must use a screen having the property of generating a charge when the light strikes it (Fig. 16-4).

Producing a charge across a surface that represents a photographic image is one thing. Converting this charge to a transmissible signal is another problem. In theory, we could connect a million wires to various points on the surface systematically, with a million high-impedance inputs to amplifiers to represent a million picture elements. That is fairly obviously an uneconomical solution, even if the elements would yield enough output to amplify in that way. The solution eventually devised for camera picture tubes uses an electron gun similar to that found in the receiver picture tube to scan the charged screen. The screen then yields secondary emission, similar to that discovered in the design of tetrodes. This secondary emission is proportional to the charge voltage built up at the point on the charged screen momentarily being scanned by the gun. The signal collector picks up this emission and amplifies it as picture signal, in just the same way that picture signal modulates the picture tube at the other end of the transmission chain.

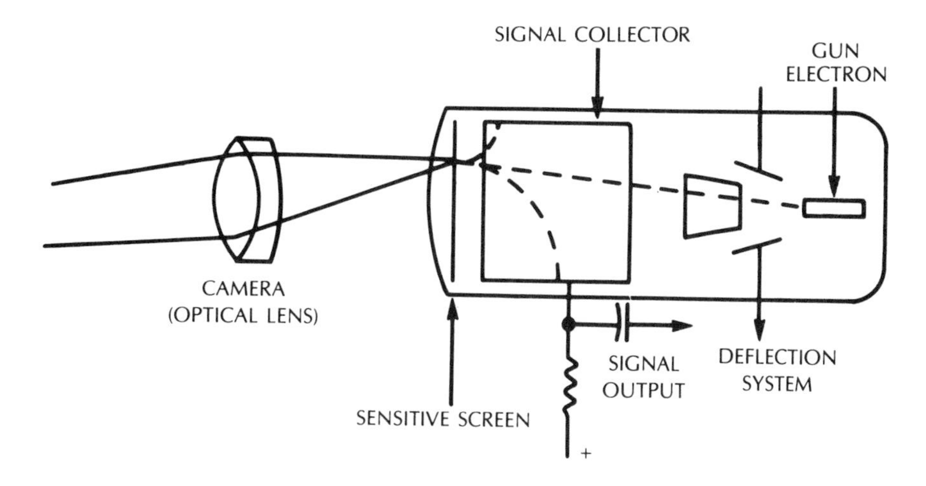

Fig. 16-4. Operational diagram of a modern TV camera tube.

The fluorescent screen used in picture tubes and lighting tubes is a direct energy converter: electrons striking it cause it to radiate light. Equivalent to the photoconductive cell in modulating an external source of power for this mode of conversion is the Kerr cell, with its attendant polarized filters (Fig. 16-5). A Kerr cell rotates the axis of polarized light.

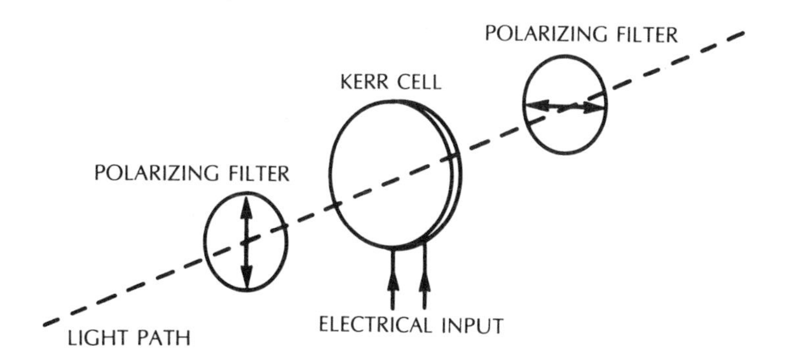

Fig. 16-5. Diagram of an early light modulator using a Kerr cell.

To understand this, we need to know what polarized light is. Most light used in everyday activities is unpolarized, which means that individual rays making up the light are randomly polarized in all directions. Light, as a form of energy radiation, is identical to the electromagnetic waves described in Chapter 8, with the difference that wavelength is infinitesimally shorter and frequency correspondingly multiplied.

Light uses wavelengths where molecules of matter resonate. Lasers are devices that generate truly polarized light, using molecules of various substances as their resonant frequency to generate energy that radiates as light in the same way that one or other form of dipole radiates radio waves of very much longer wavelength and lower frequency. The intensity of laser beams and the capability for retaining such a narrow beam over hitherto unheard-of distances is due to their perfect polarization, making them very clean. This perfection is possible because they are generated in a polarized state rather than having to be polarized from a random source. This makes the device tremendously more efficient: all the energy generated is utilized.

Before the advent of the laser, the only way to produce polarized light was to use a filter which had the effect of selecting only those rays that possessed the desired polar orientation more strongly than other orientations, while completely rejecting orientations at right angles to the desired direction. Thus, a polarizing filter stops a major portion of the rays.

When a voltage is applied to the plates, the Kerr cell rotates the angle of polarization as light passes through it. With no rotation, when the voltage applied is zero, a second polarizing filter cuts off all the light passed by the first filter because its axis of polarization is at right angles to the first (Fig. 16-5). When voltage is applied to the Kerr cell it rotates the axis of polarization from that selected by the first filter, so some of the light can now pass through the second filter, depending on the amount of rotation, which in turn is controlled by the applied voltage.

Some of the first television receivers employed a Kerr cell as a means of modulating light, which was then optically projected onto a screen. The device was quite inefficient and incapable of the speeds of modulation needed for high resolution television, so the cathode ray tube, used today in essentially the same form, was adopted, although many improvements have been made since the early models.

PIEZOELECTRIC DEVICES

Two more energy-conversion devices have achieved varied use in electronic applications: piezoelectric and Hall effect devices. Piezoelectric materials, mostly crystals of one kind or another produce voltages in certain directions when they are subjected to mechanical strain (distortion of shape). They also exhibit similar mechanical strain when voltages are applied in a converse manner. Deriving the required electromechanical interaction in an efficient manner is usually a matter of building up a compound crystal assembly.

Figure 16-6 shows the basic form of distortion produced in a single crystal, with a voltage of one polarity applied between the upper and lower faces.

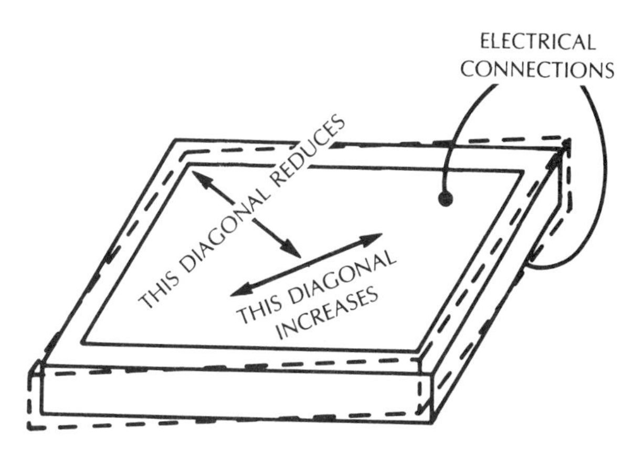

Fig. 16-6. Relationship between voltage and mechanical strains in a basic piezo crystal.

Applying a reverse voltage will produce a reverse distortion: elongation along the other diagonal. If a piece is cut diagonally from such a crystal, it will stretch when voltage of one polarity is applied and shrink when the reverse is applied (Fig. 16-7). Cementing two such pieces together, so that with the same polarity applied one of them stretches while the other shrinks, yields an assembly that bends when voltage is applied. Such an assembly is known as a "bender" crystal.

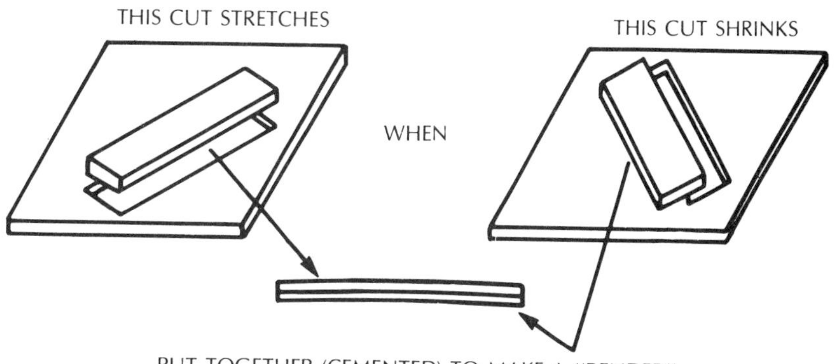

Fig. 16-7. Showing how two crystals, differently cut from the basic material and cemented together, make a "bender" crystal.

Using such a crystal is one of the more efficient ways of constructing a transducer around piezoelectric material. Where the moving coil transducers in Chapter 8 have considerable compliance and need a "stiffness" control to prevent them from "running away" or exceeding a safe degree of movement, the piezo type has the reverse problem: the transducers themselves are inherently stiff, and require some means to make them more pliable, or to enable them to work over a meaningful amplitude (the basic piezo effect is very small). Using the bender type assembly is a first step toward this. Apart from the voltage relationship to strain or distortion of the crystal, the piezo device is essentially an insulator. The impedance of a transducer as reflected at the plates by which it is connected is essentially that of a capacitance. This, too, makes it quite different from other transducers when coupled into electronic circuits.

HALL EFFECT DEVICES

Hall effect devices use materials such as indium arsenate or indium antimonide, which have properties that differ from simple electromagnetic effects. If a current is passed through the substance in one direction, with a magnetic field through it at right angles to the direction of current flow, a voltage is induced in a third direction, mutually at right angles to the other two (Fig. 16-8).

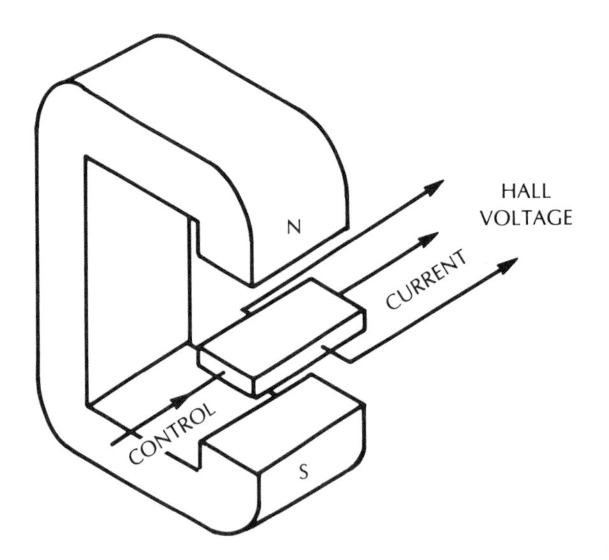

Fig. 16-8. The basic electrical properties of a Hall effect device.

Like the piezo effect, this one is quite small, measurable only with precision equipment. Such precise measuring equipment identifies a minute but consistent effect of this nature that previously known electromagnetic effects cannot explain. Development of the correct alloys has made the device at least demonstrable and may eventually find it more widespread use, just as non-ferric materials have made good magnets that replace ferric types, when at an earlier stage in the technology they were a mere scientific curiosity.

When the author first became active in this fascinating field, the solid-state diode had been discarded as "primitive"—as indeed its catwhisker form was! Sophisticated equipment of the day used only thermionic vacuum diodes. There was a distinct tendency to think of the vacuum as a "holy place" where electronic action could take place. But the solid-state principle on which the catwhisker diode had functioned came back and is more widely used at the present writing than the vacuum tubes that were the mainstay of electronics in those days. The semiconductor has proliferated into all kinds of devices that were undreamed of in those days. The Kerr cell had a short life of usefulness, since it was superseded by cathode ray tubes. But the principle on which the Kerr cell worked is as valid as ever: it could find a new application, either in its old form, or in a more sophisticated design. Or projection television might come to use a laser beam, modulated and deflected.

The Hall effect is still somewhat in the novelty category, as of the present writing, although some minor applications are being researched. The piezo effect was also a novelty for many years, until ways were found to build an assembly that made it an inexpensive alternative to the moving-coil transducer for microphones and loudspeakers.

NEWER DEVELOPMENTS IN SOLID-STATE

Solid-state is a rapidly advancing art. In later chapters we will show how advancement in overall systems, and in the components necessary for them to succeed, progress. Here we will just list a few items, to show how the detail developments occur.

Photo-electric cells have been known, in their varieties, for a long while. This leads to the question: if light can affect the electrical properties of a device, shouldn't it be possible to produce a device in which electrical potentials or currents affect their light properties? Two such devices, which had not been heard of when the first edition of this book was written, are in everyday use now, quite commonplace and inexpensive.

Light-Emitting Diode

First of these was the Light-Emitting Diode (LED). This device is built like a layer diode, except that by restricting the area of the layer that emits light, it can be made to exhibit characters, and is used for readouts on computers and other devices, such as clocks, watches, etc. The layer that changes its state from nonconducting to conducting when it is caused to emit light, which is usually colored, is restricted to the shape of the character it is to emit.

By building 7 of these segments into the shape of a figure 8 (Fig. 16-9) selectively energizing them can be used to produce all of the numerals in the decimal system. A somewhat more complicated grouping can be made to display alphabet characters as well as numerals. A typical light-emitting diode requires somewhat less than 2 volts, with a current of a few milliamps, to energize it, per segment.

The color with which it glows depends on the chemical content of the light emitting layer, just as with the fluorescent screen used in its vacuum-tube counterpart. It seems highly probable that it's only a matter of time before such light emitting elements can be combined in such a way as to produce a full-color picture screen that does not require a vacuum, or occupy more space than a relatively thin layer, somewhat like that used for polaroid type pictures.

Fig. 16-9. How 7 elements can be put together, of either LED (light emitting diode) or LCD (liquid crystal diode) to produce numerical read-out units.

Liquid Crystal Diodes

Next in this category came the liquid crystal diode (LCD) which uses a crystalline substance in liquid form, in place of the light-emitting substance. It is also polarized the opposite way: applying a reverse voltage, rather than a forward voltage, as used for the LED, causes the molecules of the crystal to reorient themselves, so it changes from being transparent to being opaque.

This has the advantage that virtually no current is required, just sufficient reverse voltage, which is quite small, to effect the change. LCDs can be formed in all the same shapes or configurations as LEDs and thus provide the basis for computers and clocks or watches which read continuously, instead of only when energized.

Miniature devices, such as wrist watches or hand-held computers, are powered by tiny batteries. If LEDs are used, leaving the display on rapidly uses up the small capacity of the battery. So when LEDs are used, the device is usually provided with a "read" button, which energizes the readout only for a short period, in order to conserve battery power. If the device is line operated, of course, this is not necessary.

On the other hand, since LCDs use a negligible amount of current, they can be left on all the time. The same tiny batteries will last a year or more, in continuous operation, because the varieties of FETs used in the calculator chip, plus the energizing of the display, all together, consumes extremely little. Virtually all the energizing has to do, is to change the charge applied to some extremely tiny capacitances.

Electrets

Another form of solid-state material was derived by applying thought to the similarities and differences between electricity and magnetism. Magnetism, as pointed out in Chapter 7, is essentially bipolar: there is never a North pole without a South pole. The magnetized and unmagnetized state of "permanent" type magnets has been explained as being due to orientation of its molecules. So the question arose, "Could not the same thing be done in a material composed of highly charged electrical molecules?"

This led to the search for, and development of, such materials, which at this stage are used principally in the "electret" (a word formed by combining "electric" with "magnet") condenser microphone. A condenser microphone, up till that invention, required a polarizing potential of thousands of volts, but virtually no current to energize it. By building an electret layer into the microphone, the polarizing is built in.

It should be stressed that an electret produces only potential, with virtually no current available. Thus it is not suitable as a power source. Thus we now have a material that, in electrostatic terms, corresponds with a magnet. The next question, on which science is now working is: can we also produce a material that converts the equivalent of current electricity into the realm of magnetism, thus getting away from the bipolar property that has always been regarded as part of the essential nature of magnetism?

Part of the problem is the extreme differences in magnitude. As was pointed out in Chapter 8, the reason the connection between "static" and "current" electricity was so long being discovered is the difference in magnitude: static electricity deals in thousands, even millions of volts, and microscopic current, while current electricity reverses that. Magnetism as it has always been known, corresponds with "static" electricity.

Pixel Displays

Until quite recently, the only way of displaying graphics electronically was by means of a cathode ray tube, which we discussed in Chapter 15. Various means were found to make the displays present colored pictures. But common to all of them was the need for a rather complex scheme of scanning the picture. The common TV picture uses some 525 lines to the frame, each line traversing the picture from left to right.

This means that the picture is made up of 525 lines of picture elements, vertically. The picture aspect ratio is 5:4, which means that if it is 12 inches high it will be 15 inches wide. So to give equal definition horizontally, each line needs to have about 656 elements. This makes a total number of picture elements of about 350,000, which have to be scanned sequentially, line by line.

Positioning of each element is achieved by two scanning frequencies: the line frequency, which scans from left to right across the picture 525 times per frame is 15,750 lines per second; and the vertical frequency which, to reduce visible flicker covers the frame in two interlaced sweeps, goes at 60 sweeps per second.

Either way, the elements are presented, in proportional brightness and color, at a rate of 30 $\times$ 350,000 elements a second, or about 10.5 megahertz. As video channels don't extend that high, some resolution is inevitably lost.

The newer approach, which is being worked on in a variety of systems, uses a flat plate to originate the display, somewhat like the LCD (in fact, LCD is one system being developed). The interesting possibility, which will soon be a reality, is that greater resolution will be possible with less rigorous frequency requirements.

Consider a single point of white on black, or black on white, where the point occupies only one "pixel" (a new word for picture element). A horizontal scan and a vertical scan, made in succession for transmission, but applied to the conversion device simultaneously, can pinpoint the single element in just two sweeps: one 525 elements long, and one 656 elements long. Total transmission time, 1181 elements, instead of 350,000! Figure 16-10 shows this, applied to a single pixel.

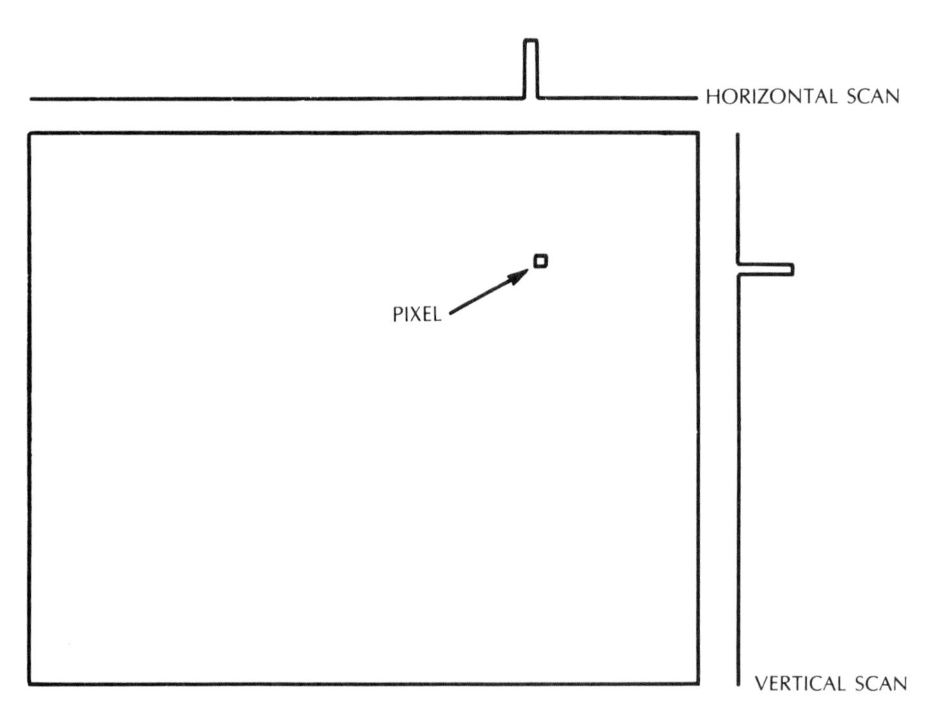

Fig. 16-10. How the new "flat-screen" solid-state display is controlled.

To build a complete picture, each scan will integrate the brilliance of the sweep, as it covers that scan. The receiver then combines the information so that each element gets the illumination determined by the combined information of the two sweeps. With modern digital technology, it will easily be possible to far exceed present picture quality, first in black and white and then, by transmitting color information as well, in color.

When this has been developed and become operational, the old cathode ray tube TV will seem as archaic as the Nipkow disc, used for the earliest transmissions before the cathode ray tube was perfected, seem now. It will, of course, revolutionize video as we know it today, based on the system that has been used for several decades to transmit television pictures.

Examination Questions

1. For 525-line television, using CR tube video, the theoretical bandwidth needed extends to over 10 megahertz, based on a frame frequency of 30 per second. Which way will definition or resolution be reduced by limiting the bandwidth to 5 megahertz?
2. Changing to a pixel type display, what bandwidth is needed to allow similar fidelity to that requiring 10 megahertz for CR tube video, assuming color is cared for by making 3 scans every frame, at the rate of 30 per second.
3. What bandwidth would be needed to raise the frame frequency to 60 per second, in question 2?
4. Assuming the same channels previously used for CR tube video are converted to pixel transmissions, requiring 3 scans per frame, and 60 frames per second, what will be the resolution vertically and horizontally? (Assume this allows a 5 megahertz bandwidth).

17
Instrumentation

Chapters 10 and 11 introduced the concepts on which the earlier methods of measurement were based. But today's instrumentation has become far more sophisticated since those days. In those days, a distinction was made between direct and indirect forms of measurement.

Direct measurement, in those days, referred to the application of a meter, such as a voltmeter, on which the voltage was read off by direct indication, as introduced in Chapter 3. Indirect measurement required the use of a bridge (Fig. 3-7) or a potentiometer (Fig. 3-13). It was called "indirect" because the person making the measurement had to adjust the instrument until a balance, or "null" was obtained, after which the desired quantity was read off the dials of the instrument, or the setting of the calibrated components.

In those days too, precise measurement of voltage involved the use of a "standard cell" of some sort, whose voltage could be relied upon as a reference. However, such standard cells were not as reliable as could be hoped, because they only held their value if no current was taken from them.

The precise manufacture of semiconductors has led to a whole new set of "standards," which are more reliable in everyday use, if not so reliable in the "basic" sense. Most significant of these is the Zener diode, which when manufactured to close tolerances, so as to be temperature and current independent, do not lose their reference value as readily as the old "standard cell" could. And they can be built into instruments, so the voltage appears whenever the instrument is switched on.

When a Zener diode is used as a reference, it takes the form shown at Fig. 17-1. If the "standing current" supplied through the resistance is 1 milliamp, then very nearly 1 milliamp can be drawn from the composite reference, before it "quits." Current can be passed "into" the reference until the Zener's dissipation is in danger.

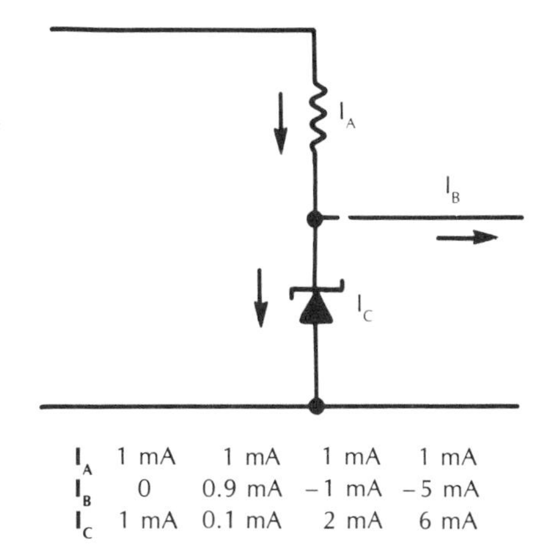

I_A	1 mA	1 mA	1 mA	1 mA
I_B	0	0.9 mA	−1 mA	−5 mA
I_C	1 mA	0.1 mA	2 mA	6 mA

Fig. 17-1. Combinations of current in a simple zener diode circuit.

However, modern voltage measuring equipment takes very little current, even while it is adjusting for balance. And when balance has been obtained, no current flows to or from the reference. The balance detector takes the form of an "op amp" (abbreviation for operational amplifier), a much-used device in modern electronics. An op amp has two identical inputs which, if they are FETs, take no current, but merely compare voltages (Fig. 17-2).

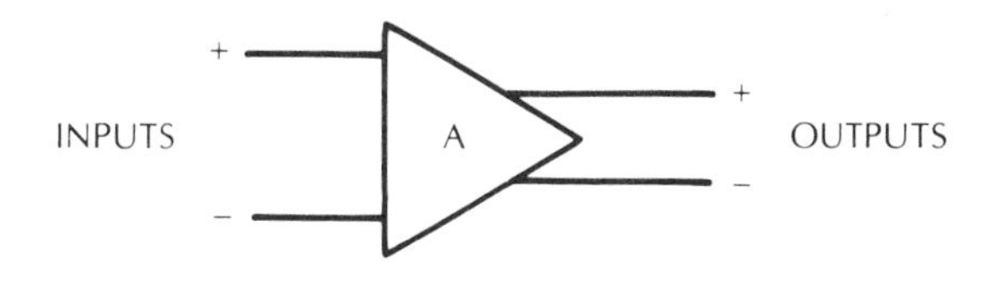

Fig. 17-2. Basic symbol for integrated circuit "op amp" (operational amplifier).

An op amp is the linear equivalent of a flip-flop, with a tremendous amount of gain. Like the flip-flop (in its basic configuration) it has two inputs and two outputs. Only when the two input voltages are precisely equal, are the two output voltages precisely equal. Any imbalance between the input voltages is amplified by a very great factor, at the outputs.

If, for some reason, an op amp is required to have a precise amount of gain—say 1000 times—this can be achieved by using negative feedback, a device first developed for use with audio amplifiers. In fact, op amps are now used in audio amplifiers, with negative feedback to control the gain quite precisely.

FUNCTIONS OF NEGATIVE FEEDBACK

Understanding what negative feedback can do requires careful thought, with a little mathematics. First assume that a small unit of input, say 1 microvolt, is applied to one input, and nothing to the other input. If the gain is a million, the output will be a million microvolts, or 1 volt. Let us call the gain A. If 1 microvolt is also applied, in the same polarity, to the other input, the output will be zero.

Now, if a resistor network is arranged to apply precisely one thousandth of the output back to the negative input, in the same phase as the original input, a 1 microvolt *difference* between inputs will still produce 1 volt output, which will produce 1 millivolt (1000 microvolts) at the second input. So now it will require 1001 microvolts at the first input to produce the 1 volt output (Fig. 17-3).

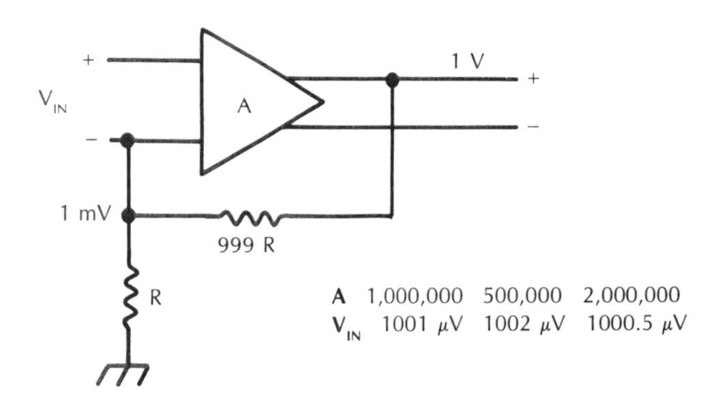

Fig. 17-3. Use of resistors to stabilize gain of an op amp used as a straight, linear amplifier.

When we say that the uncontrolled gain of an op amp is a million, that is a "rough" figure: it is not precisely a million, and could vary over a fairly wide range. Let us suppose it could vary from 500,000 to 2,000,000.

With a gain of 500,000, it would take an input difference of 2 microvolts to produce a 1 volt output. With a gain of 2,000,000, it would take an input difference of 0.5 microvolt to produce a 1 volt output. Both would produce a 1 millivolt signal to the second input. This means that, with such feedback applied, when the uncontrolled gain is 500,000, it would take 1002 microvolts to give 1 volt output, and when the uncontrolled gain is 2,000,000, it would take 1000.5 microvolts to give 1 volt output.

Without feedback, the change in necessary input, for the same output, would be 4:1. But with feedback, the change is only 1.5 microvolts in 1 millivolt, or slightly over 0.15% change. For all intents and purposes, the variation has been eliminated. Now let's write that in algebraic form.

We have used the symbol A for uncontrolled gain, which varies, in the example we have used, from 500,000 to 2,000,000. We use the symbol B for feedback fraction, which is 1/1000, or 0.001. So the feedback signal is A times

B, times the uncontrolled input necessary to produce that output. And that uncontrolled input is necessary, over and above the feed back input, to get the output. So we can write, using A_f as the symbol for gain with feedback:

$$A_f = \frac{A}{1 + AB}$$

Now, in the example we just used, AB varies from 1000.5 to 1002. So no great error is made, if we neglect the "1", in which case we can write

$$A_f \simeq \frac{A}{AB} \simeq \frac{1}{B}$$

This will be correct within a small fraction of 1%, in the case we just used.

Negative feedback was introduced first in analog audio amplification, for a variety of purposes that still serve usefully. As the foregoing formula shows, it has the effect of stabilizing gain. In discussing the effects of feedback, two quantities are important. First is the "feedback fraction," which we gave the symbol "B" and which in textbooks on that subject is usually given the Greek letter beta (β) as a symbol.

When large amounts of feedback are used, the feedback fraction controls the amount of gain with feedback.

The other important quantity is the "feedback factor," written as 1 + AB in the foregoing equations. This is also what is usually meant when the expression "amount of feedback" is used. Unfortunately, both its technical name and the colloquial expression tend to obscure the fact that it is not a constant, but a variable. If it were not variable, in some way or other, there would be no need of it.

This is first seen in the fact that if gain were already fixed and absolutely stable, there would be no need for feedback to stabilize it. But the same is true of its other uses.

In a so-called "linear amplifier," without feedback, the gain can vary over different portions of the transfer characteristic, as shown in Fig. 17-4. Feedback reduces this variation, just as it does variation in gain due to any other cause. So it has the effect of reducing distortion, or non-linearity, due to the fact that the transfer characteristic is curved. And it reduces such effects by the feedback factor.

The problem with this statement, often overlooked, is that the feedback factor is not a constant, although it is often stated as if it is. A particularly forceful example of this error occurs when a "linear" device is operated beyond the end of its linear characteristic, that is, into saturation, cut-off, or both. In either of these regions "A" rapidly disappears altogether, making the feedback factor collapse to zero. Thus feedback can remove, or reduce distortion, only where there is an effective feedback factor.

Figure 17-4 illustrates this with a hypothetical transfer curve and the effect of feedback on the waveform. Slight curvatures are greatly improved, but once the forward gain "drops dead" the distortion is more sudden.

Another thing feedback can do is to modify impedances. This can best be seen as improving regulation, although the property may be used for other reasons. Suppose that, without feedback, a linear amplification device with 1 millivolt input produces 10 volts out, without any load connected (Fig. 17-5).

Now, suppose connecting a 500-ohm load to the same output causes the output to drop to 8 volts. This can be explained, using Thevenin's theorem (Chapter 3) as being due to the device having an internal resistance, or impedance, of 125 ohms. This is arrived at by assuming the "open circuit" voltage is still 10, and that 2 volts must be dropped in the internal resistance or impedance, to leave 8 volts across the 500-ohm load.

Next, suppose we apply feedback that, without the load connected, would reduce gain by a factor of 10. This means 10 millivolts would now be needed to get the 10 volt output. The internal input is still 1 millivolt, but an "AB" of 9 reduces the gain so the external input needs to be 10 millivolts. Now, with the load connected, 1 millivolt internal input will still produce 8 volts output, reducing the AB feedback to ⅘ of 9, or 7.2 millivolts, instead of the former 9 (without the load).

But when feedback is applied, we don't have access to the internal input: that is purely a theoretical figure. So 8.2 millivolts in would produce 8 volts out. But we would maintain the 10 millivolts in, as the load was connected, so by proportion, since the amplifier is linear, the output would be $10/8.2 \times 8$, or 9.756 volts. This leaves 0.244 as the voltage dropped in the modified "internal impedance," which now figures out to $0.244/9.756 \times 500$, or 12.5 ohms.

The feedback by a factor of 10 ($1 + AB = 10$), reduces the internal impedance by the same factor: from 125 ohms to 12.5 ohms. Note that we started out by making the feedback factor 10, without the load connected. That is the factor by which the internal impedance is reduced. With the load connected, the feedback factor is 8.2, not 10, because it takes 8.2 millivolts in, with feedback, to produce the same output as 1 millivolt produces without feedback.

That result is arrived at, because the 500-ohm load is treated, in that analysis, as part of the external, not the internal circuit. Using an alternative method, which regards the 500-ohm load as part of the internal circuit, the parallel output impedance without feedback is 500 ohms in parallel with 125 ohms, which is 100 ohms. With feedback, this becomes 500 ohms in parallel with 12.5 ohms, which is 12.195 ohms. And a reduction from 100 ohms to 12.195 ohms is a ratio of 8.2, the feedback factor with load connected.

Either way of calculating produces the correct result, provided you maintain consistent references.

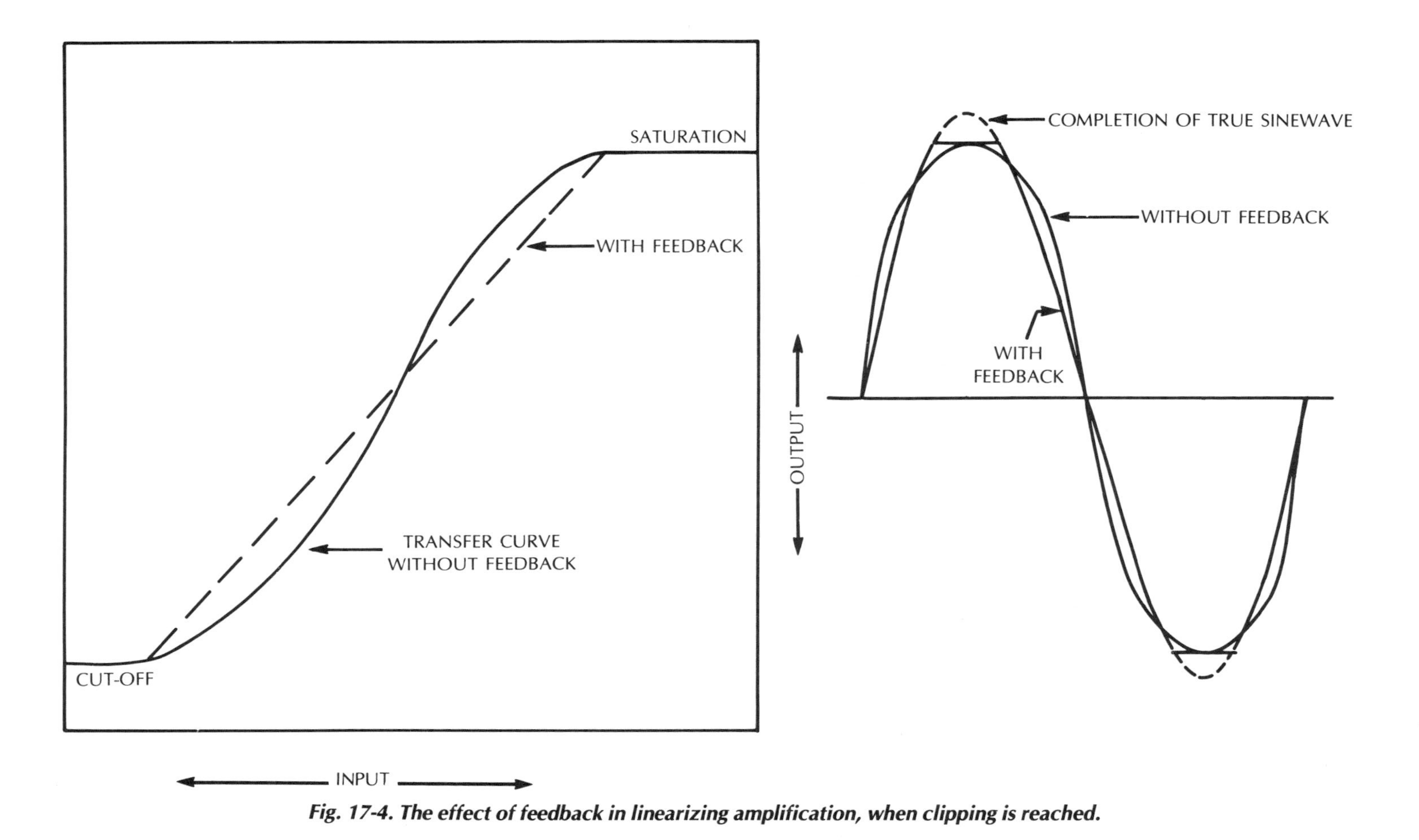

Fig. 17-4. The effect of feedback in linearizing amplification, when clipping is reached.

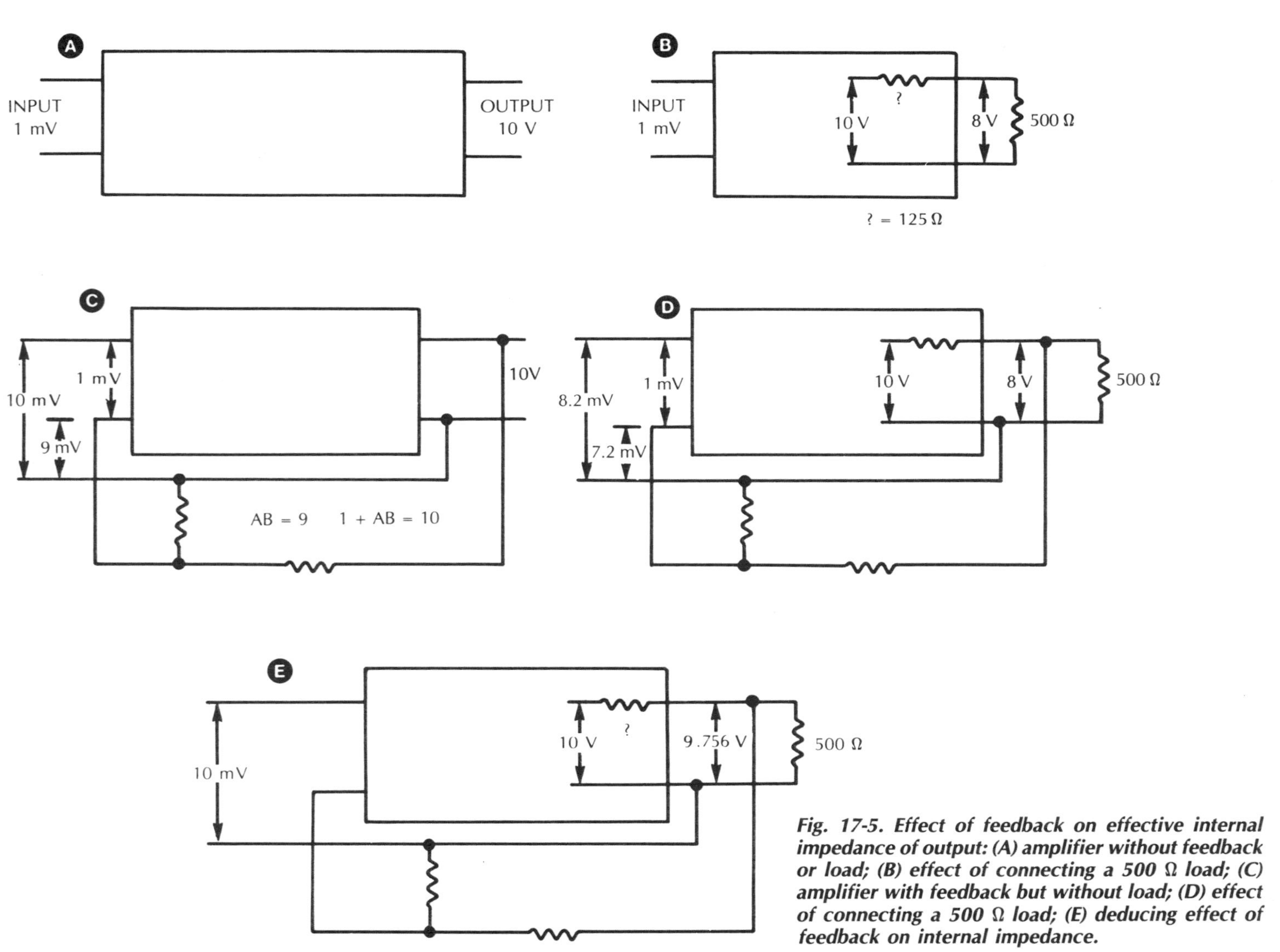

Fig. 17-5. Effect of feedback on effective internal impedance of output: (A) amplifier without feedback or load; (B) effect of connecting a 500 Ω load; (C) amplifier with feedback but without load; (D) effect of connecting a 500 Ω load; (E) deducing effect of feedback on internal impedance.

Feedback can have other effects most of which have now become somewhat redundant, but you can pursue that in books on the subject, if your area of electronics requires it. This chapter is primarily concerned with instrumentation, and the important use of feedback in this connection is to control gain to a very precise degree, so that output of an instrument amplifier is a precise multiple of its input.

DIGITAL INSTRUMENTATION

In earlier times, the essential part of a good instrument, whatever it was to measure, was an accurate indicating meter, usually of the moving coil type (Fig. 7-23, 8-10). Such an instrument, to indicate within 1% or 2% accuracy, required extremely critical design and manufacture, as well as calibration. That was quite apart from the care needed for a person to read such an instrument to that degree of accuracy.

The advent of digital read-out instrumentation resulted in far greater accuracy less expensively, and with far greater ease of reading. The essential parts of a digital instrument are a standard of reference, which is built in, and a counter that provides the figures you read (Fig. 17-6).

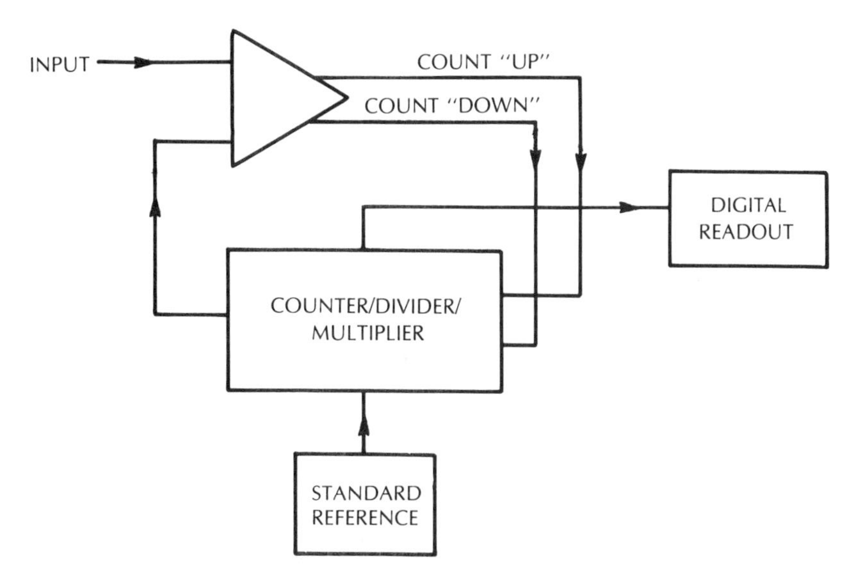

Fig. 17-6. Basic system for producing an instrument with digital readout.

There are two differences between the type of counter used in Fig. 17-6 and that shown in Fig. 14-21. That counter was only for "internal" purposes: the number it counts to, before resetting, is built in, and nobody outside needs to know what it is. There are either 6 or 12 of them to the package, to cover the octave of source frequencies, in either two packages or one package. But counters used for instrumentation must be arranged to give readouts, and to be started and stopped by external signals.

They must also be arranged in decade blocks, instead of straight binary. This means that every block of 4 binary elements makes up a decade element (Fig. 17-7).

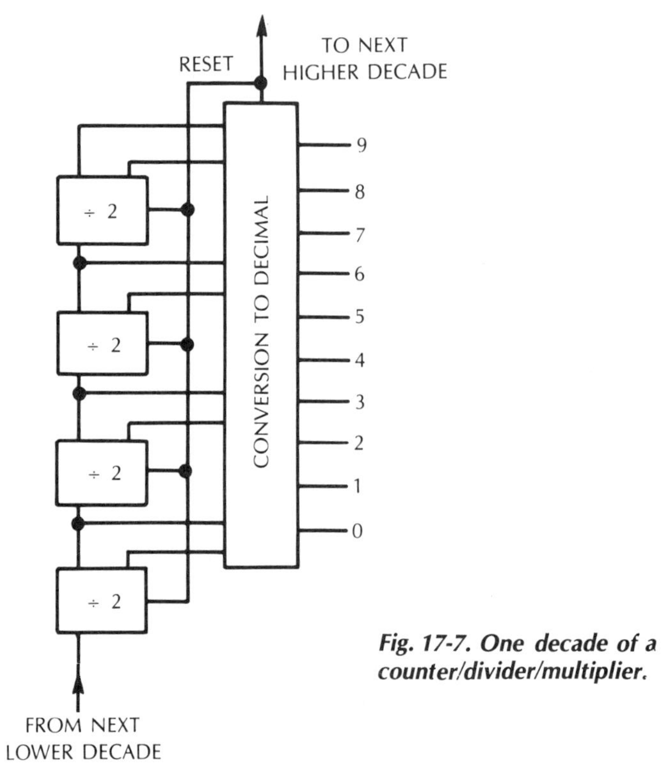

Fig. 17-7. One decade of a counter/divider/multiplier.

Theoretically a variety of binary-to-decimal conversions is possible, but two main ones have been used: the straight binary, and the biquinary (Fig. 17-8). With each of them, only 10 of 16 possible binary combinations in the 4 elements are used so, if some other combination appears, it represents an error in the system.

STRAIGHT BINARY	DECIMAL	BIQUINARY
8421		5421
0000	0	0000
0001	1	0001
0010	2	0010
0011	3	0011
0100	4	0100
0101	5	1000
0110	6	1001
0111	7	1010
1000	8	1011
1001	9	1100

Fig. 17-8. Two systems of conversion between binary and decimal.

Whichever system is used, when the count reaches the binary equivalent of 9, the next 1 resets that decade to zero, and passes an input pulse to the next higher decade.

Important elements in all such systems are conversion matrices. If the device is a calculator, its input comes from a decimal type keyboard, which needs converting to binary sets of impulses, for handling in the calculator itself. A calculator cannot handle decimal numbers directly. But the binary elements it uses are in decade blocks, so transfers can easily be made, without counting the entire number, pulse by pulse.

So, when a decimal number is "punched in" on the keyboard, the first number entered goes into the units decade. Then, when another number is entered, the first number is transferred to the tens decade, and so on, up the scale, until all the decades provided in the instrument are occupied. Now, if another number is punched in, when there are no more decades to take the first number punched in, the computer generates an "error" signal, to let the operator know that the computer cannot handle it. Only by clearing the error signal, can the operator start over.

The next matrix, or set of matrices, converts the binary information in the counter back to decimal, at each place. It takes the 4 outputs from each decade of binary, and converts it to a decimal output, which requires a separate "connection" for each number in the decade: ten of them. If this is to become a readout, then it goes to another matrix which converts the same information to a 7-element output that will cause the appropriate figure to appear (Fig. 17-9).

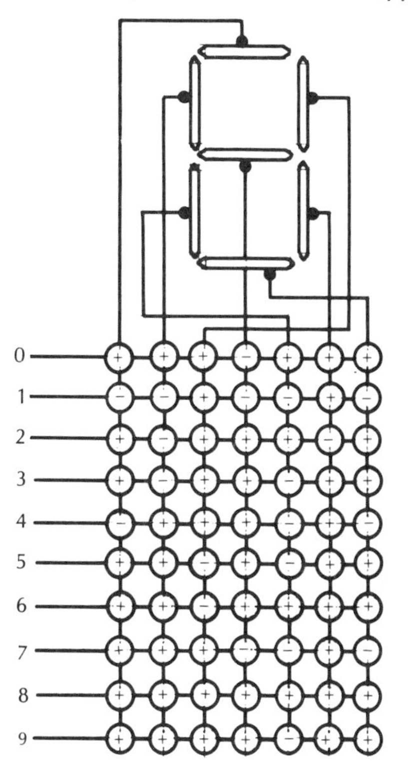

Fig. 17-9. A matrix for producing a numerical readout from a decimal output (e.g., as in Fig. 17-8).

In the alternative, if a printout is required, instead of a display, different conversion may be used. If it uses a mechanical typewriter type element, the decade outputs will be used to activate the equivalent of typewriter keys. Or if an electronic printing system is used some other appropriate matrix generates the signal required to print the required characters that way.

So, to make a measurement, of whatever, the original quantity is converted to voltage (in the case of a voltmeter, it starts as a voltage, of course), which is fed into the input of a comparator, which will be an op amp with whatever protection is necessary against possible overload. The counter then starts counting units from the standard reference, until a balance is achieved.

This is then held by signals delivered to the counter. If the counter voltage is lower than that to be measured, the counter receives an instruction to count up. If the voltage is higher than that to be measured, the counter receives an instruction to count down. As you may observe, sometimes a readout flickers between two adjacent digits in the last place, because the actual value is between the two.

FREQUENCY MEASUREMENT

This is one measurement that has always been most accurately made by digital counting. The actual cycles are counted for a specific time, not more than a second, unless a very low frequency is being counted. Often a tenth or a hundredth of a second may be used. When the count is completed, stopped by a signal from the time measurement, the instrument reads out the result, and the counter and time measurement are restarted.

OTHER MEASUREMENTS

Almost anything—any quantity—can be converted into a voltage, which is then measured, and the readout expressed in terms of the quantity being measured. But perhaps the most useful conversion is a stress gauge, which can be applied to many other things, for instance, weighing. Stress gauges that produce a voltage corresponding to stress were not very reliable, or accurate, until the advent of the FET type op amps.

When a stress gauge had to operate a moving coil type meter, although the most sensitive type meters were used,they still took a few microamps from the stress gauge, which is basically a voltage output device, any loading of which reduces its accuracy. Though loading can be compensated for in the calibration, recovery and other factors reduce its accuracy, making it difficult or unreliable to read.

An FET input takes no current, but is completely voltage operated, so the stress gauge becomes an extremely accurate device when used with an FET type op amp as an input to a comparator, followed by a counter.

But electronic measurements, using digital processing have many more advantages. For example, using the old moving coil instruments there was the problem of what to measure on ac, especially if the waveform was not sinusoidal.

For a sinusoidal waveform, the rms value could be obtained by actually measuring the rectified average, and applying the "form factor" as a correction.

But when such an instrument was used on a non-sinusoidal waveform, the reading obtained was meaningless, unless you described it as "as taken with a nominal rms meter based on rectified average!" With electronic digital metering, squaring and square root are functions that can be programmed into the microprocessor as a computer. This means that the digital information obtained from the counter can be continuously squared, and averaged at any desired rate, such as every tenth of a second, or whatever suits the measurement best, and then the square root extracted for the final readout.

Another problem frequently encountered with the indicator type instrument was one of interpretation. An example of this is the VU meter—used for telling the level of an audio recording or transmission. The level is constantly changing, and it has peaks and an average value, even at a given instant, over a few cycles. Digital solves this by making it possible to preserve and read absolute peak values, resetting every so often, or to read an average value, also taken over any desired interval.

Further than this, sometimes the actual readings are not the most important thing, but some relationship between different readings, or the same reading taken at different times. A microprocessor, of the type used for making electronic instruments can do anything an application may require.

Examination Questions

1. At different points along the transfer characteristic of a device to be used for a linear amplifier, the gain figures are: 5, 80, 120, 180, 2. Using a feedback fraction of 0.075, calculate the gain with feedback at each point on the characteristic, with feedback connected.
2. Using a feedback fraction of 0.015, calculate the gain at the same points as in Question 1.
3. The nominal load for the device of Question 1 is "Z". The measured source resistance at the same points on the curve is: 30Z, 6Z, 5Z, 4Z and 0.1Z. Calculate the source resistance with feedback at the same points, using the same feedback fraction as Question 1.
4. Using the feedback fraction of 0.015, calculate the source resistance values at the same points again.

18
System Design

If you will refer again to Fig. 17-6, you will notice that we have developed a way of representing whole functions by "boxes" on which we write what they do. Each of these boxes, in itself, is a system. The whole schematic, made up of many such boxes, is also a system. *System design* refers to the method of thinking, of working ideas out, that breaks down the overall problem in this way.

System design, at its outset, consists of thinking through all the component parts of what has to be done, and then designing a system to do that, with subsystems to do each part. If that were all it was, it would be easy. But as any design engineer knows, sometimes you put all the parts together, and it doesn't work, because the whole system occupies itself doing something its designer did not anticipate.

The answer to this, is to approach systems design by considering all the possible things a system, subsystem, or set of systems, could possibly do. In other words, we have to learn to consider what could possibly "go wrong." Perhaps this can best be shown by a few historical examples. We will choose them to illustrate the principal causes of such unanticipated results.

In the last chapter we introduced feedback as a concept, which is very widely used, and is much simpler in its application than it seemed when it was first introduced. But in systems design the concept has another importance, because it constitutes the difference between "open" and "closed" systems, which we shall come to later.

CONDITIONS OF STABILITY

The simple concept of feedback was illustrated by the algebra introduced in Chapter 17. Gain is modified, and other properties similarly changed, by the "amount" of the feedback. This oversimplifies the situation.

When amplifier designers first applied negative feedback to improve amplifier performance, they got a surprise. In those days, individual tubes did not have much gain, perhaps between 10 and 20. So to get a gain of, say 10,000 or more, required several tube stages coupled together.

The tube amplifier produces a signal at the plate which is the reverse of that causing it at its grid. So 2 stages produce an output in phase with input, 3 produce reversal again, and so on. Figuring out how to get "negative" feedback was not too difficult: just a matter of counting stages, and possibly finding a way to reverse the signal if necessary.

Varying the amount of feedback applied could easily be achieved by changing values in the feedback resistor network. The theory at that point was that, if we can apply an amount of feedback equivalent to about 10 times the internal input, the performance of the amplifier would be improved 10 times. Distortion would be reduced to one tenth, and so on.

But the experimenters found that they could not apply a feedback bigger than the internal input, before the amplifier went into oscillation. Had they made a mistake in figuring whether the feedback was positive or negative? They knew that positive feedback could cause oscillation, because it is a way whereby the amplifier provides its own input. However, checking and double checking showed that the feedback should be negative.

How did it oscillate? It did one of two things. Either it squealed at a very high frequency, perhaps even too high to be audible, or else it made a sound like "motor-boating"—an oscillation at very low frequency. Apparently what calculated out to be negative at "ordinary" frequencies, turned around to become positive feedback at these extreme frequencies, very high or very low.

The algebra doesn't "tell it all." Coupling capacitors produce a phase shift in the signal at the low frequency end, as do transformer inductances, resulting in an advance in phase at the low frequencies, and a delay at the high frequencies. Any one coupling capacitor or transformer inductance for the low frequency end produces an ultimate phase shift of 90°, when amplitude has dropped to zero (Fig. 18-1). And the self-capacitance of circuits results in a high frequency loss, with phase shift, in the reverse direction, where any one "bunch" of self-capacitance reduces output to zero when phase shift reaches 90°.

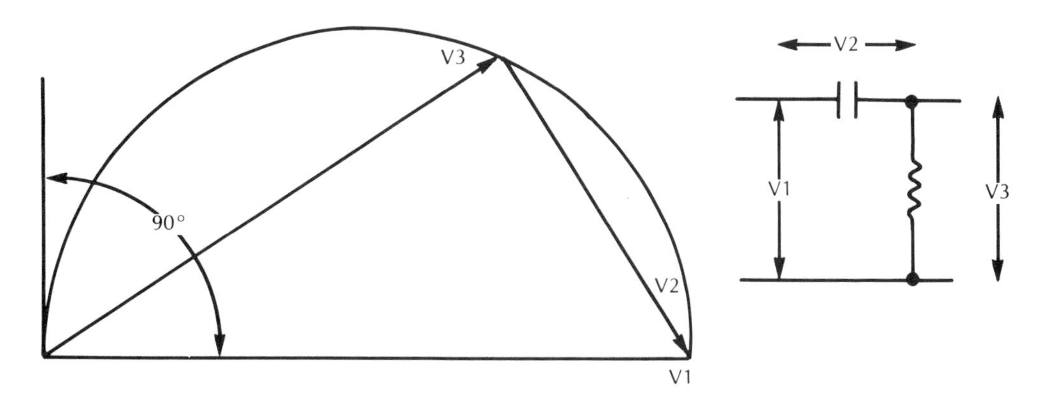

Fig. 18-1. Vector diagram for single stage coupling at low frequencies, which yields a semicircular locus.

So in a single stage, or with a single coupling element, phase shift can never get big enough to reverse the direction of feedback. But two stages, or two reactances contributing to phase shift in the same direction, produce an ultimate phase shift of 180°, which is reversal, but only when transfer, or gain, has dropped to zero (Fig. 18-2).

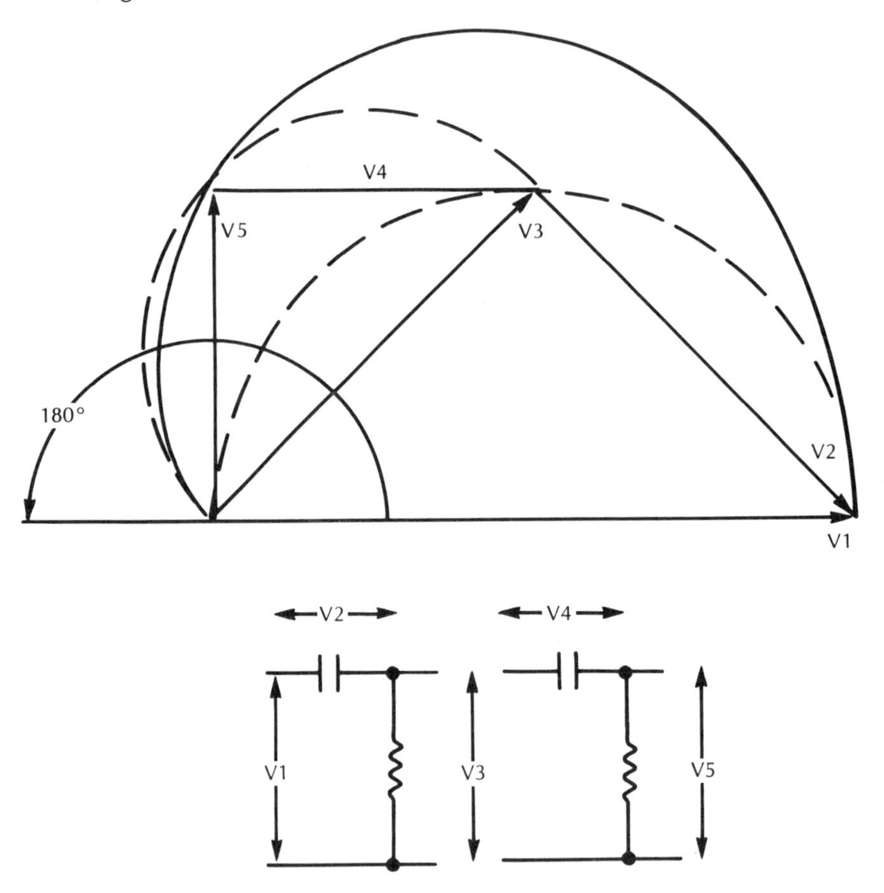

Fig. 18-2. Vector diagram for one frequency, and developed locus curve for two identical stages at low frequencies.

However, although two stages cannot reach the point of oscillation, they can result in a response that peaks, or shows an increase in gain, before it falls off, at one end of the frequency response (Fig. 18-3).

Those early tube amplifiers had three or more stages, and sometimes as many as five or more. If they had identical cut-off frequencies at one end of the response or the other, very little feedback was necessary to cause oscillation. A 3-stage system, with identical cut-offs, reaches 180° phase shift when each of them contributes 60°, which corresponds with a drop in amplitude to one half, or a combined effect of one eighth (Fig. 18-4A).

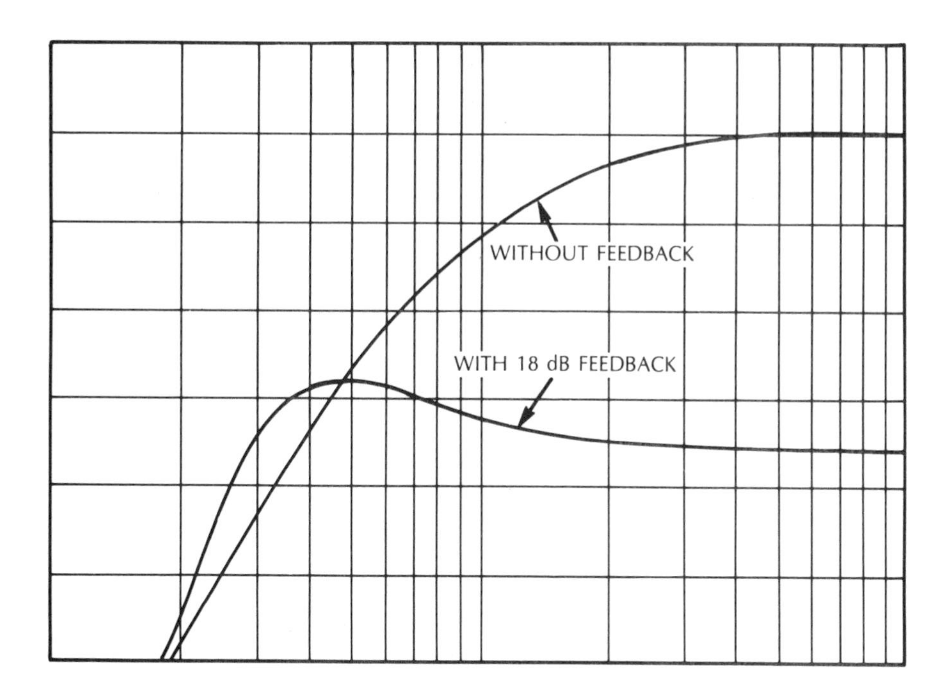

Fig. 18-3. Effect of feedback on two-stage system with identical low-frequency cut-offs.

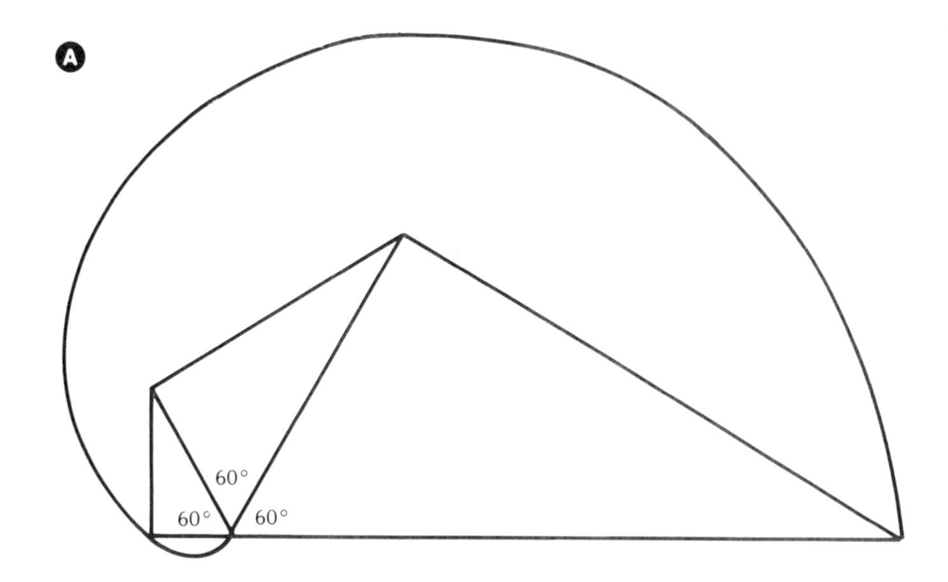

Fig. 18-4A. Making a 3-stage system work. With identical low-frequency cut-offs the amount of feedback is seriously limited (18 dB will produce oscillation).

This means the amplifier will oscillate when AB reaches 8, a feedback of less than one tenth gain reduction. A 5-stage system, which was common in those days, makes the 180° change when each contributes 36°, which represents an amplitude drop of only 0.8 multiplying out to 0.35 (approximately) for 5 stages. So such a system will oscillate before AB reaches 3. This was a serious restriction on the use of negative feedback.

In those days, a solution to that was to "tailor" the cut-offs so they did not all occur at the same frequency. If one cut-off became effective, before the remaining stages, then amplitude or gain would be reduced much more, before the others took effect, and phase reversal would occur when gain had dropped considerably lower than when identical cut-offs were used (Fig. 18-4B).

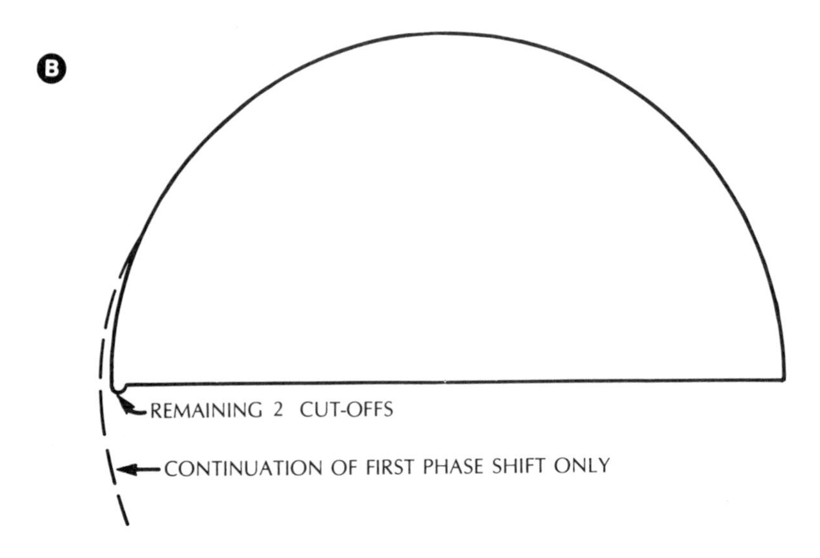

Fig. 18-4B. By staggering cut-offs, having one act much earlier than the other two, much more gain can be achieved.

The kind of diagrams shown in Figs. 18-1 through 18-4 are called Nyquist diagrams, named after the man who first used them to predict the stability or performance of a system. While they were primarily developed for application to amplifiers, the principle can be applied to systems much more widely. Nyquist distinguished 3 kinds of system, based on the form the diagram took (Fig. 18-5).

If the curve "zeroed in" without encompassing the phase reversal point at all, the system was inherently stable: it would never oscillate. It might peak before roll-off, but it would not oscillate. If the curve went beyond 180° but, by the skillful use of extra reactances in the circuit, was brought back, so the curve did not in fact encompass the "–1" point, it was "conditionally stable."

This meant it would not oscillate unless something "triggered" it, but once oscillation started, it would not quit. This was a not uncommon condition in those days. Experimentation would find a fairly complicated circuit that seemed stable, until something "upset" it, when it would go into oscillation and the only way to stop it was to switch it off.

If the curve enclosed the "−1" point, the system would oscillate under all circumstances, and was called "unstable." As soon as it was switched on, oscillations would build up till saturation limited the oscillation.

That explains the various conditions for stability and instability, in terms of geometry or algebra. The algebra involves operator "j" introduced in Chapter 10. But even that doesn't tell it all. It is all based on the assumption that the quantities "A" and "B" are both constant, except for their phase shift elements, which add "j" components that are frequency dependent. It does not take account of changes that occur due to variations in the magnitude of quantities such as "A" over the excursion produced by the signal within the system.

We have already covered the fact that A can virtually disappear. But even fluctuations in the value of A present terms in the algebra that are not easily introduced. For example, the same algebra can be applied to show how negative feedback affects distortion, perhaps using the symbol "D" for the distortion component. Now distortion contains a different frequency, or frequencies, from those present in A and B, a fact that the algebra does not show.

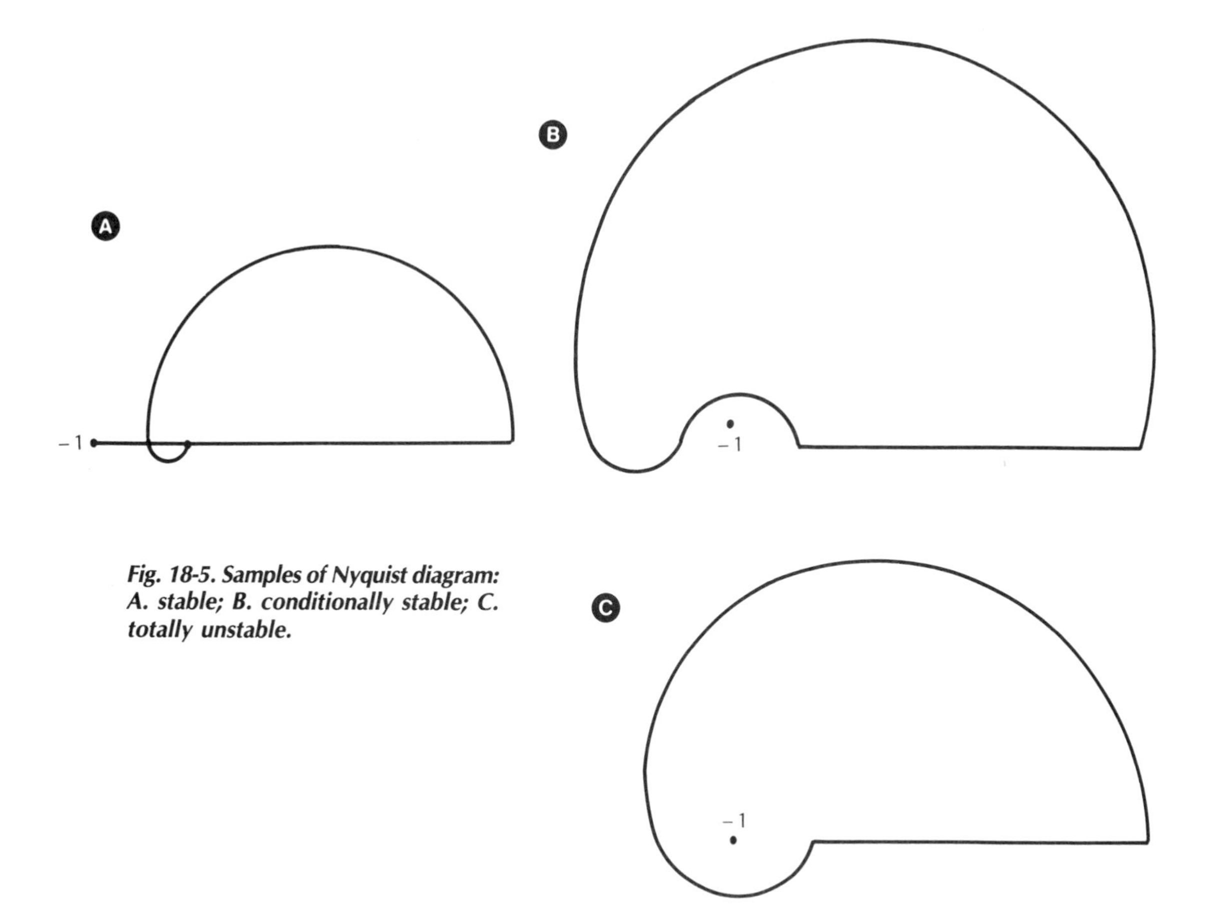

Fig. 18-5. Samples of Nyquist diagram: A. stable; B. conditionally stable; C. totally unstable.

The algebra will produce an expression containing terms in "D" multiplied by various factors, some positive, some negative and, by making the positive and negative coefficients of D equal, one would assume that distortion then disappears, by being reduced to zero. What this assumption neglects, is that the coefficients of A and B relate to the original signal frequency, while the coefficients of D relate to other frequencies, usually derivative of the same coefficients as applied to A and B. And not necessarily the same derivative.

Thus it should be evident that 4th harmonic cannot cancel 2nd harmonic, for example. Of course the algebra could introduce subscripts to indicate such differences in order, and thus keep the terms separate rather than combining them wrongly.

Such unaccounted for elements are related to time, in that they occur at different points on a "signal" waveform and are thus separated in time, just as effects due to changing frequency, represented in the Nyquist diagram, are related to time. But the latter follow the unique relationship between time and frequency, while the former will do virtually the same, whatever the frequency of the waveform.

The point to understand here, is that many factors contribute to a total consideration of conditions of stability. System design must not omit any possible consideration that could affect the performance of the system. Perhaps one tragedy of recent years will illustrate the need for this.

When the author came to America, he had been associated with system design for electrical distribution systems, many years earlier. New York Con Edison people invited him to review the East Coast distribution system. First thing he noticed was that the system had no protection against high frequency surges moving around the system, so he asked what protection there was, and what could prevent a wide scale "black out" perhaps covering several states by trigger effect.

He was told it couldn't happen, because the system distributed electricity at a low frequency, 60 Hz, so radio frequencies had no business on the lines. He pointed out that all that was necessary was the installation of "king size" rf chokes at strategic points in the system—not very costly, but a lot safer. But the Con Edison officials insisted that was unnecessary, because they were not distributing rf! The East Coast blackout is now history, and could have been avoided, had its designers had this much background in systems design.

OPEN AND CLOSED SYSTEMS

Again, in the early days, the distinction was clear. An amplifier without feedback was an open system, and one with feedback was a closed system. The concept is that signal can go from input to output of an amplifier without feedback, but when feedback is added, the output is also connected to the input. Connecting the feedback closes the system. But as we get into more advanced systems, the distinction needs more careful attention.

Take a single, automated power station as an example. Assume it is fired by some kind of fuel-fired boiler, which drives turbines which, in turn drive alternators. The speed and output, both voltage and current, are monitored to control the fuel feed to the boiler, and the steam feed to the turbines is also controlled by the system. At this point, it would appear that the system is closed: several different kinds of feedback are used to maintain correct voltage and frequency, and to balance the current demand.

But now, put that automated power station on line with other automated power stations, and we have a new situation. From the viewpoint of this new, larger system, each of the individual automated stations is an open system. Under such circumstances, it was early found that stable operation can only be reliably achieved by making one station responsible for frequency, another for voltage, and each of the others, if there are more than two, for load distribution between them.

Now, the station responsible for frequency control must reference its output against a precision clock, while the other stations use that station's output as reference for their own speed, or frequency, which will also affect load distribution, at least as regards phase of load. The station responsible for voltage control must reference its voltage to a standard reference voltage, while other stations take their reference voltage from the station that controls the system for voltage.

Finally, load distribution must be coordinated between them, according to the distribution of the load's demand, so as to avoid producing circulating currents in the system, and concentrate on supplying the consumers. Now, having added all those controls to the overall system, over and above the simple closed systems in each station, that left individual stations looking like open systems, we again have a rather complex but finally a closed system, that we will not attempt to diagram here.

Let us take a simpler application. Suppose we want to design an effective room heating system, or perhaps a space with controlled temperature, which is the same problem with a more scientific application. The simplest arrangement would be an electric heater with a thermostat "suitably placed" to control the heater. Those words "suitably placed" are the problem: where do you put the thermostat?

First we might think of putting it on the other side of the room from the heater, so that it would only cut the heat off when the room was warmed all the way across. The only problem with that is, the intervening space may be too hot by then, so the far side might go on rising in temperature, after the heater has been cut off. Then the side where the heater is might drop considerably below the "cut in" temperature, before the thermostat "feels it."

Fairly obviously, such an arrangement would vacillate in temperature more widely than the thermostat, even after several times of cycling. In fact, it might even cycle over a wider range, after the system had been on for a while. This is an unstable closed system.

Putting the thermostat closer to the heater would probably cause it to cut out before the entire room had reached the control temperature. But it is also possible that such a position would gradually bring the whole room closer to the same temperature, although not so quickly as the first method, for the initial rise, which would keep the heater on longer.

If uniformity of temperature is the objective, a solution may be the installation of several heaters in different parts of the room, separately controlled, so heat can be delivered where it is most needed. Now, although each heater and thermostat makes up a closed system, the question is, do they interact? Each helps to create the environment in which the others function.

While the behavior of a single system may vary according to the ambient temperature of the day, a number of such systems in the same room creates a little different context for each. All of them will be cycling on and off, as determined each by its own control. The interaction could have various effects, according to the way in which units' being on affected convection currents in the room.

Some placements might tend to cause the units to come on and go off, almost together, with some having somewhat different on/off ratios. Other placements might tend to make one unit come on when others are off, and vice versa. Which is the more desirable? This is a question a system designer must face, and know how to go about resolving.

DESIGN APPROACHES

Whatever it is we are designing, many good inventions have been made by the "experimenters" approach. First figure out a way to do it, then make up a circuit, which probably won't work the first time, and change various circuit components until a set of values is found which make it work. The method is called "trial and error," or "hit and miss." And many professional development laboratories do very little differently.

The more satisfactory approach is to analyze each part of the circuit, and find the theoretical parameters on which it depends. Components parameters are listed in their spec sheets, issued by the manufacturers, but often the precise parameter a designer needs to know is not listed. In this case, it is well worth the time to devise a test circuit that will measure that parameter, in the form he wants it. We showed some examples of this in Chapter 14.

Such measurements should test as large as possible a batch of samples, to get average and variation details of the parameter in question. They should also test different aspects of the parameter, its limits, for each sample, and how its value varies with different circuit components.

A couple of changes to the bistable circuit of Fig. 14-14 produces what is known as a multivibrator circuit (Fig. 18-6). You may be using transistors of the general-purpose type. But you need to know something about them, before you complete your multivibrator design. First look at the theory of how such a multivibrator works.

When each transistor suddenly starts conducting, it pulls its collector from supply voltage, where it was when it was cut off, down close to zero. As the other transistor was conducting, hopefully saturated, its base was close to zero before this happened. But this sudden change switches the other transistor from saturation to cut off.

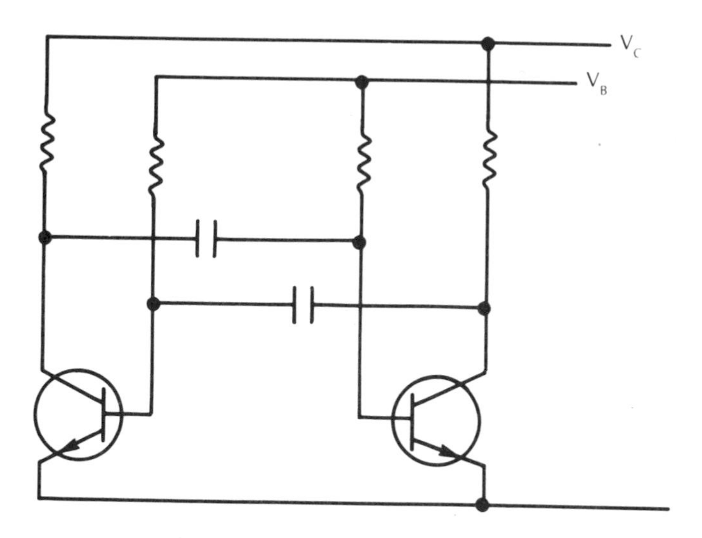

Fig. 18-6. A multivibrator with voltage control of frequency.

The rising collector voltage of the transistor just cut off, charges the other coupling capacitor, and this charging current further increases the base current of the transistor just switched on.

The waveform at either collector is a slightly modified square wave. The collector pulls the voltage down as a near-perfect square wave. But when that transistor cuts off, the return is not so perfect, rising with the time constant due to the coupling capacitor and collector resistor. As the other side of that capacitor is momentarily grounded through the other transistor base, that base resistance can be neglected. It is only important to see that the initial current, in this switching, does not exceed the transistor's momentary base current rating. At the first instant, the base current will be the same as the collector current to the first transistor had been a moment before.

Assuming the collector side of the capacitor reaches supply potential by the time switch-on occurs—it may not, this must be checked—when switching occurs the base of the transistor just cut off will go to a reverse potential equal but opposite to the supply potential. If that exceeds the safe reverse potential for the emitter-base junction of the transistor type being used, something must be done to change that.

First, however, see how changing the base supply voltage, V_B controls switching-time spacing, which in turn controls frequency (Fig. 18-7). When V_B is equal to V_C, switching occurs after approximately 0.7 t, where t is the time constant of the capacitor and base supply resistor. When V_B is, say ¼ of V_C, the point at which switching occurs, on the discharge curve, changes.

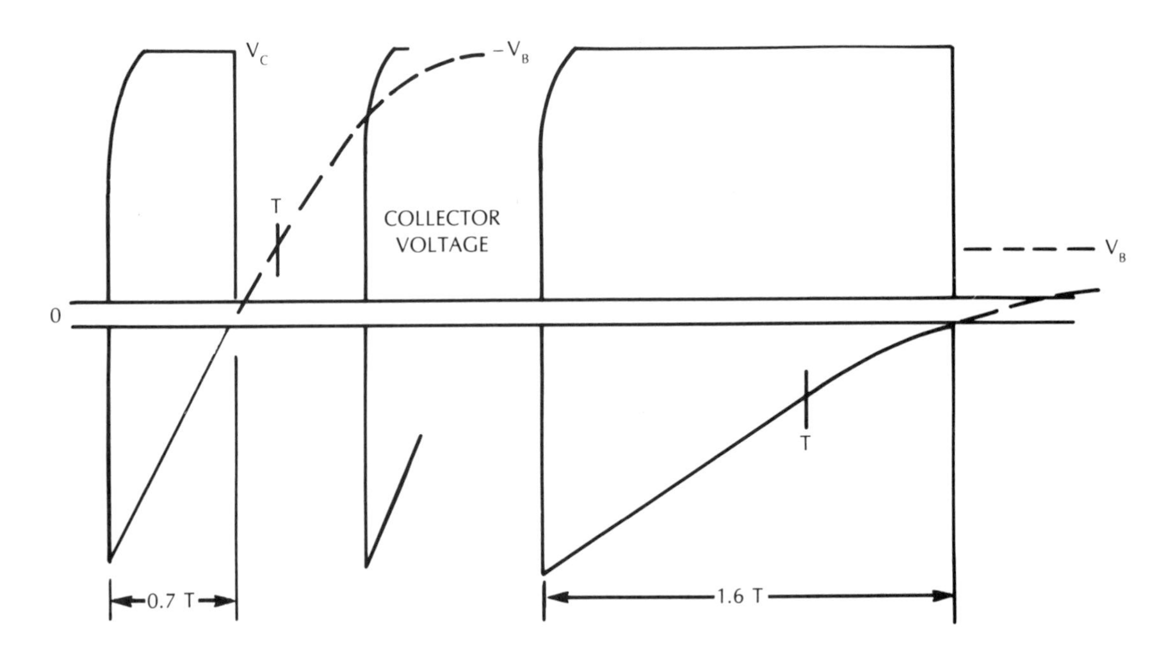

Fig. 18-7. How change of voltage on base supply (Fig. 18-6) changes frequency.

Where they are equal, the voltage across the base resistor at the beginning is double the voltage where switching occurs, so that happens when $\epsilon^t = 0.5$, which makes $t = -0.7$. That's where we got the 0.7 t. But when $V_B = ¼\ V_C$ the voltage across the base resistor at the beginning is 1.25 V_C, or 5 V_B. So switching will occur when $\epsilon^t = 0.2$, making $t = 1.6$. So the period will be nearly 2.3 times what it was with the higher value of V_B, or frequency will be reduced by that factor.

Note that, assuming the collector voltage rises to full collector supply voltage, the base goes to the same value of reverse voltage at switching. If this exceeds the rating of the transistor used, something must be done. Figure 18-8 shows two possibilities, which serve to illustrate the attendant design changes. At A, the collector is connected to a potentiometer tap point between resistors R1 and R2. C shows the attendant waveforms for one side of the multivibrator (assuming they are both the same). Both collector and base voltages behave as if a lower supply voltage were used, that at the junction of R1 and R2. Performance can be predicted by using Thevenin, which means the equivalent collector resistor is R1 and R2 in parallel.

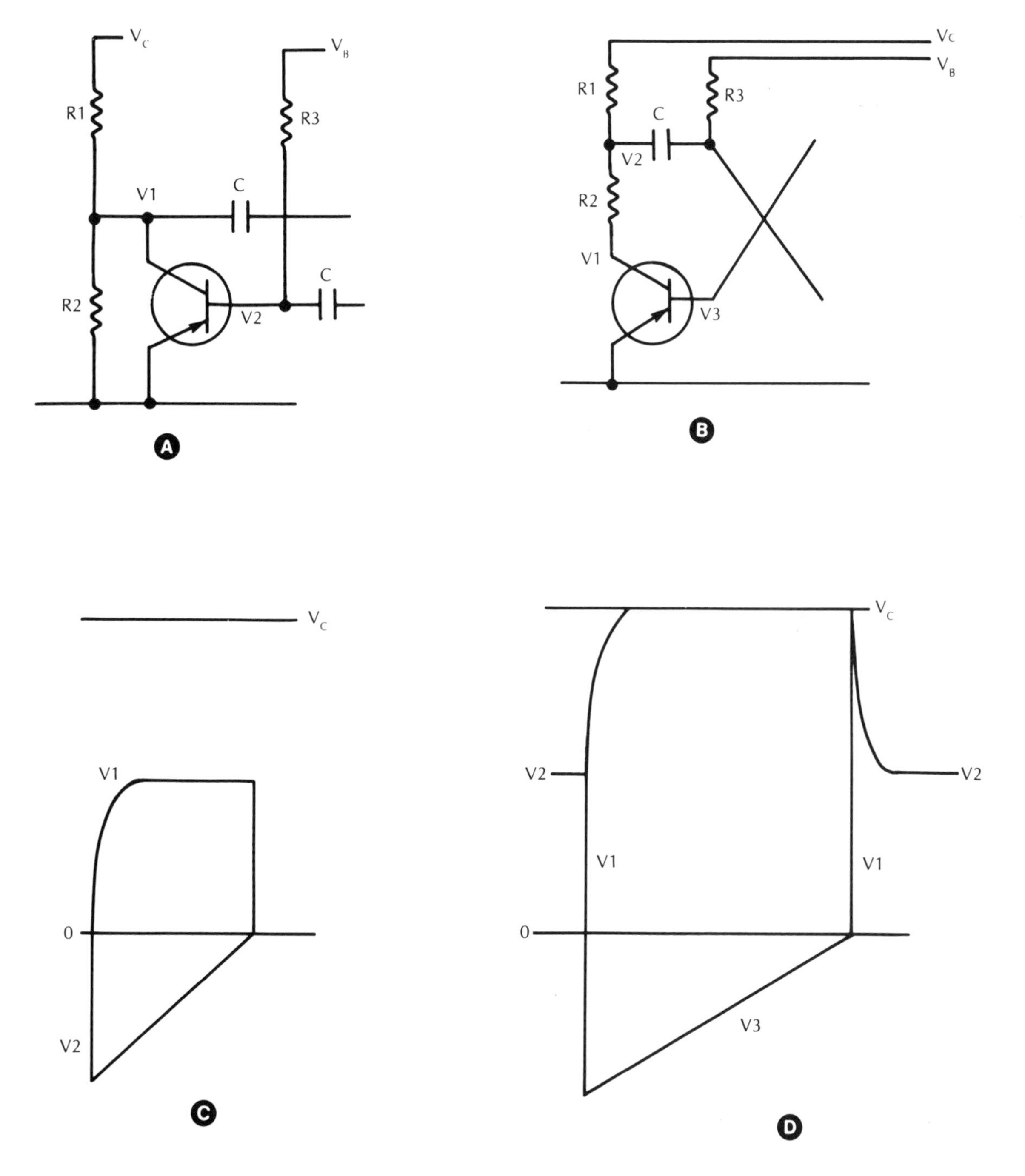

Fig. 18-8. Two ways of overcoming the limitation of action with lower base control voltages: A. tap down the collector connection, with a voltage divider, which reduces output level (C); B. tap collector resistor for cross-coupling capacitors, which reduces amplitude in a different way (D). This step may be necessary also, if reverse emitter-base voltage of the transistor used is less than supply voltage.

The arrangement at Fig. 18-8B effectively taps down the point at which the capacitor is connected, leaving the whole collector resistance (R1 + R2) in the collector. D shows the effect in the circuit. V_1, the voltage at the collector, is similar to that with no tap, except that the rise, at cut-off, to V_2 is instantaneous, while from there both follow the time constant due to R1 and C.

At turn-on, V_1 comes down all the way immediately, but V_2 falls to its value by a time constant consisting of R1 and R2 in parallel, with C. The base voltage only goes to an opposite voltage equal to the potential to which V_2 has risen before switch-on occurs.

SYSTEM BUILDING

We have already referred to how decimal numbers are entered into a digital system, as each digit is added, going from left to right, preceding entries being moved over to the left, to "make room" for the next digit as it is entered. That is a simple process that involves an instruction to a "shift register."

Now we come to consider how such a system can perform addition, subtraction, multiplication and division. Addition, of course, is the simplest, involving adding each corresponding digit of the decimal number, and carrying to the next higher order of decimal, wherever necessary, just as we learned to do in school.

For subtraction, we learned to do it in one of two ways in school: either as an answer to a question in addition, e.g., to subtract 5 from 8, we say 5 plus what makes 8? Or else we count down, just as addition was counting up, except that this too, like addition, is done by "places," with "borrowing" happening, instead of "carrying" as in addition, where necessary.

To draw out schematics of how this is done can look complicated, so we describe it here, because it's easier to understand that way. But when we get into multiplication and division, it's more complicated, so we will start there, using addition and subtraction as the basic elements from which those more complicated processes are built.

And it will be easier to follow, by looking at steps that happen at an extremely high speed inside the computer, rather than drawing out the schematic of the "hardware" that does it. Take multiplication first. Suppose we have to multiply 365 × 49. The computer takes 365 and multiplies it by 4, with a register shift to make that 40, giving 14600. Now it multiplies 365 by 9, without the register shift, adding it in as it goes, until the total is 17,885. Now when we say it multiplies, what we mean is successive addition.

So to multiply by 40, the computer quickly goes 3650, 7300, 10,950, 14,600. Then to add in the 9× part, it will go 14,600, 14,965, 15,330, 15,695, 16,060, 16,425, 16,790, 17,155, 17,520, 17,885. The same sequence can be followed with any multiplication. It's an exercise in programmed successive addition for the computer.

Division extends the subtraction operation. It needs one more element: a sensor, to know when it has subtracted enough of the divisor to tell it to move to the next place downwards (Fig. 18-9). Then suppose you program it to divide 800 by 21. The sequence would go like this (Fig. 18-10).

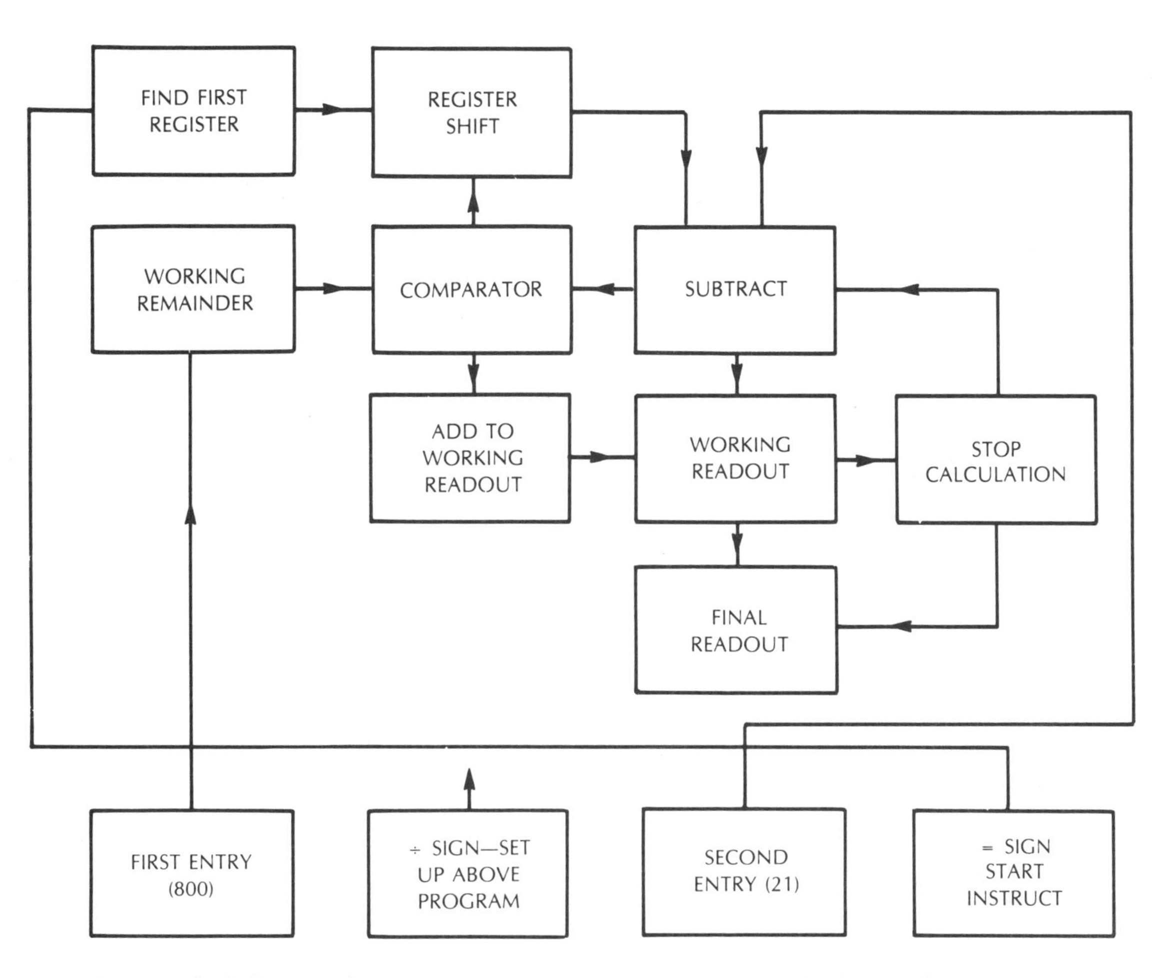

Fig. 18-9. Block diagram of operation of division, as performed by one of today's pocket computers.

The first possible division gives a result in the tens place: 21 divided into 800. Subtract 210, remainder 590, readout 10; subtract 210, remainder 380, readout 20; subtract 210, remainder 170, readout 30. Now the sensor tells that 170 is less than 210, so applies the register shift.

Continuing with 170, subtract 21, remainder 149, readout 31; subtract 21, remainder 128, readout 32; subtract 21, remainder 107, readout 33; subtract 21, remainder 86, readout 34; subtract 21, remainder 65, readout 35; subtract 21, remainder 44, readout 36; subtract 21, remainder 23, readout 37; subtract 21, remainder 2, readout 38. Now it puts in a decimal point and shifts again. But the sensor can't find any 21s in 20, so it shifts again:

2.00, subtract .21, remainder 1.79; readout 38.01; and so on till it reaches 0.32, subtract 0.21, remainder 0.11; readout 38.09; shift again, and proceed, 0.11, subtract 0.021, to get the next place. Because it works so quickly, in very little time, the readout will register 38.095238, with something still to carry. At this point, some computers will just quit, while others will round out.

remainder	subtract	readout	remainder	subtract	readout
800	210	10	0.11	0.021	38.091
590	210	20	0.089	0.021	38.092
380	210	30	0.068	0.021	38.093
170	register shift		0.047	0.021	38.094
170	21	31	0.026	0.021	38.095
149	21	32	0.005	register shift	
128	21	33	0.005	0.0021	38.0951
107	21	34	0.0029	0.0021	38.0952
86	21	35	0.0008	register shift	
65	21	36			
44	21	37	0.0008	0.00021	38.09521
23	21	38	0.00059	0.00021	38.09522
2	register shift		0.00038	0.00021	38.09523
2	register shift		0.00017	register shift	
2	0.21	38.01	0.00017	0.000021	38.095231
1.79	0.21	38.02	0.000149	0.000021	38.095232
1.58	0.21	38.03	0.000128	0.000021	38.095233
1.37	0.21	38.04	0.000107	0.000021	38.095234
1.16	0.21	38.05	0.000086	0.000021	38.095235
0.95	0.21	38.06	0.000065	0.000021	38.095236
0.74	0.21	38.07	0.000044	0.000021	38.095237
0.53	0.21	38.08	0.000023	0.000021	38.095238
0.32	0.21	38.09	0.000002	no more register	
0.11	register shift		read out final		38.095238

Fig. 18-10. Succession of steps, using the schematic of Fig. 18-9, used in performing the division: 800 ÷ 21.

In that instance, the digit 8 leaves a remainder. But if that remainder was 11 or more, the next digit would be more than 5, so some computers, sensing this, without going to the next place—because there's no room in the register—will round out to the next higher figure in the last place.

For instance, if we had programmed 900 ÷ 21, we would get 42.857142, but when the 0.000021 is subtracted from 0.00006, to get a readout of 2, the remainder is 0.000018, which is more than half of 0.000021, so it "rounds out" the reading to 42.857143.

That may seem complicated, compared to the way you did division, but for computers, that is just the beginning. Today, even a pocket computer can calculate such quantities as ϵ^x to whatever its readout capacity is. The tabulation at Fig. 18-11 shows how this is done, in columns to illustrate the progress through various stages of the computer. To understand this, you need to know that

$$\epsilon^x = 1 + x + \frac{x^2}{2!} + \frac{x^3}{3!} + \frac{x^4}{4!} + \frac{x^5}{5!} + \frac{x^6}{6!} + \frac{x^7}{7!} + \frac{x^8}{8!} + \frac{x^9}{9!} + \ldots$$

We gave that example, because it illustrates how such programs work. It involves several "temporary" memories, that follow the calculation through. For example, in this case it starts with a 1. Not all programs do, but this one does. The first column is derived by successively multiplying by the input quantity x = 0.4736. It repeats that for each successive term, then remembers it for the next term, and multiplies it again.

Term	Numerator	Denominator	Term value	Cumulative
1	1	1	1	1
2	0.4736	1	0.4736	1.4736
3	0.22429696	2	0.11214848	1.58574848
4	0.10622704	6	0.01770451	1.60345299
5	0.05030913	24	0.00209621	1.60554920
6	0.02382640	120	0.00019855	1.60574775
7	0.01128418	720	0.00001567	1.60576342
8	0.00534419	5040	0.00000106	1.60576448
9	0.00253101	40320	0.00000006	1.60576455

Fig. 18-11. Computer program for calculating ϵ^x, when x = 0.4736.

The second column is factorial notation, successively 1 × 1 × 2 × 3 × 4 × . . . and so on. As each term is calculated, that is stored in another temporary memory, and multiplied by the next digit for the following term. Then the term value is calculated by dividing the first by the second column at that "term level," and adding it into the cumulative memory.

As soon as the term value gets too small to register within the number of digits the system can hold, another sensor stops the process and delivers the readout. If the value of x had been negative, then each term in the first column would have alternated between negative and positive, and the memory would have included this information, so that the cumulative memory would also alternately subtract and add.

In other words, computers merely follow rules, either the ones built in, or programs you feed into them.

WHAT IF?

This is an important question to ask, both for system designers, and those who use them. For instance, your computer may have a button, "y^x." This means that if you hit "5" "y^x" "3" it will evaluate 5^3. Now the question should be, "how does it do it?" The "simple arithmetic" way would be 5 × 5 = 25, then 5 × 25 = 125. But is that what it does?

In the first column of Fig. 18-11, that method is used to get successive values of 0.4736^x. But if the same method is used with the y^x function, you may find limitations. That method will work only if x is an integer. Give it y = 5 and x = 3.2, for example, and it would not know what to do; its programming doesn't provide for that: it could do 3, or it could do 4, but it doesn't know 3.2. So you try feeding that in. What does it do? The designer should ask himself what it should do. It should display some kind of "error" signal, to indicate that something doesn't "compute."

However, there are series that enable such quantities to be calculated. Let's try it. The result reads 172.4662. That sounds reasonable. Try something to which you know the answer, such as 32 raised to the power 1.2, which should be 64, because 2 is the fifth root of 32, and 64 is 2 to the 6th. If 64 comes up, you know it uses a reliable method.

INTERFACING

Relatively simple programs (you may not think so, but they are in a way, because they are simply mathematical) can be built into computers, even of the pocket variety. But next we get to microprocessors, which we have to be able to instruct what to do. And we humans don't understand computer "language." For that matter, many people don't understand scientific notation, the language of mathematicians. Computers can handle that as easily as the simple notation we're all used to.

So we need various ways to "interface," and computers need the same thing to perform their microprocessing operations. To understand a bit about this, we need to understand what their own requirements are.

We have seen the need for memories, both temporary and permanent, or cumulative.They also need memories for "instructions." Instructions tell it what to do, in what order. When we hit ϵ^x, for example, we don't need to know the whole program tabulated in Fig. 18-11. The computer "knows" that, and started calculating when you complete the programming by hitting the ϵ^x button.

More than that, computers or microprocessors now handle alphanumeric characters; letters as well as numerals. They respond to a keyboard very like a typewriter keyboard. This means they must recognize words. What does this entail?

First, a pure calculator needs decade "bits" each of 4 elements, to represent decimal numerals. A typewriter keyboard has some 92 characters, and one for a microprocessor has more than that. It may have a "print" button, that instructs it to print out whatever it has stored in a program, or generated as a result of the "work" it has done. And other instructions that relate to graphic displays, interaction with other units, connection to a "modem," and so on.

Maybe it could "get by" with 128 bits to each "byte." But the standard adopted is 256. Data buses also come in multiples of that. 256 is 2^8, so that requires 8 conductors for parallel transfer, instead of series, just as 4 are sufficient for decimal digits. But by going to 16 conductors, the data bus can handle 2^{16} or 65,536. Going to 20 conductors will handle 2^{20}, or 1,048,576, commonly known as a megabyte, or a million-bit byte.

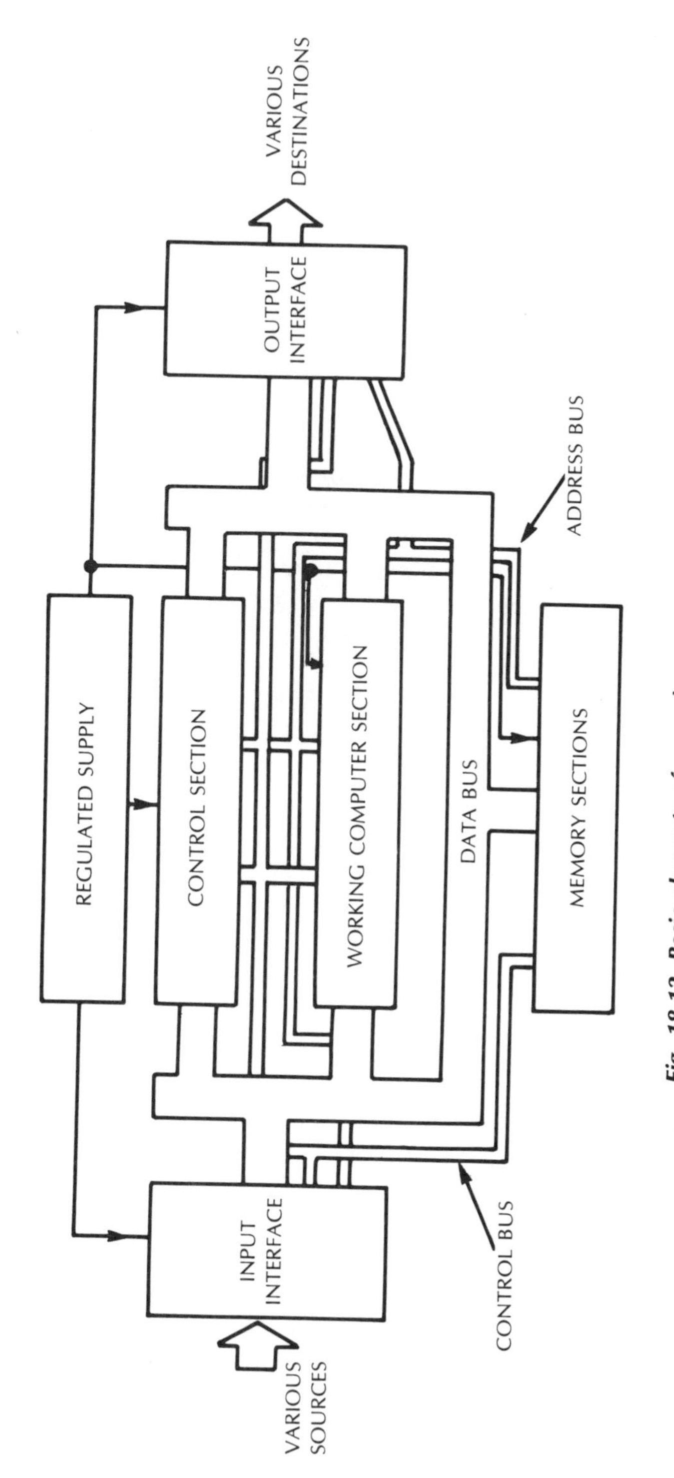

Fig. 18-12. Basic elements of any microprocessor or computer.

Why would we be considering such large numbers? Rather obviously, 16 conductors will handle the same information in half the time that 8 conductors will, or 1/16 the time that it could be handled sequentially over one line.

A computer may operate a printer, perhaps as fast as some thousands of characters a minute—much faster than the best typist you ever saw. But that is a snail's pace compared to the speeds that computers work at. So the interface must sometimes store data, and feed it out, or take it in, at these snail's paces.

Modems used have available only frequencies in the approximate range from 300 to 3000 hertz. A number of different frequencies can be used, a few of them simultaneously, over the same channel. Even so, only a short burst of a tone, at least a few cycles, is needed to identify it unambiguously. So the fastest a modem can handle data is at a rate of a few milliseconds a bit, like the printer, a snail's pace compared with the computer's lightning speed.

All these transfers must be programmed, which requires two more elements, in addition to the data handled, to make it all work. The data must "know" where it's going, from where to where. This is accomplished by addressing circuitry: an address bus tells the sender of the data, and the receiver of the same data, when to use the data bus for that purpose. Then control buses tell the various elements what they are supposed to do with it, when they send or receive it. In the calculator we discussed earlier, control circuits are operated by the button that calls, for example, for ϵ^x.

Figure 18-12 gives some idea of the elements that must go to make up any computer or microprocessor. To understand in detail requires a more advanced course in these wonderful devices.

Examination Questions

1. From Fig. 18-4A the amount of feedback necessary to produce instability in a 3-stage coupling corresponds with an F of (a); which represents (b) dB feedback ($= 20 \log_{10} F$).
2. For a 4-stage coupling with identical cut-offs, projecting the result from Question 1, with equal triangles, the feedback necessary to produce instability represents an F of (a), and (b) dB feedback.
3. Long before instability sets in, the circuit shows peaking. If n is the factor by which one cut-off is nearer the pass-band than the others, the formula for F at which peaking commences, for a 3-stage network, is $F_p = \frac{(2 + n)^2}{2\,(2n + 1)}$; and for a 4-stage network, $F_p = \frac{(n + 3)^2}{6\,(n + 1)}$. Using these formulas, find the values of F_p when $n = 10$.
4. Rearrange the formulas in Question 3, to give n in terms of F_p.
5. Using either the formulas derived in Question 4, or some other method, find the values of n needed for an F_p of 10.
6. Assuming the collector resistors in Fig. 18-6 are 1 K and the base resistors are 10 K, using transistors with a beta of 50, find the value of coupling capacitors needed for the frequency to be approximately 1000 Hz when the voltage control is at supply voltage.
7. In Question 6, neglecting contact potential effects, what is the lowest control voltage (as a fraction of supply voltage) that can be used, and what will be the frequency at that voltage?
8. Estimate the rise time on each half period of the square wave, in the example used in Questions 6 and 7.

19

Successive "Generations"

This chapter will show the student the nature of the steps by which electronics advances. The easiest way to show this is by using some examples representative of the process. Examples that can be visualized are better than abstract ones, although the same principles apply.

ELECTRONICS IN MUSIC

The earliest electronic organs, long before synthesizers, aimed at duplicating what pipe organs did, just as the earliest calculators aimed at duplicating what mechanical calculators had done for centuries. To achieve this, designers adopted various approaches.

First, because an organ pipe is a resonator, excited to speak by admitting air from the chest, the earliest electronic organs used oscillator circuits where the electrical supply to the tubes that made them oscillate was keyed (Fig. 19-1). Such an organ involved a tremendous tuning job.

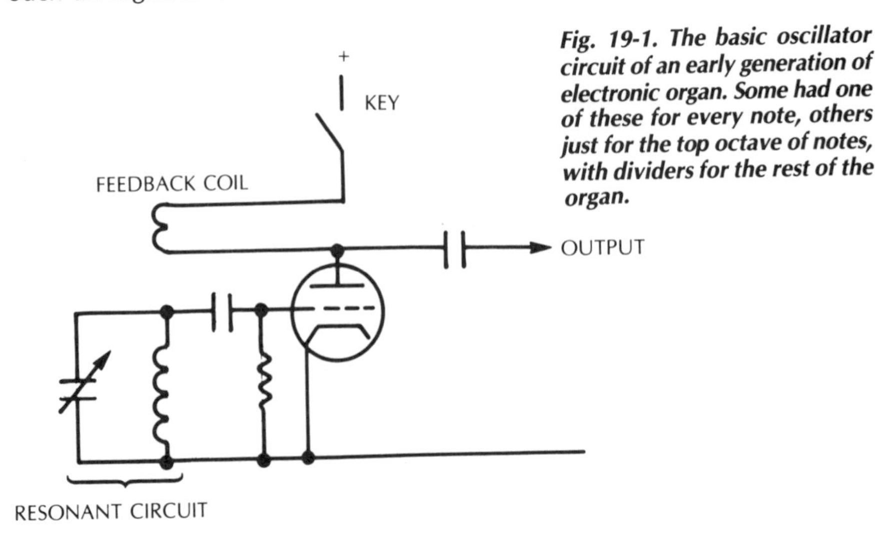

Fig. 19-1. The basic oscillator circuit of an early generation of electronic organ. Some had one of these for every note, others just for the top octave of notes, with dividers for the rest of the organ.

This led to efforts to simplify electronic organ design, which also reduced the cost, because it took less components. A very popular approach in those days was to make oscillators for the 12 notes in the top octave the organ played, then use dividers to produce all the lower octaves (Fig. 19-2).

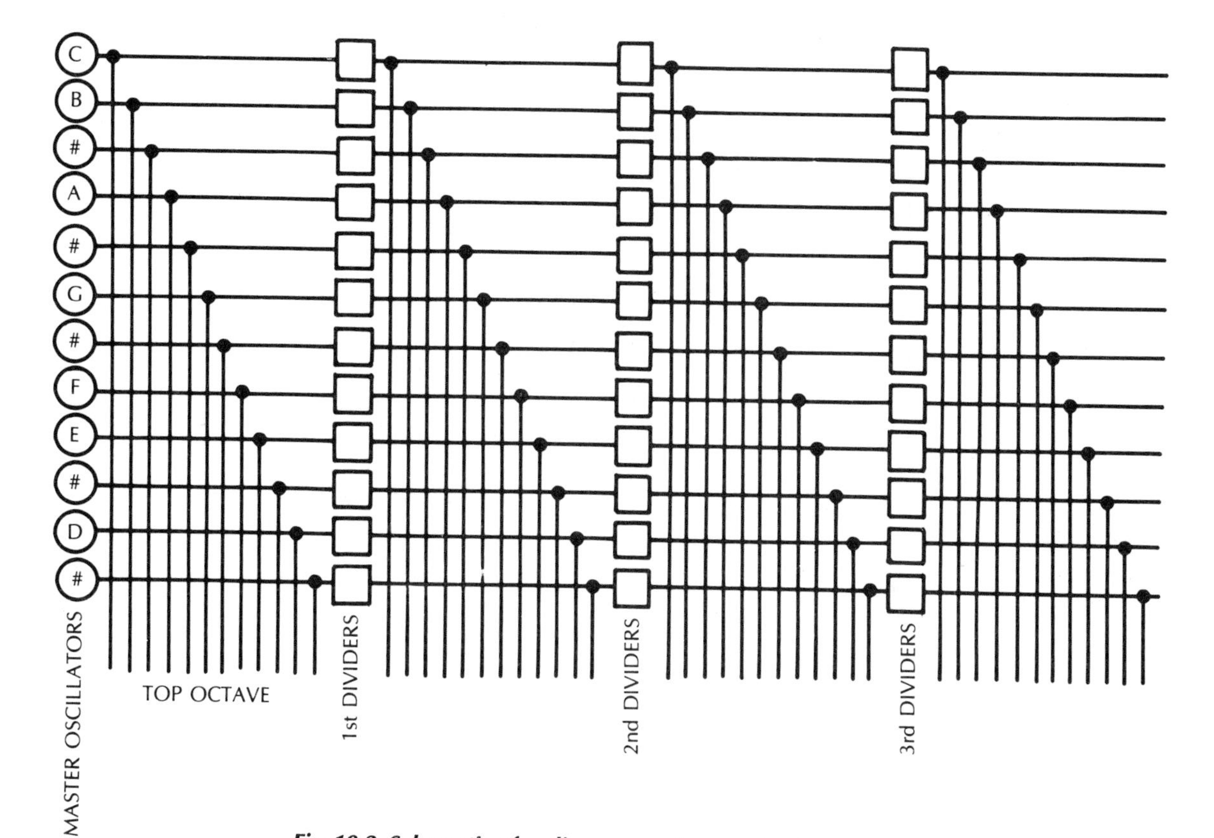

Fig. 19-2. Schematic of earlier organs, that required a master oscillator for the top octave of notes.

With this arrangement, the oscillators must be kept running continuously, and the production of voices involved filter circuits that changed the wave shape, and keying that controlled the attack and other characteristics of the particular organ stop represented (Fig. 19-3).

It was during this phase of electronic organ design and manufacture that Dr. Harry Olson of RCA Labs did some invaluable original work on synthesizers. He analyzed the characteristics of musical sounds by which they were identified, then sought to reproduce them electronically. In those days, the only active devices available to him were tubes, and rather bulky ones at that.

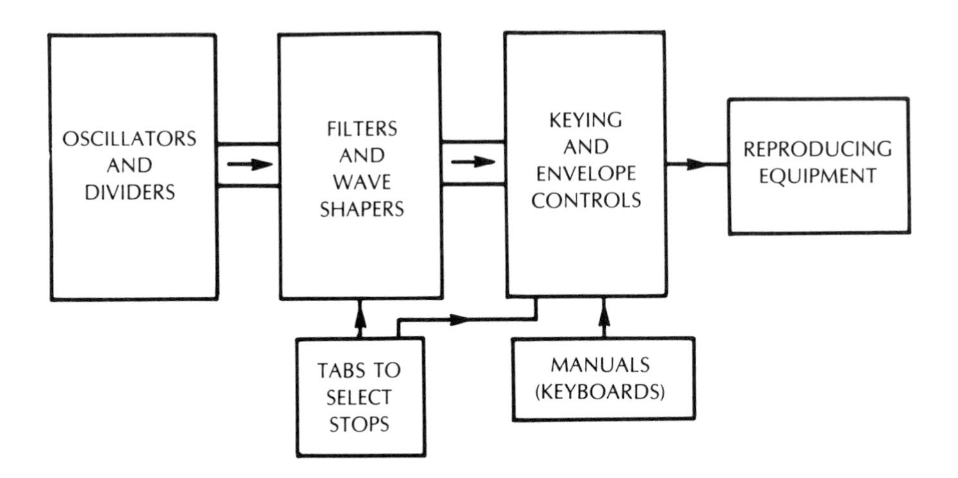

Fig. 19-3. Generalized schematic of an electronic organ in those days.

With the aid of phonograph recording, to preserve sounds once generated and to enable them to be combined (at first he could only generate one note at a time), he put together some quite primitive sounding (by today's standards) synthesized sounds, that "showed it could be done," just as the early television experiments showed that television was possible, although they came far short of modern video.

All the work thus far continued to use analog generation: some kind of oscillators which generated a waveform, which could be controlled in amplitude, growth and decay rate, and such things as glissando (gliding tone) effects.

The first step forward from this was the development of the function, or waveform generator, as distinct from the oscillator. Any kind of oscillator had a "natural" frequency, at which it oscillated, requiring build-up and decay times. If the oscillator was kept running continuously, as in many organs, because of the need to feed the octave divider networks, then build-up and decay had to be achieved by other means. This did not alter the fact that the signal started from an oscillator whose frequency was controlled in some way.

The function or waveform generator produces a wave shape, as opposed to a frequency. The shape can be square, triangular or, by modification, sinusoidal or some other shape (Fig. 19-4). It has far greater flexibility and accuracy to its designed shape production. For a while, organs continued to use oscillators while the newly introduced synthesizers, in which Bob Moog led the design, used function generators.

One thing that the use of function generators made possible was voltage control of both frequency and amplitude, which if possible with oscillators, was far more difficult to do and less precise in its result.

At the same time this was happening, digital applications of electronics were moving ahead in calculator and computer design. Integrated circuits were putting whole circuits onto a single microchip, which was a completely new concept.

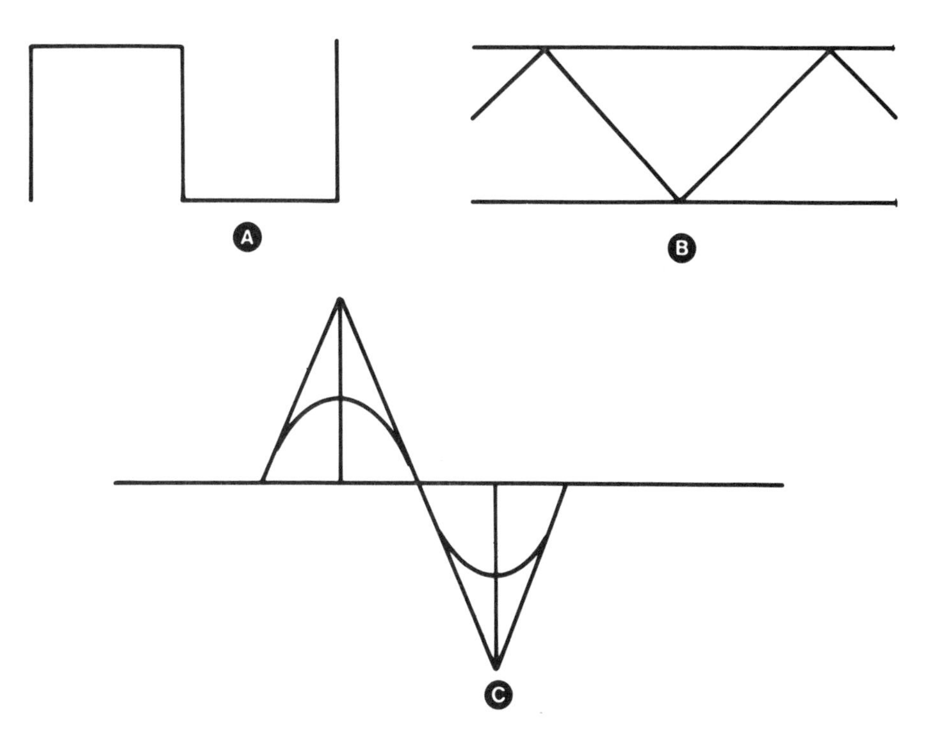

Fig. 19-4. Waveforms associated with the new class of generators, which do not use a resonant circuit: A. square wave output; B. triangular wave, which times the square-wave transitions; C. further shaping to make a sine wave output.

First, chips accommodated op amps, various gate circuits, dividers and bistable devices for other purposes. During this phase, quite a lot of individual chips were needed to make a whole instrument, but already the change was enabling a lot "more" organ to be made for the same money.

Then, in calculators and microprocessors, LSICs began to appear: large scale integrated circuits. The ones we mentioned in the previous paragraph were integrated circuits, but now many of these were combined on a single chip that performed some complete function, so that the single chip became the "guts" of a 4-function calculator, or of a watch or clock, later including calendar functions as well.

The same kind of preprogramming that could make a watch keep track of the number of days in each month of the year, could also be used to derive all the top octave notes of an electronic organ, as described in Chapter 14, using a high frequency master oscillator (instead of 12 of them) that operates in the megahertz region.

By then, electronic organs had their "guts" in an extremely small space. In the days when they used tubes, finding room for all the tubes, and providing them with adequate ventilation was a problem. The manuals (keyboards) and stop tabs that controlled them were much smaller than the circuits they controlled, although they were smaller than the equivalent pipe organs. But these new electronic organs had all the master frequencies on one small chip, all the waveforms necessary for an almost infinite variety of percussion effects on another, and so on.

To find another whole system, we need to go back to some of the earlier "hybrid" systems. The early Hammond organs used magnetic tone wheels that were rotated at the correct speeds mechanically. Other types that did not long enjoy a market used etched capacitive tone wheels, on which was engraved a variety of waveforms. But with the advent of the highly sophisticated circuitry that can be built into chips as microprocessors, an adaptation of this principle that has far greater flexibility came on the scene, initiated by Allen.

This system computerizes waveforms, rather than synthesizing frequencies. It follows the principles already finding application in digital audio. Before the advent of digital audio, artificial reverberation had a variety of choices: what was commonly called the "bed-spring," which drove a spring from one end and picked up the same analog signal at the other end, after a predetermined delay time (Fig. 19-5).

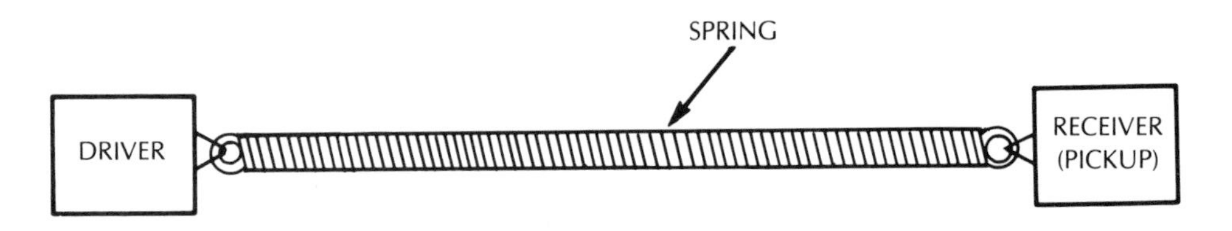

Fig. 19-5. An early basis for artificial reverberation, nicknamed the "bedspring" for obvious reasons.

As you've probably deduced, that was called the "bed-spring" system, because of the characteristic sound of the "reverberation" it produced. An alternative was the tape delay, which recorded the sound on tape, and played it back after a short delay (Fig. 19-6).

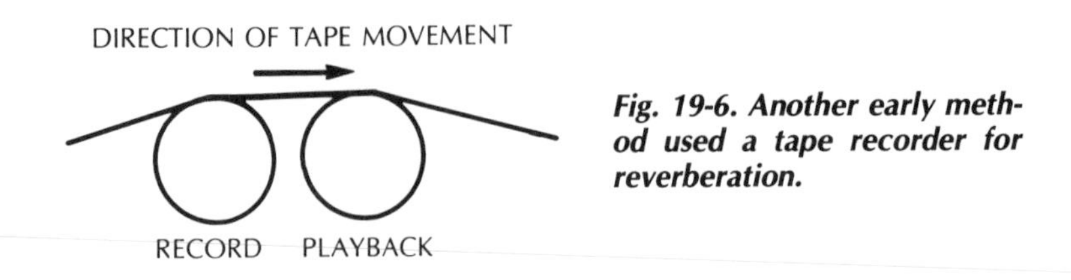

Fig. 19-6. Another early method used a tape recorder for reverberation.

Another alternative was an acoustic delay line, which actually transmitted the sound along a pipe of dimensions similar to the building being simulated, but wound into a coil for convenience (Fig. 19-7).

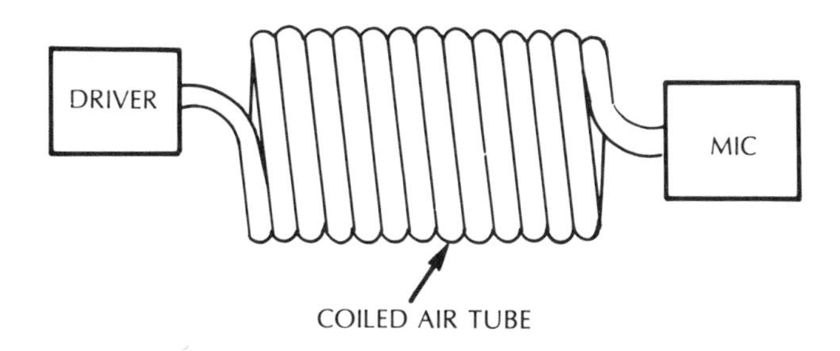

Fig. 19-7. An acoustic tube was yet another early way of getting simulated reverberation.

All of these used analog signals. Digital audio made a simpler system possible. A physical delay was no longer necessary, with its rather too "definite" impression. Real reverberation has quite a variety of delay times between repetitions, due to different paths that sound waves take in the auditorium. Digital audio takes the waveform, rather than its analog, and converts it to a succession of ordinates (Fig. 19-8).

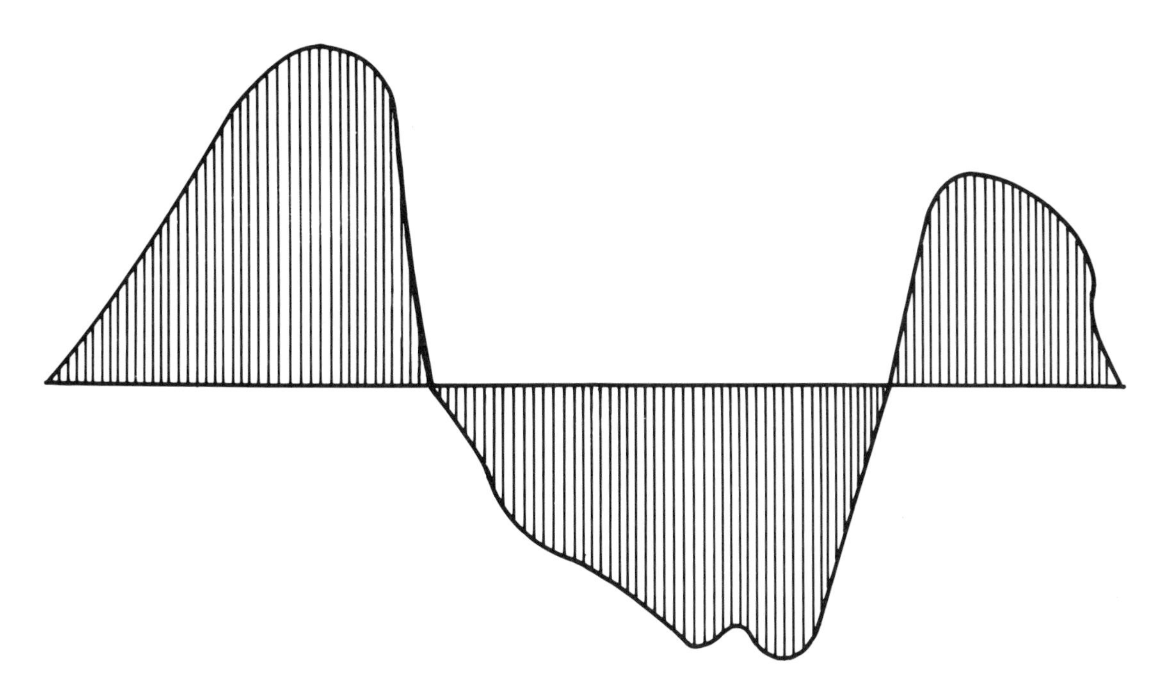

Fig. 19-8. Computer technology and digital representation of audio waveforms brought a whole new approach. Each ordinate of the waveform is stored in a computer memory for a predetermined length of time, then "read" out to produce reverberation effect.

Each of these ordinates can then be delayed by a microprocessor, and repeated at any sequence of intervals and variations in relative amplitude before recombining. This involves numbers of memory banks, similar to those used in microcomputers, into which successive elements are fed, stored, retrieved and recombined, to get extremely realistic reverberation, with virtually all the characteristics of the real thing.

In the latest form of organ, virtually the same elements are rearranged to generate predetermined waveforms, at any of the frequencies corresponding to the notes of the organ.

VIDEO GAMES

Video games simply take the basic video system used for transmitting television pictures, or recording and playing them back with video tape, and program computers to generate the "pictures," in an almost endless variety of ways. To see some of the possibilities, let us consider some of the possibilities in a system designed to play chess, since that is an old and complex game, compared to some of the more modern ones.

We will not get into the electronics necessary to present the status of the game as a video picture. We are concerned with the computer elements that program the game itself—that will be enough to show what we mean here. To get some idea of the possibilities it sometimes helps to consider how people—human beings—have done the same things for centuries.

A chess game uses 8 pieces, of 5 different kinds, and 8 pawns, for each player. The pieces are identified as king, queen, bishops (2), knights (2), and castles (2). They all have different ways of moving, which the computer must "remember" just as a human player does. And the pawns have different possible moves, too.

Human players can be broadly categorized into two types, although probably no player is exclusively one or the other: one type analyzes all the possibilities at every step in the game, for a few moves ahead, and selects the move that seems most likely to be successful at that point in the game; the other type has usually played a lot of games and memorized combinations of moves that "work," so at each move he bases what he does on previous experience.

The most successful players watch their opponents, to learn their strategy, and then utilize that information to develop a better strategy. Then there are the formula players. A chess player doesn't take long learning that position is everything in a chess game. But part of position depends on how many pieces you have on the board. A game starts with all 8 pieces and 8 pawns on both sides, and finishes with perhaps half a dozen pieces in all, when one side or the other achieves checkmate.

One aid to assessing "position" in terms of pieces "left on the board" is to set values on the various pieces. The king, of course is indispensable, since the game is over when the other side would "take" the king, which then becomes "checkmate." The queen is the most valuable piece on the board. After that, different players set different value systems, which is one thing that makes a difference in their playing style.

Because the bishops move only diagonally, and one bishop on each side starts on squares of opposite colors and can never change the color squares over which "he" moves, bishops have far greater value when you have both of them. If you lose one bishop, say the one that moves on the white squares, all your opponent has to do, is get all his pieces on the white squares, and your remaining bishop is useless.

The same is not true of castles, which move "on the square"—forward, backward or sideways. And knights are "something else again," with their own peculiar set of moves, which require a certain skill to be able to utilize. And the skilled player can use them differently with whatever other pieces he has left. So, a simple "points value" for each piece remaining on the board is hardly appropriate and the good player values them in a much more complicated way than that.

Now, let us apply this to the design of a computer to play chess, or at least see what some possibilities are.

Consider the opening move, as illustration. The player can move any one of his 8 pawns, either 1 or 2 spaces forward. That gives 16 possible moves of a pawn for the first move. None of the pieces, except the knights, can be moved, until the pawns are moved "out of the way." But each knight has 2 possible opening moves, before any pawns are moved. This brings the total number of possible opening moves up to 20.

Now, of course, the first player has 20 possible opening moves, and the second player has 20 possible opening moves, making 400 possible combinations of opening moves, one for each player. Obviously, a few of these are better than others, but we are proceeding on the basis of pure information theory, which is one way a computer could tackle it.

After the first two moves, one for each player, the number of choices for the next move varies considerably, according to the combination of positions. Usually, based on pure information theory, there will be more choices now. It is not difficult to see that, before many moves have been played, the number of possible choices runs into billions!

Now, how would you design a computer to handle that? First, of course, is the problem of designing the computer so it can compute all the possible moves at any point in the game. Then, if you are going to "play the computer" rather than merely using the computer to record a game played between two people (which is simpler to do) you need to give the computer some way of "developing skill" in the game. This is coming pretty close to designing a machine that can "think."

To see what we mean, it should be pretty obvious that designing a computer to play tic-tac-toe is much simpler than designing one to play chess. In the sense that we are developing in this chapter, they are a "generation" apart.

SENSORY PROJECTION

Another way to measure advancement can be compared to understanding how human sight and hearing functions. For the video game designer, the principles of perspective—the representation of three-dimensional space in two dimensions, with a sense of movement and timing—needs to be built in. This uses the same kind of computer functions built into a mathematical calculator—not just the 4-function kind.

But now consider synthesizing sound effects. To do this, you need to think about how human hearing normally functions, because the computer must deliver "signals" that will "fool" human hearing. A lot of the effects involve motion, so concentrate on how you perceive motion aurally.

When you listen to traffic passing, the most obvious effect is called the "Doppler" effect: the fact that sound from an object approaching is higher in pitch than when the same object is receding. How close it comes to you, at its closest point, can be simulated by the rapidity with which the pitch changes. Speed of motion is conveyed by the amount of pitch change, between approaching and receding. How close it comes to you is conveyed by the rapidity of that change.

But all of that gives you no indication of the direction in which it approaches and leaves. If you are walking along a road, you can tell whether an approaching vehicle is coming in front or from behind you. At least you usually can. How do you do this? Most people will simply ascribe that to binaural perception: the fact that you use two ears.

That can easily explain movement from left to right, or right to left, but it does not nearly so easily explain movement from front to back or back to front. The simple "binaural" theory cannot explain how you know when a sound comes from in front of you or behind you. You need more "information" to tell that: what is it?

Listen more carefully in the future. You will find, for one thing, that movement often helps. If you hear an automobile motor running somewhere, it is not so easy to pinpoint as an automobile in motion. Why is that? Because, although you may not realize it, your hearing is picking up a lot more beside that automobile sound. A blind man will tell you, because he uses a lot more of the information than a sighted person does.

Most of the time, you readily see what you hear, so you don't bother to think about how the two identify with one another. You will find that locating sounds, and more especially the movement of their sources, depends on the neighborhood. A fence, a wall, even iron railings, all affect sounds, even though they do not make any of their own. A blind man can "hear" these "soundless" objects, often from nothing more than their effect on the sound of his own footsteps.

Whether you realize it or not, you probably do some of those same things. After all, those characteristics of sound are not specially made for the blind man to hear: they are equally there for everybody, whether they pay attention or not. And just as your own footsteps are affected by objects in the surrounding neighborhood, so are all other sounds. And this is what, without realizing it, you depend on to tell where those other sounds are coming from.

You don't realize that you can "hear" the brick wall, the wooden fence, the iron railings and many more things that emit no sound of their own, but your hearing faculty is conscious of the effect these objects have on sound from objects that send out sound. Have you ever noticed that when you are driving, and enter a tunnel, you can "hear" the tunnel? But the tunnel itself doesn't make any sound: it just alters the other sounds you hear.

These are the things that the designer who wants to synthesize effects must recognize and program his computer to simulate. We are moving into new generations of realism.

SUCCESSIVE GENERATIONS

We have discussed development in several areas. The purpose of this chapter is to give the student the ability to look ahead. Early counters, for example, merely duplicated electronically what mechanical calculators had done for years, which in turn had mechanized what human fingers could once do on an abacus. Each could be regarded as a generation step ahead.

Then electronic calculators could be programmed to perform multiplication and division, and then square root—a slightly more complex program than division. Next, all of those capabilities became the building blocks with which to program calculators that would calculate mathematical quantities, like ϵ^x, logarithms, sines, cosines and tangents. Each uses the earlier programs, addition, multiplication, division, incorporated into new programs.

With video games that use graphics, first they used displays that could be activated in two dimensions, then to gain realism, the computer stored 3-dimensional information, and finally concepts of velocity and acceleration, programmed to change the other three.

In the chess game, the rules are programmed in, after the graphic representation of the chess board with its pieces has been generated. All it can do at that stage is to sort of "police" the game, making sure all the pieces make the correct moves. From there it is a short step, technologically, to enable it to predict all possible moves at any given point in the game.

That requires piece-recognition features, in turn depending on a pretty detailed memory, and memory for the moves each piece can make, then an exploratory circuit that can run each "possibility" out, likely or unlikely.

At that stage, only a chess player would know the likely choices from the unlikely choices. To predict possible moves ahead would require an impossible amount of computer memory. But a computer can "learn" chess, much like many players learn the game: by remembering previous games, and how they "came out." That involves nowhere near as much memory capacity as storing all possible future moves. That capability would mean the computer could store every chess game that has ever been played, or could ever be played, including some mighty silly ones.

Computers and microprocessors are always having new capabilities built into them. You can now program a microprocessor to proofread a manuscript you have just typed. It has a floppy disc "dictionary" against which it checks every word you put into its memory as part of the article or story you want it to type. If it cannot find any word in its dictionary, it notes that word and calls it to your attention in an appropriate way.

Then you have the option of entering it into the dictionary for future use, correcting it (if it is misspelled) or if it's a person's name, or something not important to be in the dictionary, you can tell the computer to ignore it: it did its job calling it to your attention.

A computer can also analyze data. For many elections now past, computer predictions have accurately forecast the winner, after only a relatively small part of the votes have been counted. This it does by manipulating far more data about previous elections than any single human brain could handle, at least in that way, based on all the known "variables" that contribute to the habits of people in the district being predicted.

Computers can analyze design data. Where in earlier times, the computer merely checked data, or provided help to the human designer, it can now take the parameters that go into a design, and produce the design itself. Other computers can take a design and program tools to make the parts. Then further refinement can almost let human designers out of the picture altogether.

SPEED

We have already touched on this aspect. Always research keeps coming up with faster ways of doing the same thing. For instance, the area of gate design, which we covered in Chapter 14, shows how speed can be improved. RTL gates were slow. DTL gates were faster, but TTL gates give the best speed, thus far. Attention to operating conditions can make them even faster.

But speed can also be expedited by putting more in parallel, less in sequence. Use of "broader" data buses is a help here. But for complex mathematical uses, consider the following: using the radian measure of angles, the following series will calculate the various quantities shown:

$$\epsilon^{x} = 1 + x + \frac{x^2}{2!} + \frac{x^3}{3!} + \frac{x^4}{4!} + \frac{x^5}{5!} + \frac{x^6}{6!} + \frac{x^7}{7!} + \frac{x^8}{8!} + \frac{x^9}{9!} + \ldots \quad (19\text{-}1)$$

$$\epsilon^{-x} = 1 - x \quad \frac{x^2}{2!} - \frac{x^3}{3!} + \frac{x^4}{4!} - \frac{x^5}{5!} + \frac{x^6}{6!} - \frac{x^7}{7!} + \frac{x^8}{8!} - \frac{x^9}{9!} + \ldots \quad (19\text{-}2)$$

$$\cosh x = 1 + \frac{x^2}{2!} + \frac{x^4}{4!} + \frac{x^6}{6!} + \frac{x^8}{8!} + \frac{x^{10}}{10!} + \ldots \quad (19\text{-}3)$$

$$\sinh x = x + \frac{x^3}{3!} + \frac{x^5}{5!} + \frac{x^7}{7!} + \frac{x^9}{9!} + \frac{x^{11}}{11!} + \ldots \quad (19\text{-}4)$$

$$\cos x = 1 - \frac{x^2}{2!} + \frac{x^4}{4!} - \frac{x^6}{6!} + \frac{x^8}{8!} - \frac{x^{10}}{10!} + \ldots \quad (19\text{-}5)$$

$$\sin x = x - \frac{x^3}{3!} + \frac{x^5}{5!} - \frac{x^7}{7!} + \frac{x^9}{9!} - \frac{x^{11}}{11!} + \ldots \quad (19\text{-}6)$$

Now, examining those series, you will note that they all use different combinations drawn from the same family of terms. Calculating the numerical value of successive terms is identical, whichever quantity is needed, so a central unit can perform that operation, as shown in Fig. 18-11. But then the outputs can be rotated through 4 separator collectors, or cumulators, as follows:

$$\Sigma_1 = 1 + \frac{x^4}{4!} + \frac{x^8}{8!} + \frac{x^{12}}{12!} + \frac{x^{16}}{16!} + \ldots \qquad (19\text{-}7)$$

$$\Sigma_2 = x + \frac{x^5}{5!} + \frac{x^9}{9!} + \frac{x^{13}}{13!} + \frac{x^{17}}{17!} + \ldots \qquad (19\text{-}8)$$

$$\Sigma_3 = \frac{x^2}{2!} + \frac{x^6}{6!} + \frac{x^{10}}{10!} + \frac{x^{14}}{14!} + \frac{x^{18}}{18!} + \ldots \qquad (19\text{-}9)$$

$$\Sigma_4 = \frac{x^3}{3!} + \frac{x^7}{7!} + \frac{x^{11}}{11!} + \frac{x^{15}}{15!} + \frac{x^{19}}{19!} + \ldots \qquad (19\text{-}10)$$

Then, as needed in other parts of the calculations, the following simple formulas will give the other quantities, using addition or subtraction elements only:

$$\epsilon^x = \Sigma_1 + \Sigma_2 + \Sigma_3 + \Sigma_4 \qquad (19\text{-}11)$$

$$\epsilon^{-x} = \Sigma_1 + \Sigma_3 - \Sigma_2 - \Sigma_4 \qquad (19\text{-}12)$$

$$\cosh x = \Sigma_1 + \Sigma_3 \qquad (19\text{-}13)$$

$$\sinh x = \Sigma_2 + \Sigma_4 \qquad (19\text{-}14)$$

$$\cos x = \Sigma_1 - \Sigma_3 \qquad (19\text{-}15)$$

$$\sin x = \Sigma_2 - \Sigma_4 \qquad (19\text{-}16)$$

From those basic trigonometrical quantities, all others can be calculated quite simply, using the 4-function part of the calculator. Thus time can be saved by storing the 4 quantities, (equations 19-7 to 19-10) in the initial readout from the "x" input.

Another thing this little exercise shows is that there are often many approaches to handling a particular type of problem. Which is the fastest in any given instance may well depend on the type of problem.

Examination Questions

The following statements are mathematically equivalent:

$$y = \epsilon^x \qquad x = \log_\epsilon y$$

Writing

$$z = \frac{y - 1}{y + 1}$$

$$\log_\epsilon y = 2z + \frac{2z^3}{3} + \frac{2z^5}{5} + \frac{2z^7}{7} + \frac{2z^9}{9} + \frac{2z^{11}}{11} + \ldots$$

and

$$\epsilon^x = 1 + x + \frac{x^2}{2!} + \frac{x^3}{3!} + \frac{x^4}{4!} + \frac{x^5}{5!} + \frac{x^6}{6!} + \frac{x^7}{7!} + \frac{x^8}{8!} + \ldots$$

1. Use the first series to evaluate $\log_\epsilon 2$, using the following format:
   First z = ________.
   Individual term: Sum so far: Run this tabulation out to 8 decimal places, taking as many terms as necessary.
2. Use the other series, with the same format, and the value obtained from this program, to find ϵ^x, which should evaluate back to 2.

20
Hybrid Systems

Electronics people are apt to think in channels. A classic but little known example of this occurred around 1959. The space program was just getting started, and there had been a few missions into space, some manned, many unmanned. The American technologists were working feverishly to complete a solid-state video camera with which they could photograph the moon.

Meanwhile, the Russians sent up Lunik III which photographed the far side of the moon and sent pictures back to earth. Had the Russians really "stolen a march" on American technology to achieve that? No, they had put together an off-the-shelf Polaroid camera (American made) with an even older vintage radio picture transmitter, and added their own "Rube Goldberg" device that pulled the film from the Polaroid, timed it, peeled it and wrapped the picture on the radio picture transmitter drum.

That was, in truth a "hybrid system." And American technologists didn't do it first, because they were thinking that the whole thing must be electronic.

IN NATURE

Analog and digital are not new with electronics. They both occur in nature as do hybrid systems. Human sight and hearing are both examples of this. Light and sound propagation are both analog. Light is optically focused by the human eye into an image on the retina of the eye, where it is converted into sort of a multi-channel digital signal, to be sent along the optic nerve bundle to the brain, for analysis and recognition.

Television and video use a different system from that used to communicate the image from the retina to the brain. By use of scanning, elements of the picture can be sent in extremely rapid succession over a single channel, instead of the extremely large number of channels in the optic nerve.

Sound waves impinge on the tympanic membrane—the eardrum—there to be transmitted via the three ossicles to the window of the cochlea. Thus far, everything is analog. Now, the cochlea, like the retina of the eye, converts analog to digital. First it analyzes the analog content, which it converts into discrete digital signals, nerve impulses, sent along the auditory nerve bundle, where the auditory center of the brain "processes" the information to tell us what we "hear."

Once again, audio signals in electronics make a conversion quite different from that used in the human ear, and for a similar reason to the difference already noted in video. Audio signals, whether analog or digital, use a single channel, or very few channels at most, compared with the enormous number in the auditory nerve. Each of these are examples of "hybrid systems," where the hybridization is between analog and digital.

Hybridization calls for use of some information theory, either to understand how it is done, or how to do it right. Sound waves vary in frequency and intensity. How much information is necessary to specify the frequency content and precise amplitude of various frequencies in a composite signal or waveform? That is an analog analysis (Fig. 20-1).

Light waves also possess different frequencies and intensities which in this case determine the color and intensity of the resulting point in a scene, or "picture." But a whole picture is made up of many such points. A painting, or a color photograph, is an analog representation of such a scene, but it does not present the same kind of analysis that Fig. 20-1 does for sound. Vision contains essentially another dimension.

But to transmit, record, or otherwise "handle" either sound or vision requires conversion to a digital representation of these analogs. And the kind of digital used may differ, according to the system used. In each instance, the digital used to communicate the "information" from the ear or the eye to the interpretive faculty of the brain uses a very large number of "channels" (nerve fibers) along which simple one-dimensional impulses are conveyed, at a relatively slow speed. Both the auditory and optic nerves take about a millisecond, maybe more, to transmit the impulse from the ear or eye to the brain.

Electronic digital can use impulses very much closer together, and can convey them very much faster than individual human nerve fibers can, perhaps a million times faster. So by use of appropriate "coding," a single channel can convey as much as the whole auditory or optic nerve. Where the human nerves use hundreds of thousands, maybe millions of nerve fibers, each conveying one element of the overall signal, simultaneously with all the other fibers, the electronic equivalent sequences that many elements along a single channel, completing the "information" in no more time than it takes the human equivalent to communicate over its millions of channels.

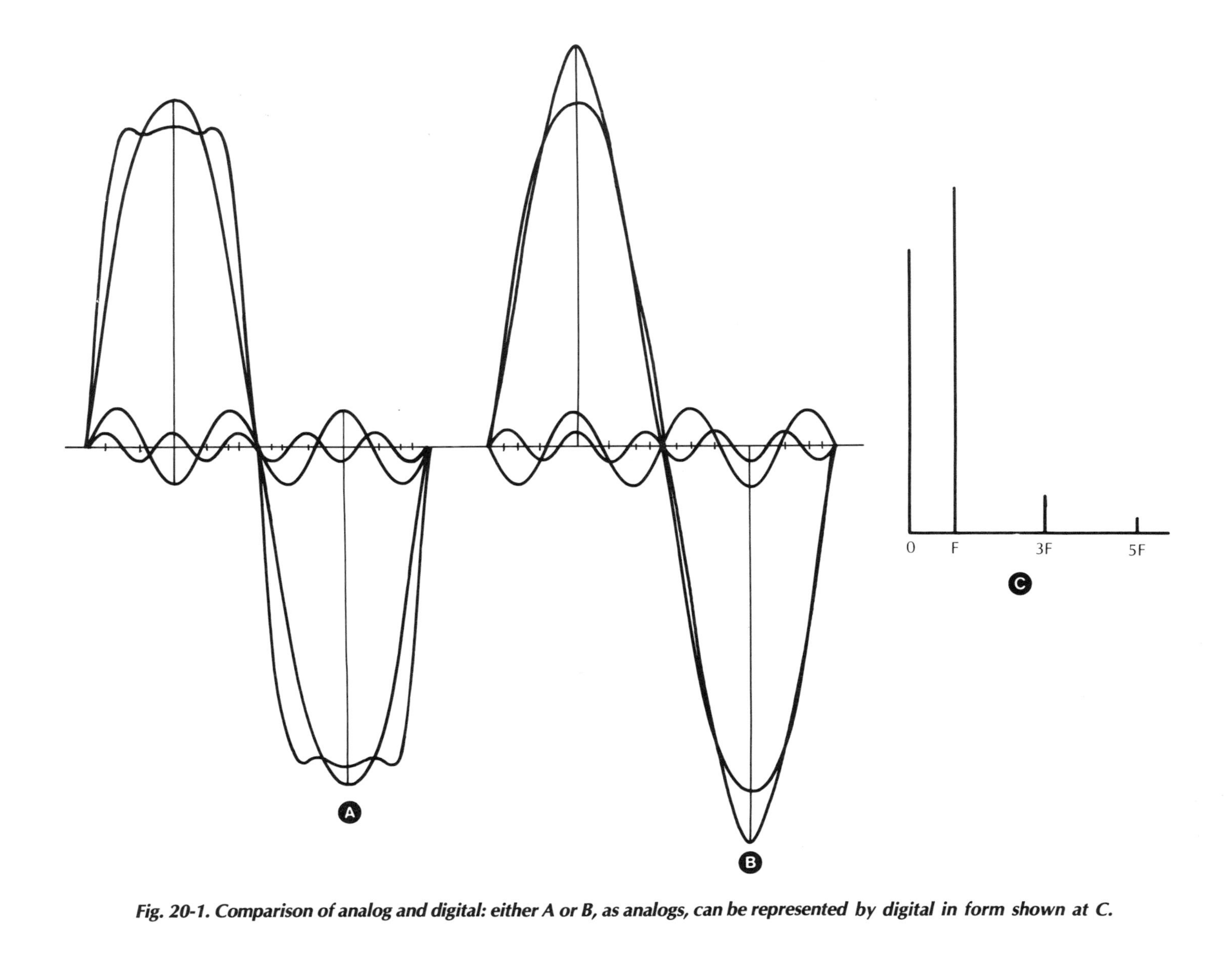

Fig. 20-1. Comparison of analog and digital: either A or B, as analogs, can be represented by digital in form shown at C.

But now, that single electronic channel can carry the same information in either analog or digital form. In video, the analog form is a waveform that represents the picture by its precise shape. In audio, the analog form is a waveform that represents the sound by its frequency content (Fig. 20-2).

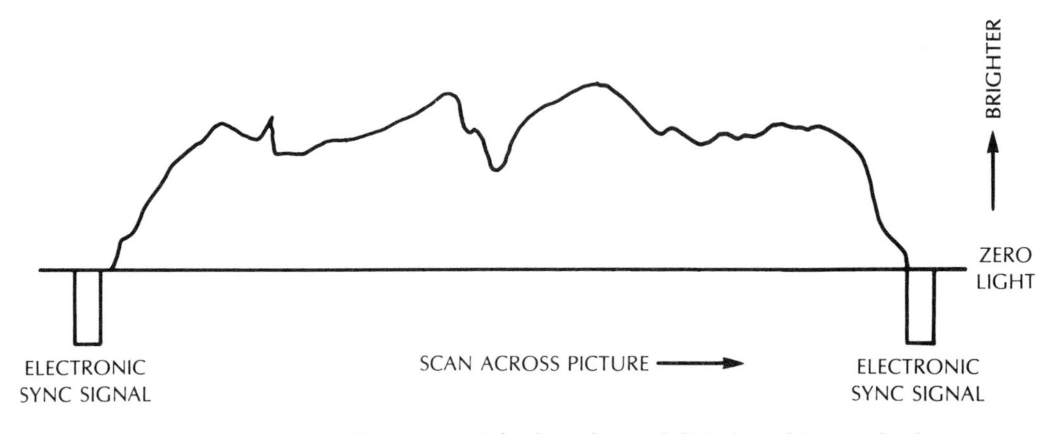

Fig. 20-2. Scan across a video picture is both analog and digital: in this case the form must be the same.

Converting video to digital means that each point on the composite waveshape must be converted to a digital representation of its amplitude. Converting audio to digital gives the system designer an option. He can convert either the instantaneous frequency content into digital information, or he can convert the waveform to digital, as with the video.

That briefly introduces us to aspects of hybrid systems as a whole. But hybridization gets to much lower levels than this. Within a system, the term often applies to mixing "integrated" elements with "discrete" elements. An integrated element, usually referred to as an "IC" (for Integrated Circuit) performs some complete function. The larger units, called LSICs (Large Scale Integrated Circuits, may contain a whole system, such as a calculator, a clock, a large portion of an organ, or whatever.

IN MACHINES

In the design of electronic equipment, hybridization is more than blending analog and digital, or converting from one to the other. It comes down to whether circuits are integrated so that various combinations of inputs result in prescribed forms of output, or whether they are discrete, designed for specific jobs, with more variations than digital provides, and interfaced with integrated units as other components of the same overall system.

For example, in the electronic organ that processes its output digitally, the digital output would be unrecognizable as music without converting to analog. While there are integrated forms of linear amplifier, this conversion can be made by an integrated section.

A function, such as logarithmic, that we introduced as being characteristic of the forward conduction of a diode in Chapter 11, may be required in a calculation or conversion. Using that characteristic of a diode is an analog property. A diode so used must be regarded as a discrete element. A linear input voltage yields a logarithmic output current.

How precisely can such a function be used, that way? Such diodes are very temperature sensitive in that characteristic. A change in temperature of a fraction of a degree can quite considerably modify the conversion from linear to logarithmic, or vice versa. And even if temperature is tightly controlled to minimize this effect, the accuracy is extremely limited, compared with the conversion provided by a digital computer, which will produce a conversion, either way, correct to 8 places of decimals, or whatever capacity is built into the calculator. The analog device may yield 2-figure, or at best 3-figure accuracy.

Possibly the most important distinction between integrated and discrete is in the way you think about them. This is best illustrated with reference to the changes that have occurred as electronic circuitry has moved from discrete toward integrated. The earlier part of this book deals with discrete components. The engineer who designed systems using only discrete components had to know, and think about, how those discrete elements behaved in the context of other components.

A tube, or a transistor, was an amplifying device, whose properties the designer had to deal with directly. He had to consider what resistors and other components he should use, for plate or collector load, whether to use plate or cathode coupling, or in the case of transistors, collector or emitter coupling, and how to bias the stage to get the proper operation.

In a multistage amplifier, he had to repeat this for every stage, and to think about how the coupling between stages interacted. Then, when he came to system design, he had to think about how the amplifier fitted into the whole system, or conversely, he had to design the amplifier to fit into the system in the way he wanted it to. The designer finished up being conversant with the whole system, and with its many parts, in their elemental terms.

In an intermediate stage, the designer relinquishes some of the smaller detail, so he can focus on the larger parts of the system. An amplifier becomes an amplifier—he no longer cares what is inside it, only what it does. Mixers, adders, subtracters, multipliers, and all other kinds of functions each do their thing according to specification, and the designer doesn't bother with what's inside any of them.

They become just "black boxes" that perform a certain function—except that they're not big enough to think of as boxes any more, a whole bunch of them can be housed in a "chip" that will sit on the tip of your little finger. But schematically, you draw them as "boxes," not bothering with how they function, only with what they do.

But these boxes are usually built into a system that has to interface with something not built into this integrated form, such as an electric motor that it is the purpose of all this electronics to control. So this requires some discrete, rather than integrated components. And at this point, the designer has to change his thinking, to consider the "workings" of such discrete elements.

The building of a digital system involves essentially a knowledge of mathematics, of an appropriate level, or at least of the mathematical processes. For instance, to design the integrated circuit that calculates logarithms, the designer of that chip needs to know how logarithms are calculated, which is fairly complex mathematics. But once the designer has put together a chip that will do that, and it is in production, nobody else needs to know how logarithms are calculated: the chip does it for them, once it is provided with the necessary supply, which can now come from a very tiny battery.

So, the logarithm calculator becomes an element in the system—a sort of "black box" on the schematic. But if it is to be included in a calculator that will also calculate trigonometrical functions, these may be built onto the same chip, and externally all that is required are different "pins" or inputs, that initiate the different programs necessary to make those different calculations.

As we have already said, a program to perform such a calculation is dealing with a converging mathematical series. The calculator starts in on the series and keeps going, until further "terms" in the series become too small to make any difference: then it stops and "reads out" the result. That is a relatively simple "decision" the calculator has to make.

In other instances the program that would normally be selected may use a value that results in a series that does not converge, in which case the computation would never end. In such a case, the computer must have a decision-making element programmed at the beginning, to select a series that will converge with satisfactory rapidity. An alternative is to design into the system an approximation formula that arrives at an adequately accurate solution, such as correct to 8 decimal places, with a limited number of terms.

This distinction can be illustrated in a different context, that is more visual.

TIME OR FREQUENCY BASED

This is mentioned, because it is an area where another "bridge" or form of hybridization can be used. Fourier analysis is, in mathematics, what corresponds with frequency analysis in electronics, such as for audio. A square waveform can be analyzed into an infinite series of frequencies, represented in the form.

$$\frac{4A}{\pi}\left(\sin \omega t + \frac{1}{3} \bullet \sin 3\,\omega + \frac{1}{5} \bullet \sin 5\omega t + \frac{1}{7} \bullet \sin 7\omega t + \frac{1}{9} \bullet \sin 9\omega t + \ldots\right) \qquad (20\text{-}1)$$

Figure 20-3 shows how this series approaches the ultimate square form, one term added at a time.

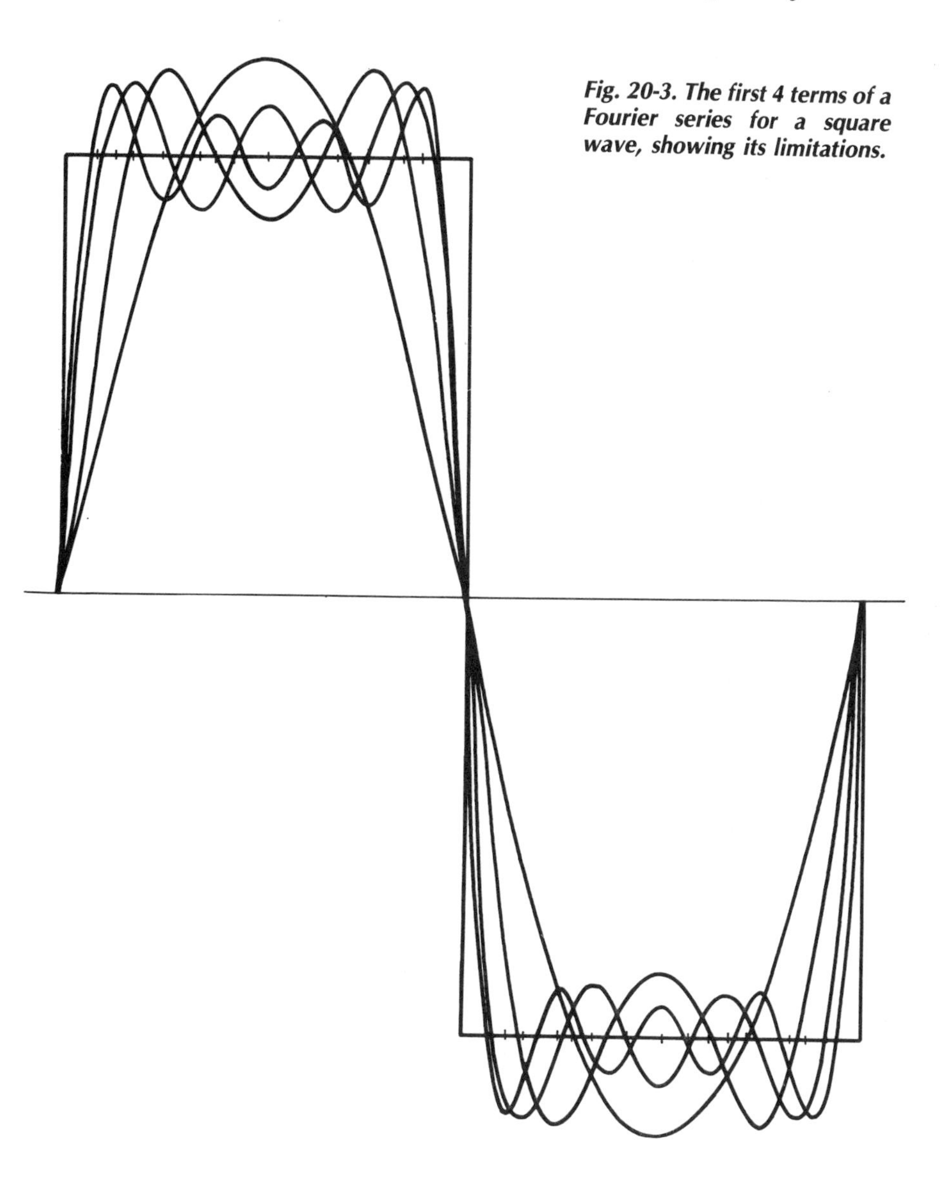

Fig. 20-3. The first 4 terms of a Fourier series for a square wave, showing its limitations.

You can see that, using this approach, an infinite series of frequencies will be needed to reach an ultimate square form and anything short of that, which any practical system must be, will have a "ripple" on it, corresponding with the "last" frequency added.

Note that the coefficients in the series of equation 20-1 are simple and can go on indefinitely, gradually coming closer to the true square form, but never reaching it.

An alternative way to generate a square wave is by viewing the system as referenced against time, rather than frequency. This is much simpler. Each "edge" of the square is a sudden change of value and between these "edges," the value holds steady. In particular, such a signal can be generated digitally. But there is still a slight problem. That theory assumes the change of value occurs in infinitely short time. In any practical system, the change must take some finite amount of time, however short.

Where a system uses digital elements, that is the only thing that matters: the speed of the change from one state to the other, and it is measured in microseconds, nanoseconds, or whatever. But if any elements in the system handle analog, then the overall waveform must be analyzed, because that is what linear systems must handle.

One analog equivalent of a completely flat-topped wave, which is the most obvious way in which the Fourier synthesis of Fig. 20-3 departs from square, produces the waveform shown at Fig. 20-4.

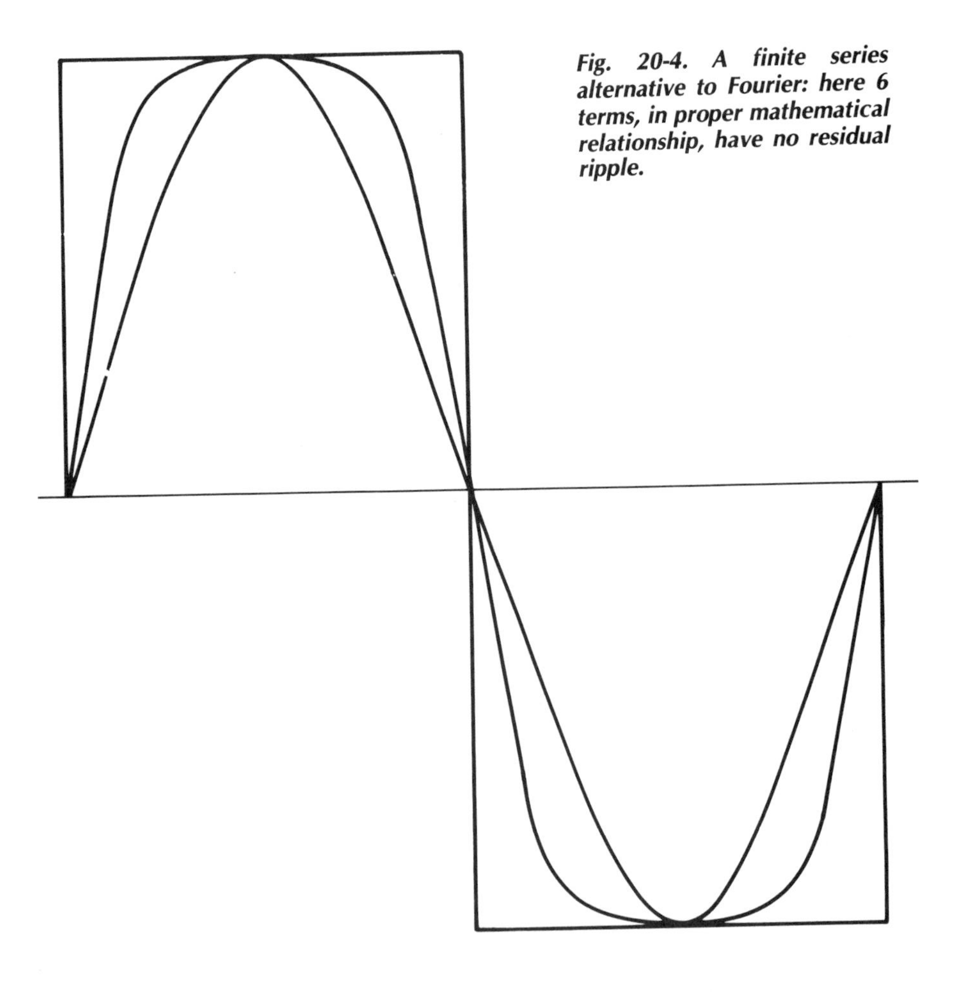

Fig. 20-4. A finite series alternative to Fourier: here 6 terms, in proper mathematical relationship, have no residual ripple.

For comparison, the following are terms using decimal approximate coefficients:

Fourier 4-term series:

ωt	$3\omega t$	$5\omega t$
1.2732396	0.42441318	0.25464791
$7\omega t$	$9\omega t$	$11\omega t$
0.18189136	0.14147106	0.11574905

6-term finite series:

1.2213364	0.29079437	0.08723832
0.02073669	0.00323105	0.00024033

Comparing the rise times of Fig. 20-3, which includes only 4 Fourier terms, and Fig. 20-4, which uses 6 terms of finite series, it is at once obvious that the Fourier gets a more rapid rise with fewer terms. But this is at the cost of the very severe "ripple" along the top. The 6-term finite series merges into a "perfect" flat top which has a defect in the opposite direction.

Actually, it does not achieve flatness at all. It would be truer to say that it has absolutely no "overshoot." Another way to approach this is with reference to the transfer characteristic concept. Fig. 20-5 illustrates this. At A is a complete linear transfer characteristic, so that a sine wave input (or any other form) produces a sine wave out (or equivalent form). At B is an ideal square wave generating characteristic.

At zero line of the input the output switches from one polarity to the other. Actually, in the input waveform is immaterial, because the only place where it affects the output waveform, is where it crosses the zero line. The remaining parts of Fig. 20-5 represent the top right part of the transfer charateristic, the bottom left part in each case being a negative duplicate of it.

At C is the transfer characteristic, of purely theoretical significance, that would duplicate the waveform produced by the first four terms of the Fourier series (up to 7th harmonic). It is thus a half-transfer characteristic representation of the same thing as Fig. 20-3. At D is the transfer characteristic corresponding to the synthesis of Fig. 20-4.

Comparison between C and D shows two extremes. E shows a transfer characteristic that includes 5 terms, in which the roots that determine where the stationary points occur have been chosen to limit variation from level to 2% (or $\pm$ 1%). This represents a series whose coefficients are approximately:

$$1.12574 \sin \omega t + 0.365706 \sin 3\omega t + 0.157857 \sin 5\omega t + 0.080879 \sin 7\omega t + 0.025213 \sin 9\omega t$$

These coefficients are noticeably larger (after the first) than those in the finite series based on no overshoot at all, but smaller than the Fourier.

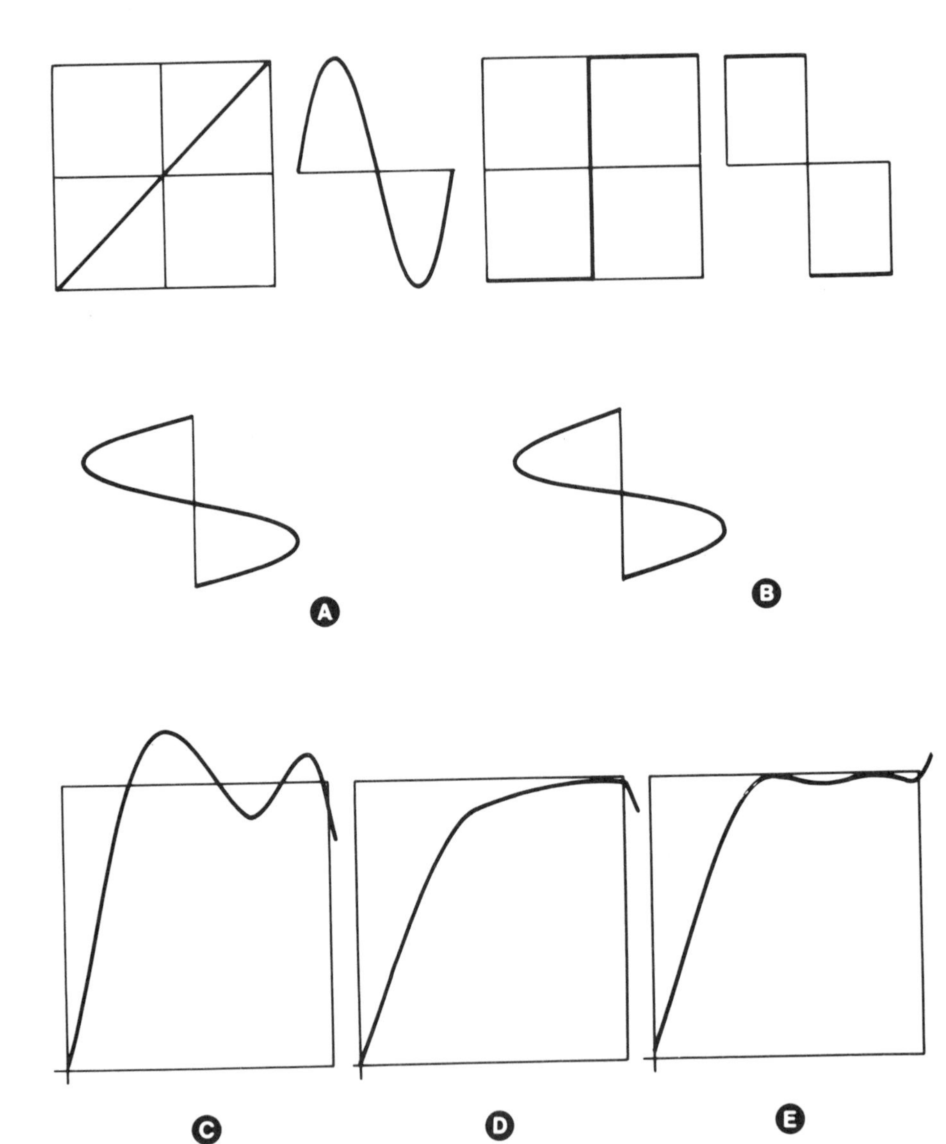

Fig. 20-5. Viewing the same thing from a transfer characteristic perspective: A. linear transfer with a sine wave; B. a perfect square wave conversion (only possible in theory); C. transfer characteristic to represent the Fourier series in Fig. 20-3; D. transfer characteristic to represent the limited finite series of Fig. 20-4; E. a compromise transfer curve that improves on both criteria.

The question for modern electronics is how to predict such responses, or how to design to a specific requirement. Such algebraic expressions, or in this instance, trigometrical ones, are continuous functions, and thus essentially analog, even if they are plotted by using digital calculators at discrete values along the curve.

If the purpose is to display a curve, a computer will program calculation of the complete function at successive values, using intervals close enough to give an appearance of continuity. But for design purposes, an approach that more nearly approximates digital is used: the first step is to program the computer to find the "stationary points:" where the maxima and minima occur, and to set those points to meet such deviation requirements. Then a different program can plot the complete display.

As a philosophy for systems design, the basis for modern systems is developed by what at one time would have been called "parts manufacturers." The electronic systems man puts together a system like a school boy at one time built something with an erector set, or a tinker toy. They are different systems of construction, which the boy—or the electronic engineer—first grasps as a set of parts, then builds whatever it is he wants.

Normally speaking, the parts of an erector set do not fit with the parts of a tinker toy set: they are completely different. The same is true of different microprocessor systems. Each manufacturer has designed, and they will continue to do so, a set of parts with a certain versatility with which they can be assembled into a variety of systems.

Generally speaking, just as with the boy constructing toys for his own amusement, it is best to work within the pattern used by the manufacturer in designing his "parts" for the system. But sometimes one system can do a job that either another one cannot do, or that it can do only in a much more cumbersome way. That is when it may be well to consider the possibility of hybridizing between systems.

To do that, you need to know considerably more about how the units of a system interconnect, than is necessary when constructing a system from parts that were made to be compatible. The tendency is to follow basics only so far. You understand, for example, how semiconductors function up through the material in this book to about Chapter 14. Then you want to take such details "as read." Since the manufacturer's designer has taken care of all those intricacies, why do you have to bother with them?

Well, for most of the time you don't. But you are always in better shape for knowing such basics. And when it comes to introducing new systems, you are much better prepared, than the type who just wants to know what he can do and what he can't do. By understanding those basics, you can answer those questions, instead of having to ask them. This capability also enables you to anticipate problems and avoid them.

EFFICIENCY ASPECTS

This can be considered in relation to the development of analog power, such as for driving a high power loudspeaker. It is always important, but this will introduce another kind of hybrid system. Figure 20-6 shows the basic elements of a "Class A" output stage, much used in the old days of tubes, but still used with transistors, for lower powers at least.

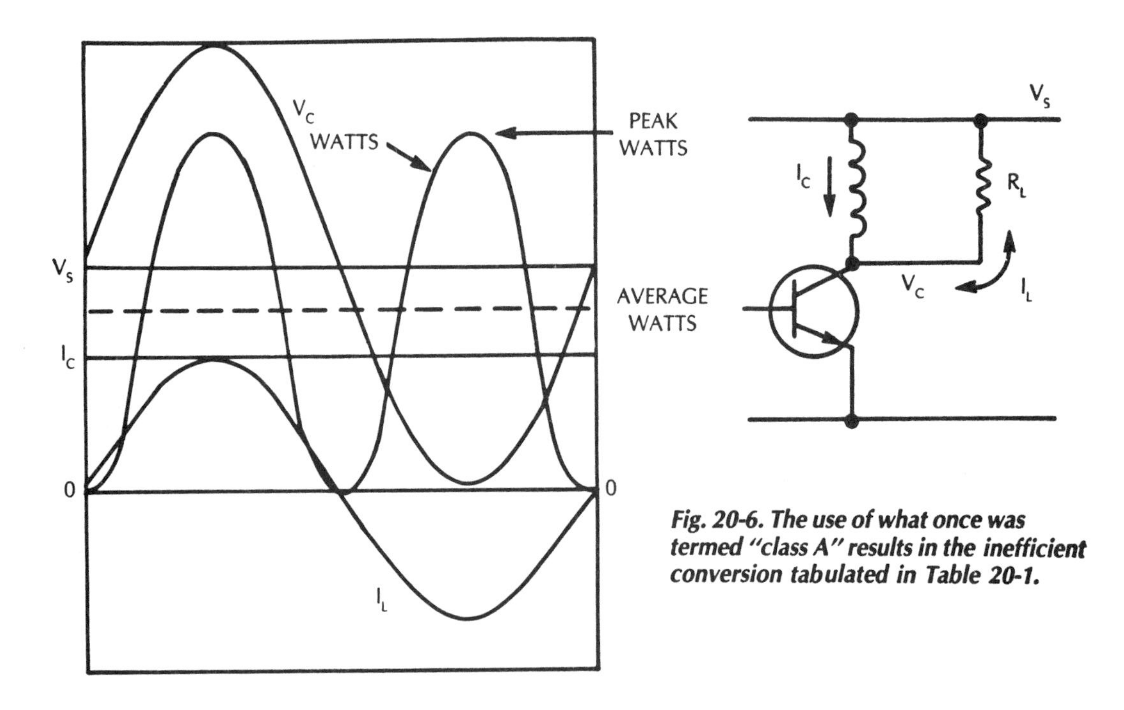

Fig. 20-6. The use of what once was termed "class A" results in the inefficient conversion tabulated in Table 20-1.

You need to realize that at all times, the supply voltage, V_S, is divided between the output device, shown as a transistor, and which in Table 20-1 we call a "valve," and the output load, designated R_L. The term valve is appropriate here, because it controls the flow of power, from the source, V_S, to the load. In this arrangement, for simplicity, an inductance carries a constant current, I_c from the supply to the junction of the valve and the load.

POWER	VALUE	LOAD	TOTAL	EFFY.
FULL	1	0.5	1.5	33⅓%
½ POWER	1	0.25	1.25	20%
¼ POWER	1	0.125	1.125	11⅑%

Table 20-1. Watts in Circuit of Fig. 20-6.

V_c is the voltage at that junction, produced by the input control to the valve, in "ways known to the art" as the patent specs say. Since the inductor carries a constant current, variation in the valve current must send positive and negative alternations, I_L, to the load. This power delivered to the load is the instantaneous product of V_c and I_L. This produces the watts curve shown, which has a peak value and an average value.

The average power in the valve does not change, being the product of V_s and I_c. Table 20-1 shows the power (average) for different power levels in the load: maximum (full), and half and quarter that amount. Adding the two powers together, and expressing the load power as a percentage of the total (supply) power gives the efficiency. Note that the maximum power is only half that the valve must dissipate all the time. In the days of tubes, it was lower than that.

Now, the next step from this was called "Class B." It is illustrated at Fig. 20-7. In it two valves share the job of supplying the load with power, each for half the time, or half the cycle, one positive, the other negative. Here, V represents the voltage produced across the load. I_1 represents the current supplied by one valve, I_2 that supplied by the other. The power curve, W, is shown for I_1. The other valve will have a duplicate power curve, in the other half cycle.

This circuit raises the maximum efficiency from the 33⅓% of Fig. 20-6 to 78%. You should realize that both these figures are theoretical values, only to be reached by "perfect" valve devices, so practical values are a little lower. But note that efficiency drops off at lower output levels, with actually more power being dissipated in the valves at lower levels, than at full level.

This led to another kind of hybrid system, illustrated in Fig. 20-8. These are waveforms associated with pulse width modulation. The little waveforms in circles are the changing shape of the pulse waveform, at a "carrier" frequency that is many times that of the signal. The modified sine wave could be an audio waveform, "carried" by the ultrasonic frequency shown sampled in the circles.

This system operates by having the output valves always either "on" or "off." So they have either zero current, or zero voltage, at every instant. When they are "on" the full voltage is passed to the load, with whatever current accompanies it. When they are "off" the load is momentarily disconnected. Like the Class B system, separate "valves" supply the two halves of the waveform.

Notice that this system produces an inherent discontinuity, between no pulse at all, and a minimum pulse width, and again, between the maximum pulse width, and "full on" with no separation between pulses. That is why the output is not a sine wave: we would like it to be.

Figure 20-8B shows how this defect can be eliminated. Instead of both pulses being off, at the zero line, both are limited, by appropriate circuitry, to little "spikes" that don't go full height (shown in the circle pointing to the zero line). Small inputs reduce one set of spikes, and increase the other, alternately. Then, when one set of spikes disappears, the other grows to full height—fully "on" for a brief part of the carrier cycle—then the width grows till it fills the time space, and finishes up with little inverted spikes, to eliminate the other discontinuity.

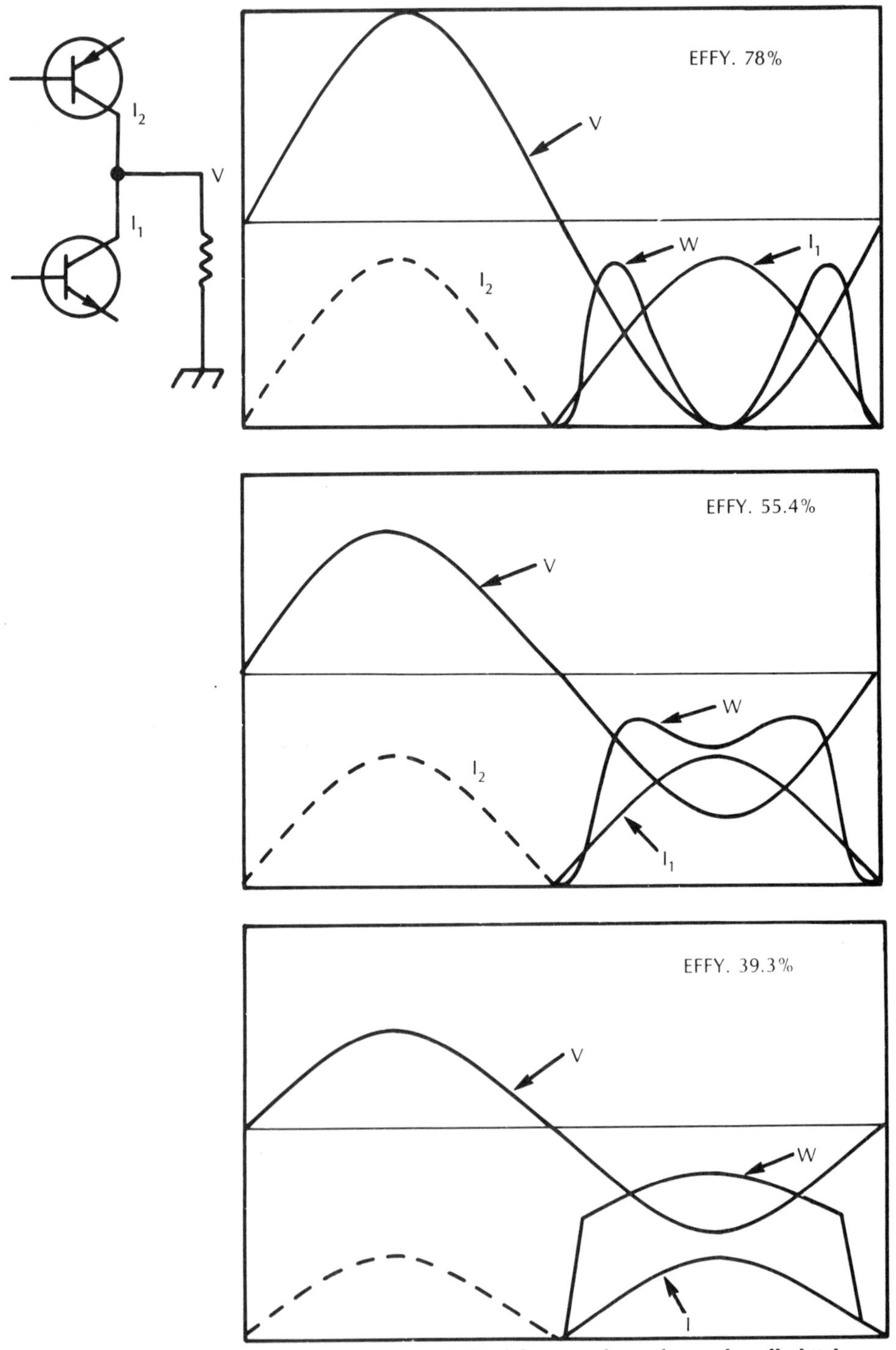

Fig. 20-7. Use of separate outputs for each half of the waveform, formerly called "class B" raises possible efficiency from a maximum of 33⅓% to 78%, with dramatically reduced levels of efficiency at all but maximum output level.

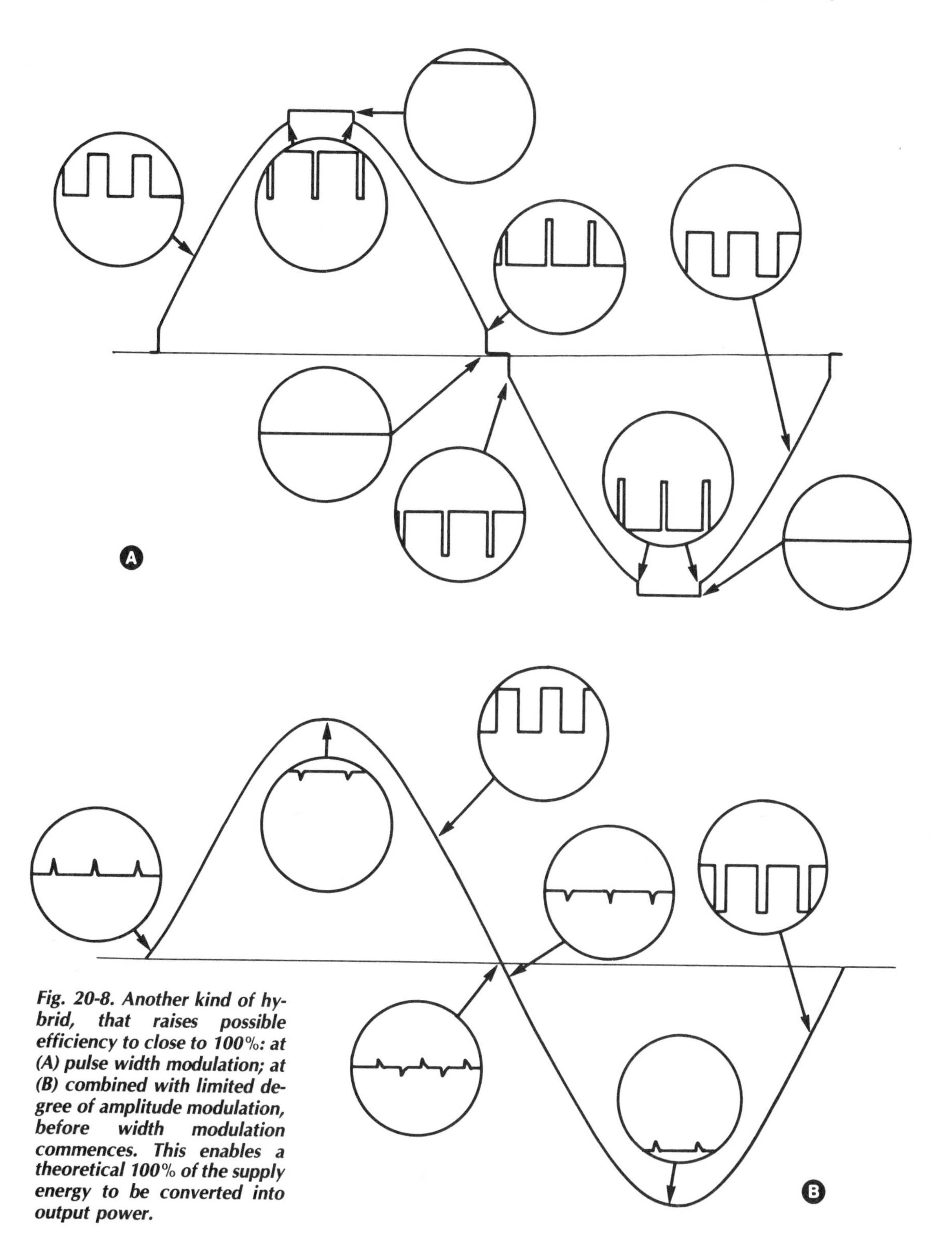

Fig. 20-8. Another kind of hybrid, that raises possible efficiency to close to 100%: at (A) pulse width modulation; at (B) combined with limited degree of amplitude modulation, before width modulation commences. This enables a theoretical 100% of the supply energy to be converted into output power.

We showed this system (without giving details of its implementation, which can take many forms) to illustrate another concept of hybrid. The output valves in such a system are essentially two-state, like those used in digital system. Except for the tiny spikes, which only appear at certain quite limited intervals, the output valves are always either "on" or "off" just as in all other two-state systems.

But to eliminate the discontinuity inherent in such systems, a tiny bit of a different form of modulation is introduced. It's not analog, either, but it uses properties of the electronics not unlike analog.

This is not to be confused with analog to digital, and digital to analog conversion in systems. There, any particular stage is either analog or digital, never both. And if it finishes up with analog, necessary to drive a loudspeaker, the efficiency problem is still with us. With this system, efficiency can be made very high, close to 100%, not only at full output, but at lower outputs too. But to get from a digital system to this kind of output, would normally require a digital to analog conversion first.

In this chapter, we have aimed to get you to think in terms of different kinds of system, which combine analog and two-state properties in different ways.

Examination Questions

The following questions are based upon a signal that by analysis consists of the following component frequencies: 10 volts of 1 kHz; 1 volt of 2 kHz; 0.5 volt of 3 kHz; 0.25 volt of 4 kHz; 0.1 volt of 6 kHz; and 0.05 volt of 9 kHz.

1. Assuming that these frequencies are possible selections from the band beginning at zero, over which any frequency at 100 Hz intervals may appear, in any amplitude down to that of the 9 kHz component, and up to the 1 kHz component, and that the analysis is to be sampled every millisecond, find the bandwidth needed to convey this information over one channel.
2. Assuming the precise amplitude within the range specified in Question 1 is to be conveyed, as a waveform conversion to digital, sampled at intervals corresponding to 16 kHz cycles, what bandwidth will be needed to convey the information in this form, over one channel?
3. Assuming the frequencies named are selected from 8 discrete possible frequencies, at a selection of 8 possible levels, and that data on this synthesis is to be conveyed every 8 milliseconds, what bandwidth is needed to convey this data in its simplest form?

21
On Keeping Up with Change

Electronics is undoubtedly the most rapidly advancing branch of technology today. The first edition of this book would have been outdated before it came from the printer, if it had attempted to convey to the student the latest in "basics." But it did not. It provided the student with the foundation for keeping himself up to date with changes, for which reason the book remained in print much longer than most.

We could make this second edition much more "up to date" by giving details of what, as this is written, is the latest development, but which will probably be obsolete within a couple of years. But knowing the real basics that underlie all of these developments will enable you to keep up, even though this book makes no attempt to tell you the "latest."

Electronic technology has advanced by what have come to be called "generations." Biological generations have overlaps—there are no precise years which can be considered the province of each generation—and there are individuals whose age comes between the ages of others belonging to successive generations. Thus a person 30 years of age comes between the generations of a father and son, whose ages are 40 and 20 respectively.

The same is true in the advance of electronic technology. Let us trace a few of them, so you can see what this means. Back in the "primeval past" experimenters found out many of the things we have covered somewhat cursorily, but necessarily in the early chapters of this book. Then came radio. Several experimenters theorized that such waves were possible and set out to find them, or produce them, with success.

At that stage, even finding that they had been successful presented problems: how do you know something is there, that you cannot see or hear? Accidentally, someone discovered what later proved to be the rectification properties of what later was formalized into a diode, but then took the precarious form of a crystal and catwhisker. Phone transmission had to wait for the vacuum tube to be invented.

Because telegraphy came first, radio was known for years as "wireless telegraphy," even after "telephony" would have described it more accurately. Had the diode been perfected first, and progress then moved on to the transistor, before anyone had invented a tube, who knows how different this development might have been?

In those days tubes were quite large. A receiving tube, which did no more than a single transistor (of which there are now thousands in the average microchip, that can sit on your little finger tip) was as big as a light bulb. A good receiving set used from 3 to 5 of these. Building a modern computer from such primitive elements, if it could be done, would occupy a whole city, and the time taken for what today's computer takes microseconds to do, would have lasted all year.

Then there were big tubes, used for transmission, to get the much higher power needed. A couple of these (they were used in pairs) could heat your house as a byproduct. Today's transmitters have gone to solid state, and size, as well as heat, has reduced enormously.

During this era, Dr. Harry Olson, of RCA labs worked on an electronic synthesizer that would look primitive indeed alongside today's music synthesizers. And it sounded crude too. It occupied a whole lab to itself, and took many run-throughs to synthesize just one note, or short phrase of music. It could never have been played as modern musicians play theirs today. But it was a beginning: the proof that it could be done. Those who saw the possibilities set about doing it, as newer technology became available.

The start of the "revolution" came with the transistor, a device that began to replace the vacuum tube, in a package about ¼" round by ¼" high, instead of the size of a light bulb, and that quickly showed it could do far more than tubes ever could, although it was limited in some respects at first. Even then it was the beginning of a "size" revolution, because the "guts" of that ¼" × ¼" package were already so tiny you needed a microscope to see them. That size was only needed to make connections.

The next step was the development of monolithic chips into integrated circuits (ICs). A whole new technology had dawned. New parameters came into use, and digital technology really came into its own, where before everything has been analog. An example of this follows.

The original simple flip-flop—its full name was multivibrator, or bistable device, according to use—was characterized by extremely high speed switching from one bistable state to the other. In each stable state, one transistor was on and the other off (refer back to Fig. 14-14). When it flipped (or flopped) for a moment it was an amplifier with high gain and very high positive feedback, which didn't quit till it came to rest in the other stable state.

Newer units with a similar purpose contain many more components, which do not all function simultaneously during its active moments. This results in a more successive transition from one stable state to the other. One outcome of this is shown in Fig. 21-1. Refer back to Fig. 18-6 for a moment. This circuit generated an approximation to a square wave, based on the bistable operation referred to in Fig. 14-14.

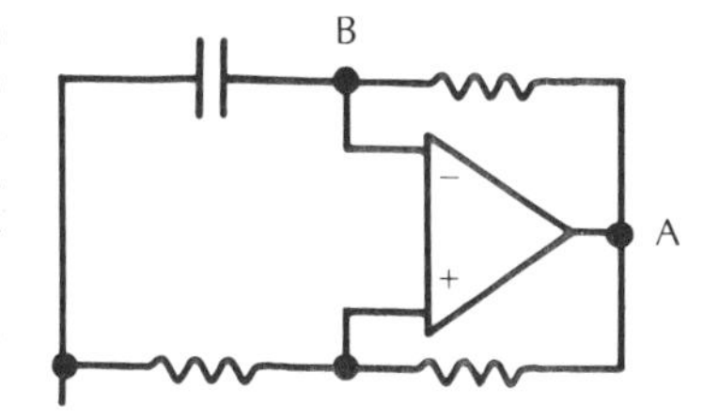

Fig. 21-1. Variations in the use of an op amp as a generator: (A) the theoretical use as a square/triangular wave generator; (B) the intrusion of slew rate limits the squareness of the wave, and the sharpness of the triangular form; (C) proportioning the time, so the slew rate occupies a critical part of the waveform, the intended triangular form closely approximates sinusoidal; (D) going beyond the limit represented in (C), the former square wave output becomes triangular, and the B waveform departs from sinusoidal in the opposite direction.

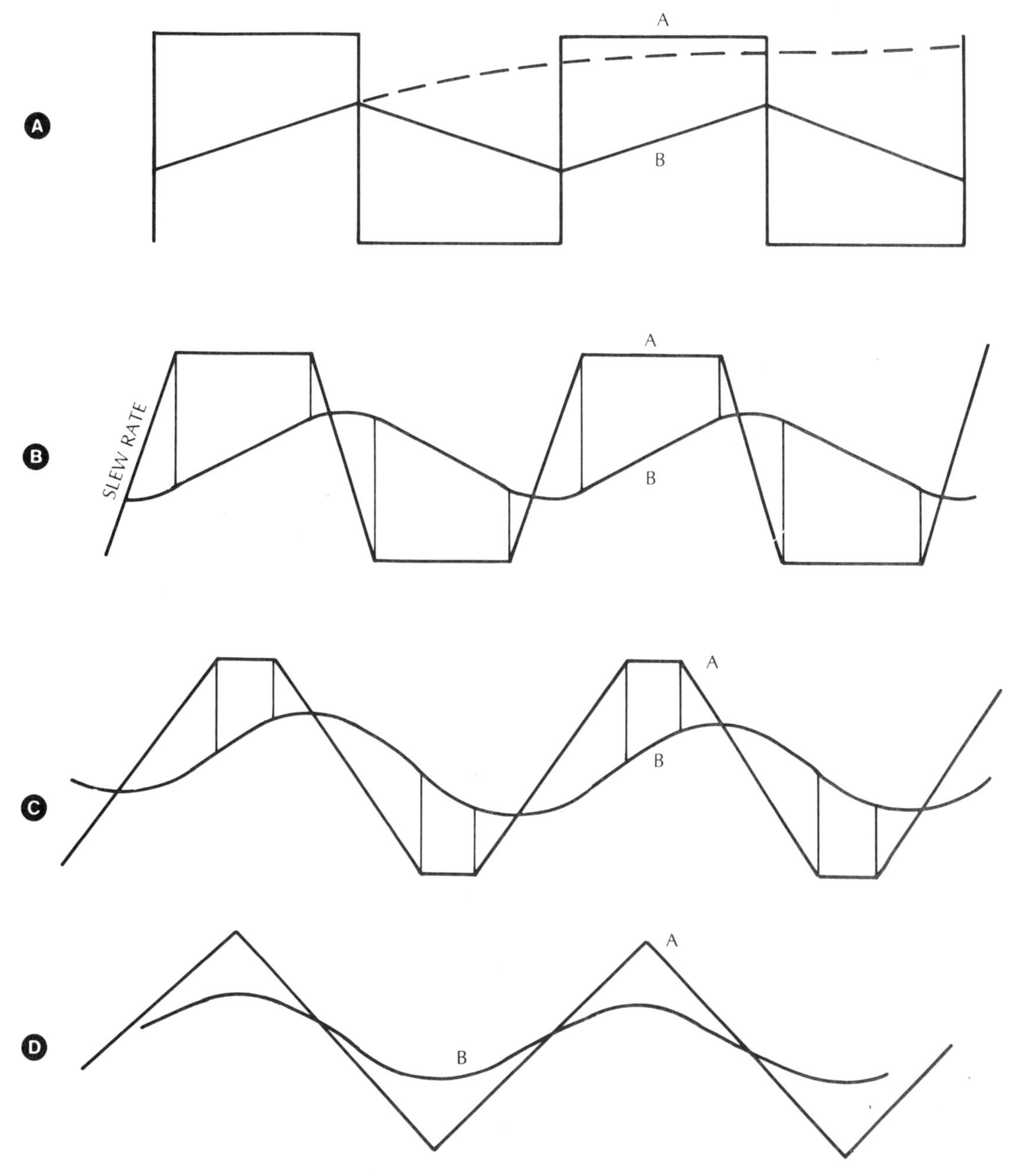

The limitations were noted in Chapter 18, but the transition was very fast. In Fig. 21-1 is shown a circuit that combines the actions into one package, so the same capacitor serves the purpose of the two capacitors in Fig. 18-6. As we have explained earlier, the triangle is a symbol for an op amp—operational amplifier. It is assumed to have the necessary plus and minus supply voltages connected. It has two inputs, an inverting (−) and a non-inverting (+). This refers to the relationship between input and output (at the point of the triangle).

The op amp is a differential amplifier (a simplified one was shown at Fig. 13-24) with a very high gain (more than 70 dB, or 3000:1 is typical) so that, unless the two inputs are within a millivolt or so of each other, the output "hits the rail"—goes to supply positive or negative, as the case may be. Thus, if the + input is positive of the − input, the output goes to supply positive. And if the + input is negative of the − input, the output goes to supply negative.

You can reason this either of two ways: referencing the + input, the output goes to the same polarity by which it differs from the − input; referencing the − input, the output goes to the opposite polarity by which it differs from the + input. Then, when the output is not "on the rail," meaning that the two inputs are very close together in voltage, a change on the + input produces a much amplified version of the same change at the output. Correspondingly a change on the − input produces a much amplified, but inverted change at the output.

Now, to explain Fig. 21-1. At (A) is the theoretical operation. First, the output being on the + rail, the capacitor charges until the − input reaches the same voltage on the + input, then the output switches to the − rail. This changes the voltage on the + input, the two resistors acting as a voltage divider, so now the capacitor discharges until the − input drops to the same voltage as that on the + input.

The square wave appears at the output, and the triangular wave, synchronized with it, appears on the − input, marked as A and B in Fig. 21-1. That's the theoretical operation, and pretty much how it operates, provided you don't push the frequency too high. As frequency goes higher, you find that the "verticals" of the square wave are no longer vertical (B). That is characteristic of this type of device.

This is something we didn't run into in the more primitive bistable units, so it has a name: *slew rate.*

It has an absolute value that cannot be "beaten" in that particular device. There was something like it in the device of Fig. 18-6: each transistor has a cut-off frequency, that may be specified in various terms. The one that counts here is the frequency at which its β (beta) drops to unity. This limits the speed at which either transition takes place. But each way, the transition takes place simultaneously.

In these more complicated devices, a cut-off frequency is less relevant. The transition is better and more accurately described as a rate of voltage change, usually given in volts per microsecond (V/μsec). The "workhorse" 741 op amp (which just may be out of date by the time you read this) has a slew rate of typically 0.5v/μsec. More recent developments have pushed this to as high as 50v/μ (and maybe even higher by the time you read this).

But now look at what happens when you "push" into the slew rate. At (B) you can see that, as well as making the sides of the square wave slope, the "points" of the triangle wave get rounded. Push a bit further, and the "B" waveform becomes a very close approximation to a sine wave. Push even further, and the "A" waveform becomes triangular, because the two slew rates bump into one another, and there's no "flat portion" to the "square wave!" Then the "B" waveform distorts in a different way from that when the slew rate is a relatively small part of the waveform.

These are the kind of changes you need to accommodate to, with progress. Now let's look at another example.

TOUCH DIALING

This is a good example, because it shows several aspects that must be considered in making progress in systems design. Fig. 21-2 shows the layout of a "push-button" phone, which replaces the older dial system. Now let's consider the problems involved in making such a transition. At first, people on telephone exchanges that used dialing equipment could not use a push button phone. They had to stick with their old-fashioned dial phones.

That was possible, of course, but sometimes inconvenient. If subscribers didn't need it, communication between exchanges that used the different systems did. So it had to be developed, to make any intercommunication, or transition possible. Had this not been done, people on dial-phone exchanges could have called only other people on the same kind of exchange, and people with touch tone dialing could only talk to other people on touch tone dialing. You can see how inconvenient that would have been.

So a way had to be devised to translate data, entered by either dialing or touch tone, from one to the other, so a different system could respond to it at the other end. How is this to be done? Let's take a closer look at the touch tone system, which after all must be responsible for achieving compatibility: you cannot, all at once, change millions of dial systems already in use.

The touch system uses 7 frequencies (the system is designed for 8, but one is "spare" for the time being). Four lower frequencies determined which row a button is pushed in, and three (later four) higher frequencies tell which column a button is pushed in. Thus every button pushed initiates two tones, one from the lower four, one from the higher three (or four).

At the other end, if it is a touch system, sensors pick out the two frequencies, and make the appropriate circuit selection, usually with an electronic selector, rather than the old mechanical dial. If it connects into a dial system, each pair of tones will "instruct" the dial where to go, the remaining "digits" being stored, if the data comes in too fast to handle, until they can be sequenced.

Conversely, where a dial code comes into a touch system, the reverse conversion must be applied: each number round the dial will generate the proper two tones for that number.

One more thing is shown in Fig. 21-2. Each button must "send" two frequencies, which means that, either by contacts, or an electronic equivalent, it must make those connections. If it used electronic switching, there may be no problem, but mechanically made contacts may pose a problem: the two frequencies may not be connected exactly simultaneously. They need to be, for the system to work correctly. So another function is included: a button depressed signal.

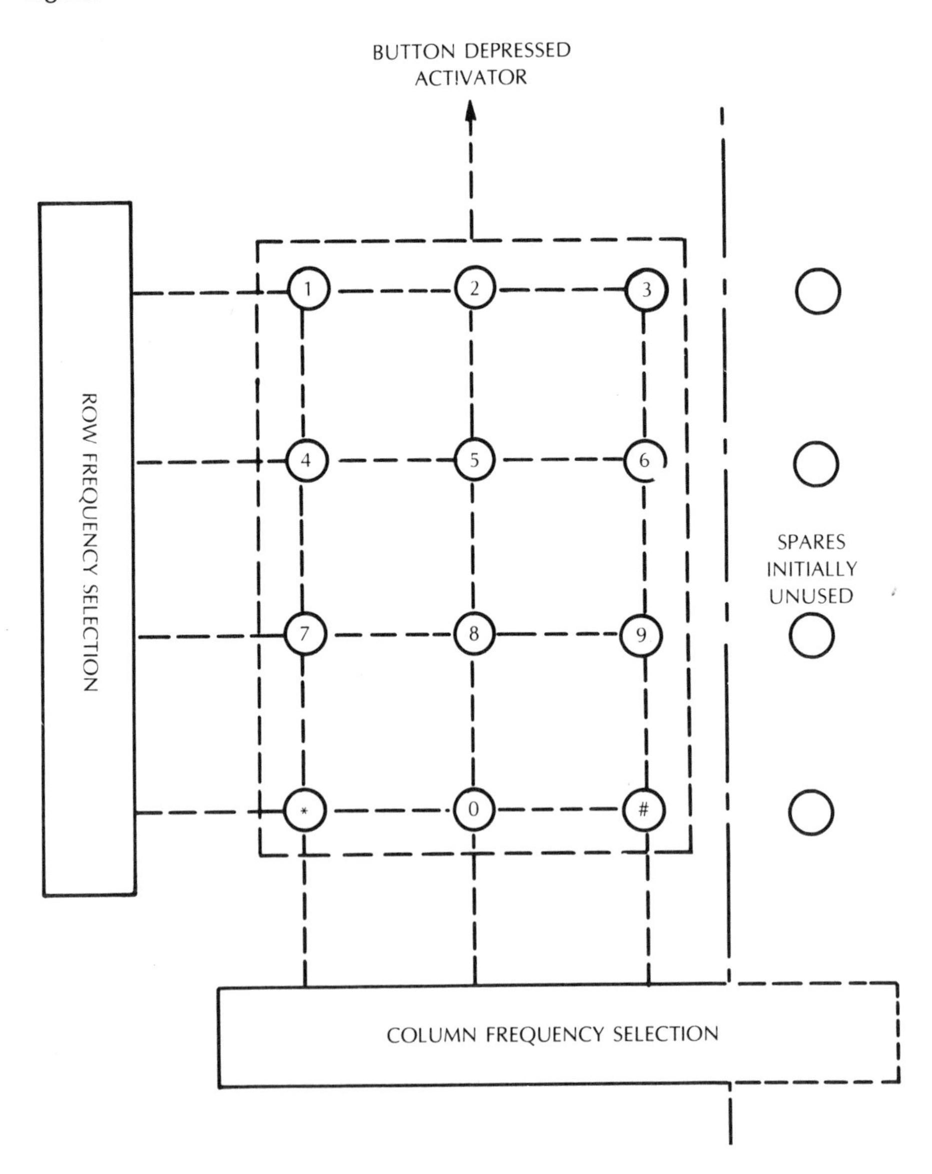

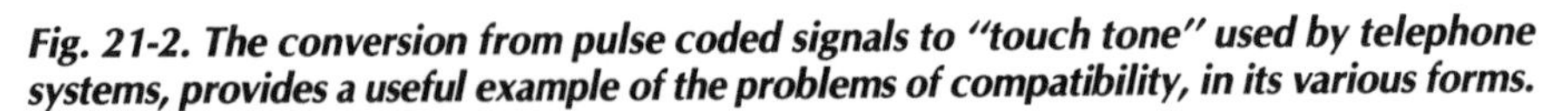
Fig. 21-2. The conversion from pulse coded signals to "touch tone" used by telephone systems, provides a useful example of the problems of compatibility, in its various forms.

In the mechanical version, this is the last contact to be made. It means that the other two will not be activated, although the contacts are made, until the last contact is made. This format is useful beyond this simple precaution: it provides signal separation in making stored-signal transfers. You remember we said that, in transferring to a dial system, the dial works more slowly? Well because of this, numbers entered by touch tend to "stack up" during transfer to a dial system.

Many telephone systems now store the whole number—however many digits it has, and that usually depends on the numbers used—before beginning to transmit any of them. This is how some telephone instruments are able to redial the last number you dialed, by pressing only one, or maybe two buttons: the instrument routinely stores the whole number before transmitting it.

You will recognize that the early digits tell how many are to follow it. Certain 3-digit number sequences are area codes. When these are entered, the system "looks for" however many more digits (usually 7 in the United States) are needed to complete the number. But if the first three digits are 011, dialed or pressed quickly enough not to allow the time delay to initiate the next step (such as calling the operator) this enters the international code section. After that, more digits will select the country, more again the city or area within the country, and the final however many digits are the number you are calling there.

Your local storage facility cannot know all the numbers in the world, to tell how many digits a particular call should have. That may not emerge until you have reached a point half-way round the world. So the system has the delay just mentioned, built in. It will store numbers you either dial in, or touch in, until you make a pause of more than a built in amount of time: then it "assumes" (actually it is completely non-intelligent, so this word is only an analog) that you have finished entering, and proceeds to follow out what your numbers "ordered."

If, in the course of doing this, something does not fit the requirements further along the line, a message is triggered back, that puts on a recording that tells you the connection cannot be completed as entered. If it reaches a "busy" intersection, that message will be sent back, urging you to "try again later."

In other points in a system, usually as part of an older system, you will be instructed to pause until you hear some signal. For instance, if you are calling some remote country, you may need to establish contact there, before you transmit the number you want. The system will then send an appropriate "ready" signal back, for you to complete your dialing, or touch-toning.

RULES FOR KEEPING UP

Those are illustrations that are, of course, already past. It is impossible to predict what has not been discovered yet! But you need to be ready for whatever comes. The basic rule should be obvious, but unfortunately is not: connect the new with something you already know about. The reason this is not more often done, is because too often people don't know anything about something that is quite familiar to them.

If you know how an existing system works, it is natural to ask how a new system works differently from the one you already know. But too often we just know how to use a system, without knowing how it works. The telephone systems we just gave a quick run-through is a good case in point. You know what and how to dial, with no knowledge of what that accomplishes, beyond hoping to hear the right voice on the other end.

You have seen, perhaps, one of the old-fashioned rotary selector exchanges, where hundreds of electro-mechanical switches make the connections. If you have, you can picture that happening. But with these electronic "switches" activated by tones—if you've seen the system, you haven't seen the "switches"—they're inside incredibly small microchips! You haven't seen the storage units that "hold" the number till you're ready for it to "go." Your phone may have one of those inside, but you can't see it, it's too tiny.

So it's much more difficult to know how things work today. You have to be interested in that, rather than merely what to do to make it work. Even those who design such systems most usually don't know how they work. All they know is a succession of "pinouts"—how to connect them to make them do what they're supposed to do, whether that's a clock to tell time and date, or a computer to perform the 4 functions, with memory and mathematical functions perhaps. Connect it up right and it does it: that's all many people know.

But when something new comes along, if you can connect up how it's been changed from whatever preceded it, that's a big help to understanding how it can be used. That's in the general sense.

Now, in the more specific sense, whether or not you can do that, what you need to ascertain about a new system can be put under two headings, with scores of subheadings.

- What can the new system or device do that the old one couldn't? Or how has it been improved over the old one?

- What can the old one do that the new one can't?

There's usually something that had to be sacrificed.

After these basic questions come questions of speed, reliability, compatibility with earlier devices (or lack of it), and a whole lot more. At a more "nitty gritty" level—that of the components within—many chips are designed to need ground, a positive supply and a negative supply, which totals three supply connections. Some will accept only two, provided the associated components will accept the same. Microprocessors use zeros (0) and ones (1), or "highs" and "lows." A low, usually at or close to negative supply, is a zero. A high, usually at or close to positive supply, is a 1.

Then, as we have described, a recent development is the 3-state device. In these the output can be "low" "high" or "don't care!" When the circuit is in use, it is required to transmit a low or a high. But it is also permanently connected to a data bus, or some other bus, to which many more like it are also connected. So when each unit is not in use—i.e., when it is another unit's "turn,"—it has its output go "high impedance." This means that whatever some other unit puts on the bus will not affect this one.

This chapter was intended to give you an outlook that will help you digest how things work, in a little greater depth than most people do. It will pay you to know this some time. Today's trend results in more and more people asking "Tell me what to do now." What will happen when nobody knows enough to answer the questions everyone else is asking? We hope we have got you started on the road to being one of the people with more answers than questions!

Examination Questions

1. An op amp slew rate is given as 10v/μ. It is to be used in the circuit of Fig. 21-1 to generate a square wave, using a supply voltage of +10 and −10. How much of each square wave cycle is taken up with lack of verticalness due to slew rate?
2. At what frequency will the output wave become triangular, with no vestige of the square wave left?
3. At a frequency of 1 MHz, what will be the amplitude of the output wave?
4. A typewriter keyboard has 44 keys, and uses 2 shifts, one for capitals (from lowercase) and the other to allow the same keyboard to be used for microprocessor functions. If this keyboard uses parallel binary (each with high or low choices) how many circuits will it need?
5. If, for transmission over a distance, using "touch-tone" type signals, with two tones for each "character," how many tones would be needed?
6. If, instead of two tones, each "character" uses three tones, how many tones will be needed then?

Glossary

accelerator tube A device used to progressively accelerate electrons, beyond the velocity attained due to the supply voltage used.

admittance The reciprocal of *impedance*. A measure of the current taken divided by the voltage associated with it.

alpha A factor in current transistors expressing collector current as a fraction of emitter current, typically between 0.99 and 1.

ambient temperature The temperature surrounding a device or system. Room temperature.

amplification factor Any output quantity, divided by the corresponding input quantity (e.g., voltage/voltage or current/current) when other quantities are held constant. Originally applied to tubes about voltage amplification, it was then given the symbol μ (Greek letter mu).

amps Short for *amperes* the standard measure of electrical current, subdivided into milliamps, microamps, nanoamps, etc.

analog Generally, something that qualitatively represents another quantity; in modern electronics, continuously variable quantities, as opposed to *digital*.

analog IC An integrated circuit chip that by its design functions in variable quantities, but which may also be used in digital circuitry.

anode In any device the positive electrode, to which electrons are attracted within the device. Specifically, in British technology, the plate (q.v.) of a valve (q.v.).

audio Electrical or electronic equipment handling signal that represents sound.

base The central electrode of a current transistor, between emitter and collector. In a chip, the substrate on which the device is built.

beam tetrode A vacuum tube with only two grids, thus called a tetrode (having 4 electrodes) but with additional beaming electrodes that give it characteristics more like a pentode, in fact superior from some viewpoints to a pentode.

beta (β) A factor in current transistors obtained by dividing collector current by the corresponding base current, and equal to $(1 - \alpha)$ (see Alpha). Typically, according to type, somewhere between 10 and 10,000.

B-H A product used to rate a permanent magnet: B represents the magnetix flux density, H the magnetizing force associated with it. *B-H-Max* refers to the point on its demagnetization curve at which the B-H product is a maximum.

bias Any supply used to establish a required operating condition, in the absence of signal. May refer to voltage or current, or to the circuit that provides it.

binary The system of numbering using base 2. Or any system characterized by two state operation, otherwise called digital.

biquinary A system of numeration that uses 2s and 5s to build up a synthetic decimal system. Of great use in calculators that use the "rounding out" since absence of the '5' binary digit in the next place drops that place, while its presence raises the last presented place by 1.

bit In computer and microprocessor language, a bit represents a single binary element, which may be present or absent.

bridge circuit Any electrical circuit using two dividers in such a way that a measuring element across their junctions can detect a null.

byte A group of *bits* that represent a single, but maybe complex, piece of information in a computer or microprocessor system.

calibration Any process for using a known quantity to establish the accuracy, or determine the significance of, another instrument, value, or piece of equipment.

capacitance A form of *reactance* in which change of voltage is accompanied by current, but no current accompanies a steady (d.c.) voltage.

cathode The part of a device from which electrons are emitted or flow, usually the negative electrode.

CAT scan Computer Analyzed Tomography. A system for examining internal anatomy, using successive intensified X-rays (low-level X-rays using a fluorograph intensifier) and a computer to develop a picture of a certain plane within the anatomy, virtually obliterating data originating in other planes.

channel (1) a path along which signal or data can travel; (2) the conducting path in a field-effect transistor (FET) type of device.

chip A colloquial term for silicon substrate on which an integrated circuit (IC) is built or assembled.

circuit board A board on which an electrical circuit is etched, into which components are inserted to complete the circuit.

class of operation Classified as *Class A* in which complementary circuit elements are active throughout the entire signal cycle or period; *Class B* in which separate elements handle opposite polarities of signal; and *Class AB* using Class A for small signals and Class B for bigger ones.

C-MOS A series of devices designed to utilize a form of operation where little change in voltage and or current (at a given point in the circuit) results in complete reversal (from on to off, or vice versa). It has the advantage of providing a low-current, faster operating device.

coercivity The magnetizing force needed to reduce magnetism in a permanent magnet, or in an oxide tape, to zero.

coil An electrical conductor wound into parallel turns, so as to multiply the effect of current flowing in it, producing magnetization.

collector The electrode of a current type transistor to or from which current flows as controlled by the current flow between the other electrodes (base and emitter). The collector-base junction is normally reverse polarized, so no current flows without such control to permit it.

color code Any system of identifying values of components by the use of colors. The most used one is a code for resistor values.

conductance Reciprocal of resistance. A measure of the current conducted by an electronic element, divided by the voltage that causes it; where conductance is a component of admittance, it is that part of the admittance that is strictly proportional to applied voltage, instant by instant.

conductivity A measure of the specific resistance of a material, usually measured between opposite faces of a centimeter cube of that material.

contact potential Particularly in silicon devices, the forward potential needed to initiate current flow; this may vary, depending on the magnitude of current necessary to activate the associated circuit.

current The quantity of electricity flowing, as a rate of flow, as distinct from the potential that causes it to flow.

current electricity The study of electrical effects due to current flow, as distinct from that due to potentials and fields.

cut-off (1) in frequency, the point at which the behavior of a device departs from its normal operation to diminish its performance; (2) in electronic devices, such as transistors, the condition in which no current flows.

decimal A system of numbering using the base ten (the one in everyday use).

deflection The means used to move the passage of an electron beam.

dielectric In some senses, another name for insulation, but usually with specific reference to concentrate the electric field due to a voltage applied across the material.

digital Literally, something whose operation depends on counting; more specifically in electronics, a system that uses devices with two-state (q.v.) capabilities.

digital IC An integrated circuit in which the elements always operate with a two-state limitation.

diode A two-electrode semiconductor, which normally conducts current one way but not the other (although there are exceptions). However it may be characterized by other capabilities, such as emitting light (LEDs, light emitting diodes) or becoming opaque (LCDs, liquid crystal diodes).

dipole A device for emitting or receiving polarized electromagnetic radiation, or an element of such a device.

discrete components Single elements that have to be connected to other parts before they become operational, or that are separately fabricated (see *integrated circuits*).

dissipation Means for removing heat generated in the operation of a device, or the process of doing it. Also a measure of the heat so removed.

drain The electrode of a field effect transistor (FET) device that corresponds with the collector in a current type device.

duty cycle A measure of the relative time a device is operational, as compared with the time it is not.

electret A device using a substance that produces a permanently electrified or charged effect, similar to the way a permanent magnet is permanently magnetized.

electric field Influence exerted by the presence of voltage difference, or lines of electric potential.

electrode Any terminal of an electrical device having properties that are used by circuits to which it is connected.

electromagnet The use of electric current in a coil, usually with magnetic material within or associated with the coil, whereby electric current produces a magnetic effect.

electromagnetic radiation Any form of radiation characterized by possessing both electric and magnetic fields, each of which is at right angles to the other, each of which by its change sustains the other, in such a way as to propagate in a third direction, mutually at right angles to the directions of the two fields.

electromechanical transducer Any device for the interchange of electrical and mechanical energy, converting either one to the other.

electromotive force A source of electric potential, that tends to drive electric current. Abbreviated to EMF it signifies a voltage, where the voltage causes a current, rather than vice versa.

electron gun The part of a cathode ray tube, or other tube or device, that generates a beam of electrons and issues them in a converging ray.

electrophorus A simple device that uses electrical charges to demonstrate an effect.

emitter The electrode of a current type transistor that is normally in its conducting mode with the base, so that base-emitter current controls collector current.

fan-out A term applied to *integrated circuits (ICs)* referring to either input or output connections, that specifies the number of inputs the same terminal can receive, or the number of outputs the same terminal can supply, connected at the same time.

farads The basic measure of capacitance, represented if a change of 1 volt per second produces a current of 1 amp; normally microfarads and small subdivisions are specified or used.

feedback The closing of a circuit, by connecting the output of a system back to its input in some way.

feedback factor Essentially the amount by which the gain of a system is reduced or changed, as a result of connecting feedback, or the amount by which its connection affects other properties or characteristics.

feedback fraction The fraction of output fed back to the input.

FET (field effect transistor) Any of a variety of devices in which electric field within a solid-state material controls current flow.

filament In the earliest vacuum tubes, an electrically heated wire, called a filament, was the source of electrons.

Fleming's rules Convenient rules for remembering the electromagnetic relationships in motor or generator settings.

focussing Means used for focussing an electron beam, so it converges to a point.

form factor The relationship between the rms and rectified average values of a waveform. For a sine wave it is 1.11, for a square wave 1, and for a triangular wave 1.5.

frequency The repetition rate for a cyclical signal, measured in the number of complete cycles per second called *hertz*.

full wave Applied to *rectification,* a conversion from ac to dc in which both half cycles of the ac are utilized.

function generator A type of signal generator which is time-referenced, rather than frequency referenced, generating the shape of the wave consecutively, rather than as an oscillation (see *oscillator*).

fundamental frequency The basic frequency of a signal, to which all others relate. Usually it is the lowest frequency and the related frequencies are multiples of it, see *harmonics*.

fuse A protective device intended to prevent overload from destroying the equipment, or parts of it.

gate (1) a device for handling a variety of inputs and delivering an output indicating certain things about the relationship between the input signals present; (2) the electrode of a *FET,* or similar device, which controls the current flow between other electrodes.

gauss The unit for measuring the density of a magnetic field produced by a magnet.

gilberts The unit for measuring total magnetomotive force in a magnet.

grid A wire mesh electrode used in thermionic vacuum tubes, used to control the flow of electrons through it, by the voltage applied to the grid.

ground A connection that becomes a reference zero for voltage, whether or not it is actually connected to physical ground. In British publications, this is called *Earth*.

group velocity The velocity at which the energy in a wave travels, as distinct from the rate at which its change of state travels (see *phase velocity*). Group velocity can never be greater than *propagation velocity*.

half wave Applied to rectification, a conversion from ac to dc in which only one half or polarity of the waveform is utilized.

harmonic motion A rhythmic motion at a natural frequency, which means the form of the motion is basically sinusoidal.

harmonics Frequencies related to a fundamental frequency, usually as direct multiples. Thus a frequency three times that of the fundamental is usually termed the "third harmonic" (or in music, the "second overtone").

heater In thermionic tubes, an electric element associated with the cathode whose purpose is to raise its temperature to enable it to emit electrons.

henry The unit of inductance. If a change of one ampere per second produces one volt EMF, the inductance has a value of one henry.

hertz The unit of frequency, being the number of complete cycles per second. Formerly called "cycles per second," the symbol was ~, now replaced by Hz.

hysteresis In the magnetization cycle of a magnetic material, the lag of magnetization behind the magnetomotive force producing it, results in energy loss, which is termed hysteresis. Also applied to similar phenomena in other fields.

impedance The relationship between voltage and current in a device, measured in ohms. If a current (usually ac) of 1 amp produces a potential of 1 volt, the impedance is 1 ohm.

inductance A device that produces an electromotive force whenever current changes (also called an *inductor*). The commonest form utilizes a coil so the current produces a magnetic field, the changing of which generates an EMF. The unit of *inductance*, using the word as a property, is the *henry*.

induction coil The earliest kind of transformer, in which fluctuation of a current in the primary winding was effected by use of an interrupting device.

influence machine Any of a variety of machines that uses electrostatic effects to generate the high voltages associated with static electricity.

instrument multipliers Additional components to metering circuits, to extend the range of the instrument by some multiplier factor.

insulation Material through which electric current will not pass.

integrated circuits (ICs) Complete electronic circuits, often consisting of a great many elements, in which the whole entity is manufactured at once, rather than being assembled from separate parts.

ionization Separation of a gas into positive and negative ions, before it eventually "breaks down" to allow an electrical discharge to pass through it.

J-FET An adaptation of the junction transistor to *FET* elements, in which the role of collector and emitter (in current transistors) is reversible, or in FETs the role of the source and drain. Applied to IC technology in which the inputs to the IC are J-FETs.

keeper A piece of magnetic material put in place between the poles of a magnet, to prevent self-demagnetization when the working magnetic circuit is removed.

LCD (liquid crystal diode) A form of diode that is translucent until an application of reverse voltage renders it opaque.

leakage In a transformer or other device incorporating mutual inductance, some of the magnetic flux linkages from one coil do not engage the other; the linkages that fail to do so are called leakage. The effect has the properties of an inductance, which is called *leakage inductance*.

LED (light emitting diode) A form of diode that emits light (usually colored) when a forward current is applied to it.

linear amplification In general, amplification in which states intermediate between those used in a two-state system are utilized. Specifically, when steps are taken to make the relationship between output and input strictly proportional, such a system is called linear.

linear IC Chips, notably op amps, that fit the previous definition. Because of the very high gain built into them, negative feedback can make their operation extremely linear. However, even if such ICs are not used for linear purposes, the name still applies, to distinguish them from other ICs designed for two-state operation.

load In general, the impedance connected to the output of an amplifier or other device, which causes it to produce a combination of voltage and current as output, rather than one or the other.

magnetic field Primarily the influence of a magnet that produces forces with other magnetic objects due to that field; when combined with electric field in electromagnetic radiation the important feature is the rate at which the field changes.

magnetic flux The field from a magnet (see previous entry) is measured by intensity and called *flux*.

magnetic potential The capability of a magnet, or a magnetic source, such as a current flowing in a coil, to produce magnetization, analogous to voltage or potential with electricity.

magnetomotive force The production of magnetic potential, such as by current in a coil of wire.

maxwell The unit of magnetic flux. *Gauss* (q.v.) is a unit of flux *density*, i.e., maxwells per square centimeter, while maxwells are used to measure the total magnetic field from a given source.

micro- A prefix meaning the millionth part of.

milli- A prefix meaning the thousandth part of.

MOS-FET A type of integrated circuit utilizing the advantages of both the MOS and FET technology (see C-MOS).

mutual inductance Applied to two coils whose magnetic fields interlink, so that changing current in one of them will induce an EMF in the other, as well as itself.

nano- A prefix meaning the one billionth part of.

ohm The unit for measuring resistance, reactance or impedance: if a current of 1 amp produces a voltage of 1 volt, the resistance, reactance or impedance is 1 ohm.

Ohm's law Simply states that the relationship between voltage, current and resistance is linear.

oersted The unit of magnetic potential, per centimeter length (i.e., gilberts per cm).

op amp (operational amplifier) One of the commonest IC units in modern electronics, it is a unit, multipurpose, differential (having 2 inputs) amplifier. A single IC may contain a number of op amps.

oscillator An electronic device that produces a natural frequency, determined by electrical elements such as inductance and capacitance.

parallel connection Connecting elements so they each share the same voltage, but the current divides between them.

parallel data The use of a number of conductors, in a *data bus* (or other system *bus*) each of which carries one element of the data, all at the same instant in time.

pentode A thermionic tube consisting of a cathode, plate and three grids, making five electrodes in all.

permanent magnet A magnet with high retentivity, so that its magnetism remains after the magnetizing force is removed.

permeability A measure of a magnetic material's readiness or capability of being magnetized. Compares magnetization present when the material is in place with what would be present in its absence.

phase velocity The rate at which a certain condition of a wave passes along a waveguide tube or circuit, independently of the conveyance of actual energy along with it. Phase velocity can exceed propagation velocity.

pica- A prefix meaning the same as micro-micro-.

piezo-electric A property that converts mechanical stress into electrical potential, and vice versa.

pixel In a matrix system, such as may be used for constructing a picture, a pixel is an element where two coordinates intersect, or the electrical counterparts that produce an element of image at that point.

plate The common name for the *anode* (q.v.) of a thermionic tube.

polarizing (1) the orienting of an electromagnetic wave so that its component fields are in precise positions, mutually at right angles; (2) the part of a circuit that provides necessary voltage to render it operative.

potential Generally, another name for voltage, without regard to whether such potential exists as a cause or an effect.

potential divider Networks in series designed to apportion potentials applied to them in specific proportions.

potentiometer Name for a component (possibly variable) used as a potential divider.

power factor In a dc circuit, power is the product of voltage and current; in an ac circuit, because voltage and current may not coincide in timing or phase, power can be less than that product. Power factor converts the product of voltage and current to actual power in the combination.

preferred values A set of values designed to make selection and availability easier.

printed circuit (PC) A common name for an etched circuit board.

progagation velocity The speed at which a wave propagates, or travels.

pulse A concentration of energy into a short time, with spaces between such concentrations.

push-pull Before the advent of complementary symmetry circuits, similar devices (rather than complementary) had to serve both halves of a waveform. Such devices were said to operate in push-pull.

Q value In a resonant circuit, this is an efficiency factor, obtained by dividing the reactive component(s) by the resistive component. The higher the numerical Q number, the more efficient the resonance effect.

reactance A circuit element in which, on an ac signal, voltage and current do not occur together, and which is thus characterized by zero power factor. The two kinds of reactance are inductance and capacitance (q.v.).

rectification The process of switching alternate halves of an ac waveform so that they all pass in the same direction.

rectified average One way of measuring an ac waveform, by rectifying it and then measuring the average value of that.

reflector In electromagnetic radiation, a device that concentrates or modifies the radiation in the same way a mirror does light.

reluctance The relationship between magnetic potential and magnetic flux in a magnetic circuit, analogous to resistance in an electrical circuit.

remanence The magnetism remaining in a magnetic material after the magnetizing force has been removed.

resistance The relationship between voltage and current, primarily in a dc circuit, but also applied to the in-phase, or power components in ac circuits.

resistivity Reciprocal of conductivity. Generally, the resistance of a centimeter cube of the material, measured between two opposite faces.

resonance A condition when reactances of opposite kinds balance one another, at what thus becomes a natural frequency of the circuit.

retentivity The magnetic density remaining after the magnetizing force has been removed, when no magnetizing force is present.

reverberation Sustaining sound by means of acoustic reflections in a building or auditorium; or simulating such effects electronically, by various means including the introduction of delay times.

rms (root mean square) A measure of an ac waveform, obtained by squaring a virtually infinite number of samples of the whole wave, taking the mean, or averages of these squares, then taking the square root of that average.

saturation A condition in a semiconductor when further change in the control circuit produces no further increase, or an immeasurable increase, in the circuit controlled.

sawtooth A waveform characterized by uniform rate of change in one direction, with a very rapid return to its starting point.

semiconductor A device or material characterized by allowing electrical conduction in one direction only.

serial data The presentation of individual bits in every byte of data, in sequence over a single channel, rather than in parallel over many channels.

series connection Connection elements so they share the same current, but have different voltages across them.

shunt A component connected in parallel, usually to bypass some of the current.

signal A general name for changes that pass through a system and are handled by it, regardless of what they represent, such as audio, video, or virtually anything.

sinusoidal A waveform shaped like a sine wave, and thus representative of a natural oscillation.

slew rate In ICs, the maximum rate at which an output voltage can change, specified in volts/μsec.

solid-state Earlier electronics relied on behavior of elements in evacuated spaces, such as vacuum tubes. Modern electronics uses materials in "solid" form, without evacuation, and thus is termed solid-state.

source In a FET type device, the electrode to which the gate potential is referenced, and from which the current, of whichever polarity, may be regarded as originating.

split load A circuit in electronics, where an amplifier stage has part of the load in each "leg" of its output circuit, for instance collector and emitter.

standard cell A reference for voltage formerly used, in which the components of the cell were carefully specified. However, all standard cells become unreliable when connected to even the smallest current drain, so they are little used now solid-state devices provide better references (e.g., zener diodes).

static electricity The study of electrical effects due to the presence of electrical charges, in which current flow plays little effect.

substrate In IC production, the base layer, usually silicon, on which all the component elements are built.

superconductivity A condition at temperatures close to absolute zero, at which conductors lose all resistance.

temperature coefficient A measure of the rate at which a quantity changes its value as temperature changes.

temperature rise Increase in temperature due to the operation of a device.

tetrode A thermionic tube consisting of a cathode, anode and two grids, making four electrodes in all; any other four electrode device.

thermionic tubes Electronic devices contained in evacuated tubes, or tubes containing inert gas at low pressure, and consisting of a heated cathode and other electrodes.

time constant In a simple resistance-reactance combination, the time it would take for the quantities (voltage and current) to reach their ultimate values, if the change continued without changing from its initial rate.

tomography The use of X-rays with multiple exposures, superimposed or so combined that details in a certain plane are emphasized, and details outside that plane are neutralized, like a picture out of focus. Focusing that works with visible light cannot be used with X-rays, so use of this method is an alternative that yields a similar result.

transconductance The relationship between current in a second circuit, and the voltage causing it, as a control, in a first circuit. See *conductance*.

transducer A device for converting electrical or electronic signals into mechanical or acoustic ones, or vice versa.

transformer A device applying mutual inductance in its most efficient manner, so that circuits with different voltage/current requirements can be matched.

transistor Primarily a device consisting of two diodes with a common base, so arranged that one influences the other in a prescribed way.

triode A thermionic tube consisting of a heated cathode, a plate or anode, and a grid, making three electrodes in all. Other three-electrode devices.

tri-state devices In digital circuitry, particularly computers and microprocessors, devices whose output, like the two-state types, may be high or low, but in addition a third state is possible: high impedance, which means high or low may be determined elsewhere, without affecting the tri-state device that is in this third state.

two-state A method of operation, in which no intermediate states can occur: each device in the system is either "on" (conducting) or "off" (nonconducting) or the voltage at a terminal is either "high" or "low."

vacuum tube A tube containing electrodes for an electronic device, from which the air or gas which would otherwise occupy them has been evacuated, as a necessity for its proper operation.

valve The British term for "*tube*."

variable-mu A special type of tube, or other device, capable of having its gain varied over a wide range.

vector representation A method of showing the behavior of circuits by use of geometry.

video The electronics and other equipment that handle the presentation of pictures by the scanning method originally developed for television.

voltage A measure of electrical "pressure"—the force needed to drive current in a circuit.

voltage divider A network that divides voltage in prescribed proportions or manner.

volts The standard measure of voltage.

watts The standard measure of power: volts times amps equals watts, unless a power factor (q.v.) is involved.

waveguide A sort of pipe which confines electromagnetic propagation of prescribed frequency along its inside.

zener voltage The voltage at which a diode begins to conduct when a reverse voltage is applied to it. Diodes that use this property are called *zener diodes.*

Answers to Selected Examination Questions

CHAPTER 2

1. RED WHITE BLACK GOLD
   GREEN BROWN BLACK GOLD
   BROWN BROWN BROWN GOLD
   ORANGE GREEN BROWN GOLD
   YELLOW BROWN BROWN GOLD
   YELLOW GRAY BROWN GOLD
   YELLOW ORANGE RED GOLD
   RED RED RED GOLD
   BROWN BLACK RED GOLD
   ORANGE GREEN BROWN GOLD
   BROWN BLACK BROWN GOLD
   BROWN BLACK BLACK GOLD

3. 650,000 ohms 5 percent
   95,000,000 ohms 5 percent
   (95 megohm)
   2,700,000 ohms 10 percent
   (2.7 M)
   8.2 ohms 10 percent
   3600 ohms 5 percent
   6.8 ohms 20 percent
   4.7 ohms 20 percent
   470,000 ohms 20 percent
   84 ohms 5 percent
   10 megohms, 5 percent

5. 27.55 and 30.45
   48.45 and 53.55
   104.5 and 115.5
   332.5 and 367.5
   389.5 and 430.5
   456 and 504
   4085 and 4515
   2090 and 2310
   950 and 1050
   332.5 and 367.5
   95 and 105
   9.5 and 10.5

7. 617,500 and 682,500
   2.43 M and 2.97 M
   3420 and 3780
   3.76 and 5.64
   79.8 and 88.2
   90.25 M and 99.75 M
   7.38 and 9.02
   5.44 and 8.16
   376,000 and 564,000
   9.5 M and 10.5 M

9. 30 microamps

11. 1 volt (approx)

13. 23.55 ohms

15. 43.35 ohms

17. 3.11 K and 4.29 K

19. 2.925 K and 3.69 K

21. 6.79 K

23. 6.111 K and 7.469 K

CHAPTER 3

1. V1 equals V2 equals 4 volts; V3 equals V4 equals V5 equals 8 volts; I1 is 400 mA; I2 is 267 mA; I3 is 333 mA; I4 is 222 mA; I5 is 111 mA. 400 + 267 is 667 mA (two-third amp, more exactly); 333 + 222 + 111 is 666 mA (also representing two-third amp).

3. The internal resistance of the battery is 0.3 ohms.

5. Current is 1.67 (or one and two-thirds amps); voltages are 1.367 and 1.133; 1.367 + 1.133 or 2.5

7. 3 K

9.

4 volts	16.7 ohms
6.7 volts	22.2 ohms (or 6 and two-thirds volts)
3 volts	20 ohms
15 volts	66.7 ohms
27 volts	132 ohms

11. Before connecting R5: V1 is 6 volts, V3 is 9 volts, I1 equals I3 equals 150 mA; V2 is 9 volts, V4 is 6 volts, I2 equals I4 equals 100 mA. After connecting R5: I5 is 15 mA; V1 is 6.36 volts; I1 is 159 mA; V3 is 8.64 volts; I3 is 144 mA; 159 is 15 + 144; V2 is 8.46 volts; I2 is 94 mA; V4 is 6.54 volts; I4 is 109 mA; 94 + 15 is 109; V5 is 2.1 volts; 6.36 + 2.1 is 8.46; 6.54 + 2.1 is 8.64.

13. 50 ohms; 99,950 ohms; 0.05005 ohms

15. If 27 K and 39 K are close tolerance, the voltage will be close to 59, as stated; source resistance is 27 K × 39 K divided by 66 K or 16 K. A 1-volt drop means the meter resistance is approximately 58 × 16 K, which is close to 1 meg (allowing for error in reading). Meter is probably 10,000 ohms per volt.

17. The error suggests 22 K is low and 33 K is high. If 22 K is 17.6 K and 33 K is 39.6 K, the voltage calculates to 69.2. Source resistance of 17.6 K with 39.6 K is 12.2 K. For this to drop 9 volts, the extra resistor must be 12.2 K × 60 divided by 9 or 81.5 K. An 82 K resistor will almost certainly come close.

19. Resistance for 50% low calculates to 947 ohms; for 20% high, to 729 ohms; a mean of 820 might do the job; a more sophisticated circuit could use a different resistance each way, separated by diodes.

CHAPTER 4

1. 400 K, 1 milliwatt; 125 K, 5 milliwatts; 3.3 K, 3 watts; 1.35 K, 0.39 microwatts; 13.2 ohms, 18.1 milliwatts
3. (a) $V = \sqrt{WR}$ (b) $I = \sqrt{W/R}$
5. 70 degrees C
7. 80 degrees C, 13 watts
9. 28.4 K
11. Nearly 10 percent
13. 65 degrees C
15. 100 milliamps (average), 10 watts; 2 amps (average during one-twentieth), 20 watts; 300 volts.

CHAPTER 5

	RECTIFIED AVERAGE	RMS	FORM FACTOR
1.	7.5 V	7.906 V	1.054
3.	8.333 V	8.660 V	1.039
5.	7.5 V	8.165 V	1.089
7.	6.4 V	7.097 V	1.109
9.	1 V	2.582 V	2.582

CHAPTER 6

1. 1023
2. 4

3. 999; 4095.

4. 10

5. 0000000000000000
0000010101010110
0000100101100010
0001000100010001
0000000110010011
0000011100010100
0001000001000100
0000001110000000
0000100001010001
0001000010010100

6. 1023

7. ¾

8. (a) 1, 2, 4, 8; (b) 3, 5, 9; (c) 7.

9. 3 and 9.

10. (a) 31; 4064

CHAPTER 7

1. 62,500, 90; 190,500, 108; 64,500, 91.44; 212,850, 8.89.

3. 80,625, 31.75; 80,625, 44.45; 80,625, 16.51; 80,625, 4.0005.

5. 8470, 1.272.

7. 247.5 oersteds.

9. 9.675 square centimeters or 1.5 square inches; 4.24 centimeters or 1.67 inches.

11. 29.2 square centimeters or 4.5 square inches; 2.12 centimeters or .83 inches.

CHAPTER 8

1. Triangular waveform, 4,167,000 maxwells peak (8,333,000 maxwells peak-to-peak).

3. 183.6 kilograms

5. 100 watts (200 volts at 0.5 amp)

7. 54 cents a week

9. Half a meter (50 centimeters)

CHAPTER 9

1. 50 watt-seconds, 250 watt-seconds, 125 watt-seconds; 12.5 watt-seconds

3. 5 watt-seconds, 1.25 watt-seconds, 12.5 milliwatt-seconds, 2.5 milliwatt-seconds

5. Maximum current
7. 1884 ohms, 1570 ohms, 1327 ohms, 94.25 ohms
9. 16.7 millisecond, 72.7 microseconds, 25 milliseconds, 22 milliseconds
11. 950 micromicrofarads, −2.3 percent per degree C

CHAPTER 10

1. 5 ohms, 1700 ohms, 340 ohms, 125 ohms, 410 ohms
3. 4.00 volts, 7 mA through the resistance, 24 mA through the reactance. 17.0 volts, 0.26 amp through the resistance, 1.68 amp through the reactance.
5. 5000 Hz (more precisely 5,003.5 Hz); in series 47.43 K, in parallel 23.7 K
7. 0.5 henry, 10 ohms
9. 2250 Hz, 3380 Hz

CHAPTER 11

1. 79 milliamps
3. 92 ohms, 10.87 milliamps
5. Half-wave rectifier with a current shunt on the meter for dc readings: (1) 10 V dc (2) 10 V ac (3) 10 V dc with an ac peak greater than 10 V

CHAPTER 12

1. 19, 11.5, 39, 99, 65.67, 124, 249, 199
3. 1.01, 0.92, 0.97, 1.00, 0.995
5. 1.14 megohms
7. 51.5; 515 millivolts (0.515 volt)
9. 207 K

CHAPTER 13

1. 1.8 K, 0.196 volt
3. 52 microamps, 10 K, 0.75 volt
5. 3.15 K, 5.57 K; 7.94, 4.9 volts

7. 3 volts, 9 milliamps; 40 K (39 K could be used); 4 volts, 2.4 volts (based on 40 K)

9. 4 mfd (approx), 8 mfd, 1.6 mfd, 6 mfd, 7.5 mfd, 22 mfd

CHAPTER 14

1. 650 milliamps, 5 volts

3. 13.3 kHz

5. 640 milliamps

7. 280 microseconds

9. 100 nanoseconds

CHAPTER 15

1. Plate load 220 K in parallel with 1 meg is 180 K. Amplification factor, 1.25 × 80 is 100. Stage gain is 100 × 180 K divided by 260 K or 69.3 (69 or 70 acceptable)

3. 257

5. 800 ohms, 245 volts.

CHAPTER 16

1. Horizontally.

2. 3.5 kHz.

3. 7 kHz.

4. Over 12,000 vertically, by 15,000 horizontally.

CHAPTER 17

1. 3.64, 11.43, 12, 12,414, 1.74.

2. 4.65, 36.4, 42.86, 48.65, 1.942.

3. 22Z, 0.86Z, 0.5Z, 0.276Z, 0.087Z.

4. 28Z, 2.73Z, 1.8Z, 1.08Z, 0.097Z.

CHAPTER 18

1. (a) 8, (b) 18 dB.
2. (a) 4, (b) 12 dB.
3. 3.43, 10.7 dB; 2.56, 8.17 dB.
4. $n = 2(F_p - 1) \pm \sqrt{4 F_p^2 - 6 F_p}$

   $n = 3(F_p - 1) \pm \sqrt{9 F_p^2 - 12 F_p}$
5. 36.44, 54.93.
6. 0.072 μF.
7. about 1/10, 290 Hz.
8. 72 μsec.

CHAPTER 19

1. 1/3.

			cumulative
$2z$	=	.666666666	
$2z^3/3$	=	.024691366	.69135803
$2z^5/5$	=	.001646090	.69300412
$2z^7/7$	=	.000130642	.69313476
$2z^9/9$	=	.000011290	.69314605
$2z^{11}/11$	=	.000001026	.69314707
$2z^{13}/13$	=	.000000096	.69314717
$2z^{15}/15$	=	.000000009	.69314718

2. ϵ^x

			cumulative
$1 + x$	=	1.69314718	
$x^2/2!$	=	.24022651	1.93337369
$x^3/3!$	=	.05550410	1.98887779
$x^4/4!$	=	.00961813	1.99849592
$x^5/5!$	=	.00133336	1.99982927
$x^6/6!$	=	.00015404	1.99998331
$x^7/7!$	=	.00001525	1.99999856
$x^8/8!$	=	.00000132	1.99999988
$x^9/9!$	=	.00000010	1.99999998
$x^{10}/10!$	=	.00000001	1.99999999

CHAPTER 20

1. 1.8 MHz.
2. 150 kHz.
3. 1 kHz.

CHAPTER 21

1. 4 μsec
2. 250 kHz.
3. 5 volts.
4. 8
5. 23
6. 18

Index

A

absolute temperature, 24
ac circuits, 73-97
  examination questions for, 96
ac collector loading, 253
ac current, measurement of, 80
accelerator devices, 316
acceptor base, 244
admittance, 192, 235
air-core inductors, 179
alphanumerics, 112, 362
alternating current, 73
alternation, 73
amplification, 104, 241
  linear, 249-271
amplification factor, 303
amplifiers, operational, 268
amplifying valve, 297
amplitude, 95
analog-to-digital conversion, 107
analog and digital bases of operation, 98-114
  examination questions on, 114
analog vs. digital, 104, 381
anisotropic magnets, 138
anode, 314
antennas, 156
anti-phase dipoles, 160
artificial reverberation, 370
attenuation, 48
avalanche diode, 283

B

back emf, 153
balance, 149
ballast, 311
band-pass filters, 204
band-reject filters, 205
beam tetrode tube, 305
  typical curves for, 306
beam-type thermionic tubes, 313
beta tolerance deviation, 259
BH product, 128
BH-max measurement, 128
biasing, 255, 268, 306
bibber, 283
binary logic circuitry, 108
binary-to-decimal conversions, 104, 105, 342
biquinary systems, 109
bistable circuit, 285
bits, 102, 112
branching circuit, 30
breakdown voltage, 214
bridge circuits, 32, 198, 334
  capacitance, 198
  effects of spurious elements on, 200
  frequency-dependent, 203
  frequency-independent, 198
  instrument multipliers and, 40
  meter measurement of, 39
  potentiometers and, 42
  practical features of, 207
  Thevenin's theorem and, 35
  voltmeters and, 40
buses, 288, 362

C

calculators, 362
capacitance, 169
  Ohm's law and, 172
  relationship of current and voltage in, 172
  resistance combined with, 178
capacitance bridge, 198
capacitive reactance, 171
capacitors
  electrolytic, 180
  mica, 180

reactance in, 166
reservoir, 229, 230
temperature coefficients for, 181
types of, 179
carrier frequency, 391
cathode ray tube (CRT), 313
electrostatic deflection in, 314
magnetic focusing in, 314
pixel displays vs. 331
cathode resistor
grid bias using, 307
cathodes, 298
ceramic magnets, 129
BH-max point and curve of, 138
chips, 7
choke smoothing, 228
chokes, 310
circuits, 31
branching, 30
bridge, 32, 198
direct current, 30
Kirchhoff's law and, 30
matched, 49
power factor of, 189
pulse, 69
resonant, 195
test, 277
tuned, 181
voltage divider, 44
Wheatstone Bridge, 32, 33
clamp diode, 283
class-B waveforms, 392
clipping, 250, 339
clock frequency, 292
closed system design, 352
CMOS, 243
coercivity, 120
coils, 16, 179
critical rating for, 64
electromagnetic wave generation in, 158
inductance in, 153, 166
leakage in, 154
magnetic field generated by, 143
magnetism induced by, 118
reactance in, 166
windings of, 63
collector
common, 238
grounded, 238
collector load resistance, 237
collector voltage-current curves, 241
common base, 238
common collector, 238
common-emitter stage, 234, 237
compound semiconductors, 233-248
examination questions for, 247
computer languages, 362
computer programs, 362
computers, 375
basic elements of, 363
conditionally stable curves, 350
conductance, 192, 235
conductor resistance, 18
conductors, 6
comparison of copper and aluminum as, 19
effects of cooling on, 62
heat, 57
magnetic fields produced by, 143
movement in magnetic field of, 145
skin effect in, 182, 183
super-, 25
temperature coefficients affecting, 24
trans-, 49
windings of, 144
conservation of energy, 145
contact potential, 216, 217
control ratio, 312
cored inductors, 180
counter, 293
coupling capacitors, 347
critical rating, 64
Crookes, Sir William, 313
crystals, piezoelectric effect of, 326
current, 3, 30, 146
capacitive relationship of voltage to, 172
changes in distribution of, 59
frequency of, 77
inductive relationship of voltage and, 171
input of, 47
linear amplification and, 251
magnetic fields force produced by, 144
magnetism induced by flow of, 118
output of, 47
relation of temperature rise to, 59
current bibber, 283
current distribution, effect of heating on, 59
current swing, 263
current-amplifying transistors, 233
varieties of, 240
cutoff frequency, 273, 349
cyclotron, 317

D

Darlington circuit, 267
data buses, 362, 377
dc stabilization, 307
decade blocks, 342
decimal logic circuitry, 108
decimal-to-binary conversions, 104, 105, 342
decoupling capacitor, 262
delay time, 274, 276
demagnetization, 124
demodulator, 113
dependent variables, 98
dielectrics, 6
digital audio, 371
digital instrumentation, 341
digital systems
analog vs., 104
understanding of, 112
digital-to-analog conversion, 107, 381
diode transistor logic (DTL), 290, 376
diodes, 16
avalanche, 283
basic characteristics of, 210
clamp, 283
contact potential of, 216
examination questions on, 231
germanium, 211, 216
layer, 329
light emitting, 329
liquid crystal, 330
logarithmic properties in, 212
ratings for, 230
silicon, 216
temperature coefficients for, 227
vacuum, 296
zener, 334
diodes and simple semiconductors, 210-232
dipole antenna
anti-phase, 160
basic relationships in, 156
dummy reflector, 161
pickup/radiation pattern of, 160
reflectors and, 159
direct current, 30
direct measurement, 334
discrete elements, 382
dissipation
effect of pulsing on, 68
energy, 168
heat, 56

power, 53
thermionic tubes and, 304
distribution density, 126
donor base, 244
Doppler effect, 374
DTL logic gate, 289
duty cycle, 279
dynamo, 4

E

effective resistance, 17
efficiency, 390
electrets, 330
electric motors
efficiency in, 151
reversibility in, 151
voltage, rpm, current, and torque in, 150
electrical devices vs. electronic, 7
electricity
current, 2
static, 2
study of, 1
electrochemical devices, 1
electrolysis, 309
electrolytic capacitors, 180
electromagnetism, 2-8, 142-165
early experiments in, 4
energy transfer and balance in, 149
examination questions for, 165
generation of electromotive force by, 147
generation of mechanical force by, 149
inductance and, 153
left-hand and right-hand rules of, 146
radiation of, 156
radio transmission and, 6
relationship between electric and magnetic fields in, 158
reversibility in, 151
rules for, 143
spatial relationships in, 143
transducers and, 152
waves generated by, 157
electromotive force (emf), 115
generation of, 147
electron gun, 313
electronvolt, 315
electron-multiplier, 313
electronic applications, 366-378
electronic organs, 366
artificial and simulated reverberation in, 371
bedspring system in, 370
general schematic of, 368
hybrid systems for, 370
use of digital audio components in, 371
electronics
definitions of, 1-8
increasing performance speed of, 376
keeping up with changes in, 395
sensory projection use of, 373
successive generations in, 375
video game use of, 372
electrons, 296
electrophorus experiment, 5
electrostatic deflection, 314
electrostatic generator, 4
emitter, grounded (common), 234
energy transfer, 149
equipotential law, 126
equivalent parallel networks, 193
equivalent series, 192
examination questions, answers to, 416-423

F

Faraday, 2
feedback, 307
ac and dc, 257
negative, 336
feedback factor, 337
feedback fraction, 337
field, 146
field effect transistor (FET), 243, 335, 344
basic construction of, 244
filters
band-pass, 204
band-reject, 205
droop and peak in, 206
high-pass, 204
low-pass, 204
polarizing (light), 325
reactance, 204
flip-flops, 110, 284, 285, 335, 396
fluctuation, 73
flux, 146
flux density, 115
form factor, 85, 90, 345
Fourier analysis, 384
frequency, 74
organ notes, 294
ranges of, 96
reactance and, 173
frequency based, 384
frequency measurement instrumentation, 344
frequency-dependent bridges, 203
frequency-independent bridges, 198
friction, 151
function generator, 368
fundamental mode, 75
fuses, 61

G

gain, 268, 346
gas-filled tubes, 309
gates, 106
gauss, 117
generators, 4
germanium diodes, 211, 216
germanium transistors, 240
gilberts, 117
glossary, 404-415
grid-bias, cathode resistors for, 307
grid current biasing, 308
schematic of, 308
grounded base, 238
grounded collector, 238
grounded emitter, 234
group velocity, 164

H

half-wave rectifiers, 221
Hall-effect devices, 326, 328
harmonic mode, 75
harmonics, 74
waveform effects caused by, 86-95
heat, electronic equipment and, 65
heat flow, 57
heating effects, 56
henrys, 167
Hertz, 74
high-pass filters, 204
high permeability region of magnetism, 119
high-frequency cutoff, 273
hybrid systems, 379-398
digital electronics in, 380
efficiency aspects of, 390
examination questions for, 394
machine, 382
pulse width and amplitude modulation, 393
hysteresis loop, 130, 131

I

impedance, 35, 188
calculation for, 191
reciprocal of, 192

independent variables, 98
indirect measurement, 334
indium antimonide, 328
indium arsenate, 328
inductance, 153, 166
 mutual, 154
 Ohm's law and, 170
 Q values of, 202
 relationship of voltage and current in, 171
inductance-resistance combination
 relationship of voltage and current in, 175
induction coil, 3
inductive reactance, 173
inductors, 16
 air-core, 179
 complicated nature of, 201
 controlling gas discharge with, 310
 cored, 180
 temperature coefficients for, 181
 types of, 179
initial domain of magnetism, 119
input impedance, 260
input resistance, 220
input source resistance
 effects of, 239
instrument multipliers, 40
instrument rectifiers, 217
 calibration of, 218
 input resistance in, 220
 low-voltage scales of, 219
instrumentation, 334-345
 digital, 341
 examination questions for, 345
 frequency measurement in, 344
insulators, 6
integrated circuits (ICs), 396
integrated elements, 382
interconnection, 47
interfacing, 362
internal impedance, 338
 effect of feedback on, 340
internal resistance, 36
ionization, 3, 296, 309

J

j operator, 191

K

keeper, 124
Kerr cell, 325, 326, 329
Kirchhoff's laws, 3-52, 70, 80, 234
 examination questions on, 50
 interconnection problems and, 47
klystron tube, 318, 318
 cross-section of element arrangement of, 318

L

lambda, 159
large scale integrated circuits, 369
lasers, 326
layer diodes, 329
leakage flux, 154
leakage inductances, 155
left- and right-hand rules, 146
light-emitting diodes (LEDs), 329
linear accelerator tube, 316
linear amplification, 249-271, 337
 Darlington circuit in, 267
 effect of feedback on, 339
 examination questions for, 270
 input for, 251
 negative feedback and, 336
 operational amplifiers in, 268
 removing ac feedback in, 257
 split load stages in, 264
 voltage-gain stage for, 258
linear scales, 81
linear transistors, 273
linearity, 250
 ac collector loading for, 253
liquid crystal diodes (LCDs), 330
loading, 151, 206
 ac collector, 253
logarithmic properties of diodes, 212
logic circuitry, comparisons of binary and decimal, 108
logic gates, 106
 DTL, 289
logic systems, two-state, 284
low-pass filters, 204

M

machine hybrid systems, 382
magnetic amplifiers, 132
magnetic fields, 16, 126, 143
 force due to current in, 144
 moving conductor in, 145
 relationship of voltage to, 148
 waveforms and, 148
magnetism, 2, 115-141
 basic quantities of, 115
 calculations for, 134
 de-, 124
 domains of, 119
 examination questions for, 140
 fields of, 126
 hard vs. soft materials of, 118
 high permeability region of, 119
 initial domain of, 119
 relationship between measurement quantities for, 117
 saturation point of, 119
 study of, 1
 unit cube representation of, 116
 use of coil to induce, 118
 using current to produce, 118
magnetomotive force (mmf), 115, 142
magnetron, 313, 318
 cross-section of, 319
magnets
 anisotropic, 138
 BH product of, 128
 BH-max measurement of, 128
 calculations for, 134
 ceramic, 129
 charging of, 121
 design requirements for, 134
 early concepts and calculations for permanent, 125
 effect of combining ac and dc on, 133
 hard vs. soft materials for, 120
 keepers in, 124
 leakage in, 135
 optimizing performance of, 122, 128
 permanent, 121
 permanent, BH product and BH-max curve for, 128
 permanent, practical operating condition of, 123
 permanent, relationships in field of, 126
 poles of, 125
 removing magnetic circuit from, 123
 soft material, 130
 work area gap of, 121
 working air gaps for, 134

Marconi, 6
matched circuits, 49
maxwells, 117
mechanical force, 145
  generation of, 149
mercury-vapor rectifiers, 312
meters, polarized vs. nonpolarized, 81
mica capacitors, 180
microchips, development of, 368
microprocessor, 7, 375
  basic elements of, 363
modem, 113
modulator, 113
Moog, Robert, 368
MOSFET, 243
motion, 146
motor-boating, 347
motors, 4
moving coil meter, 152
multimeters, 221
multipliers, instrument, 40
multistage amplifier, 383
multivibrator, 355
music synthesizers, 366, 367
mutual conductance, 303
mutual inductance, 154

N

negative feedback, 346
  functions of, 336
negative temperature coefficients, 64
networks, 47
nonlinearity, 250
npn type transistors, 240
null points, 37
Nyquist diagrams, 350, 351, 352

O

Oersted, 1
oersteds, 117
Ohm's law, 3, 14, 30, 34, 80
  capacitance and, 172
  inductance and, 170
  problems using, 15
ohms, 9
Olson, Dr. Harry, 367, 396
open system design, 352
operational amplifiers, 268, 335, 397
  gain and bias in, 268
  integrated circuits using, 335
  operation of, 269
organ note frequencies, 294
oscillation, 73, 347
oscillators, basic circuit for, 366
oscilloscopes, 77

P

parabolic reflector, 162
parallel I/O, 110
parallel resistors, 20
parallel values, 192
pendulum, 73
pentode tube, 301
  typical curves for, 302
period, 74
permanent magnets, 121
  BH product and BH-max curve for, 128
  curves of early, 129
  early concepts and calculations for, 125
  magnetization and demagnetization cycle of, 121
  practical operating condition of, 123
  relationships in field of, 126
permeability, 116
perpetual motion, 145
phase relationships in waves, 87
phase shift, 347
phase velocity, 164
photoconductive cells, 324
  TV camera use of, 324
photosensitive devices, 322
photovoltaic cells, 322
  characteristics of, 323
  cross-section of, 323
piezoelectric devices, 326
pithball experiment, 4
pixel displays, 331
plate resistance, 303
pnp type transistor, 240
polarity, 125
polarizing filters, 325, 326
potentiometer, 42, 334
  measurement of low voltages with, 45
power calculations, 53-72
  examination questions for, 71
  heating effects and, 56
  units for, 55
power dissipation, 53
power factor, 189
power supply rectifier, 217
pressure, 3
processors, 7
pulse circuits, 69
pulsing, 68
Pythagorean theorem, 186

Q

Q values, 202

R

radiation, electromagnetic, 156
radio, 6
ratio arms, 200
reactance, 166-184
  bridge circuits and, 198
  circuit, 185-209
  circuits, examination questions for, 208
  effect of frequency on, 173
  examination questions for, 184
  filters for, 204
  power flow in circuits with, 187
  reciprocal of, 192
  time constants and, 174
  waveforms associated with, 185
reciprocals, 192
reciprocity, 142
rectified average readings, 84
rectifiers
  full-wave, 217
  half-wave, 217, 221
  instrument, 217
  mercury-vapor, 312
  power supply, 217
  ratings for, 230
  silicon controlled (SCR), 245
reflectors, 159
  parabolic, 162
regulators, 215
reluctance, 115
reservoir capacitor, 229, 230
resistance, 9-29
  effects of low temperatures on, 25
  capacitance combined with, 178
  collector load, 237
  conductor, 18
  effect of tolerances on, 22
  effective, 17
  effects of input source, 239
  examination questions for, 28
  factors affecting, 20
  input, 220
  low, measurement of, 39
  measuring changes in, 68
  Ohm's law and, 14
  other values of, 16
  parallel, 21
  plate, 303
  reciprocal of, 192
  series, 20
  specific, 18
  temperature coefficients affecting, 24